Geology of the Ambleside district

The rugged mountains of Scafell and Bow Fell, the picturesque valleys of Wasdale, Eskdale and Great Langdale, and the gently undulating areas around the lakes of Coniston Water and Windermere are the heart of the English Lake District. The villages of Ambleside and Grasmere and their neighbouring fellsides are long established as peaceful retreats for poets, writers, walkers and climbers. By contrast, 19th century Coniston was the centre of a thriving copper mining industry. The economy today is dominated by tourism, though the principal land use is hill farming and some slate is extracted from a number of quarries. The geology of this scenically very beautiful part of Cumbria, wholly within the Lake District National Park, is explained in this account and illustrated on Sheet 38 Ambleside of the 1:50 000 scale geological map of England and Wales, available as separate solid and solid with drift editions.

The geology of the Ambleside district chronicles the rifting of the microcontinental terrane of Avalonia from Gondwana early in the Ordovician, its drift northwards across the Iapetus Ocean and collision with Laurentia in the Silurian leading to the climactic Acadian orogeny during the Early Devonian. Repeated glaciation during the Quaternary has modified the original landforms of the district, but only fragmentary evidence survives of events prior to the last, Late Devensian, regional ice-sheet glaciation.

The oldest rocks of the district, Tremadoc or Arenig siltstone and sandstone belonging to the Skiddaw Group, crop out only in a very small area in the west of the district. They are related to the more extensive tracts in the northern part of the Lake District, and in Black Combe and Furness to the south.

The 8 km-thick, Llandeilo to early Caradoc, Borrowdale Volcanic Group was erupted from subaerial volcanoes within an extensional basin at the margin of Eastern Avalonia. It is a calc-alkaline suite that was generated above a subduction zone. The lower part of the succession was dominated by the effusion of andesite lavas. This was succeeded by voluminous dacitic and rhyolitic ignimbrites, associated with the formation of calderas. Volcanotectonic collapse produced lacustrine basins. The volcanic rocks are underpinned by a major, late Ordovician granitic batholith, which is exposed in the western part of the Ambleside district as the Eskdale and Ennerdale intrusions. The volcanic and intrusive rocks are host to a wide variety of mineral veins, predominantly with ores of copper or iron.

The overlying Windermere Supergroup of Ashgill to Ludlow age comprises some 5 km of marine sedimentary rocks. Upper Ordovician strata are of relatively shallow-water facies and include limestone, calcareous sandstone and siltstone. In the Silurian part of the Supergroup, Llandovery strata are dominantly hemipelagite; this lithology is then interbedded with progressively more turbiditic sand, silt and clay upwards through the Wenlock and Ludlow.

Much of the district is covered by glacial deposits formed during the Devensian glaciation. Thick till is present mainly in river valleys. Lacustrine deposits are locally extensive around the lakes and strata in Lake Windermere form the type area for the Late Devensian Interstadial. Extensive spreads of hill peat cover the volcanic rocks in the southwest.

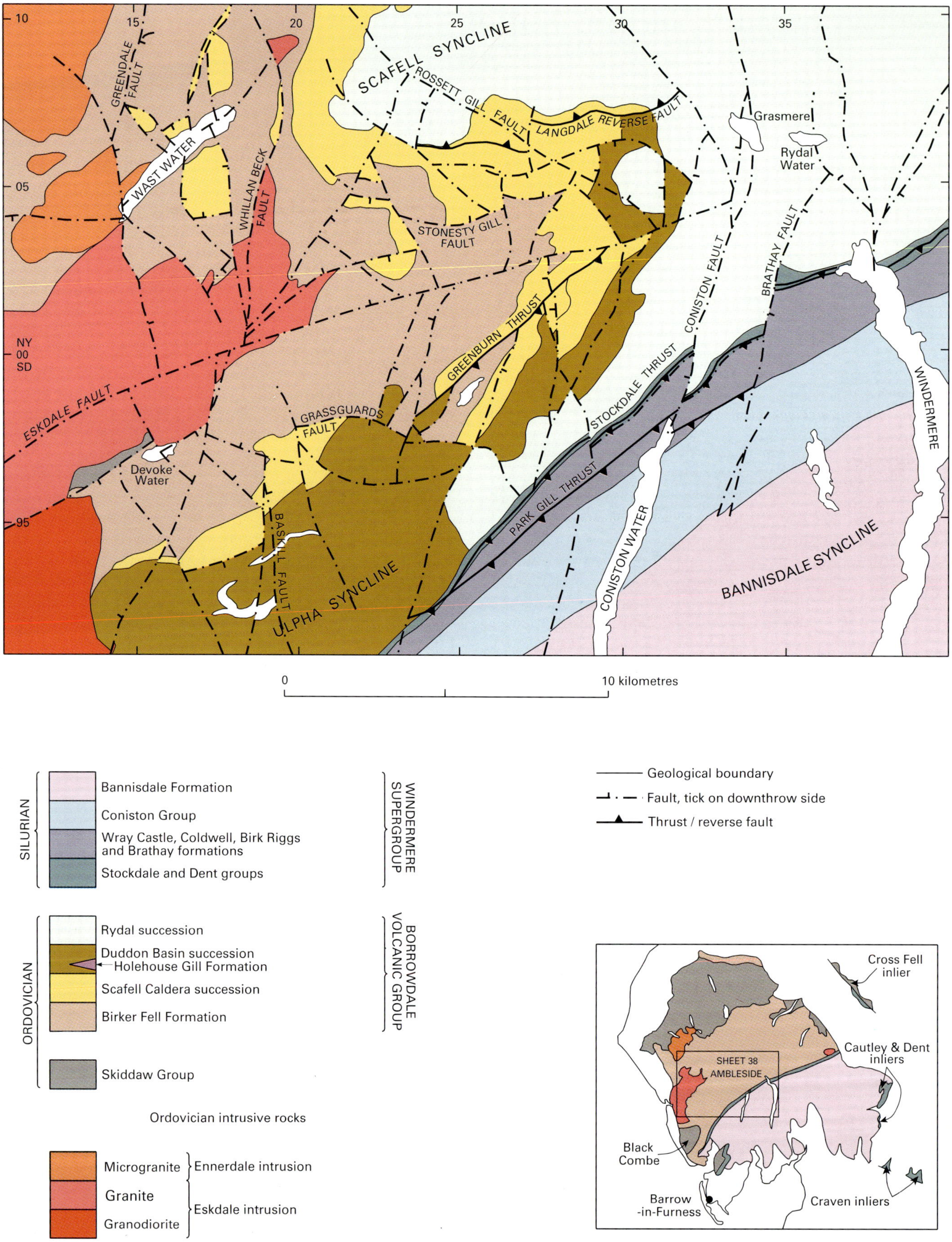

Figure 1 Summary geological map of the Ambleside district.

BRITISH GEOLOGICAL SURVEY

D MILLWARD
E W JOHNSON
B BEDDOE-STEPHENS
B YOUNG
B C KNELLER
M K LEE
N J FORTEY

Geology of the Ambleside district

Memoir for 1:50 000 Geological Sheet 38 (England and Wales)

CONTRIBUTING AUTHORS

Stratigraphy and structure
P M Allen
M J Branney
B P Kokelaar
B J McConnell
J W Merritt
M G Petterson
R W Scott
N J Soper
P Stone

Biostratigraphy
S G Molyneux
A W A Rushton

Petrology and geochemistry
D C Cooper
S Hirons
B Roberts
C C Rundle

Hydrogeology
R J Marks

London: The Stationery Office 2000

First published 2000

ISBN 0 11 884547 0

Bibliographic reference

MILLWARD, D, JOHNSON, E W, BEDDOE-STEPHENS, B, YOUNG, B, KNELLER, B C, LEE, M K, FORTEY, N J, ALLEN, P M, BRANNEY, M J, COOPER, D C, HIRONS, S, KOKELAAR, B P, MARKS, R J, MCCONNELL, B J, MERRITT, J W, MOLYNEUX, S G, PETTERSON, M G, ROBERTS, B, RUNDLE, C C, RUSHTON, A W A, SCOTT, R W, SOPER, N J, and STONE, P. 2000. Geology of the Ambleside district. *Memoir of the British Geological Survey*, Sheet 38 (England and Wales).

Authors

D Millward, BSc, PhD
E W Johnson, BSc, PhD
B Beddoe-Stephens, BSc, PhD
B Young, BSc
British Geological Survey, Edinburgh

B C Kneller, BSc, PhD
Department of Earth Sciences, University of Leeds

M K Lee, BSc, MSc, PhD
N J Fortey, BSc, PhD
British Geological Survey, Keyworth

Contributing authors

J W Merritt, BSc
M G Petterson, BSc, PhD
P Stone, BSc, PhD
British Geological Survey, Edinburgh

P M Allen, BSc, PhD
D C Cooper, BSc, PhD
S G Molyneux, BSc, PhD
A W A Rushton, BA, PhD
British Geological Survey, Keyworth

R J Marks, BSc
British Geological Survey, Wallingford

C C Rundle, BSc, PhD
formerly of the NERC Isotope Geoscience Laboratory

M J Branney, BSc, PhD
Department of Geology, University of Leicester

B P Kokelaar, BSc, MSc, PhD
Department of Earth Sciences, University of Liverpool

B J McConnell, BSc, PhD
Geological Survey of Ireland, Dublin

R W Scott, BSc, PhD
N J Soper, BSc, PhD
Department of Earth Sciences, University of Sheffield

B Roberts, BSc, PhD
S Hirons, BSc
Birkbeck College, University of London

Printed in the UK for The Stationery Office
TJ000848 C6 3/00

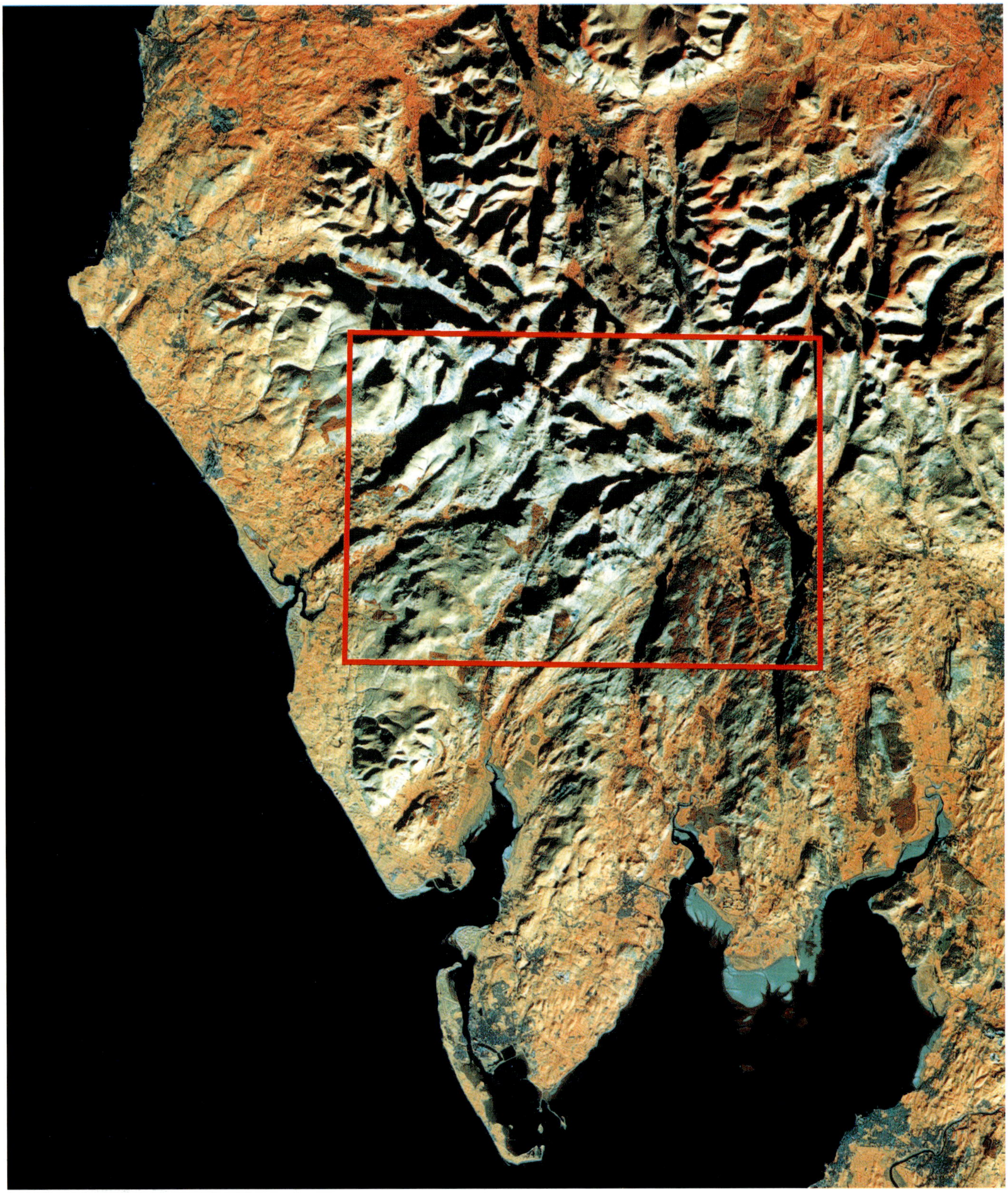

Frontispiece Landsat TM image of north-west England showing the Ambleside district.
Processed by BGS, Keyworth. Bands 4, 5, 7 used.

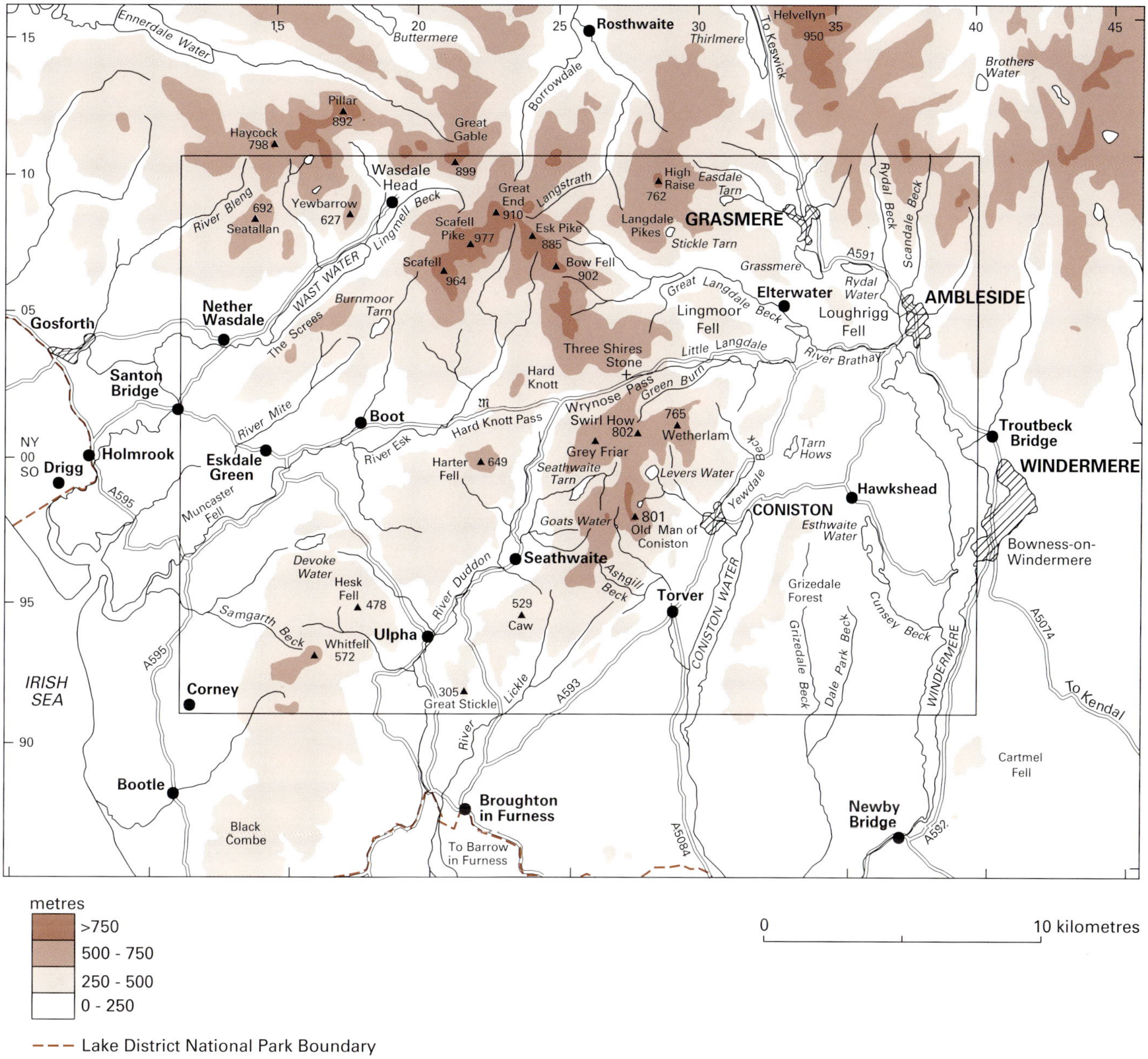

Figure 2 Physical geography of the district and the adjoining areas.

ONE

Introduction

The rugged mountains of Scafell and Bow Fell, the picturesque valleys of Wasdale, Eskdale and Great Langdale, and the gently undulating areas around the lakes of Coniston Water and Windermere form the heart of the English Lake District. The villages of Ambleside and Grasmere and their neighbouring fell sides are long-established and popular retreats of poets, writers, walkers and climbers. This now-peaceful scene contrasts with the industrial activity of the last century when Coniston was the centre of a thriving copper mining industry. The geology of this scenically very beautiful part of Cumbria, wholly within the Lake District National Park, is portrayed on the 1:50 000 Series geological map of the Ambleside district (Figure 1, British Geological Survey, Solid geology 1996; Solid and Drift geology 1998) and is explained in this account.

The mountain scenery of the northern and central parts of the district, including Scafell Pike (977 m), the highest mountain in England, has been carved out of the remnants of a 450 million-year-old volcanic province. By contrast, the lower ground in the south-east is underlain by later-formed sedimentary strata. The volcanic rocks are deeply dissected by glacial valleys that drain to the west via the rivers Bleng, Irt, Mite and Esk, to the south through the rivers Duddon and Lickle, and into Coniston Water and Lake Windermere (Figure 2). There are many dramatic examples of glacial erosional features in the district, including the U-shaped valleys of Wasdale, Eskdale and Great Langdale (cover picture). Corrie glaciers sculpted the spectacular corries that now contain Low Tarn [NY 162 093], Stickle Tarn [NY 287 077], Levers Water [SD 279 994] and Goat's Water [SD 266 977]. The imposing screes at Wast Water, which began forming after the retreat of these glaciers, testifies to the continued effect of climate on the landscape.

The economy of the district is dominated today by tourism, centred in the villages of Grasmere, Ambleside and Coniston. These are served by a road network from the north and south-east (Figure 2) that generally follows the course of the main valleys. Communication between centres in the east of the district and west Cumbria is by the narrow, winding mountain road over the Wrynose [NY 277 027] and Hard Knott [NY 231 015] passes. The former Lakeland counties of Cumberland, Westmorland and Lancashire meet at the Three Shires Stone [NY 2772 0275] on the summit of Wrynose Pass in the centre of the district. Hill farming is the principal land use and slate, mainly for ornamental and architectural work, is extracted from quarries north-east of Ambleside [NY 397 078], at Elterwater [NY 323 048], Coniston [SD 284 984; SD 278 974] and Broughton Moor [SD 255 945].

GEOLOGICAL SUCCESSION

The district lies within the Lake District Lower Palaeozoic inlier (Figure 1). A threefold geological succession within the inlier has been recognised for more than 150 years and the main groups are represented in the district. Strata of Late Palaeozoic, Mesozoic and Cainozoic age originally present have been removed by erosion and the Lower Palaeozoic rocks are now overlain by spreads of superficial deposits of Quaternary age. The geological succession in the Ambleside district is given in Table 1.

The lowest and uppermost of the stratigraphical divisions comprise marine sandstone, siltstone and shale. The lowest strata, of Tremadoc to Llanvirn age, belong to the Skiddaw Group (Cooper et al., 1995), which crops out in the northern and eastern parts of the inlier, and on Black Combe and in the Furness area to the south (Figure 1, inset). The Skiddaw Group is more than 5 km thick though the base is never seen. It probably underlies much of northern England, beneath Carboniferous rocks of the Alston and Askrigg blocks and has likely correlatives in the Isle of Man and Ireland (Johnson, 1961; Cooper and Molyneux, 1990; Cooper et al., 1995). The upper marine sedimentary succession crops out in the south-east of the district and is included within the 8 km-thick Windermere Supergroup of Caradoc (Longvillian) to Přídolí age (Kneller et al., 1994); these strata are coeval with the Powys Supergroup of Wales.

The marine rocks are separated stratigraphically and geographically by the Borrowdale Volcanic Group of probable Llandeilo to early Caradoc age. At the northern margin of the inlier the Eycott Volcanic Group is stratigraphically equivalent to the Borrowdale Volcanic Group. The three major divisions are separated by unconformities of regional significance (Moseley, 1974; Ingham et al., 1978; Millward and Molyneux, 1992). The central part of the Lake District is underpinned by a granitic batholith (Bott, 1974; Lee, 1986a), which is exposed in the western part of the district as the Eskdale and Ennerdale intrusions.

Skiddaw Group

In the Ambleside district the oldest rocks crop out as a very small inlier in the south-west, near Devoke Water [SD 142 965] (Figure 1). Cleaved, hornfelsed sandstone and siltstone belonging to the Skiddaw Group lie between the Eskdale pluton and the Borrowdale Volcanic Group. The Skiddaw Group formed the original substrate to the volcanoes. Hence abundant silty mudstone and sandstone clasts, derived from the Skiddaw Group occur within

hydrovolcanic deposits in the lower part of the Borrowdale Volcanic Group, and microflora from the Skiddaw Group have been reworked into some of the sedimentary rocks interbedded within the volcanic deposits. Subsequently, components of the Lake District batholith were intruded approximately along the contact between the Skiddaw and Borrowdale Volcanic groups, but interpretation of seismic reflection profiles suggests that lenses of Skiddaw Group also occur at depth within the Lake District batholith which has been interpreted as a cedar tree laccolith (Evans et al., 1993; 1994).

Borrowdale Volcanic Group

Remnants of the 460 million-year-old subaerial volcanic province comprise the 8 km-thick Borrowdale Volcanic Group, a succession which was produced during two volcanologically distinctive episodes. The first phase was dominated by the eruption of andesite lava flows, now comprising a single stratigraphical unit, up to 2.8 km thick and referred to as the Birker Fell Formation (Petterson et al., 1992). The second phase was dominated by explosive acid volcanism and caldera formation (Branney and Soper, 1988), the products of which, together with intercalated volcaniclastic sedimentary rocks, form a succession varying in thickness across the district from 2 to 5 km.

The Birker Fell Formation, described in Chapter 5, crops out in the west of the district, around the margins of the Eskdale and Ennerdale intrusions (Figure 1). Andesite lavas, each from a few tens of metres to about 250 m thick, dominate the sequence and are separated by thin beds of volcaniclastic siltstone, sandstone and breccia (Table 1). Some sills are present. Where dips are shallow the volcaniclastic deposits have weathered out to form prominent benches whereas the more resistant andesites typically form near-vertical crags. The resulting trap topography is locally very well developed and good examples occur at Border End on Hard Knott [NY 225 018], between Crook and Green crags on Birker Fell [SD 200 985] (Plate 1), and around Scoat Tarn [NY 160 104]. The formation also includes subordinate basaltic, dacitic and rhyolitic lava flows and sills, together with mainly andesitic and dacitic ash-fall tuff, pyroclastic surge deposits and ignimbrite. Interbedded sedimentary rocks mostly comprise tephra which has been reworked, re-deposited and lithified to form siltstone, sandstone, conglomerate and breccia. These lithologies commonly form beds 1 to 15 m thick separating the andesite sheets, but thicker lensoidal units locally reach 50 to 200 m thick.

The lithostratigraphy of the second volcanic phase is described in Chapter 6 in three successions, associated with distinctive eruptive and depositional centres (Figure 1; Table 1). The oldest of these occurs in the Central Fells and comprises a complex sequence of garnet-bearing pyroclastic rocks, but it is considered to have been erupted from the Scafell Caldera (Plate 2; Branney and Kokelaar, 1994). The first pyroclastic formation is andesitic in composition and comprises a lower, welded lapilli-tuff that buried the underlying topography, overlain by a bedded ash-fall tuff. The succeeding lapilli-tuff formations are of dacitic and rhyolitic composition and include very densely welded and lava-like ignimbrite. The final stage of this eruptive cycle includes the Scafell Dacite lava and lacustrine sedimentary rocks.

The succession in the overlying Duddon Basin is up to 3 km thick, comprises nine formations of pyroclastic and volcaniclastic sedimentary rocks and is preserved only within the Coniston Fells and Ulpha Syncline in the southern part of the district (Figure 1). A number of locally significant unconformities are present in this succession. Densely welded andesitic ignimbrite occurs at the base in the south-west of the district, succeeded by more widespread andesitic bedded tuff and lava flows. Grey mudstone of the Holehouse Gill Formation in the Ulpha area contains a marine microflora and provides the only evidence within the Borrowdale Volcanic Group for incursion by the sea. Higher parts of the succession in the Duddon Basin consist of alternating formations of

Plate 1 Trap topography within andesite lavas in the Birker Fell Formation. The prominent crags are the massive central parts of the lavas; the benches comprise thin beds of volcaniclastic rocks separating successive flows. Crook Crag from the west (D 3876).

Plate 2 The Scafell Caldera.

Thick sheets of welded acid ignimbrite overlain by sedimentary rocks are seen here within the Scafell Syncline, formed from the Early Devonian tectonic compression of the Ordovician caldera. In the crags on the left of the skyline the rocks dip to the right and on the right the sandstones exposed in the pointed peak of Bow Fell dip to the left. The bedded rocks in the foreground are the Hard Knott breccias, produced from the collapse of penecontemporaneous fault scarps within the caldera. View looking north from Hard Knott [NY 232 024] (D 4031).

volcaniclastic sedimentary rocks and welded dacitic to rhyolitic ignimbrite.

The youngest part of the Borrowdale Volcanic Group in the district is preserved mainly in the east, in the Rydal area, and is 0.6 to 2.2 km thick. Unconformably overlying both the Scafell Caldera and Duddon Basin successions, it consists predominantly of sedimentary strata overlain by pyroclastic rocks. A substantial number of andesite sills are present.

The Borrowdale and Eycott volcanic groups have been assigned customarily to the Llandeilo and early Caradoc because they lie between the Llanvirn Tarn Moor Formation of the Skiddaw Group and the middle Caradoc Drygill Shales of the Windermere Supergroup (Wadge, 1978). The Holehouse Gill Formation is probably Caradoc (Harnagian–Soudleyan) on the basis of its marine microflora (Chapter 6; Molyneux, 1988). Radiometric dates from the Borrowdale Volcanic Group in the Ullswater and Haweswater areas (457 ± 4 Ma; Sm–Nd on garnet–whole rock pairs; Thirlwall and Fitton, 1983), and for the Eskdale and Ennerdale plutons (c. 450 Ma; U–Pb on zircon; Hughes et al., 1996), which intrude the Birker Fell Formation and Duddon Basin succession, are compatible with the Caradoc biostratigraphical age of the Holehouse Gill Formation.

Borrowdale Volcanic Group rocks are typically porphyritic with abundant phenocrysts of plagioclase and alkali feldspar with clinopyroxene, orthopyroxene, olivine, biotite, iron–titanium oxide or garnet. Phenocrysts of red almandine–pyrope garnet are an unusual feature of the suite (Fitton, 1972). They are a distinctive characteristic of some tuff and dacite units in the Birker Fell Formation and are significant in the Scafell Caldera succession. However, garnet does not occur in higher parts of the succession in the district. Basalt in the Throstle Garth Member, and the Great Whinscale Dacite within the Birker Fell Formation are exceptional examples of aphyric rocks. Alteration of the primary mineralogy and geochemistry of these rocks is widespread and multiphase, resulting from such processes as contemporaneous subaerial weathering, subvolcanic hydrothermal alteration, burial, contact metamorphism and cleavage development.

Windermere Supergroup

The Windermere Supergroup crops out in the southeastern part of the district (Figure 1) as a sequence of folded and cleaved, predominantly marine sedimentary rocks, which unconformably overlies the Borrowdale Volcanic Group. Within the district the Windermere Supergroup ranges from Cautleyan (mid-Ashgill) to Ludfordian (late Ludlow) in age; it is more than 5 km thick.

The Dent Group (Table 1; traditionally known as the Coniston Limestone) at the base of the Supergroup, is of late Ordovician age and comprises 20 to 85 m of limestone, calcareous sandstone and siltstone with shelly faunas indicative of a relatively shallow water environment. A number of non-sequences and minor unconformities are present. Pockets of conglomerate occur at the base and a thin acidic tuff is present in the upper part. Though deposition was apparently continuous from the Ordovician into the Silurian, a change in lithofacies and biofacies occurred close to the boundary, near the top of the Hirnantian Stage, with the first indications of an anoxic environment.

The lower part of the Silurian succession consists largely of graptolite-bearing hemipelagite with rare shelly fossils (Table 1; Stockdale Group; c. 100 m thick), but progres-

sively more turbiditic sand, silt and mud was introduced through the Wenlock and Ludlow. Thus, above the Stockdale Group is a sequence of Wenlock and lower Ludlow strata, 600 to 1000 m thick, dominated by dark grey, finely laminated, graptolitic, muddy siltstone. The base of this unit is marked by a transition from bioturbated siltstone to laminated siltstone; the probably diachronous top is marked by the incoming, during the earliest Ludlow, of numerous sandstone turbidites which periodically swamped the laminated siltstone. However, the laminated siltstone persists as a background facies throughout the Gorstian Stage. Four formations are recognised, of which the lowest (Brathay Formation) and uppermost (Wray Castle Formation) consist almost entirely of laminated siltstone; turbiditic sandstone dominates the two intervening formations (Table 1). The cumulative thickness of laminated siltstone remains more or less constant at about 600 to 650 m, and overall variations in thickness of this unit are caused largely by the presence or absence of the interbedded turbidites. The middle and upper Ludlow is represented by about 2000 m of sandstone-dominated turbidite units (Coniston Group) overlain by at least 2.5 km of mainly thinly bedded siltstone (Bannisdale Formation).

Intrusions

Two major granitic bodies, the Eskdale and Ennerdale intrusions, crop out in the west of the district and are the exposed westernmost part of the Lake District Ordovician–Devonian batholith (Figure 1; Chapter 8). Geophysical modelling suggests that the top surface of the batholith is at a depth of less than 2 km over much of the area underlain by the Borrowdale Volcanic Group (see Chapter 3; Lee, 1986a). Evidence for such shallow depths is provided also by the unusually wide zone of contact metamorphism around the exposed intrusions (Chapter 10). Interpretation of seismic reflection profiles along Wasdale show the subsurface shape of the granites as a cedar-tree laccolith (Evans et al., 1993; 1994). U–Pb isotopic ages of 450–452 Ma on zircon from the Eskdale and Ennerdale granites (Hughes et al., 1996), indicate a Caradoc emplacement age. Many minor intrusions, of Ordovician to Devonian age, occur sporadically across the area, but with some varieties considerably more abundant around the margins of the Eskdale and Ennerdale granites.

Eskdale pluton

Two components, the Eskdale granite and the Eskdale granodiorite, comprise the Eskdale pluton. From Devoke Water [SD 152 972] north-east towards Burnmoor Tarn, the granite contact is subparallel to bedding in the adjacent and overlying rocks, rising in stratigraphical level from within the uppermost Skiddaw Group to cut the Craghouse Member, in the middle part of the Birker Fell Formation. A cupola on the roof of the batholith crops out near Wasdale Head [NY 195 090] and is within a few hundred metres of the base of the Scafell Caldera succession. The granodiorite is a discordant intrusion cutting through from the base of the Birker Fell Formation to the Duddon Basin succession within the western part of the Ulpha Syncline. The Eskdale granite has a U–Pb zircon age of 450 ± Ma (Hughes et al., 1996). Though its relationship to the granodiorite is unexposed, these intrusions are considered to be approximately coeval.

The Eskdale granite consists of three main varieties. In order of decreasing abundance these are medium-grained non-porphyritic muscovite granite, aphyric or megacrystic microgranite, and coarse- to very coarse-grained granite. Microgranite is most common in the northern part where the low-dipping contacts and inliers of hornfelsed volcanic rocks suggest that the roof zone of the intrusion is exposed. Xenoliths of country rock are extremely rare in the Eskdale granite and there is a thin zone of microgranite at the margin attributable to chilling. These features suggest that stoping was not a significant intrusion mechanism, though locally the adjacent Borrowdale Volcanic Group rocks are permeated by granitic veinlets.

Greisen, formed by metasomatic recrystallisation of the granite, occurs locally within medium- to coarse-grained and fine-grained facies of the Eskdale granite (Young et al., 1988). Close to exposed contacts of the intrusion, quartz–white mica greisens characteristically occur as near-vertical layers up to 2 m wide, adjacent or parallel to joints and quartz veins. Topaz is an abundant constituent in some greisens and fluorite is a common accessory phase. A distinctive quartz–andalusite rock is associated with topaz greisen adjacent to the contact with the Skiddaw Group on Water Crag [SD 1529 9733], near Devoke Water.

The Eskdale granodiorite is medium grained, locally rich in biotite and in the south is extensively altered to clay minerals (British Geological Survey, 1991). A marginal microgranodiorite is developed along the contact. Compared with the Eskdale granite, the granodiorite contains hornblende, has a greater proportion of plagioclase and biotite, and lacks coarse, platy muscovite. Almandine garnet (Ansari, 1983) is locally present, particularly in rock exposed in Waberthwaite Quarry [SD 1124 9430]. Mafic xenoliths and aplitic veins up to 0.15 m across are common.

Ennerdale intrusion

The Ennerdale intrusion, commonly referred to as the Ennerdale Granophyre, lies to the west and north-west of Wast Water where it forms three outcrops, covering an area of about 16 km^2 (Figure 1). Within the district the intrusion is in contact with the Borrowdale Volcanic Group except at the southern end of Wast Water where there is an unexposed, possibly faulted, contact with the Eskdale granite. Gravity modelling and seismic reflection studies (Lee, 1989; Evans et al., 1993) suggest that it is a relatively thin tabular body, less than 2 km thick, and underlain by denser, less silicic plutonic rocks. The age of the intrusion is 452 ± 4 Ma (Hughes et al., 1996).

Pink, leucocratic, fine-grained, slightly porphyritic granite, commonly with granophyric texture is the dominant lithology. Biotite was probably the main mafic component though it is now altered. Dioritic rocks occur locally, adjacent to the margin of the intrusion. The Bleng diorite, exposed on the flanks of the Bleng valley [NY 125

085], is associated with a prominent magnetic anomaly (Chapter 3). The diorite contains conspicuously acicular pseudomorphs of chlorite and sphene, probably after amphibole, giving rise to the local name 'needle rock' (Rastall, 1906). At the margin of the intrusion south of Nether Wasdale around Mecklin Wood [NY 117 025] are dolerite and hybridised dioritic, granodioritic and melanocratic granitic rocks.

MINOR INTRUSIONS

Aphyric ***basalt and dolerite*** dykes, 0.5 to 6 m wide, are abundant within the Borrowdale Volcanic Group around the margins of the Eskdale granite from Eskdale to Wasdale. A few dykes cut the Eskdale granite, for example near Boot [NY 179 007], whereas others are hornfelsed by the granite. East of Burnmoor Tarn, in the upper Esk valley, dykes are orientated west-north-west to east-south-east, but around Wast Water the trend is more north-westerly. Macdonald et al. (1988) subdivided the suite into a high Fe–Ti tholeiitic, and a calc-alkaline, group.

Splintery, aphyric to sparsely microporphyritic ***rhyolite*** dykes, up to 8 m wide, are locally abundant within the Borrowdale Volcanic Group and the Eskdale pluton. North-west of Wast Water they form conjugate north-east and north-west sets, and south of Devoke Water they generally trend west-south-west. The rocks have a fine-grained to cryptocrystalline felsitic texture with fan-shaped, variolitic to spherulitic intergrowth textures. The dykes are genetically linked with the Ennerdale and Eskdale plutons and so the Rb–Sr isochron ages of 428–436 Ma (Rundle, 1992; Al Jawadi, 1987) are likely to have been reset.

The distinctive ***quartz–feldspar granite porphyry*** dyke of the Wasdale area and a geochemically variable suite of ***microdiorite–microgranite porphyry*** minor intrusions cropping out in the Scafell area, in the Duddon valley and in the Windermere Supergroup between Windermere and Coniston lakes may have been linked with the Skiddaw–Shap intrusive episode. This is based on Rb–Sr isochron ages of 390–395 Ma, close to the 395–400 Ma age of emplacement of the Skiddaw and Shap granites (Al Jawadi, 1987; Rundle, 1992). The quartz–feldspar granite porphyry is up to 35 m wide on Kirk Fell. Plagioclase and alkali-feldspar are present as phenocrysts in the microdiorite–microgranite porphyry suite which occurs as small bosses, stocks and dykes intruded along major faults. Garnet is present in some of these intrusions (Beddoe-Stephens and Mason, 1991). Other garnet-bearing and non-garnetiferous andesite and dacite minor intrusions, together with prominent sills intruding the volcaniclastic rocks, are related to the Borrowdale Volcanic Group.

Sparse, uncleaved, ***lamprophyre*** dykes are restricted to the west of the district where they intrude the Eskdale granite and Borrowdale Volcanic Group. In neighbouring districts lamprophyre dykes intrude Silurian strata of the Windermere Supergroup and are probably Early Devonian in age (Macdonald et al., 1985).

Quaternary deposits

The dominant deposit resulting from the Late Devensian glaciation is ***till***, now forming extensive, featureless spreads (British Geological Survey, 1998; Chapter 12). On major interfluves in the west of the district the till is generally less than 5 m thick and is penetrated by small exposures of solid rock, whereas in the east, the till is mostly confined to valley bottoms where more than 10 m occur locally. Till formed from materials within and on top of the glacier is heterogeneous, permeable, a metre or so thick, and composed of very poorly sorted, crudely stratified gravels with large boulders up to several metres in diameter and intercalations of silty sand, silt, and clay. The subglacial till is homogeneous, and typically overconsolidated, more clay rich, matrix supported and relatively impermeable. In the south-east of the district, mainly over the Windermere Supergroup, the till has been extensively moulded into drumlins.

Moraines, accumulated during the Dimlington and Loch Lomond Stadial corrie glaciations, which followed the retreat of the regional ice sheet (Sissons, 1980), are formed of ***Morainic deposits***. These highly variable and permeable lithologies include complex intercalations of matrix- and clast-supported diamicton and boulder gravel with beds of sand, silt and clay. A single arcuate ridge commonly marks the former position of a glacier front and hummocky moraine on the up-valley side of this represents supraglacial debris that was deposited as the glaciers retreated. Small areas of ***Glaciofluvial deposits***, deposited from meltwaters, occur locally and glacial meltwater channels are present in the west and south-west of the district.

River and lake sediments, along with spreads of peat, scree and head are of postglacial, Flandrian age (Chapter 12). ***Alluvium*** underlies present-day floodplains and forms ***River terrace deposits*** above the present floodplain; ***Alluvial fan deposits*** occur downstream from any sudden decrease in the watercourse gradient. These deposits consist mainly of gravel, sand and silt. Locally, organic-rich silt and peat are associated with the river deposits. Many rivers have built deltas into the lakes. Alluvial fans, predominantly of poorly sorted pebble and cobble gravel, occur notably in Wasdale, Eskdale, the Duddon valley and Great Langdale. The westernmost 2 km of the River Esk within the Ambleside district are tidal and where the alluvium is considered to have been deposited in an estuarine environment it is classified on British Geological Survey geological maps as 'undifferentiated' ***Marine deposits*** (British Geological Survey, 1998). These probably contain a higher proportion of mud and silt and less gravel than the river alluvium farther upstream.

Lacustrine deposits occur at the bottom of present-day lakes and tarns and within sites of former lakes and abandoned meanders on river floodplains. Much of the evidence of Late Devensian and Flandrian climatic variation is from the pollen assemblages found within sediments in the present-day lakes and tarns. Up to 40 m of sediment are found in Lake Windermere (Howell, 1971) which is designated the type area for the Late Devensian (Windermere) Interstadial (Chapter 12).

The rugged relief and relatively good drainage in central and eastern parts of the district has restricted blanket ***peat*** development there, but it covers more of the relatively smooth terrain farther west, notably on the water-

shed between Eskdale and the Duddon valley (British Geological Survey, 1991; 1998). Generally, the thickness increases uphill to a maximum of about 3 m though greater amounts occur in some poorly drained depressions on both high and low ground. Reed-swamp or fen peat is forming in poorly drained lowland basins today, for example, adjacent to Little Langdale Tarn [NY 305 033] and Elter Water [NY 335 042].

Head deposits are generally less than 1 m thick and probably formed under periglacial conditions from the downhill movement of rock regolith. Head deposits have not been mapped extensively though a thin veneer is probably widespread. The rockhead surface may be cambered, particularly on steep slopes where bedding or cleavage planes dip steeply into the hill side. Foundered strata occur locally on some of the major crags where rock masses have become detached and moved slightly down slope.

Aprons of ***scree*** are present below most crags of volcanic rock and occur locally where the Eskdale granite is exposed. The Screes at Wast Water [NY 146 038 to NY 176 064] are a particularly impressive example which probably began to accumulate during deglaciation and continue to grow today.

Structure

Geological structure of the district is dominated by the Scafell and Ulpha synclines, which affect the Borrowdale Volcanic Group, and by a major south-facing monocline, the steep limb of which incorporates the north-western margin of the Windermere Supergroup outcrop (Figure 1; Chapter 9; Johnson et al., 1979; Kneller and Bell, 1993). The monocline is Acadian (Early Devonian) in age, but the broad, open Scafell and Ulpha synclines, with approximately north-easterly and east-north-easterly trending axial planes, originated as Caradoc volcanotectonic structures (Branney and Soper, 1988) subsequently compressed during the regional, Acadian event. The Ulpha Syncline in the south-west of the district is cut by the Eskdale granodiorite and truncated by the pre-Ashgill unconformity at the base of the Windermere Supergroup; it is thus unequivocally of Ordovician age. Two areas of differing structural style affect Windermere Supergroup strata. In the north-west a broadly homoclinal region of steep south-east to south-south-east dip (the steep limb of the monocline) contrasts with an area of widespread minor folding situated farther south-east and forming the axial region of the Bannisdale Syncline.

A spaced to slaty and penetrative cleavage of Early Devonian age (Soper et al., 1987) affects all rocks of the district, including the Eskdale granite. It is best developed in the steep limb of the monocline affecting both volcaniclastic sedimentary rocks of the Borrowdale Volcanic Group and mudstone of the Windermere Supergroup. Cleavage is generally steeply inclined and its strike is relatively consistent across the district, trending 060° to 070°. However, the direction of dip varies systematically in association with the main structures of the district. Many of the cleaved rocks have proved suitable for working as slate.

The complex fault pattern within the district has a polyphase evolution. Many faults within the Borrowdale Volcanic Group originated through extension and volcanotectonic collapse contemporaneous with volcanism (Branney and Kokelaar, 1994). In parts of the Central Fells these faults are widespread, spaced 100 to 500 m apart and have vertical displacements of up to several hundreds of metres. Some faults have been inactive subsequently whereas others have undergone re-activation and mineralisation; for example, substantial late reverse movements occurred on the Langdale Fault. Some of the fault re-activation resulted from Acadian thrusting, Late Palaeozoic and Mesozoic extensional episodes and possibly Cainozoic uplift, but at each stage new faults may have developed also.

North-east to south-west-orientated thrusts and a north-north-east-trending swarm of normal and wrench faults are characteristic of the monocline. The principal thrusts, the Greenburn, Stockdale and Park Gill thrusts, are north-west directed. Their magnitude is illustrated by the cumulative displacement of at least 4 km across the Greenburn thrust zone.

The Coniston Fault Zone is one of the most substantial fracture systems within the central Lake District, traceable through the entire Borrowdale Volcanic Group from St John's in the Vale, east of Keswick, via Grasmere to Coniston, a distance of some 40 km (Moseley, 1993). Ordovician displacements are inferred from the volcanic lithostratigraphy. Substantial throw on the fault at the basal Windermere Supergroup unconformity diminishes upwards. The Coniston Fault Zone has a similar orientation to the boundary faults of the Lake District Lower Palaeozoic inlier with the Permo-Triassic East Irish Sea Basin and the Pennine faults of the Vale of Eden and was probably re-activated during Permian extension.

The east-north-east-trending Eskdale Fault cuts the Borrowdale Volcanic Group and Eskdale pluton in the western part of the district and is a prominent lineament on satellite images of the Lake District (e.g. British Geological Survey, 1992, fig. 4). The fault terminates eastwards at the Coniston Fault, but extends west of the district. North of the Eskdale Fault rocks generally dip shallowly northwards, whereas south of the fault dips are steep and to the south-east.

Mineralisation

The metalliferous mineral veins of the Lake District are mainly characterised by carrying major copper, major lead–zinc, or major iron (haematite) mineralisation; small but locally significant concentrations of antimony, arsenic, bismuth, cobalt, manganese, molybdenum, nickel and tungsten also occur. Lead is almost invariably accompanied by silver, and small amounts of gold have been reported from some deposits. Baryte is locally a substantial component of some veins. Iron ore, principally haematite, is locally common. Suggested ages of mineralisation range from late Ordovician for the copper-bearing veins to Early Devonian for the tungsten-bearing veins, Early Carboniferous for the main lead–zinc mineralisation and Cretaceous or Cainozoic for the haematite mineralisation.

In the district the Borrowdale Volcanic Group and the granitic intrusions host many mineralised veins (Chapter 11). Metalliferous veins are very rare in rocks of the Windermere Supergroup. The extraction of copper ores from veins within the volcaniclastic succession of the Borrowdale Volcanic Group in the Coniston, Tilberthwaite and Greenburn areas made the Lake District one of Britain's most important copper mining centres in the 19th century. Significant quantities of haematite have been mined from veins cutting the Eskdale granite.

OUTLINE OF THE GEOLOGICAL HISTORY

Evolution of the Avalonian continental margin

The Lake District lies immediately south of the line of closure of the Iapetus Ocean which, in Ordovician times, separated the continents of Laurentia to the north from Avalonia–Gondwana to the south. The ocean was destroyed by plate tectonic processes and the line of collision of its margins, the Iapetus Suture, now underlies the Solway Firth. Scotland formed part of Laurentia along with Greenland and north-east North America, whereas the Lake District lay on a microcontinental terrane, Eastern Avalonia, that comprised southern Britain and adjacent parts of continental Europe. Faunal and palaeomagnetic evidence (Cocks and Fortey, 1982; 1990; McCabe and Channell, 1990; Trench et al., 1991) show that in the Early Ordovician Avalonia was attached to Gondwana in a high southern latitude. Avalonia then rifted from the supercontinent, drifted north and eventually collided with Laurentia in tropical latitudes during the Silurian (Soper and Woodcock, 1990). The Lower Palaeozoic geology of the Lake District records this history and three main stages are recognised (Cooper et al., 1993). Firstly, the marine Skiddaw Group strata were deposited at the continental margin of Avalonia and then uplifted to form the subaerial basement for the second, volcanic phase that followed. During this phase volcanism was associated with caldera collapse forming the Borrowdale Volcanic Group on the Avalonia margin, above the subduction zone in which the Iapetus Ocean was destroyed. In the third stage Windermere Supergroup sedimentation and subsequent orogenic deformation were effects of the collision of Eastern Avalonia with Laurentia.

Pre-volcanic uplift

The marine turbidites of the Skiddaw Group were deposited in deep water on the margin of Eastern Avalonia, from late Cambrian to early Llanvirn times (Cooper et al., 1995). Regional uplift from a deep oceanic to a subaerial environment took place prior to building of the first volcanoes within the Llandeilo to Caradoc Borrowdale Volcanic Group. The absence of conclusively pre-volcanic tectonic structures in the Skiddaw Group (Hughes et al., 1993) indicate that large-scale terrane collision or accretion was an unlikely cause of this dramatic change. Uplift may have been the inevitable consequence of the generation of large volumes of magma that preceded volcanism at the continental margin. Whatever the cause, the uplifted Skiddaw Group was eroded considerably before volcanism commenced (Millward and Molyneux, 1992; Cooper and Hughes, 1993). North-west of the Ambleside district the probably alluvial Latterbarrow Sandstone lies at the base of the Borrowdale Volcanic Group and illustrates that the subaerial environment was established before the first volcanic eruptions (Allen and Cooper, 1986).

Volcanism at the continental margin

Following uplift of the Skiddaw Group a major volcanic province formed on the margin of Eastern Avalonia above the subduction zone in which the Iapetus Ocean was being consumed. Volcanism was relatively short lived during the time interval from the Llandeilo until the early Caradoc. Lavas and voluminous pyroclastic rocks were of basaltic to rhyolitic composition (Fitton et al., 1982; Beddoe-Stephens et al., 1995), and this phase of magmatism culminated in the emplacement of granitic laccoliths later in the Caradoc (Evans et al., 1993; Hughes et al., 1996). The volcanoes had low relief and were probably located within an extensional basin. The marine sedimentary rocks of the Holehouse Gill Formation testify to at least one incursion by the sea, but the volcanoes were essentially subaerial and their products were subject to continual reworking by processes involving explosive volcanic activity, mass wastage, and erosion by wind and water. Because of the ease with which unconsolidated subaerial pyroclastic deposits may be reworked, resulting sedimentary deposits may be relatively over-represented in the sequence. In general, preservation of the volcanic province is attributable to the extensional tectonic regime and successive episodes of caldera collapse associated with the cataclysmic eruption of very large volumes of tephra.

Low-profile andesite volcanoes

Volcanism began in the Ambleside district when large volumes of water, possibly from lakes, very shallow seas or an aquifer, came into contact with a near surface body of magma which was then explosively erupted to form tuff cones or tuff-rings. The unconsolidated cones were eroded and re-deposited locally. Following this establishment of magma channel-ways to the surface, volcanism in the Birker Fell Formation (Table 1) was dominated by eruptions of andesite block-lava flows that built up plateaux (Petterson et al., 1992). In many examples the top and bottom of the lava flows are remarkably regular and subparallel over considerable distances; laterally persistent bedded sedimentary units of river and sheet-flood deposits commonly occur conformably between the lavas. Simple and compound aa basalt lavas were erupted locally. The andesite lavas probably emanated from many vent sites and pyroclastic eruptions periodically filled valleys with ignimbrite and volcanic mudflow deposits (Petterson et al., 1992). At the end of the effusive phase the terrain had a relative relief of no more than 150 m and this was ultimately buried by the widespread andesitic ignimbrite at the base of the succeeding Whorneyside Formation.

CALDERA-RELATED MAGMATISM

The eruption of andesitic pyroclastic flows and phreatoplinian ash of the Whorneyside Formation heralded a major change to explosive volcanism. Thick ponded silicic ignimbrites and lava domes, followed by lacustrine sedimentary strata are typical of a 'caldera cycle' (Branney and Soper, 1988). The diversity of pyroclastic facies within the sequence records rapid alternations in eruption style between magmatic and phreatomagmatic activity, changes in the location, geometry and number of vents, and rapid changes in highly irregular caldera-floor topography during volcanotectonic collapse. Many of the pyroclastic and sedimentary formations thicken into the Scafell and Ulpha synclines, suggesting that these structures were probably initially sags resulting from caldera collapse and basin extension respectively (Chapter 9; Branney and Kokelaar, 1994; Branney and Soper, 1988; Davis, 1989; Channell and McCabe, 1992).

Within the Central Fells of the district, the earliest pyroclastic succession was associated with the development of the Scafell Caldera (Branney and Kokelaar, 1994). Vents for the Whorneyside eruption are believed to have been sited to the north-west of the Central Fells, in the adjoining Keswick district. Subsidence associated with that eruption allowed water to interact with the magma causing a change to phreatoplinian activity that deposited an air-fall ash blanket throughout the district (Branney, 1991). The succeeding silicic pyroclastic flow deposits of the Airy's Bridge Formation probably followed soon after and incremental, piecemeal subsidence created increasingly large fault scarps, many of which were the sites of avalanching and landsliding. The waning stage of this phase of volcanism is represented by the eruption of small-volume pyroclastic flows and effusion of volatile-poor lava domes, such as the Scafell Dacite. The broad depression that resulted from caldera formation became the site of lacustrine sedimentation.

A complex sequence of pyroclastic, effusive and sedimentary rocks is preserved in the Duddon Basin. The events that produced these rocks probably occurred over a relatively long time span compared with the Scafell Caldera volcanism which was probably fairly continuous. In the Duddon Basin a pattern of extensive and thick, welded ignimbrites succeeded by sedimentary rocks indicate that calderas probably were formed, though the geometry of these structures is less well understood than is the case in the Scafell Caldera. Phreatoplinian ash eruptions, associated with the effusion of andesitic lavas followed the first of the major ignimbrite-forming events. A narrow fault-controlled structure within the south-west of the district allowed marine conditions to be established locally with the deposition of the Holehouse Gill Formation.

Subsidence associated with eruptions in the Scafell Caldera and Duddon Basin produced a widespread basin in which fluvial and lacustrine sedimentation dominated (Table 1). Contemporaneous volcanic activity continued in a reduced form, with the influx of eruption-generated gravity flows of breccia, and beds of ash-fall andesitic tuff. Sills of basaltic andesite and andesite were emplaced into the unconsolidated, wet sediments and some of the sills may have broken surface to become extrusive locally. Thin silicic pyroclastic deposits presaged the return to explosive volcanism with the succeeding Lincomb Tarns Formation. The final products of the volcanic episode are not preserved, probably because emplacement of laccolithic elements of the Lake District batholith gave rise to greater uplift in the west of the district resulting in erosion to form the south-westward overstepping relationship seen at the unconformable base of the overlying Windermere Supergroup.

Foreland basin development and Acadian deformation

As the magmatism waned, marine conditions became established across the eroded and thermally subsiding volcanic pile; they were to last for more than 40 million years during the late Ordovician to the Early Devonian. A marked increase in subsidence and the corresponding increase in sedimentation rate during the Ludlow epoch has been associated with a foreland basin migrating southward across the Lake District during the final stages in the closure of the Iapetus Ocean (Kneller, 1991; King, 1992). This was initiated when the northern margin of Eastern Avalonia collided with the margin of Laurentia and a thrust sequence was propagated south-eastwards preceded by a flexural basin (Kneller, 1991; Hughes et al., 1993; Kneller et al., 1993b).

The Ashgill rocks were deposited during cyclic periods of deepening water. The earliest deposits were of beach and shore-face origin, with clastic material derived from the underlying Borrowdale Volcanic Group; these gave way to carbonate-rich sedimentation with restricted clastic sediment supply. Succeeding storm-dominated environments were followed by sedimentation during a period of maximum water depth. Reworked silicic pyroclastic material was brought in by turbidites from contemporaneous rhyolitic volcanism to the south-east. Minor erosion in the district resulted from a fall in relative sea level, but a subsequent period of deepening seas saw renewed carbonate sedimentation, followed by the accumulation of silty muds.

The abrupt change to relatively deep, marine shelf environments during the Llandovery resulted from the eustatic sea-level rise following the Hirnantian glaciation (Brenchley, 1988). The fine-grained strata of the Stockdale Group consist of graptolitic laminated black mudstone with subordinate blue-grey mudstone containing a sparse shelly benthos.

During the early Wenlock hemipelagic silts were deposited in an anaerobic deep shelf environment to form the Brathay Formation. A succeeding series of small, coalescing turbidite fans formed the Birk Riggs Formation, but their development ceased abruptly in the late Wenlock when two episodes of eustatic regression re-introduced aerobic conditions during which the carbonate-rich Coldwell Formation was deposited (Laufeld et al., 1975; McKerrrow, 1979; Scoffin, 1971).

As the foreland basin developed subsidence rates accelerated from the end of the Wenlock, and during the Ludlow thick spreads of turbidite sands were deposited to form the Coniston Group. Sediment dispersal was predominantly from north-east to south-west (Kneller, 1990a;

1990b) with detrital sandstone grains derived from a rapidly uplifted and geologically heterogeneous source area, probably an earlier orogenic belt (McCaffrey, 1991). Interruptions to this system resulted in periodic low sand input; deposition of the silty Latrigg Formation may reflect a period of relatively high sea level, but the subsequent Moorhowe Formation silts may have resulted from a change in the pathway and sorting mechanisms of sediment prior to final deposition. A change from sand-rich facies to mud-rich facies during deposition of the lower part of the succeeding Bannisdale Formation was diachronous, migrating progressively towards the south-south-east with time.

Sedimentation continued into the Early Devonian and the Windermere Supergroup rocks may have been buried beneath an overburden of at least 8 km to produce anchizonal metamorphic conditions (Merriman et al., 1995). Final inversion of the foreland basin and development of typical 'slate belt' structures, folds, cleavage and associated faults appears to have been a climactic event during the Acadian orogeny in the Early Devonian. Cleavage formation was synchronous with emplacement of the Early Devonian Shap granite and its associated dykes (Boulter and Soper, 1973; Soper and Kneller, 1990).

In terms of large-scale regional structure the south-east-facing monocline of the southern and central Lake District has been proposed as the consequence of thrusting south-eastwards above a ramp dipping gently north-west (Kneller and Bell, 1993). In this interpretation the ramp lies beneath the central Lake District and continues south-eastwards as a flat detachment beneath the Windermere Supergroup. A displacement up the ramp of about 20 km is accommodated by back-thrusting within the monocline and by shortening within the Windermere Supergroup strata.

Upper Palaeozoic to Cainozoic evolution

There is no direct evidence from the Ambleside district of Upper Palaeozoic to Cainozoic sedimentation and tectonics. However, evidence from neighbouring districts gives clues to a history associated with the formation of block and basin structure in northern England during the Carboniferous, development of the East Irish Sea Basin in the Mesozoic and the opening of the North Atlantic Ocean from the mid-Cretaceous (Chadwick et al., 1994; 1995).

The Acadian orogeny resulted in at least 8 km of uplift for the Lower Palaeozoic rocks but the resulting Devonian molasse deposits, mainly formed of eroded Windermere Supergroup lithologies, are preserved only locally as the Mell Fell Conglomerate (Wadge, 1978) in the northern Lake District. During the Early Carboniferous the Lake District massif formed a relatively stable, buoyant block, adjacent to rapidly subsiding basins (Chadwick et al., 1995). The massif only became submerged later and accumulated a thin sequence of limestone and siliciclastic deposits. The influence of the structural blocks in the north of England waned during the later Carboniferous and deltaic, lagoon-swamp conditions prevailed in the Westphalian producing the Coal Measures now preserved in west Cumbria.

Uplift and erosion of the Lake District massif during the latest Westphalian and Early Permian exposed the Borrowdale Volcanic Group rocks which were then eroded and re-deposited as alluvial fans (Brockram) in west Cumbria at the eastern margin of the East Irish Sea Basin. Extensional faulting at the western margins of the Lake District and Pennine blocks was the initial control on subsidence and sedimentation, but this was replaced by thermal subsidence from the Late Permian into the Early Jurassic. During this time sedimentation across the Lake District block was dominated first by aeolian and fluvial systems, then by marine conditions. From the Mid-Jurassic to Early Cretaceous the district was apparently once again emergent (Cope et al., 1992).

The propagation of sea-floor spreading into the North Atlantic region by mid-Cretaceous times produced widespread shelf conditions and a relatively uniform Upper Cretaceous chalk sequence was probably deposited across the Lake District block (Chadwick et al., 1994). By the end of Cretaceous times the district probably reached its maximum Mesozoic burial. Regional, flexural uplift across the Lake District block of approximately 1750 m (Chadwick et al., 1994) then occurred from about 65 Ma (Lewis et al., 1992) resulting in the erosion of all Carboniferous and younger strata and exhumation of the underlying Lower Palaeozoic rocks.

Glaciation

Climate in northern Britain began to deteriorate from the mid-Miocene, with the growth of the Antarctic and Greenland ice sheets, but the region was not glaciated fully until about 750 000 years ago. Cold episodes of ice-sheet growth were separated by warmer periods with similar or even milder climatic conditions to those of the present day. Repeated glaciation has modified the pre-Quaternary, radial drainage topography of the Lake District by widening, straightening and deepening pre-existing river valleys and breaching watersheds. Only fragmentary evidence survives of Quaternary events prior to the last, Late Devensian, regional ice-sheet glaciation.

The last glacial episode began about 120 000 years ago, with intermittent full glaciation of the Lake District continuing until the Dimlington Stadial, about 25 000 years ago. Glacial outwash and till accumulated in Lake District valleys with most local ice deflected southwards by the main ice-stream emanating from Scotland. Amelioration of the climate resulted in the Lake District becoming almost completely deglaciated by the onset of the Windermere Interstadial at about 14 000 years BP (Pennington, 1978) and by 13 000 years BP summer temperatures were similar to those of the present day (Lowe and Walker, 1984). Sand and gravel were laid down from meltwaters and ice-marginal moraines and glacial meltwater channels were formed.

Vegetation in the early part of the Windermere Interstadial was dominated by grasses, sedges and herbaceous plants that thrived on disturbed soils. Open-habitat conditions were succeeded by a shrub vegetation of juniper, willows and crowberry, which later developed into birch woodlands in the lowlands (Pennington, 1977).

The Loch Lomond Stadial saw another deterioration of the climate from about 11 000 to 10 000 years BP (Pennington, 1977). Corrie glaciers reformed and tundra vegetation was re-established in the surrounding areas. Periglacial activity was widespread and included the formation of mountain-top blockfields, solifluction lobes, screes and protalus ramparts that developed at the feet of perennial mountainside snow patches (Ballantyne and Harris, 1993). The patchy vegetation cover left soils vulnerable to erosion so that rivers carried high bed-loads and formed gravelly river terraces.

Retreat of the corrie glaciers around 10 300 BP was initiated by the abrupt amelioration in climate at the beginning of the Holocene Interglacial. Since then the landscape has continued to be modified through rock fall, soil creep and debris flow and the re-colonisation of the terrain, firstly by a shrub assemblage and subsequently by mixed deciduous woodland on lower ground (Pennington, 1964; 1970).

Widespread clearances of the forests began in the late Iron Age (Pennington, 1975a; Barber et al., 1994), when arable agriculture was increasing, and the landscape was almost completely cleared of trees during Roman times. The later influence of human activity on the environment of the district is recorded in the lakes. Atmospheric pollution from the beginning of the industrial revolution has increased the concentrations of certain heavy metals such as zinc in the lacustrine sediments (Mackereth, 1966; Haworth, 1985). The biota in Coniston Water is less diverse than in other lakes following pollution in the 19th century when copper mining in the Coniston area was at its peak (Davison et al., 1985). Accelerated erosion and mass transport during the past few centuries (Macklin et al., 1994) may be related to the destabilisation of mountain soils by overgrazing (Harvey et al., 1981), particularly during periods of wetter and cooler climate such as in the 'Little Ice Age' at the beginning of the 18th century.

TWO

Applied geology

The Ambleside district lies entirely within the boundaries of the Lake District National Park, 2280 km^2 of upland with outstanding natural beauty, designated in 1951 following the *National Parks and Access to the Countryside Act* of 1949. Hill farming remains the principal land use but throughout the Lake District tourism now dominates the local economy. It is the spectacular scenery which attracts large numbers of visitors and the geological foundation of that scenery is thus the principal local asset. This aspect of the Lake District has a long history but popular tourism really became established in the latter part of the 18th century. Then the developing taste for picturesque Gothic was led by the poets Thomas Gray and William Wordsworth. The latter in particular drew inspiration from the dramatic surroundings produced by the glaciation of the hard, Lower Palaeozoic rocks, and geological allusions are not uncommon in his work. Wordsworth settled at Grasmere and his work served as a magnet for others, including Lamb, Coleridge, Shelley, Tennyson and Ruskin. Ruskin's personal interest in geology led to him assembling a collection of minerals and rocks at Brantwood, his Coniston home; this collection is now housed in Coniston.

Today's visitors are still drawn largely by the same spectacular scenery and thus the unique geological characteristics of the Lake District. The National Park Authority estimate 20 million visitor-days per annum to the park. The volume of tourist traffic has become a geological agent of some importance as erosion of footpaths and development pressures of various kinds increase. This enormous influx of visitors also has an impact on such issues as water supply and waste disposal, both of which demand solutions outside the Ambleside district as here defined.

KEY ISSUES

With the local dominance of the tourist industry the conservation of the primary asset, the scenery, is likely to continue to be of paramount importance. In part, the landscape is the result of previous exploitation of natural resources, which still continues on a small scale. Slate quarrying and metal mining both have long histories with the former the only currently active extractive industry.

Major issues on which the geology has a bearing include:

- Quarrying
  - Slate
  - Building stone and aggregate
- Mining
  - Non-ferrous metals
  - Iron ore
- Drift resources
  - Sand and gravel
  - Peat
- Geothermal resources
- Ground stability
- Hydrogeology and water supply
- Waste disposal
- Seismicity
- Conservation

QUARRYING

The Neolithic stone axe industry of Langdale records one of Man's earliest use of natural resources, one of Britain's oldest stone-based industries. It has been known for a long time that sites in Great Langdale, and to a lesser extent on Scafell Pike and Glaramara, were the source of huge numbers of beautifully polished stone axes found over a wide area of Britain (Bunch and Fell, 1949; Chappell, 1987; Cummins, 1979). Though the stone from which the axes were made varies somewhat in character (Wooley, in Claris et al., 1989), the source is recognised to be a bed of very fine-grained volcaniclastic sandstone and siltstone within the Seathwaite Fell Formation (Chapter 6). Claris et al. (1989) have provided a detailed description of the working sites.

Slate

Slate was formed across wide areas of the Ambleside district during the Acadian orogeny, when an intense cleavage was imposed on many of the rocks. The most even and regular cleavage occurs in the finer-grained and lithologically uniform sedimentary rocks of the Windermere Supergroup. These form what are referred to locally as blue-grey slates. However, the district is most renowned for its green slate formed from fine-grained, volcaniclastic sedimentary rocks of the Borrowdale Volcanic Group (Chapters 4 to 6). Locally, coarse-grained pyroclastic rocks, andesite lavas and welded tuffs are also cleaved, and have been worked as poorer quality slate. The larger quarries within the district are shown in Figure 3.

Most of the commercial quarries within the Borrowdale Volcanic Group, including all those that are currently active, occur within the Seathwaite Fell Formation (Chapter 6). The remainder are found in the Duddon Hall and Lickle formations in the Duddon valley. Smaller operations occur in suitable lithologies throughout the volcanic succession. The largest quarries in the Windermere Supergroup occur principally in the Ashgill and Brathay formations (Chapter 7). All the quarries lie within the high-strain zone of the steeply dipping, south-east-

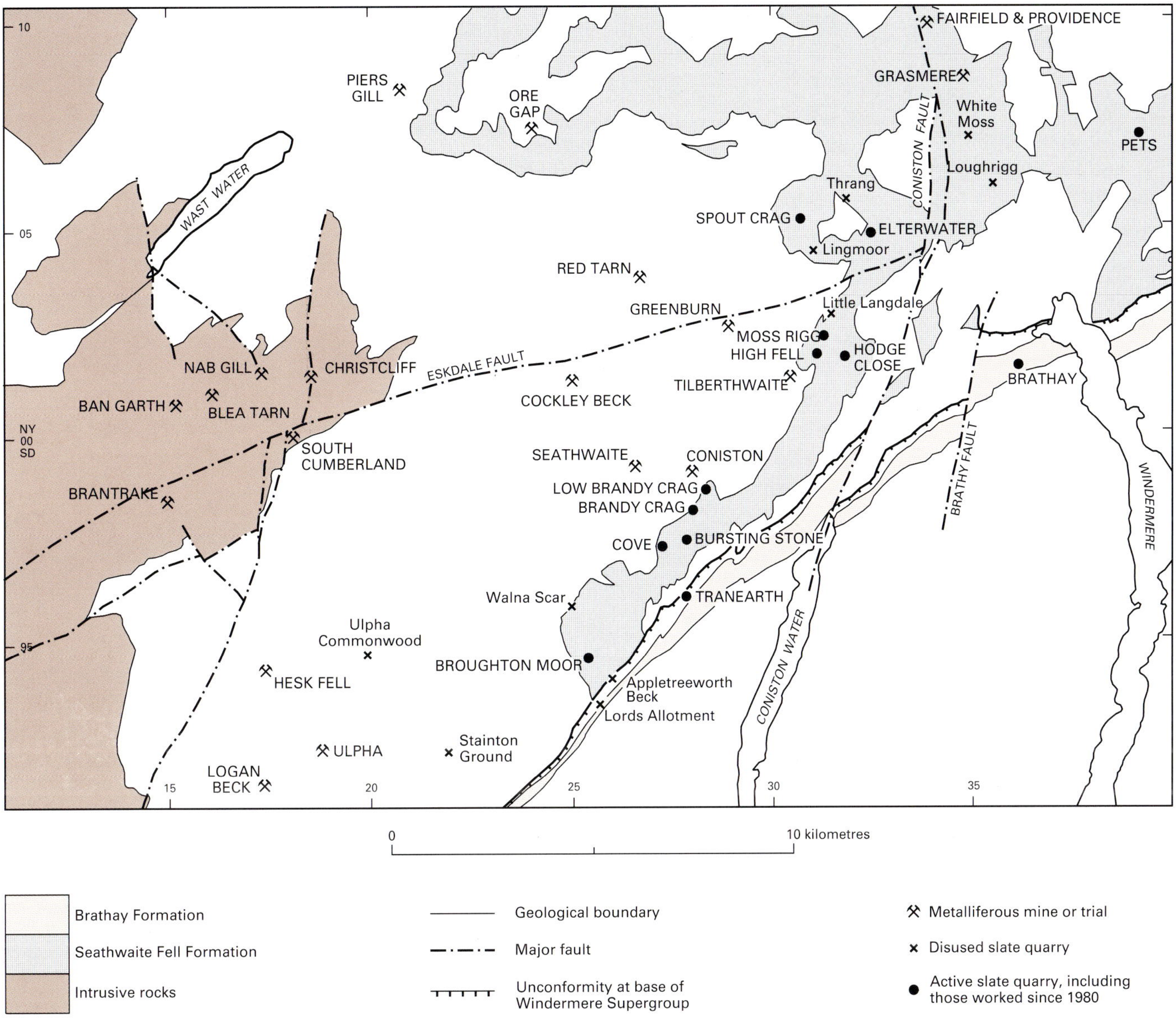

Figure 3 Location of major slate quarries and sites of metalliferous mining.

Abbreviations: Enn Ennerdale intrusion; EskG Eskdale granite; EskGd Eskdale granodiorite. Some of the mine symbols, for example those labelled as Coniston, Greenburn and Tilberthwaite, represent groups of mines.

facing limb of the monocline on the south side of the Lake District batholith.

The essential characteristic for roofing slate is the regular fissility that gives a product of uniform thickness. This is generally found in fine-grained lithologies affected by the highest strain; ideally the cleavage should be orthogonal to the bedding, which may result in attractive patterns on the slate surface from the exposure of fine detail of sedimentary structures. The natural colour and markings, variety of finishes and textures make it much sought after for internal, architectural and ornamental purposes. Colour, grain size and appearance are factors that influence the choice of finish that can be given to the cut slate. For structural and architectural products the quarrying requirement is for blocks up to 5 tonnes. The durability, resistance to weathering, dirt and abrasion, as well as the attractive appearance, make the slate a suitable material for roofing, cladding and paving.

History of the slate industry

Green slate was first worked nearly 2000 years ago by the Romans who used it for damp-proof courses, flagging and roofing in their forts. In the 12th and 13th centuries ecclesiastical buildings and houses of the nobility were roofed with slate. The introduction of gunpowder in the 16th century, growing prosperity and a desire for fine buildings, led to quarrying on a commercial scale and by 1805 some 25 000 tonnes of slate were produced annually from quarries in the Coniston and Kirkby in Furness areas (Stephens Associates, 1988). The use of compressed

air for drilling and lifting equipment early in the 19th century revolutionised quarrying, and the coming of the railways in the middle of the century overcame transportation constraints and boosted production. The first slate company, Burlington Slate Quarries, was founded in 1843. Lake District slate was widely used for prestigious building projects, including Chelsea Hospital and Kensington Palace in London. The decline of the industry began in late Victorian times, initially caused by the availability of cheaper Welsh slate and the development of alternative roofing materials.

The beginning of the 20th century saw the closure of many of the small quarries, leaving only a few commercial operations. Statistics for 1909, issued by the Chief Inspector of Mines, record slate production in Cumberland, Westmorland and Lancashire at 24 778 tonnes, which was 6.3% of the UK total. By 1975 the amount extracted annually had reduced to around 15 000 tonnes, though this represented 23.9% of the UK total (Stephens Associates, 1988). These figures reflect the growing demand for slate for architectural, structural and ornamental purposes in the UK and abroad, though with a continuing decline in the use of slate for roofing.

The last 30 years have seen further reduction and rationalisation of the industry. There are only two major companies and a handful of smaller operators active at present. The two major companies produced some 8000 tonnes of blue-grey and green slate in 1991 and 1992 from quarries in the Ambleside and neighbouring Ulverston districts, approximately 60% as roofing slate and the remainder as structural, architectural and decorative products. An increased interest amongst UK architects in the use of natural stone, and the requirement of the Lake District Park Special Planning Board for the employment of local materials in the park, are likely to ensure a continued market for Lake District slate. Significant slate resources are likely to be present as extensions of existing quarries, in currently abandoned workings and in hitherto unexploited areas.

QUARRYING METHODS

Originally the slate was quarried from hillside exposures by hand, but as skills developed and demand grew, much of the slate extracted during the last century was by underground mining. Some of these caverns are still accessible, for example 'Cathedral' quarry [NY 314 028] within the Little Langdale Quarries complex and several on the northern flanks of The Old Man of Coniston [SD 277 980]. Since the introduction of pneumatic equipment, drilling and blasting have been employed generally. Either horizontal or vertical drilling is used, depending on the inter-relationships of the cleavage and the quarry face; the aim is to split the slate along the cleavage and safely loosen it from the working surface. The growth in demand for sawn and polished slate since the 1960s increased the need for large blocks free from blast damage (Plate 3). The introduction of wire-sawing techniques overcame these difficulties. Technological improvements, such as the introduction of industrial diamonds into cutting wire, now allow wire-sawing to be used widely.

Plate 3 Sawn blocks of slate. The material shown here has been sawn from volcaniclastic rocks of the Seathwaite Fell Formation at Pets Quarry, Kirkstone [NY 392 071] (MNS 5415).

Building stone and aggregate

The vernacular architecture of the district is characterised by the use of stone obtained locally from the most convenient source. Most of the hard rocks have been employed for domestic or farm buildings, and for the construction of the many miles of drystone walls erected at the time of the Parliamentary Enclosures Act. The use of slate has been outlined above. Only for the most important or expensive buildings has stone been imported into the district. All bricks used in the district have been imported.

The Eskdale granite, at Beckfoot Quarry [NY 1642 0034], was worked for making road setts in the 1920s (Davies, 1981), and some granite for building stone may have been obtained there and at other small quarries. The largest quarries worked for building stone were those within the grey, biotite and garnet-bearing granodiorite of the Broad Oak Quarries at Waberthwaite [SD 1122 9438]. The quarries are not permanently active, though in recent years a small amount has been extracted as ornamental stone. All preparation of the stone is done outside the district.

Quarries in the Eskdale granite at Beckfoot and on the north side of Muncaster Fell at Murthwaite [SD 1153 9906] have been worked for both crushed rock aggregate and railway ballast. The history of granite quarrying in Eskdale has been reviewed briefly by Davies (1981). Very large reserves of granite, and of other hard rocks within the Borrowdale Volcanic Group, exist within the district though the comparative remoteness from large, potential markets and the situation within the National Park render any resumption of working extremely unlikely.

DRIFT RESOURCES

Sand and gravel

Small areas of sand and gravel occur both within the glacial and fluviatile deposits of the district (Chapter 12). No large or extensive deposits are known and none has been worked except for very small-scale local use.

Peat

Variable thicknesses of hill peat blanket many parts of the district and low-level peat is present locally, for example in the valley of the River Esk (British Geological Survey, 1998; Chapter 12). Some of the hill peat blanket has been dug in the past for local use as fuel. Traces of old peat diggings may be seen at a few sites on the hills above Eskdale [e.g. NY 1585 0085, NY 1694 0167 and NY 1944 0202]. In view of the very small size of the deposits the resumption of peat working either for fuel or for horticultural use is very unlikely.

MINING

Non-ferrous metals

Much of the southern portion of the Lake District metalliferous orefield is located within the district including Coniston, one of Britain's major centres of copper mineralisation. Descriptions of veins within the Coniston mining field and the associated smaller concentrations of copper mineralisation in the Greenburn, Tilberthwaite and Ulpha areas (Figure 3) are given in Chapter 11. Metal production within the district has been dominated by copper mining, though significant quantities of iron ore have been raised as well as minor amounts of lead, zinc, cobalt and nickel. Ores of other metals are present, but are not known to have been worked. The bulk of the copper production within the Coniston field was from the two large mines, Paddy End and Bonsor, located within the Copper Mines Valley, between Coniston village and Levers Water. Holland (1986) has provided a detailed historical description of the mining industry at Coniston.

Though the Coniston deposits may have been discovered and worked by the early Britons or the Romans (Dewey and Eastwood, 1925), the first documentary records of mining at Coniston date from around 1599. At this time systematic exploration and development of the veins was begun by German miners under the supervision of Danial Höchstetter who was employed by the Elizabethan 'Company of Mines Royal'. Some of the outcrop workings, such as Simon's Nick, just south-east of Levers Water, may date from this period. Such was the richness of some of the deposits near Levers Water that Höchstetter is said to have considered draining the tarn in order to secure further ore reserves.

It is likely that working continued intermittently until the English Civil War brought mining to an end (Matheson, 1986, p.31). The 18th century saw considerable mining and prospecting activity at Coniston, but the middle years of the 19th century became the heyday of the Coniston mines. During this period several new access levels, including the Bonsor Deep, Flemming's and Taylor's levels were driven, and many shafts such as the Old and New Engine, Paddy End Engine and Thriddle were sunk. Annual outputs of 3000 to 4000 tonnes of dressed copper ore were recorded in the 1850s, though by 1876 output had fallen to about 1000 tonnes. By 1895 the world price of copper had fallen to well below the economic limit for the deep workings at Coniston. Moreover, the increasing proportion of magnetite in the ore from the deepest workings at Bonsor rendered the ore almost impossible to dress by the gravity methods then available. Pumping was therefore stopped and production restricted to the removal of payable ore from remaining pillars. The tonnage of ore raised between 1895 and the mines' final abandonment in 1908 is unknown, but is likely to have been comparatively small. Whereas there are no reliable records of the grade of run-of-mine ore during its long life, Eastwood (1959, p.171) noted that contemporary reports during its final years (1891 to 1908) suggest a general average grade of 3% copper, with the Paddy End ore usually averaging 2 to 5%. Total production of dressed copper ore from Bonsor and Paddy End mines, estimated to have contained between 5 and 13% copper, amounts to about 52 000 tonnes (Eastwood, 1959). In 1855 about three tonnes of nickel and cobalt ore were raised at Coniston and in 1893 there is a record of about 12 tonnes of lead ore.

The large dumps at both Paddy End and Bonsor mines have been estimated to contain an average copper content of 0.9% with some parts up to about 1.5%. An attempt was made between 1912 and 1914 to extract copper from dump material in an electrolytic plant built on the site of the Bonsor dressing floors. The operation was abandoned in 1915 (Shaw, 1970, p.116). The most recent attempt at working the Coniston mines was an abortive attempt to reopen parts of the Paddy End Mine in the 1950s.

Rather less historical information is available for the associated centres of copper mineralisation in the district. Mining is known to have been in progress at Tilberthwaite in the 16th century, or perhaps even earlier. There is evidence for 18th century mining both there and at other centres such as Ulpha though, as with the main Coniston mines, it was the 19th century which saw the most extensive exploration and working. Some unsuccessful prospecting was undertaken in the Tilberthwaite area in the 1930s, but it was not until 1942 that the Greenburn and Tilberthwaite Mining Company went into liquidation.

Apart from the galena- and sphalerite-bearing veins of the Coniston copper mines the only other known lead–zinc veins of the district are those of Grasmere Mine. These were worked in Elizabethan times, but they are not known to have been worked since. The production of lead from these mines must have been very small.

Mineral reserves, particularly of copper, are still present in the Coniston area though they are unlikely to prove economic. Cameron et al. (1993) have identified a number of geochemical anomalies around Ulpha, Torver and in the upper catchment of the River Lickle that may indicate hitherto unrecognised sites of mainly base metal mineralisation. The present distribution of metallic elements in stream sediment is summarised in the geochemical atlas for the Lake District (British Geological Survey, 1992).

The old workings and mine dumps, may present a local problem in terms of undermining or pollution but in neither case is this likely to be significant. The historical extent of heavy metal pollution was considerable, however, and the biota in Coniston Water is still less diverse than in the other lakes following 19th century pollution at the peak of copper extraction from the Coniston mines (Davison et al., 1985).

Iron ore

The district includes a suite of haematite-bearing veins that are believed to be genetically related to the large replacement haematite ore bodies of the Carboniferous limestones of west and south Cumbria. The origin and characteristics of the district's haematite deposits are discussed in Chapter 11. Iron mining and smelting in the area may date from Roman or earlier times. Small piles of iron slag at several places in Eskdale suggest small-scale iron working, possibly in medieval times, no doubt using locally produced charcoal as fuel. No mine workings are known that can be dated positively to these early periods.

Some mining for haematite in veins in the Borrowdale Volcanic Group rocks is recorded in the 17th and 18th centuries at Red Tarn [NY 268 037] and Little Tongue Gill [NY 339 099], Grasmere. However, a more systematic exploration for ore reserves at these localities and in the Eskdale area dates from the second half of the last century when demand for home-produced iron ore increased, mainly as a result of the Franco-Prussian War. At this time deposits in the Eskdale granite, such as those at Ban Garth, Brantrake, Christcliff, South Cumberland and Nab Gill mines were developed and brought into production (Figure 3). Though some good ore was raised, only the Nab Gill Mine [NY 174 013] proved to be productive for more than a few years. The deposit here was considered sufficiently promising to warrant the construction of the original Ravenglass and Eskdale Railway (Davies, 1981). However, by the turn of the century mining had all but ceased and Nab Gill finally closed in 1917. According to manuscript notes by W C C Rose in the British Geological Survey archive, some exploration for haematite was undertaken at Red Tarn in the 1930s.

Postlethwaite (1913) made brief reference to veins of haematite up to 15 m wide near Ore Gap [NY 240 072] between Esk Pike and Bow Fell. He recorded that high-quality, mainly kidney ore was worked in about 1700. Smith (1924, p.214) quoted an old report of mining for haematite near Crinkle Crags and on Harrison Stickle, Langdale, but the sites of these workings were not given. Despite Smith's suggestion that these veins together may hold several million tonnes of ore it seems that no successful mine was established. None of these deposits are likely to be of economic interest.

Production figures have not survived for most mines and where such records are available they are incomplete. Blea Tarn Mine [NY 167 007] is known to have produced at least 300 tonnes of ore in 1874 and in 1881 the Birker Moor Mine raised more than 3 000 tonnes; in 1881 South Cumberland Mine [NY 179 000] was reported to be producing 100 tonnes of ore per week (Hibbert et al., 1940). Figures obtained from various sources indicate that total ore production from Nab Gill Mine exceeded 28 500 tonnes, though returns for some years may include ore raised by the same company at Blea Tarn.

There are no grounds for expecting the presence of further haematite deposits of economic interest, either within the known veins of the district, or in hitherto undiscovered veins.

GEOTHERMAL RESOURCES

In the early 1980s, the Lake District batholith was identified as a potential hot dry rock geothermal resource because of its large intrusive volume and indications that it might contain above average concentrations of the main heat-producing radioelements (uranium, thorium and potassium). A study funded by the Department of Energy was carried out by the British Geological Survey, in collaboration with the Open University and Imperial College, to assess the hot dry rock potential of the region (Lee, 1986b; Lee et al., 1984; 1987; Webb and Brown, 1984; Webb et al., 1987a; Wheildon et al., 1984). This was focused mainly on the Shap and Skiddaw granites where measurements in boreholes showed elevated heat-flow values of 77.8 and 100.9 mW/m^2 respectively, caused by above average radiogenic heat production of 5.2 $\mu W/m^3$ and 4.2 $\mu W/m^3$ respectively in the two intrusions.

No heat flow measurements were made in the Ambleside district and the high values observed over the Shap and Skiddaw granites are considered to be unreliable indicators of likely values elsewhere within the composite Lake District batholith. Limited heat production data are available for the Eskdale granite and Ennerdale intrusion (Table 2), but the measured values at the surface are difficult to interpret in terms of predicting the likely heat flow (and thus hot dry rock geothermal potential) over the western parts of the batholith. The data for the Eskdale granite include a rather low uranium content and consequent low heat production of around 1.9 $\mu W/m^3$. However, a detailed study of other geochemical data led Webb and Brown (1984) to conclude that intensely oxidised alteration, as shown by extensive haematisation at outcrop, had mobilised and depleted primary uranium content in the upper part of the intrusion. They thought that if haematisation and leaching of radioelements are

Table 2 Heat production data for the Ennerdale and Eskdale granites.

Intrusion	Uranium range (ppm)	Uranium mean (ppm)	Thorium range (ppm)	Thorium mean (%)	K_2O range (%)	K_2O mean (%)	Heat production ($\mu W/m^3$)
Ennerdale granite	4.0–6.4	WR 4.6	17.2–22.5	WR 21.0	0.24–4.98	1.91	2.8
Eskdale granite	1.1–6.2	WR 4.2 GS 3.3	6.1–25.1	WR 10.3 GS 11.3	1.53–5.09	4.53	1.9

Notes:

WR indicates values derived from whole-rock analyses.
GS indicates values derived from calibrated field gamma spectrometry.
Data from Webb and Brown (1984).

less severe at depth, or if radioelements there are hosted by more stable accessory mineral phases, higher uranium contents would be expected in the deeper parts of the intrusion. Thus, while the observed heat production value of 1.9 $\mu W/m^3$ may represent the upper part of the Eskdale granite, the values at depth may be as high as 4 to 5 $\mu W/m^3$, indicating that heat-flow values over the Eskdale granite may be as high as those observed at Shap and Skiddaw. The likely heat-flow values over the Ennerdale intrusion and Eskdale granodiorite are more difficult to predict than over the Eskdale granite, because seismic interpretation and gravity modelling (Chapter 3) indicate that these are laccoliths underlain by concealed components of the batholith. Subsequent mapping and geochemistry have revealed the inhomogeneous nature of the Eskdale granite at surface, but with some exceptions (Chapter 11), have confirmed the relatively low surface concentrations of uranium in the freshest rocks obtainable (Table 14).

GROUND STABILITY

Solid rock is exposed over large areas of the district; characteristically it is thoroughly indurated with very low intergranular permeability. In general, the rock should provide a sound foundation, but it is variably fractured and cleaved (see Chapter 9) and the intensity and distribution of these discontinuities may influence the design and erection of structures.

Extensive and locally thick accumulations of drift or superficial deposits are a characteristic of many of the major river valleys and lake surroundings (Chapter 12; British Geological Survey, 1998). Extensive spreads of glacial till occur on interfluves in the west of the district but these are generally less than 5 m thick. However, in the valley bottoms the till may be in excess of 10 m thick. The till is lithologically very variable as described in Chapter 12; the coarser deposits are generally permeable, whereas the more clay-rich ones are relatively impermeable and overconsolidated. The lacustrine and alluvial deposits are permeable and unconsolidated. These may contain layers and lenses of laminated clay and organic-rich material that have implications for the stability of structures.

A small number of landslips has been recorded locally in till on the slopes of the River Lickle in the south-west of the district. Details are given in Chapter 12.

HYDROGEOLOGY AND WATER SUPPLY

This mainly high, mountainous area is deeply dissected by valleys and includes the large lakes of Wast Water, Grasmere, Rydal Water, Coniston Water, Esthwaite Water and Windermere (Figure 2). The district is drained to the west by the rivers Bleng, Irt, Mite and Esk, to the south through the rivers Duddon and Lickle, and into Coniston Water and Windermere. Rainfall varies from 1300 mm/a in the south-east of the district over the generally low ground, to levels of over 5000 mm/a in the high mountains in the north.

The formations within the district do not include any regionally significant aquifers. Most of the water supplies are derived from abundant surface sources. Groundwater sources mostly consist of springs and shallow wells. The bedrock has very low intergranular porosity and permeability. As a consequence the groundwater storage and flow is restricted to joints and fractures in the shallow, weathered zone. The combination of steep slopes and low hydraulic conductivity results in small volumes of groundwater flow and randomly orientated small-scale flow systems. Water balance calculations suggest that the annual infiltration to groundwater in these rocks is less than 40 mm (Wadge, 1966). The superficial deposits have some potential; the permeability of glacial till, though quite variable, gives rise to numerous springs, and alluvium forms a modest aquifer in places.

Limited groundwater resources in the district are indicated by the sparse records of hydrogeological data. The British Geological Survey Groundwater Archive contains 51 sites in the district. About half of these are described as springs, and half as shallow wells. A small number of deeper boreholes have a maximum depth of 35 m. The shallow wells could result, in places, from the development of springs and the overall average yield is 12 m^3/d. A groundwater abstraction licence from the Environment Agency (formerly National Rivers Authority) is only required for more than 20 m^3/d and there are only 17 licensed abstractions in the district.

The highly indurated and heavily faulted Borrowdale Volcanic Group rocks have almost no intergranular flow or storage capacity. Thus, any groundwater is likely to be limited to fractures and faults, and weathered rock within 30 or 40 m of the ground surface. This gives rise to numerous springs, mostly of low yield, that can be subject to marked seasonal variation, commonly failing in periods

of drought. Of the 13 sites in the Groundwater Archive most are springs with some shallow wells and two deep wells; of these five are licensed. The largest yield is 45 m^3/d, from the Seathwaite Fell Formation. The Birker Fell, Duddon Hall and Lincomb Tarns formations also produce sources of groundwater with an overall average yield of 11 m^3/d (excluding the largest source of 7.6 m^3/d).

The Windermere Supergroup rocks have more favourable hydrogeological characteristics, though intergranular permeability and storage capacity are very low, even in the greywacke sequences, because of the high degree of cementation. Faults are more widely spaced than in the Borrowdale Volcanic Group. Groundwater quantities again depend on the presence of fractures, which are considered unlikely to be generally open below 30 or 40 m from the ground surface. There are 30 sites in the Groundwater Archive, mainly springs with some shallow wells and two deep wells. Only three formations are known to have been developed for groundwater, the Wray Castle and Bannisdale formations and the Coniston Group. These sources have an average yield of 20 m^3/d.

The intrusive, coarse-grained igneous rocks in the west of the district have eight sources recorded in the Groundwater Archive and these are all shallow wells. Information is limited to rocks of the Eskdale pluton, with an average yield of 7.7 m^3/d. The groundwater resource is limited to joints and faults and the surface weathered zone. On the higher ground the weathered zone tends to drain rapidly, and little groundwater may remain in storage.

The superficial deposits contribute significantly to the groundwater regime. The extensive cover of till in the valleys controls groundwater flow from the solid formations in the surrounding uplands. Springs form at the edge of the outcrop, associated with the solid rocks and also rise through fractures or permeable zones within the till. Shallow wells collect groundwater from springs or intersect the potentiometric surface. Of the 51 sites in the Groundwater Archive, 23 are associated with till and have an average yield of 15 m^3/d. The glacial sands and gravels have some potential for supply, though they are not always free of a substantial clay or silt fraction. The alluvium in the base of the valleys, though of more limited outcrop area than the till, forms a useful aquifer deriving the water from the solid formations below and from infiltration from the local surface drainage system. This has been exploited by shallow, dug wells and boreholes. However, the size of the resource is small and the six sites in the Groundwater Archive have an average yield of 3.3 m^3/d.

No chemical analyses of groundwater are known to have been published. Data derived from analysis of river water during baseflow conditions (Patrick, 1978) indicate that water from the granites, Borrowdale Volcanic Group and Windermere Supergroup form three chemically district groups. The granites form sodium–potassium chloride waters, whereas the Windermere Supergroup produces calcium bicarbonate water; the Borrowdale Volcanic Group waters are variable in composition. The Lake District Regional Geochemical Atlas provides data on pH, conductivity, bicarbonate and fluoride contents of surface waters.

WASTE DISPOSAL

Domestic, commercial and industrial waste from the Lake District National Park is currently disposed of at four landfill sites, only one of which partly lies within the park (Lake District National Park, 1994). None of these sites are within the Ambleside district. Closed landfill sites for waste disposal are recorded from 13 locations within the district and most of these are in alluvial deposits.

The district includes a number of abandoned quarries which offer a considerable volume for potential landfill. At the present, use of these for other than small-scale disposal of inert waste to meet local needs is unlikely.

SEISMICITY

Minor seismic events arising from natural earthquakes are not uncommon within north-west England (Musson et al., 1984; Musson, 1987; 1994) and a number of records exist of seismic events originating within the Ambleside district. The Grasmere area was the epicentre for events on 23 February 1867 and 16 May 1911, with magnitudes on the Richter Scale of 2.7 and 3.1 ML respectively. The Ambleside earthquake of 12 September 1988 consisted of two shocks, only seconds apart, of Richter Scale magnitude 3.2 and 3.0 ML. These originated from depths of 15 and 7 km respectively. Slight cracking of plaster was recorded in Ambleside and the event was felt widely in the central Lake District and as far east as Kirkby Lonsdale (Musson, 1994). The Wrynose Breast area [NY 267 027] was the epicentre of a 2.2 ML earthquake originating at a depth of 12 km on 18 July 1994 (Wright and Richards, 1995).

The effects of earthquakes originating outside the district have been felt within it. Notable among the nearer examples are the Carlisle earthquakes of 9 July 1901, the epicentre of which was about 10 km south to south-west of Carlisle, and 26 December 1979 which had its epicentre close to Longtown near the Scottish–English border.

Though there are no grounds for regarding the Ambleside district as especially prone to seismic activity, the recorded local events indicate that, along with much of Britain, earthquakes of unusual intensity possibly could occur.

CONSERVATION

There is an increasing awareness of the need to conserve important sites of geological or natural history interest. Localities of national significance are scheduled as Sites of Special Scientific Interest (SSSI) by English Nature. Within the district several sites are designated as SSSIs for their geological importance. Lately, English Nature has instituted a second scheme, Regionally Important Geological and Geomorphological Sites (RIGS), that do not warrant SSSI status, but are of value primarily for teaching purposes. Designation of sites within the district under this scheme is in hand.

THREE

Concealed geology

The concealed geology in the Ambleside district has been interpreted from gravity, magnetic and, at the western margin, seismic reflection data. No deep boreholes have been drilled to confirm the geophysical models. The pioneering work on the concealed geology of the Lake District was undertaken by Bott (1974; 1978) who carried out the first regional-scale gravity survey of the area. This revealed a belt of relatively low Bouguer anomaly values linking the Shap, Skiddaw and Eskdale granites, which was interpreted in terms of a major granitic batholith, much of which lies beneath the Skiddaw and Borrowdale Volcanic groups. At outcrop the batholith comprises five distinct acid igneous intrusions (Eskdale, Ennerdale, Shap and Skiddaw granites, and Eskdale granodiorite; Chapter 8). The concealed part of the batholith underlies an area many times greater than the exposed portions and is unlikely to be any less complex.

In the early 1980s the British Geological Survey carried out a new regional gravity survey of the Lake District with a distribution of observations considerably improved over that provided by the previous survey. These new data were used by Lee (1984a; 1986a) to define further the overall three-dimensional form of the batholith. Whilst the modelling of Bott (1974; 1978) and Lee (1984a; 1986a) provided a reasonable picture of the broad-scale deep structure of the Lake District, this was based on relatively simplified assumptions of the density distribution and was carried out without the control provided by the modern geological mapping described in this memoir. Detailed geophysical studies, including additional surveys, were therefore undertaken as part of the Lake District mapping programme to investigate the form of the composite batholith, the structure of the Skiddaw and Borrowdale Volcanic groups and the tectonic evolution of the region (Lee, 1988; 1989). Since this work was completed, very detailed geological and geophysical investigations, including seismic reflection and low-level airborne surveys, have been undertaken in west Cumbria as part of a programme of investigations into the subsurface disposal of low- and intermediate-level radioactive waste. The westernmost part of the Ambleside district has been included within the study area and important additional interpretations of the subsurface structure have been reported (Evans et al., 1993; 1994; Kimbell, 1994).

GEOPHYSICAL DATA

Gravity surveys

The widely spaced gravity survey of Bott (1974) was superseded by a regional survey of the Lake District carried out by the British Geological Survey in 1981 and 1982 (Lee, 1984a; 1986a). The regional data were supplemented by closely spaced observations along traverses and additional regional observations in the western Lake District to improve the resolution of anomalies over the margins of the exposed intrusions. A Bouguer anomaly map of the Ambleside district (Figure 4a), based on the complete dataset, shows prominent gravity lows over the outcrop of the Eskdale granite (ES) and Ennerdale intrusion (EN), and a broad low extending east-north-eastwards from the Eskdale granite over the concealed part of the batholith. Bouguer anomaly values increase sharply south-eastwards, away from the batholith, with a prominent change of gradient across the contact between the Borrowdale Volcanic Group and the Windermere Supergroup. To the west of the district the gravity low over the Eskdale granite merges with that caused by sedimentary strata within the East Irish Sea Basin (IS).

Aeromagnetic surveys

The aeromagnetic survey of the Lake District was carried out by Canadian Aero Service Ltd in 1958 and 1959. In the southern part of the Ambleside district the survey was flown along east–west flight lines 2 km apart with north–south tie lines 10 km apart. In the northern part of the area, flight lines were orientated north–south and tie lines east–west. Mean terrain clearance was 305 m. The data were recorded originally in analogue form and subsequently the posted data values along each flight line were digitised by the British Geological Survey. A contour map of total field aeromagnetic anomalies based on the digital data (Figure 4b), shows a broad magnetic low over the Eskdale granite (ES) and central Lake District, and high amplitude, short spatial-wavelength anomalies over the Ennerdale intrusion (EN). The magnetic field over the Borrowdale Volcanic Group is relatively flat, except for a few minor short-wavelength anomalies which are not resolved in Figure 4b but are visible on individual flight line profiles. A broad magnetic high (WI) occurs over the Windermere Supergroup to the south-east of the Ambleside district, and this forms part of a major magnetic ridge that extends from the southern Lake District, via the Askrigg Block, to eastern England (Lee et al., 1990).

Two detailed aeromagnetic surveys, which extend into the western part of the Ambleside district, were flown in 1990–91 (Figure 5; Kimbell, 1994 and references therein). A 'high level' survey was flown at a constant barometric elevation of 300 m along north-east to south-west survey lines 500 m apart, with north-west to south-east tie lines 1000 m apart. A subsequent low-level survey was flown with the sensor at a nominal 100 m terrain clearance along 200 m-spaced flight lines orientated north-east to south-west, with orthogonal tie lines at 400 m intervals. The detailed aeromagnetic surveys resolve local magnetic

anomalies over the Borrowdale Volcanic Group which are poorly imaged by the regional data. These are related to relatively magnetic units within the volcanic succession, particularly within the Birker Fell Formation around the northern and western margins of the Eskdale granite (Kimbell, 1994). The low-level survey also included magnetic vertical gradient, Very Low Frequency (VLF) electromagnetic and radiometric measurements.

Seismic reflection surveys

In 1990, several seismic reflection profiles were acquired across the Eskdale granite and Ennerdale intrusion in the Wasdale area as part of the Sellafield investigations (Evans et al., 1994 and references therein). The location of the seismic lines is shown on Figure 5 and the significance of the results is discussed below in relation to those of the gravity and magnetic modelling.

PHYSICAL PROPERTY DATA

The accuracy of gravity modelling is critically dependent on knowledge of density variations within the granitic batholith and within the Lower Palaeozoic rocks into which it was emplaced. New density determinations were therefore made as part of the geological mapping programme on samples from over 350 localities, together with a limited number of magnetic susceptibility and sonic velocity determinations (Lee, 1988). Values for rocks occurring in the Ambleside district are given in Tables 3–5.

Granitic intrusions

Properties for granitic rocks cropping out within the district are listed in Table 3. The Eskdale granite, its outlying cupola at Wasdale Head and the microgranite within the Ennerdale intrusion have almost identical properties, but the Eskdale granodiorite is characterised by higher density values. After accounting for the effects of weathering on the samples (Lee, 1988), the computed representative in-situ density values are 2.70 Mg/m^3 for the Eskdale granodiorite, 2.63 Mg/m^3 for the Eskdale and Wasdale Head granites and 2.62 Mg/m^3 for the Ennerdale microgranite. All of the granitic rocks have very low magnetic susceptibility values in the range 0.0001 to 0.0002 SI. In contrast, the dioritic part of the Ennerdale intrusion has a higher density than that of the granitic rocks and a much higher value of magnetic susceptibility (representative in-situ density of 2.74 Mg/m^3 and susceptibility of 0.009 SI).

No spatial variation in density has been found across any of the granitic intrusions within the Ambleside district (Lee, 1988). This is an important conclusion because the limited data available to Bott (1974; 1978) led him to interpret the western part of the Lake District batholith as a near-vertically zoned intrusion, with density values increasing from 2.61 Mg/m^3 in the centre of the Eskdale granite to 2.68 Mg/m^3 on the northern margin of the batholith. Though density values must increase towards the northern margin of the batholith to explain the shape of the negative gravity anomaly, the lack of zonation within the granitic outcrops suggests that the variation may be in the form of a series of major batholith components rather than a fine zonation.

The contrast in properties between the Ordovician intrusions (Eskdale and Ennerdale) in the western Lake District and the Devonian granites (Shap and Skiddaw) is noteworthy. The Shap granite is characterised by both higher density (2.66 Mg/m^3) and higher magnetic susceptibility (0.0085 SI) values than the western granites (Lee, 1984b). The Skiddaw granite is severely affected by hydrothermal alteration (Webb and Brown, 1984) and is characterised by density values in the range 2.58 to 2.67 Mg/m^3 and a low magnetic susceptibility value of 0.0003 SI (Lee, 1984b).

Borrowdale Volcanic Group

Analysis of density variations within the Borrowdale Volcanic Group is more complex than for the granitic rocks because each of the constituent formations comprises a characteristic mixture of four principal lithological types (basalt, andesite, silicic rocks and clastic rocks). The approach adopted, therefore, was to establish representative density values for each lithological group which could then be used to calculate representative values for each formation depending on the proportion of each group within the formation (Lee, 1988).

The results of the analysis are given in Table 4. Density values for the four lithological groups range from 2.70 Mg/m^3 for silicic rocks (lava and welded ignimbrite) to 2.88 Mg/m^3 for basalt. Values for andesite and clastic rocks lie between these values (2.78 and 2.75 Mg/m^3 respectively). Thus, formation densities within the Borrowdale Volcanic Group range from 2.70 Mg/m^3 for the Airy's Bridge Formation (100% silicic rocks) to 2.81 Mg/m^3 for parts of the Birker Fell Formation containing a high proportion of basalt. This variation is significant in the gravity modelling because density values for formations dominated by silicic rocks are similar to those of the Eskdale granodiorite (and by implication possibly to concealed components of the batholith).

The limited number of magnetic susceptibility determinations on samples of andesite from the Borrowdale Volcanic Group give values in the range 0.0004 to 0.0006 SI. The range is low compared with dioritic parts of the Ennerdale intrusion but slightly higher than typical values for the western granites. The general lack of major high-amplitude magnetic anomalies over the Borrowdale Volcanic Group (Figure 4b) suggests that these relatively low values are typical of most of the sequence. However, some more magnetic units are present, as evidenced by minor anomalies observed over the outcrop by the detailed aeromagnetic surveys. This is supported by the recent measurements on cores of volcanic material from west Cumbria which show susceptibility values generally below 0.001 SI, except within some formations that have values up to 0.01 SI (Kimbell, 1994).

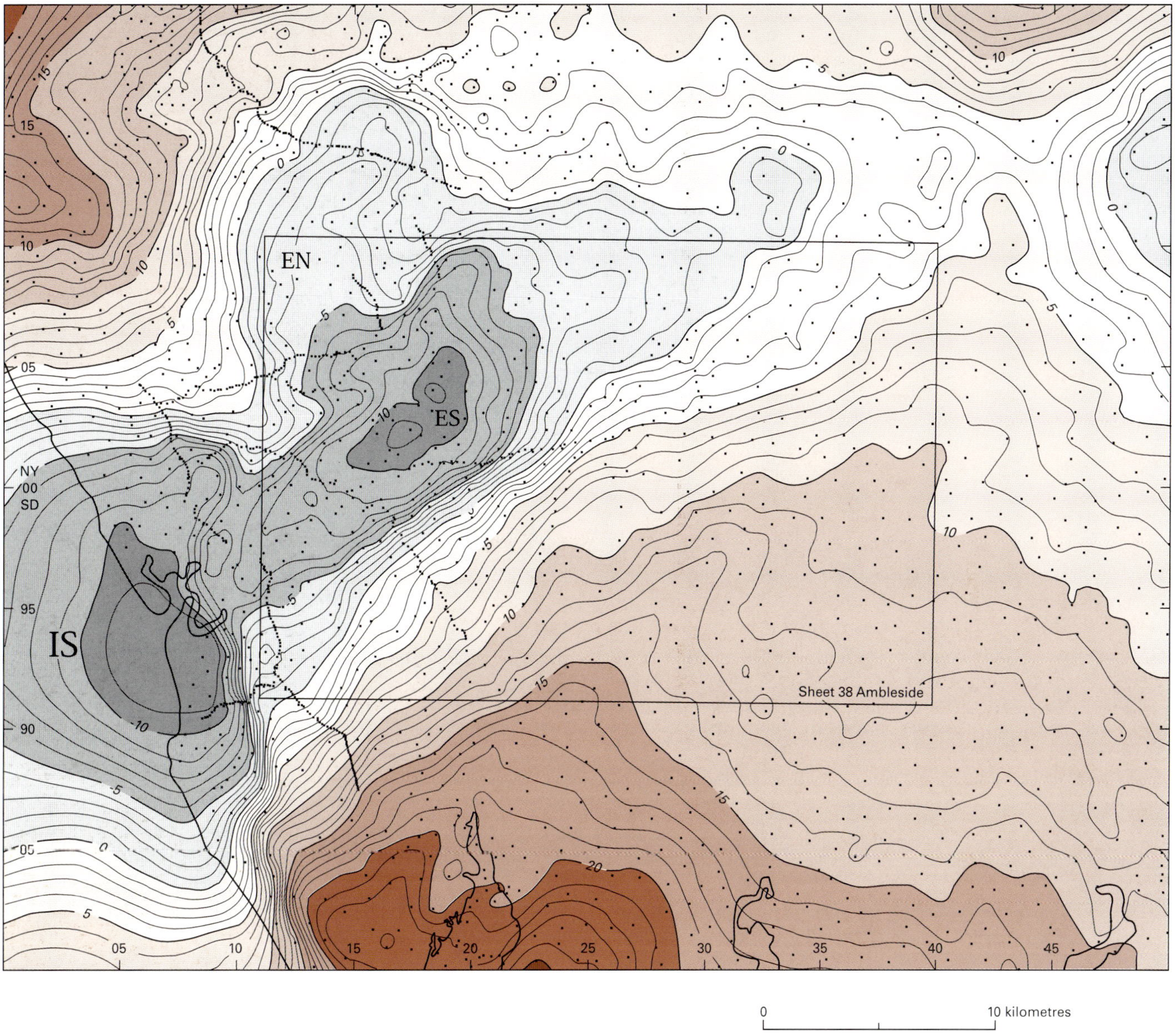

Figure 4 Bouguer gravity and aeromagnetic anomaly maps of the district and surrounding area.

· Recording station.

a. Bouguer gravity anomaly map. Anomaly values in mGal.

Anomalies discussed in the text: EN Ennerdale microdiorite; ES Eskdale granite.

Skiddaw Group and Windermere Supergroup

Though Skiddaw Group rocks crop out only within a small inlier on the southern margin of the Eskdale granite, they form the Black Combe inlier to the south of the district, and are presumed to underlie the Borrowdale Volcanic Group and Windermere Supergroup, forming the 'basement' into which the batholith was emplaced. The detailed analysis of density variations within the Skiddaw Group (Lee, 1988) showed values ranging from 2.72 Mg/m^3 for sandstone and greywacke to 2.81 Mg/m^3 for mudstone and siltstone. Five samples of hornfelsed mudstone from the Black Combe inlier had values at the high end of the mudstone range. The importance of these data for modelling gravity anomalies in the Ambleside district is that the calculated mean density of the Skiddaw Group sequence at depth is about 2.78 Mg/m^3 (Lee, 1988). The few magnetic susceptibility measurements and absence of prominent magnetic anomalies over the outcrop indicate generally low susceptibility values (less than 0.001 SI).

Density data from the Windermere Supergroup (Lee, 1988) vary from 2.69 Mg/m^3 for the Coniston Group and Kirby Moor Formation to 2.77 Mg/m^3 for the Browgill

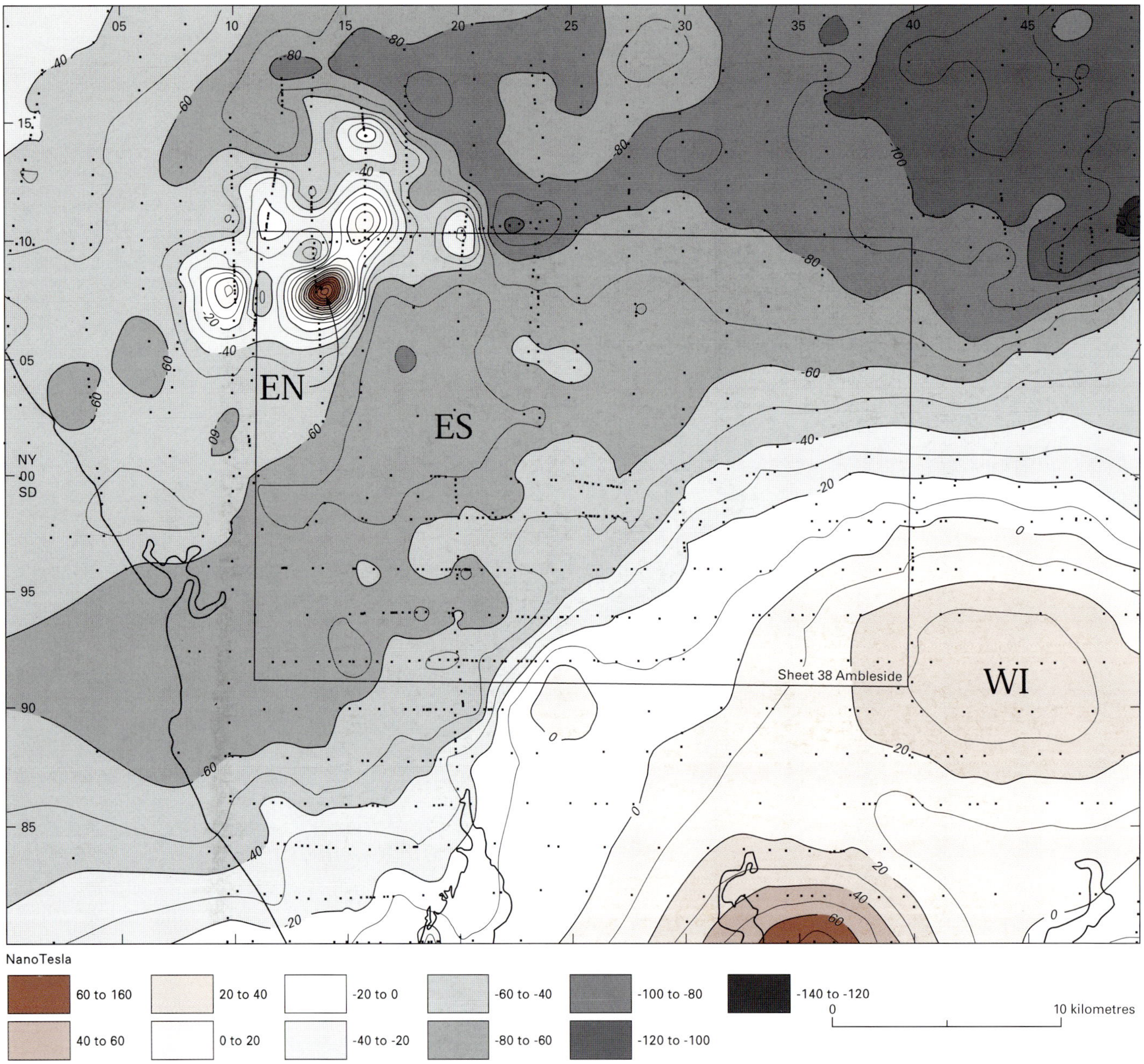

Figure 4 (continued).

b. Aeromagnetic anomaly map. Anomaly values in nanotesla.

Anomalies discussed in the text: EN Ennerdale microdiorite; ES Eskdale granite; WI sub-Windermere Supergroup.

Formation (Table 5). No magnetic susceptibility measurements were made on Windermere Supergroup rocks; the lack of magnetic signature (Figure 4b) suggests that these rocks have very low susceptibility.

STRUCTURAL ANALYSIS OF GRAVITY AND MAGNETIC DATA

Image processing and digital filtering techniques were used by Lee (1989) to analyse structural information from the gravity and aeromagnetic fields. A number of prominent linear features and residual gravity and magnetic anomalies were identified which provide important information on the form of the concealed batholith and its relationship to the wider structure of the region. The high-pass filtered gravity map (Figure 6), for example, resolves residual anomalies related to individual components of the batholith and structures within the Borrowdale Volcanic Group. The most important lineaments and anomalies in the Ambleside district, derived from the whole suite of images, are shown in Figure 7.

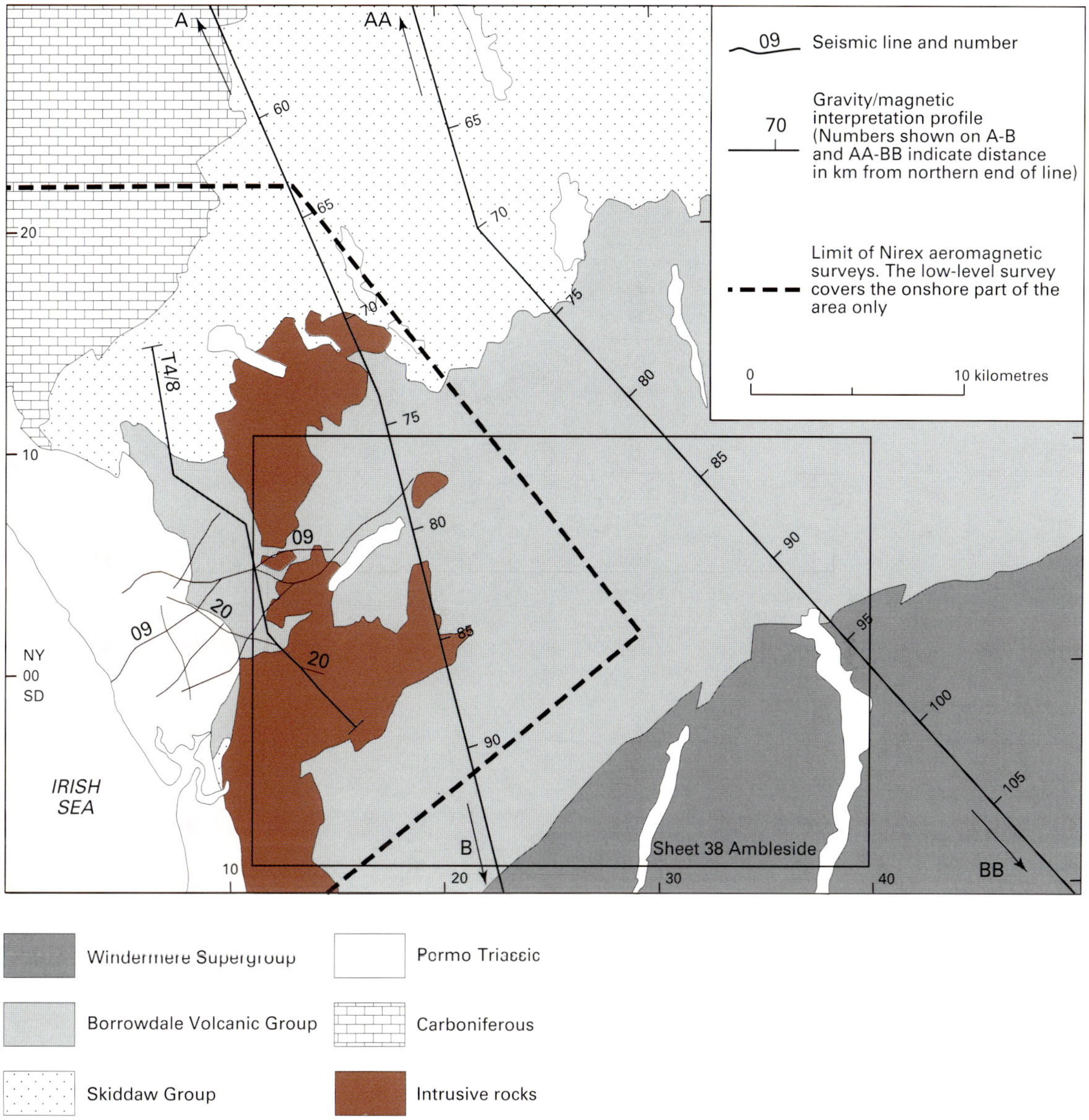

Figure 5 Location of interpretation profiles, seismic lines and detailed aeromagnetic surveys in the district and surrounding area.

Lineaments

Two sets of geophysical lineaments are observed within the district; a prominent east-north-easterly trending set and a subsidiary north-easterly trending set. Three major east-north-east-trending lineaments cross the Lake District (Figure 7; lineaments 1, 2 and 3), two of which (lineaments 2 and 3) occur within the Ambleside district.

Though lineament 1 lies outside the Ambleside district, it is mentioned as the most prominent of all the geophysical lineaments in the Lake District and the one most clearly related to fundamental structures within the Skiddaw Group. Designated by Lee (1989) as the Crummock Lineament (CL, Figure 7) it separates south-south-east- and north-north-west-facing slump folds within the Skiddaw Group (Cooper et al., 1995) and seems to have been an important structural feature from early Ordovician to Early Devonian times.

Lineament 2, which Lee (1989) named the Ullswater Lineament (UL, Figure 7), is visible as a prominent gravity feature on shaded-relief images, but the gravity gradients which form the lineament are not related to a single density contrast. At its western end the line runs close to the northern margin of the granite at Wasdale Head (residual gravity anomaly WS, Figure 6) and then eastwards, following the northern margin of the outcrop of the Airy's Bridge Formation which marks the northern limb of the Scafell Syncline. Farther east, the line is located along the southern side of the Ullswater inlier of the Skiddaw Group and there is a hint that it also extends across the Vale of Eden on to the western margin of the Alston Block. Magnetically, the Ullswater Lineament is visible as a subtle feature on shaded-relief images over the Borrowdale Volcanic Group outcrop (a weak, linear residual magnetic low occurs where the Airy's Bridge Formation crops out on the northern limb of the Scafell Syncline).

Lineament 3, the Southern Borrowdales Lineament of Lee (1989; SBL, Figure 7), lies close to the southern edge of the Borrowdale Volcanic Group outcrop, just south of the contact with the Windermere Supergroup. The

Table 3 Physical property data for the Eskdale and Ennerdale intrusions (after Lee, 1988).

A. Density	No. of samples		Grain density (Mg/m^3) Mean (SD)	Saturated density (Mg/m^3) Mean (SD)	Calculated in-situ density (Mg/m^3)
Eskdale and Wasdale granites	40		2.64 (0.01)	2.61 (0.02)	2.63
Eskdale granodiorite	26		2.70 (0.03)	2.68 (0.03)	2.70
Ennerdale granite	16		2.63 (0.01)	2.61 (0.02)	2.62
Ennerdale microdiorite	4		2.75 (0.01)	2.72 (0.03)	2.74
B. Magnetic susceptibility and sonic velocity	**No. of samples Lab**	**Field**	**Laboratory susceptibility (SI × 10^{-3}) Mean (SD)**	**Field susceptibility (SI × 10^{-3}) Mean (SD)**	**Sonic velocity (km/s)**
Eskdale and Wasdale granites	4	19	0.2 (0.1)	0.1 (0.1)	5.4
Eskdale granodiorite	0	4		0.2 (0.1)	
Ennerdale granite	1	5	0.1	0.1 (0.1)	5.6
Ennerdale microdiorite	3	4	9.0 (10.0)	9.0 (6.4)	6.1

Determinations are based on one representative sample per location. Field magnetic susceptibility values are based on portable susceptibility meter measurements at outcrop. Sonic velocity values are the mean of N measurements, where N is the number of laboratory samples.

Table 4 Physical property data for Borrowdale Volcanic Group rocks in the district (after Lee, 1988).

A. Density by lithology	No. of samples	Grain density (Mg/m^3) Mean (SD)	Saturated density (Mg/m^3) Mean (SD)	Calculated in-situ density (Mg/m^3)
Basalt (BVG1)	18	2.88 (0.04)	2.86 (0.04)	2.88
Andesite (BVG2)	50	2.79 (0.04)	2.78 (0.04)	2.78
Silicic rocks (BVG3)	53	2.69 (0.04)	2.71 (0.04)	2.70
Clastic rocks (BVG4)	43	2.77 (0.04)	2.74 (0.04)	2.75

B. Density by formation	BVG1 %	BVG2 %	BVG3 %	BVG4 %	Calculated density (Mg/m^3)
Esk Pike, Lincomb Tarns and Seathwaite Fell formations	0	10	20	70	2.74
Dunnerdale, Lickle and Caw formations	0	0	65	35	2.72
Duddon Hall and Ulpha formations	0	20	0	80	2.76
Airy's Bridge Formation	0	0	100	0	2.70
Birker Fell Formation: BFA5	10	40	0	50	2.78
Birker Fell Formation: BFA4	0	85	0	15	2.78
Birker Fell Formation: BFA3	40	35	5	20	2.81
Birker Fell Formation: BFA2	10	30	50	10	2.75
Birker Fell Formation: BFA1	0	40	10	50	2.76

C. Magnetic susceptibility	No. of samples Laboratory	Field	Laboratory susceptibility (SI × 10^{-3}) Mean (SD)	Field susceptibility (SI × 10^{-3}) Mean (SD)	Sonic velocity (km/s)
Andesite	16	47	0.6 (0.1)	0.4 (0.1)	6.4

Density values for the four principal lithological groups are based on laboratory measurements (Lee, 1988). Formation densities are calculated from the proportion of each lithology in each formation. Five divisions of the Birker Fell Formation (BFA) reflect the lithological variations both in undivided parts of the fomation and within formally defined members.

Field magnetic susceptibility values are based on portable susceptibility meter measurements at outcrop.

Sonic velocity values are the mean of N measurements, where N is the number of laboratory samples.

BVG Borrowdale Volcanic Group.

Table 5 Density data for the Windermere Supergroup.

Formation	Source	No. of locations	No. of samples	Grain density (Mg/m³)	Saturated density	Suggested representative in-situ density (Mg/m³)
Kirkby Moor Formation	Bott, 1974 Lee, 1988	1 1	20 4	2.74 2.69	2.69 2.67	2.69
Bannisdale Formation	Bott, 1974 Lee, 1984a	1 10	21 10	2.73 2.74	2.72 2.69	2.72
Coniston Group	Lee, 1984a Lee, 1988	1 1	2 5	2.72 2.69	2.69 2.69	2.69
Wray Castle and Coldwell formations	Lee, 1988	2	15	2.73	2.69	2.71
Brathay Formation	Bott, 1974 Lee, 1988	2 1	41 5	2.77 2.74	2.74 2.73	2.74
Browgill Formation	Lee, 1988	1	6	2.79	2.77	2.77
Undivided Windermere Supergroup; calculated from the proportions of each formation in the sequence						2.71
Buried Windermere Supergroup; assuming in-situ porosity values lower than those at outcrop						2.72

Note that the Kirkby Moor Formation is not present in the Ambleside district.

Samples of slate are prone to splitting which results in lower measured values for saturated density than would be expected in-situ. The recommended in-situ values listed allow for this and the differences between values from the various sources.

lineament is visible mainly as a gravity feature and is related to the density contrast between these major lithostratigraphical units. Projected to the east, the lineament is aligned approximately with the Lunedale/Butterknowle faults on the southern margin of the Alston Block.

Along the Ullswater and Southern Borrowdales lineaments, Skiddaw Group rocks are mostly concealed beneath younger rocks and there is no evidence that these lineaments are directly related to early structures within the Skiddaw Group. However, they are similar to the Crummock Lineament because they extend over several tens of kilometres and are related to structures of different age along their length. Thus they may represent major re-activated deep-seated fractures that influenced the emplacement of the volcanic succession and the granitic intrusions.

Several less extensive east-north-east-trending lineaments lie between the Ullswater and Southern Borrowdales lineaments and are related mainly to the Scafell Syncline and to topographical lows in the western Lake District. Magnetic lineament 5 lies on the northern limb of the Scafell Syncline and correlates with the southern side of the outcrop of the Airy's Bridge Formation. Lineament 6, which has both a gravity and magnetic expression, correlates with the centre of the Scafell Syncline at its eastern end and passes through the contact between the Eskdale and Ennerdale intrusions. It also runs close to the line of Wasdale which may partly account for the gravity expression. Lineament 7 has a gravity expression at its western end coinciding with Eskdale, and a magnetic expression at its eastern end, corresponding to the outcrop of the Airy's Bridge Formation on the southern limb of the Scafell Syncline.

At least three north-east-trending lineaments can be recognised on the images. The most prominent (lineament 4) appears as a gravity feature which at its north-east end lies between the Scafell and Haweswater synclines and at its south-west end lies close to the line of the strike of steeply dipping strata within the Borrowdale Volcanic Group west of Coniston. Lineament 9 lies along the contact between the Borrowdale Volcanic Group and the Windermere Supergroup in the south-west and is parallel to the axial plane trace of the Black Combe Anticline. Lineament 10, a weaker gravity feature, continues north-east of lineament 9 along its south-east limb of the Haweswater Syncline.

Residual gravity lows

Prominent residual gravity lows occur over the Eskdale granite (ES, Figures 6, 7), the granite at Wasdale Head (WS), the Eskdale granodiorite (ED) and the Ennerdale intrusion (EN). Within the central Lake District, a number of less prominent but significant lows were also recognised by Lee (1989) where the batholith is concealed by volcanic rocks. The Dunmail low (DM) lies between lineaments 2 and 4 and follows the axis of the concealed batholith of earlier models (e.g. Bott, 1974; Lee, 1984a), coinciding approximately with the Scafell and Place Fell synclines. The Rydal low (RD) lies between lineaments 4 and 10 on the south-east side of the batholith and coincides approximately with the south-western part of the Haweswater Syncline. A small low occurs to the south-west of the Rydal anomaly, between lineaments 4 and 9 in the Coniston area (CN), and an identifiable low occurs over the Ulpha Syncline (UL).

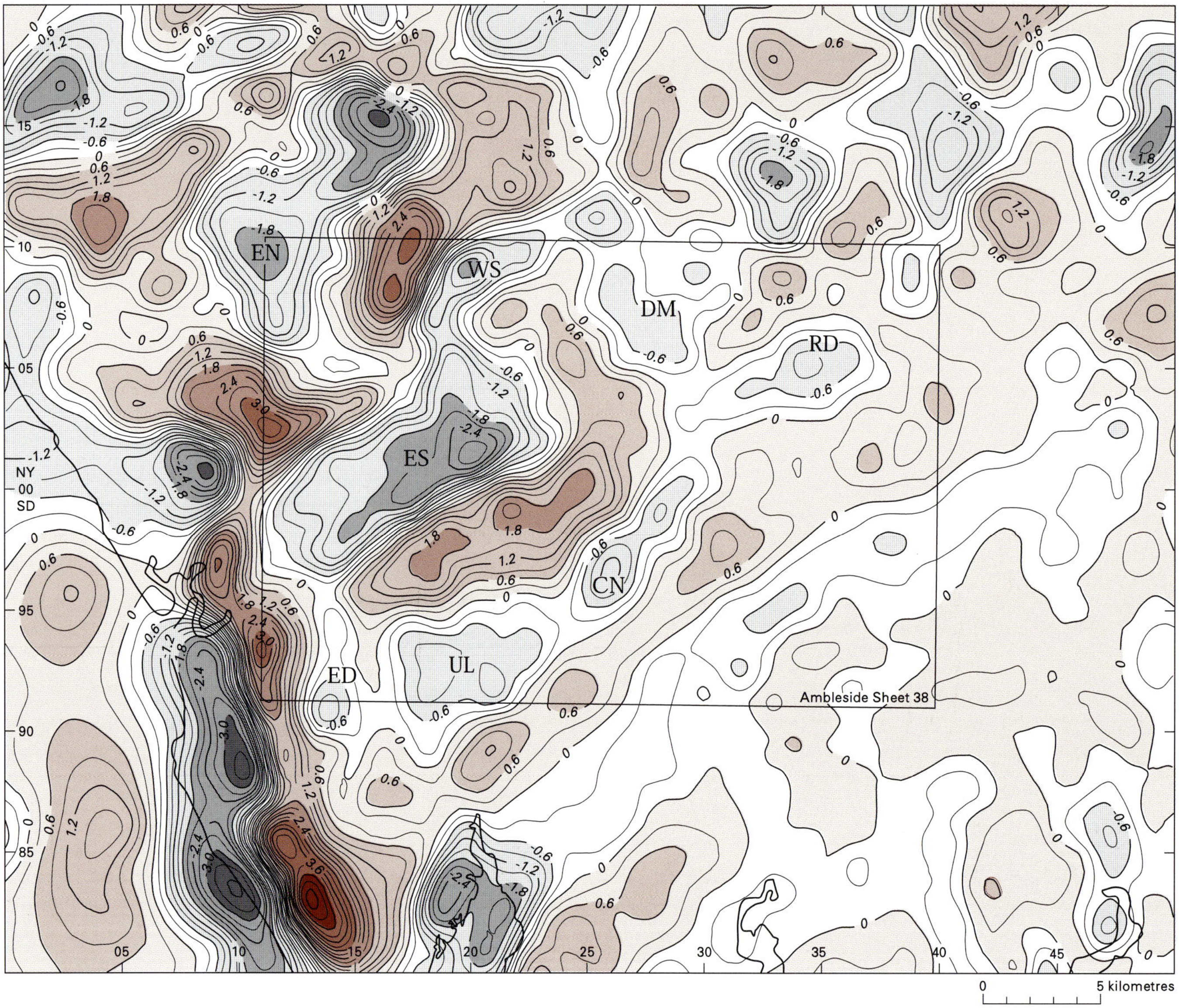

Figure 6 Residual gravity anomaly map generated by removing anomalies of spatial-wavelength greater than 5 km.

Anomaly values in mGal. Residual gravity lows: CN Coniston anomaly; DM Dunmail anomaly; ED Eskdale granodiorite; EN Ennerdale intrusion; ES Eskdale granite; RD Rydal anomaly; UL Ulpha anomaly; WS Wasdale granite.

The association of residual gravity lows with the exposed granites suggests that the other lows in the central Lake District may also be due to distinct components of the concealed batholith or to separate, high-level granitic intrusions. Alternatively, the spatial correlation of lows DM, RD and UL with the three major synclines in the Borrowdale Volcanic Group (Scafell, Haweswater and Ulpha respectively) suggests that they may be caused by thick sequences of low-density volcanic rocks (e.g. silicic pyroclastic rocks) within the synclines. The feasibility of these models is discussed below.

COMPONENTS AND SUBSURFACE FORM OF THE BATHOLITH

Lee (1989) carried out detailed, integrated gravity/ magnetic modelling along 12 profiles which resulted in a new model for the subsurface form of the Lake District batholith (Figure 8). In addition to the six granitic intrusions recognised at outcrop (Shap granite, Skiddaw granite, Eskdale granite (including the cupola at Wasdale Head), Eskdale granodiorite, Ennerdale intrusion and the Threlkeld microgranite), the model included up to

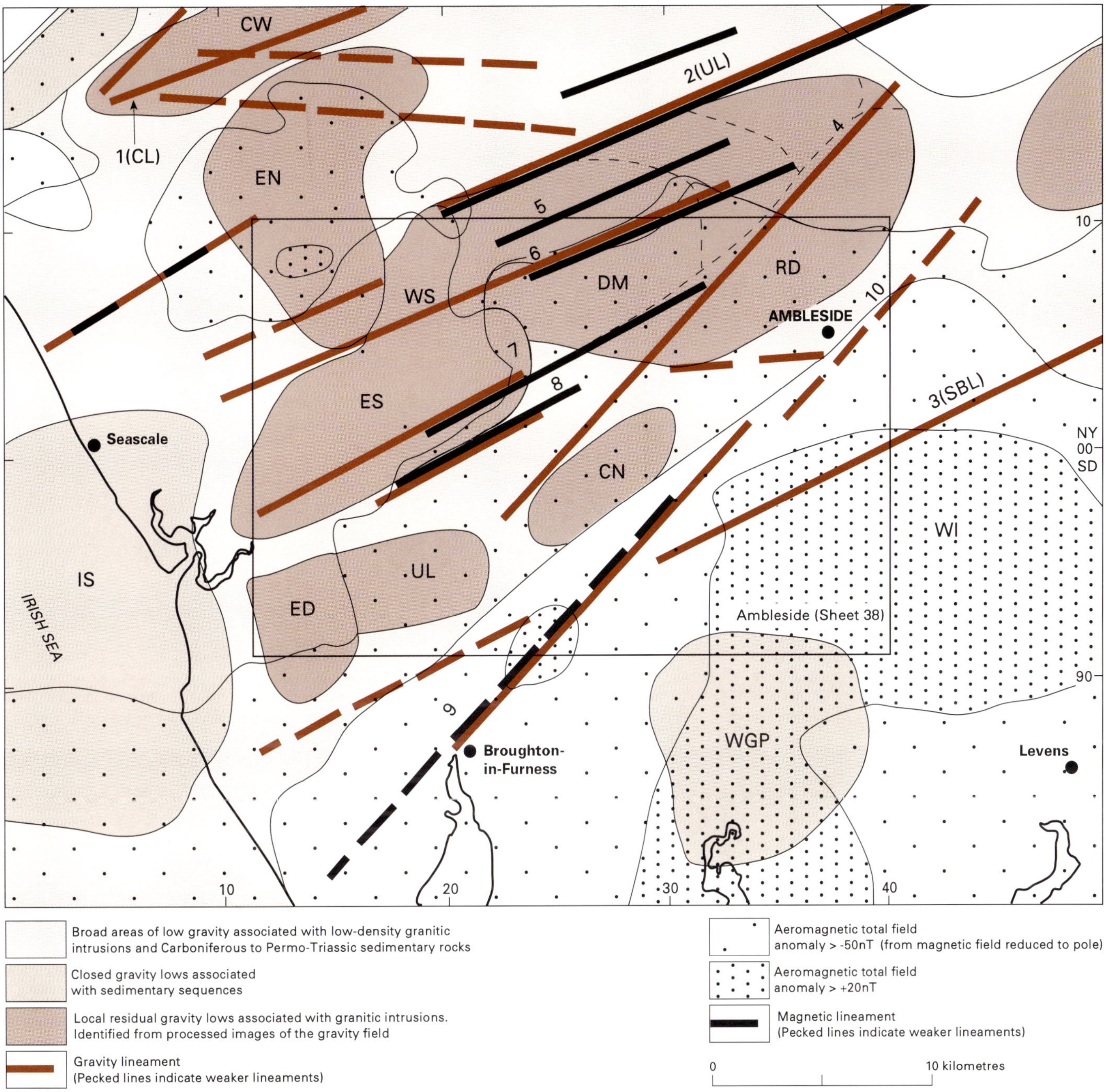

Figure 7 Principal gravity and magnetic lineaments and anomalies (after Lee, 1989).
Numbers indicate lineaments discussed in the text. Principal lineaments: 1 Crummock (CL); 2 Ullswater (UL) and 3 Southern Borrowdales (SBL). Residual gravity lows associated with exposed or concealed granitic intrusions: CN Coniston anomaly; CW Crummock Water anomaly; DM Dunmail anomaly; ED Eskdale granodiorite; EN Ennerdale intrusion; ES Eskdale granite; RD Rydal anomaly; UL Ulpha anomaly; WS Wasdale granite. Closed gravity lows associated with sedimentary sequences: IS East Irish Sea Basin; WGP Windermere Supergroup.

eight concealed bodies, some postulated as deep-seated components of the batholith and others as high-level granitic intrusions. Of these wholly concealed intrusions, the Dunmail, Rydal, Ulpha and Buttermere components occur within the Ambleside district. Additional modelling was subsequently carried out along profile AA–BB (Figure 5) (across the Ambleside and Keswick districts) to integrate the findings from the earlier modelling with the final results of the geological mapping programme.

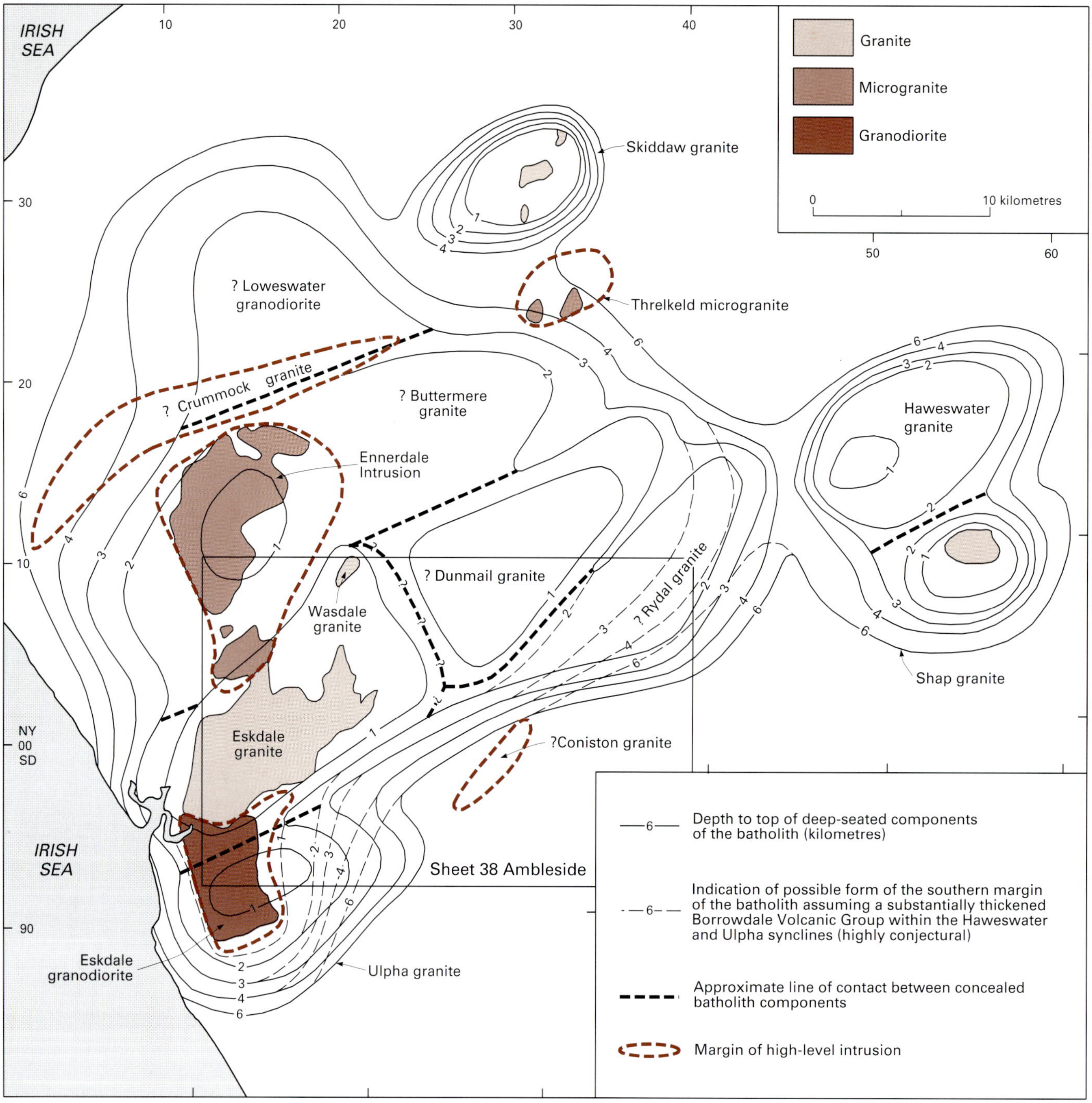

Figure 8 The 3-D form of the Lake District batholith and its various components (after Lee, 1989).

Exposed intrusions

The Eskdale granite is interpreted as a deep-seated intrusion extending to a depth of about 9 km (Figures 8, 9a). Its measured density at the surface (2.63 Mg/m^3) is sufficient to account for the whole of the negative gravity anomaly over the outcrop and there is no (gravity) evidence for a separate, underlying intrusion of different density. However, the seismic reflection data (Evans et al., 1994, figs 3–6) indicate that the western part of the Lake District batholith may be interpreted best as stacked, sheet-like (laccolithic) intrusions. Also, combined seismic, gravity and magnetic modelling along seismic line GGDG-90-20 (see Figure 5 for location) shows that a better fit is obtained to the observed gravity field on the north-western margin of the Eskdale granite if the edge of the granite is interdigitated with lenses of country rock (such as the Skiddaw Group) rather than a simple outward-sloping intrusive contact (Figure 10). It is possible, therefore, that the Eskdale granite comprises a series of closely spaced

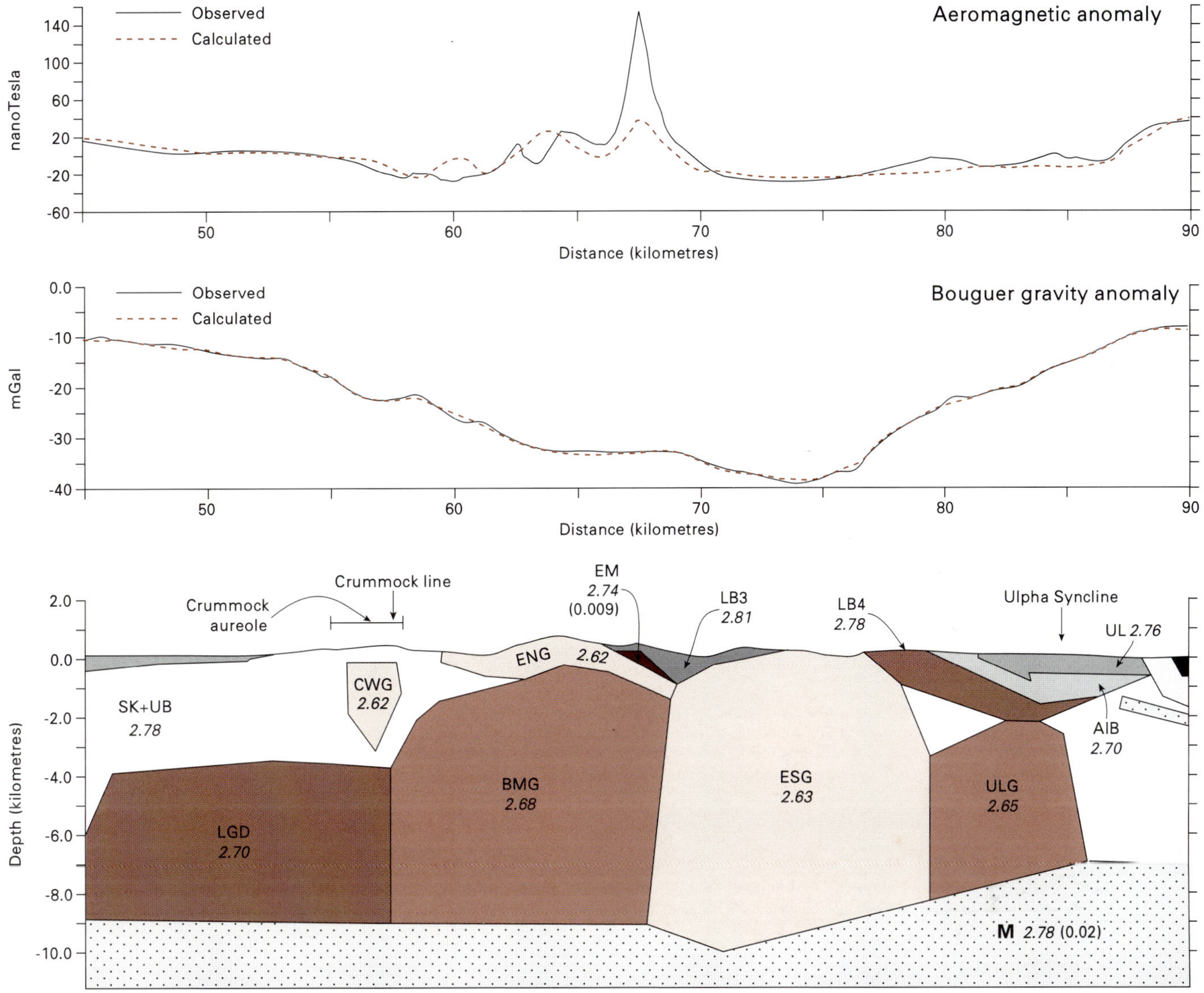

Figure 9 Integrated gravity and magnetic models (see Figure 5 for locations).

a. Profile AB (central part). Density values used in the modelling are shown in Mg/m^3 and magnetic susceptibility values (in brackets, where greater than zero) are shown in SI units.

AIB Airy's Bridge Formation; BMG Buttermere granite; CWG Crummock Water granite; EGD Eskdale granodiorite; EM Ennerdale microdiorite; ENG Ennerdale granite; ESG Eskdale granite; LB3 & LB4 Birker Fell Formation; LGD Loweswater granodiorite; M magnetic basement; PT Permo–Triassic; SK + UB Skiddaw Group and undifferentiated basement; UL Ulpha Formation; ULB undifferentiated adjacent batholith; ULG Ulpha granite.

granitic laccoliths, rather than a single pluton. Gravity models support petrographical and geochemical data (Chapter 8) which indicate that the granite at Wasdale Head is a northern, higher level off-shoot of the Eskdale granite (Lee, 1989).

The Eskdale granodiorite (Figure 8) may be interpreted geophysically either as a marginal granodioritic phase to the Eskdale granite or a discrete laccolith underlain by a separate, deep-seated component of the batholith (the Ulpha granite). However, if the granodiorite extends at depth on the southern margin of the Eskdale granite, the northern half of the granodiorite must be only 1 to 2 km thick and underlain by a shoulder of the Eskdale granite extending southwards. On balance, Lee (1989) concluded that the best fit to the observed gravity field is achieved if the Eskdale granodiorite is assumed to be a laccolith, about 1 km thick (Figure 9b). High temperature mineralisation at Black Combe (Cameron et al., 1993) supports further the concept of a granitic intrusion at depth on the southern margin of the batholith (rather than the Eskdale granodiorite extending to depth as a marginal phase of the batholith).

Gravity modelling of the Ennerdale intrusion (Bott, 1974; Lee, 1989) suggests that the surface density values

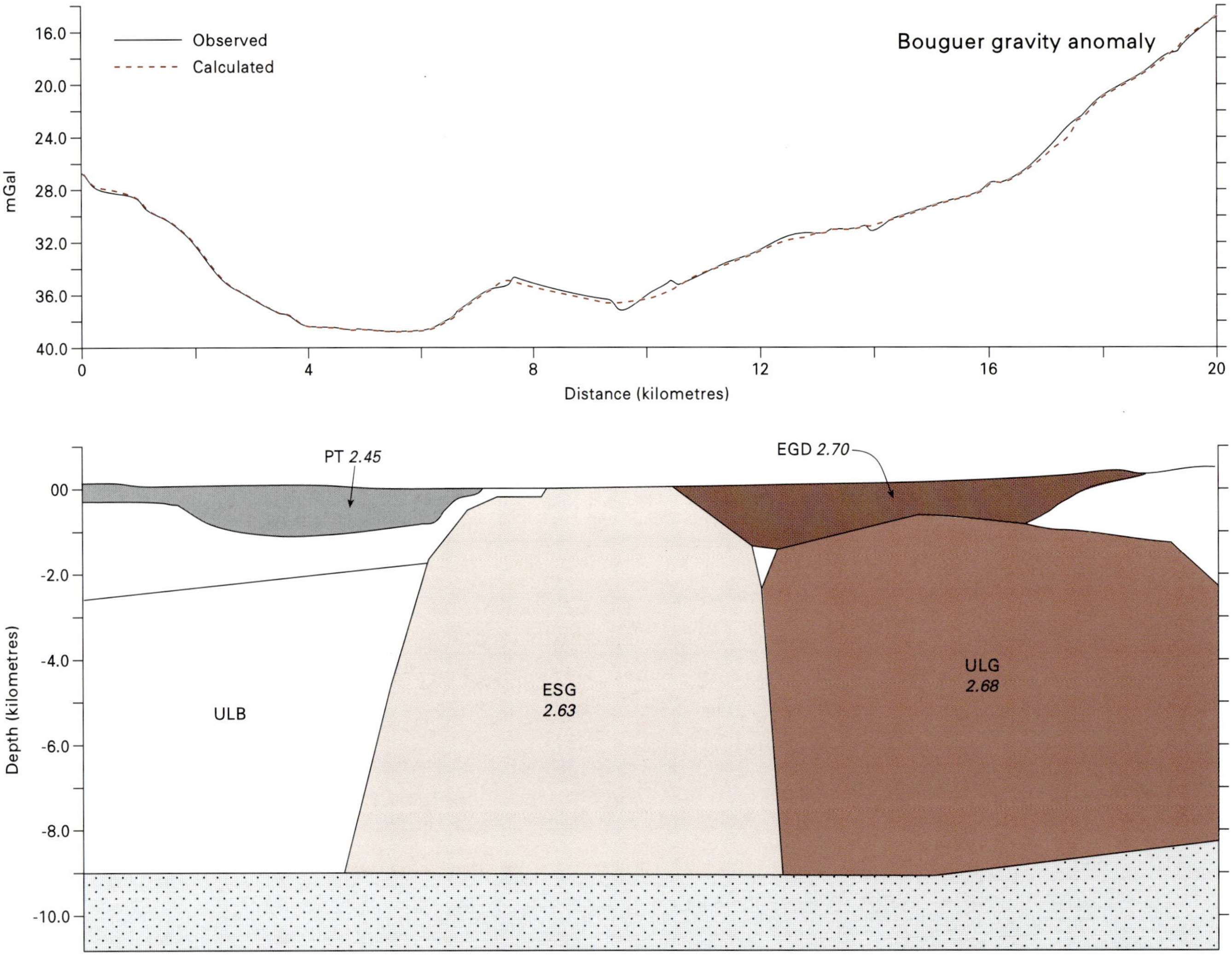

Figure 9 (continued).

b. Traverse 4/8 assuming that the Eskdale granodiorite is of laccolithic form and underlain by a separate component of the batholith, the 'Ulpha granite'. Density values used in the modelling are shown in Mg/m^3.

cannot extend to any great depth (c. 1 km thick; Lee, 1989; EN, Figures 8, 9a) and that the granite must be underlain by a separate pluton of higher density (the Buttermere granite of Lee, 1989). Concealed parts of the Ennerdale intrusion extend south-eastwards towards the Eskdale granite and the northern part of the underlying Buttermere granite is probably not in contact with the Ennerdale intrusion (Figure 9a). Magnetic anomalies over the Ennerdale intrusion are associated with bodies of magnetic hybrid rocks within it. The seismic reflection data fully support the interpretation of the Ennerdale intrusion as a laccolith of 1.1 km thick (Evans et al., 1994).

Concealed intrusions

The 'Buttermere granite' (Figures 8, 9a) was interpreted by Lee (1989) as a major, separate component of the batholith lying between the Crummock and Ullswater lineaments. A reasonable fit to the gravity field is obtained if a single intrusion is assumed to extend to around the same depth as the Eskdale granite (i.e. 9 km) and has a density between that of granite and granodiorite (c. 2.68 Mg/m^3). However, seismic reflection data indicate that the concealed batholith beneath the Ennerdale intrusion comprises a series of granite laccoliths interleaved with significant thicknesses of country rock such as hornfelsed Skiddaw Group rocks (Evans et al., 1994). Such a combination could have an overall density similar to that of granodiorite. Thus, the 'Buttermere granite' as defined by Lee (1989), may represent a distinctive batholith component comprising moderately spaced sheets of granitic composition rather than a single intrusion of granodioritic composition.

The 'Dunmail granite' (DMG, Figures 8, 11) lies to the east of the Eskdale granite and forms part of the ridge of the batholith underlying the Scafell Syncline in the central Lake District. It lies between the Ullswater Lineament and

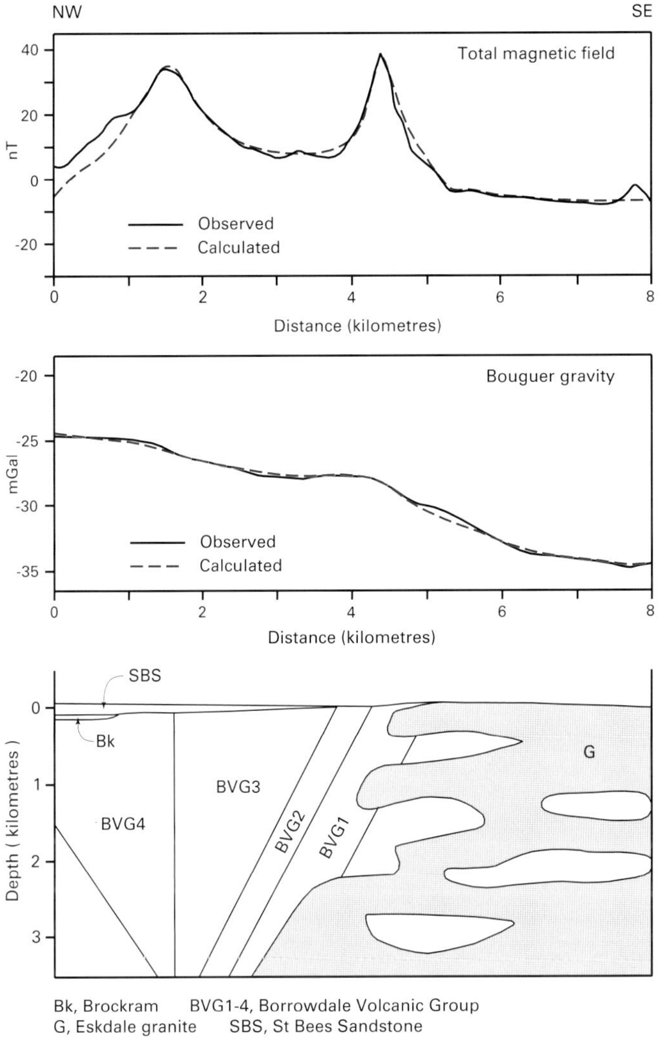

Figure 10 Gravity and magnetic interpretation along seismic line GDGG–90–20; location shown on Figure 5 (Evans et al., 1994).

Preferred model (values in brackets give density and magnetic susceptibility values in Mg/m^3 and SI units respectively): BK Brockram (2.70, 0); BVG1 Borrowdale Volcanic Group (2.78, 0.003); BVG2 Borrowdale Volcanic Group (2.73, 0.0007); BVG3 Borrowdale Volcanic Group (2.69, 0.0007); BVG4 Borrowdale Volcanic Group (2.72, 0.0035); G Eskdale granite (2.63, 0); SBS St Bees Sandstone (2.50, 0); unlabelled Skiddaw Group (2.78, 0.0004).

lineament 4 (Figure 7). The gravity field in this area may be interpreted as an eastwards extension of the Eskdale granite, but modelling as a separate intrusion of slightly higher density (c. 2.66 Mg/m^3) was considered by Lee (1989) to be marginally more realistic. In view of the implications of the seismic data for the western parts of the batholith, it is also possible that the Dunmail granite could be a batholith component with a slightly greater proportion of interleaved country rock between granitic sheets than is the case within the Eskdale granite. Interpretation of the gravity anomaly as a thick sequence of low-density acid volcanic rocks is not substantiated (Lee, 1989).

The 'Rydal granite' (Figure 8) lies between lineaments 4 and 10 (Figure 7) and underlies the south-western part of the Haweswater Syncline. Best fit models are based on a density of approximately 2.68 Mg/m^3, which is between that of granite and granodiorite at outcrop elsewhere in the Lake District. Alternatively, there may be a thick sequence of low-density acid volcanic rocks above a south-easterly extending shoulder of the Dunmail granite. However, a separate component of the batholith was considered by Lee (1989) to be more likely, given the present understanding of the bulk density and thickness of the overlying volcanic sequence. As with the other concealed intrusions, the 'Rydal granite' could be a series of stacked laccoliths.

It is possible to interpret the residual gravity low over the Ulpha Syncline (UL, Figure 7) as a 2 km-thick sequence of low-density silicic volcanic rocks underlain by a shoulder of the Eskdale granite extending to the south. However, a separate intrusion of density 2.65 Mg/m^3 ('Ulpha granite', ULG; Figures 8, 9) beneath a 2 km-deep syncline containing about 1 km of acid volcanic rocks fits the observed gravity field marginally better. Together with the models of the Eskdale granodiorite (Figure 9b) the modelling results support the existence of a separate Ulpha granite underlying the Eskdale granodiorite and the Ulpha Syncline. Stacked granitic sheets are equally compatible with the gravity data.

A small, residual gravity low in the Coniston area (CN, Figures 6, 7) may be modelled as either a high-level granitic intrusion into the Borrowdale Volcanic Group ('Coniston granite'; Figure 8), a low-density sequence of silicic volcanic rocks or a small deep-seated component of the batholith. The gravity field is not particularly well defined over the anomaly and a thick sequence of acid volcanic rocks is compatible with the surface geology. However, the presence of a small granitic intrusion above the southern wall of the batholith should not be ruled out, especially in view of the presence of the extensive copper mineralisation in the Coniston area (Chapter 11).

IMPLICATIONS FOR EVOLUTION OF THE BORROWDALE VOLCANIC GROUP AND THE BATHOLITH

The U-Pb zircon ages for the Eskdale and Ennerdale intrusions (Hughes et al., 1996) indicate that the exposed magmatic rocks of the Lake District are confined principally to the late Ordovician and Early Devonian. Timing of emplacement of the concealed parts of the batholith is thus important. The spatial correlation of the major concealed components of the batholith with the synclines within the volcanic sequence, as demonstrated by the gravity modelling, along with the initiation of the Scafell and Ulpha synclines as volcanotectonic structures (Chapter 9), strongly implies that the concealed batholith is late Ordovician and contemporaneous with the Eskdale and Ennerdale intrusions, rather than Early Devonian in age. Furthermore, the seismic reflection data suggest that each

of the major batholith components identified by Lee (1989) could comprise a distinctive suite of stacked laccoliths emplaced beneath each syncline, rather than homogeneous plutons within a multi-phase batholith.

MAGNETIC BASEMENT AND RELATIONSHIP BETWEEN DEEP AND SHALLOW STRUCTURE

The integrated gravity and magnetic modelling by Lee (1989) interpreted the broad magnetic highs in the southern Lake District (e.g. anomaly WI, Figure 4b) as magnetic basement at relatively shallow depth (5 to 6 km) beneath the Windermere Supergroup. However, in order to fit the long-wavelength magnetic anomaly across the Lake District magnetic basement must also be present beneath, and to the north of, the batholith, with its northern margin dipping beneath the Solway Basin. The magnetic basement may comprise a thick layer of pre-Skiddaw Group (magnetic) sedimentary rocks extending down to the mid-crust, perhaps similar to the magnetite-bearing rocks of Arenig age encountered in the Beckermonds Scar Borehole (Wilson and Cornwell, 1982), or magnetic crystalline basement, which may have acted as a source for the magnetite observed at Beckermonds Scar, or a combination of both (Lee, 1989).

The most realistic models of the southern Lake District show the magnetic basement as a ramp directly beneath the southern margin of the batholith, cutting through its southern flank at depth (e.g. Figure 9a). This led Kneller and Bell (1993) to adduce that the batholith had been displaced southwards above a northward dipping mid-crustal ramp during the Acadian, resulting in the formation of a monocline and associated backthrusts on the southern margin of the batholith (Chapter 9).

The relationship between the structure of the magnetic basement and thrusts on both the southern and northern margins of the batholith was investigated by the modelling along profile AA–BB (Figure 11), based on the principles established by Lee (1989) but constrained by the resurvey. The magnetic field is explained as a series of north-north-west-directed ramps segmenting the magnetic mid-crustal layer and implying Acadian imbrication of the Avalonian crust. The presence of a major crustal ramp (or series of ramps) to the north of the Lake District (North Lakes Ramp System) is supported by the presence of dipping reflectors on seismic reflection profiles across the Iapetus Suture Zone (WINCH and

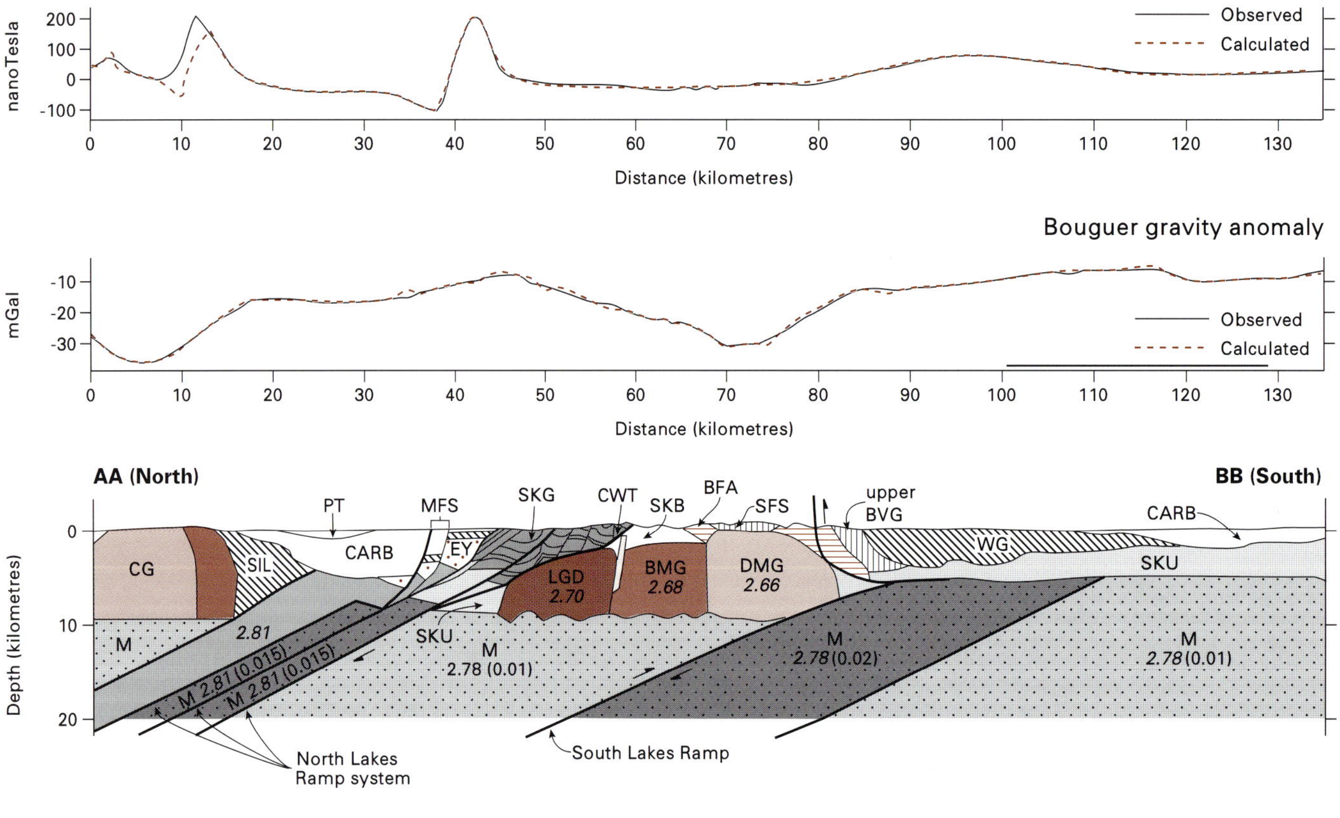

Figure 11 Integrated gravity and magnetic model along profile AA–BB (Figure 5).

Density values used in the modelling are shown in Mg/m^3 and magnetic susceptibility values (in brackets, where greater than zero) are shown in SI units. BFA Birker Fell Formation; BMG Buttermere granite; BVG Borrowdale Volcanic Group; CARB Carboniferous rocks; CG Criffel granite; CWT Crummock Water Thrust; DMG Dunmail granite; EY Eycott Volcanic Group; LGD Loweswater granodiorite; M magnetic basement; MFS Maryport fault system; PT Permo-Triassic rocks; SFS Scafell Syncline; SIL Silurian rocks; SKB Buttermere Formation; SKG Skiddaw Group; SKU Undifferentiated Skiddaw Group; WG Windermere Supergroup.

NEC profiles, Soper et al., 1992) and by modelling of magnetic anomalies along strike in the Irish Sea (Kimbell and Stone, 1995). The conclusion that there is a ramp beneath the southern Lake District (South Lakes Ramp) is difficult to avoid if a reasonable fit is to be obtained to the magnetic anomaly and is supported by north-north-west-dipping reflectors observed on seismic data along strike in the eastern Irish Sea (Soper et al., 1992).

The model suggests that south-south-east-directed thrusts in the Skiddaw Group ramped up over the concealed northern shoulder of the batholith and are probably rooted in the North Lakes Ramp System. Southward displacement of the batholith over the South Lakes Ramp produced the backthrust and the monocline, the position of which were controlled by the south-south-east flank of the batholith. The Lake District batholith acted as a rigid upper-crustal block just to the south of the Iapetus Suture Zone, over which the Borrowdale Volcanic Group is relatively undeformed. During the Acadian, upper-crustal thrusts developed on its north-north-west and south-south-east margins, and crustal delamination and imbrication occurred beneath it.

FOUR

Ordovician: Borrowdale Volcanic Group

Much of the mountainous central part of the Lake District is formed from the Borrowdale Volcanic Group. The succession, up to about 8 km thick in the Ambleside district, comprises basaltic, andesitic, dacitic and rhyolitic lavas, sills and volcaniclastic rocks that comprise a dominantly subaerial (Branney, 1988a), medium- to high-K calc-alkaline volcanic association. It is considered to have been produced as a result of subduction of the Iapetus Ocean at the continental margin of Eastern Avalonia (Fitton and Hughes, 1970; Millward et al., 1978; Fitton et al., 1982). The Borrowdale Volcanic Group constitutes the main geological division of the Ambleside district. In the west erosion has exposed the underlying Skiddaw Group and components of the Lake District batholith, and in the south-east the volcanic rocks are overlain unconformably by the Windermere Supergroup; thus a complete section through the volcanic complex is present. This chapter provides a general introduction to the Borrowdale Volcanic Group and describes its overall petrology and geochemistry; details of the lithostratigraphical units are given in Chapters 5 and 6.

The age and time span of the Borrowdale Volcanic Group is poorly constrained between the Llanvirn Tarn Moor Formation (Skiddaw Group; Cooper et al., 1995) and the middle Ashgill (Cautleyan) Dent Group (see Chapter 7), and has been assigned customarily to the Llandeilo and early Caradoc. Dark grey mudstone from Holehouse Gill in the Ulpha valley (Figure 1), first reported by Numan (1974), contains a microflora that is probably Caradoc (Harnagian–Soudleyan; Chapter 6; see also discussion by Molyneux, 1988).

Rb–Sr dates of Borrowdale Volcanic Group rocks reflect subsequent resetting of the isotopic systems (Rundle, 1992). Extensive data from the Great Whinscale Dacite (Kanaris-Sotiriou et al., 1991) gave a poorly defined Rb–Sr age of 398 ± 11 Ma. Separate transects through the dacite at Yoadcastle [SD 157 951], Crook Crag [SD 196 989], Harter Fell [SD 218 998] and Brandy Crag [SD 225 990] yielded older ages of 430 to 420 Ma, close to the Rb–Sr ages for the Eskdale and Ennerdale intrusions (Rundle, 1979), but still younger than the Caradoc biostratigraphical age for the volcanic rocks. The U–Pb zircon ages of 450 ± 3 and 452 ± 4 Ma (Hughes et al., 1996) for the Eskdale and Ennerdale plutons, which intrude the Birker Fell Formation are close to 457 ± 4 Ma (Sm–Nd on garnet–whole rock pairs) obtained from garnetiferous rocks from the lower part of the Borrowdale Volcanic Group in the Ullswater and Haweswater areas (Thirlwall and Fitton, 1983). Though there are no precise correlations between the Ambleside district and these areas, the samples dated are probably from both the lower and upper parts of the Borrowdale Volcanic Group. The Sm–Nd age is compatible with the biostratigraphical age of the Holehouse Gill Formation on the timescales of Harland et al. (1990) and Fordham (1992). On this basis a timespan of 10 to 14 Ma is available for the eruption of the Borrowdale Volcanic Group, but the duration of the volcanic episode may have been considerably shorter if time is allowed for the period of uplift prior to the onset of volcanism.

Rock classification

In this account the compositions of lavas, sills and pyroclastic rocks in the Borrowdale Volcanic Group are classified, where possible, using the schemes recommended by the IUGS Subcommission on the systematics of igneous rocks (Le Maitre, 1989). Where geochemical data have not been acquired, field appearance and petrographical characteristics such as phenocryst assemblage are used.

Components of the volcaniclastic rocks comprise pyroclasts, reworked pyroclasts and epiclasts. In most of the literature describing the Borrowdale Volcanic Group the stratiform volcaniclastic rocks have been called tuff, though many are demonstrably of sedimentary origin. In this account pyroclastic nomenclature is strictly applied. Where a non-volcanic emplacement is adduced then the rocks are classified using the terms volcaniclastic sandstone, siltstone etc., according to the Wentworth grain-size classification. In some parts of the succession interbedded pyroclastic, reworked pyroclastic and sedimentary rocks are distinguished, but in some formations, in particular those dominated by plane parallel bedding, the origins are equivocal and in such cases the clastic sedimentary terminology is used. The volcaniclastic rocks are interpreted according to their lithofacies using criteria established from modern examples (e.g. Cas and Wright, 1987).

SKIDDAW GROUP: BASEMENT TO THE VOLCANIC COMPLEX

The base of the Borrowdale Volcanic Group crops out west of Devoke Water [SD 132 959] where the volcanic rocks overlie poorly exposed, siliciclastic sedimentary rocks assigned to the Skiddaw Group. The contact is not exposed but an unconformity is presumed from evidence elsewhere in the Lake District, where an angular relationship has been demonstrated (Wadge, 1972).

In the Ambleside district strongly cleaved, grey fine-grained sandstone and siltstone crop out as narrow slivers between the Eskdale granite and the Borrowdale Volcanic Group near Devoke Water [SD 142 965]. On Water Crag [SD 153 973] faint bedding dips steeply to the south-east subparallel to the inferred contacts with the Eskdale

granite and Borrowdale Volcanic Group. A narrow wedge (up to 1.5 m) of these rocks is exposed in the adjacent Linbeck Gill [SD 1505 9714]. To the west of Devoke Water, two small exposures occur next to the margin of the Eskdale intrusion [SD 147 970 and 1415 9673] and, a little to the south, cleaved bedded siltstone is exposed near a sheep fold [SD 1421 9661]; these are considered to be part of a continuous outcrop extending south-westwards to Barnscar [SD 133 959]. The sedimentary rocks are intercalated with feldspar-phyric basalt that is not autobrecciated; because it appears to be transgressive in the eastern part of its outcrop it is probably intrusive.

In thin-section these rocks are fine grained, strongly foliated and consist entirely of quartz, chlorite and sericite. Chlorite probably replaces biotite indicating that the cleavage-related mineralogy postdates the hornfelsing and confirms that the Eskdale granite (dated at 450 ± 3 Ma) is pre-cleavage. Subtle changes in grain size in the hornfels may be relict lamination, but otherwise all original textures have been destroyed. Near the granite contact conspicuous small crystals of tourmaline are present.

These rocks were noted as 'altered ?Skiddaw Slate' on the Old Series six-inch Geological Sheet, Cumberland 83, an interpretation accepted by Firman (1957), and are now assigned to the Skiddaw Group. Mudstone, siltstone and sandstone clasts, derived from the Skiddaw Group, also occur in variable quantities in pyroclastic rocks within the lower part of the Borrowdale Volcanic Group. The tabular clasts have been strongly abraded and are imbricated in beds where they are abundant. The bed forms of the pyroclastic rocks are indicative of an hydrovolcanic origin with the clasts derived explosively from the conduit walls.

LITHOSTRATIGRAPHY OF THE VOLCANIC ROCKS

The volcanic rocks were referred to as Green Slates and Porphyries by Sedgwick (1832), and then successively as the Lower Silurian Volcanic Series of Borrowdale by the Geological Survey (Ward, 1876), Borrowdale Series by Marr (1916) and Borrowdale Volcanic Series by Green (1920). The change from Series to Group was required by modern lithostratigraphical usage (Wadge, 1978; Moseley, 1984). The primary survey (Geological Survey, 1860–90) produced the only previous complete set of maps of the district, but it was left to Marr (1900) and Green (1919) among others, to apply a formal lithostratigraphy. During the ensuing 50 years researchers, notably J J Hartley, G H Mitchell and R L Oliver, erected local lithostratigraphies for almost all of the Borrowdale Volcanic Group in the Ambleside district. Each area contained its own set of names but precise correlation with the local stratigraphies of adjacent areas was not made. Subsequent attempts at resolving the inter-area correlations have been made, though the plethora of local names was not rationalised (Mitchell, 1956a; Millward et al., 1978; Moseley and Millward, 1982). These lithostratigraphical models developed for the Borrowdale Volcanic Group were not based on new field surveys and contained much uncertainty caused mainly by the lack of fossiliferous strata, the lateral impersistence of many units, complex faulting and the presence of many sills, previously identified as lavas and ascribed to stratigraphical formations. Also, the complex isoclinal fold model for the Lake District Lower Palaeozoic rocks proposed by Green (1920) influenced strongly Mitchell's (1940) interpretation of the lithostratigraphy in the Coniston area. Mitchell's (1956b; 1963) later contributions on the Dunnerdale and Seathwaite fells had a simpler structural interpretation, but the correlations between the areas could not be resolved without remapping.

The resurvey (British Geological Survey, 1996) provides a resolution to the previous uncertainties of correlation in the Ambleside district and forms a sound basis for the volcanological interpretation of the volcanic complex. Where possible, unit names in common usage are retained, but necessarily many widely known terms are abandoned. Previous names for the newly defined lithostratigraphical units and comparison between the definitions of the base of new compared with old units are given in Chapters 5 and 6. To aid comparison with previous accounts a summary is given in the chart of Table 6. Superficially this chart suggests that the previous lithostratigraphies were basically correct, but the complex differences between old and new units are revealed by duplication of old names; for example, in part of the Seathwaite Fells, Mitchell's (1963) Kidson How Tuffs belong to the newly defined Whorneyside and Airy's Bridge formations but, in another part, to the Low Water Formation.

In their attempts at producing a unified lithostratigraphy for the Borrowdale Volcanic Group Millward et al. (1978) and Moseley and Millward (1982) noted that the succession could be divided into two distinctive parts, which are evident in the Ambleside district (Figure 12). The lower part comprises dominantly andesite lavas intercalated with volcaniclastic deposits, including thin but stratigraphically important ignimbrites. These rocks are assigned to the Birker Fell Formation, 1.8 to 2.8 km thick (Chapter 5; Petterson et al., 1992). The upper part is 2 to 5 km thick and consists of intermediate and silicic pyroclastic rocks, and volcaniclastic sedimentary rocks. Abundant penecontemporaneous sills were emplaced into the sedimentary succession. The fundamental change in eruptive process from the dominantly effusive activity of the Birker Fell Formation to the subsequent voluminous explosive activity associated with caldera formation (Branney and Kokelaar, 1994) occurs at the base of the Whorneyside Formation. This unit is widely distributed throughout the district and extends northwards into the adjoining Keswick district.

The lithostratigraphy of the volcaniclastic upper part of the Borrowdale Volcanic Group (Chapter 6) is described in three areas, the successions of which were associated with spatially and temporally distinctive eruptive and depositional centres (Figure 12). The lowest of these is developed best in the Central Fells and comprises pyroclastic rocks of the Whorneyside, Airy's Bridge and Lingmell formations. Though some of these rocks also occur in the south of the district, they were considered to have been erupted from the Scafell Caldera (Branney and Kokelaar, 1994). Overlying the Airy's Bridge Formation in the Coniston Fells and Ulpha Fells, in the southern

Table 6 Broad comparison of previous lithostratigraphical divisions of the Borrowdale Volcanic Group in the Ambleside district with that newly proposed. No direct equivalence is implied; many of the new formations have redefined bases. Some old names appear at several levels within the areas, indicating where new interpretations of the stratigraphy apply. (See text for details.)

Formations used this memoir	Coniston (Mitchell, 1940)	Dunnerdale (Mitchell, 1956a)	Seathwaite (Mitchell, 1963)	Wastwater to Duddon (Firman, 1957)	Langdale and Grasmere (Hartley, 1925, 1932)	Scafell (Oliver, 1961)
Middle Dodd Dacite						
Tarn Hows						
Esk Pike					Steel Fell Rhyolite	Esk Pike Fm.
Lincomb Tarns	Yewdale Breccia				Felsitic and basic tuffs	Lincombe Tarns Fm.
Seathwaite Fell	Yewdale Bedded Tuffs	Wrengill Andesites*	Wrengill* Andesites		Wrengill Andesites*	Seathwaite Fell Tuffs
	Wrengill Andesites*	Tilberthwaite Tuffs: White Pike Andesites*	Tilberthwaite Tuffs: White Pike and Dow Crag Andesites*		Bedded Tuffs	
	Upper Tilberthwaite Tuffs	Walna Scar Quarries Tuffs	Walna Scar Quarries Tuffs			
Lingmoor Fell					Langdale Rhyolite	
Low Water	Upper Tilberthwaite Tuffs		Kidson How Tuffs			
Lag Bank		Lag Bank Tuffs	Lag Bank Tuffs			
Caw		Caw tuffs	Caw Tuffs			
Lickle	Paddy End Rhyolites	Lickle Rhyolites	Lickle Rhyolites	Wallowbarrow Crag Group		
Dunnerdale		Dunnerdale Tuffs	Dunnerdale Group: Dunnerdale Tuffs	Duddon Hall Tuffs		
Wallowbarrow				Wallowbarrow Crag Group		
Ulpha		Ulpha Andesites (Bleamsley Bank Flows)	Ulpha Andesites	Ulpha Andesites		
Holehouse Gill						
Duddon Hall	Lower Tilberthwaite Tuffs and Dow Crag Andesites*	Ulpha Andesites (Ivenscar flows) and Duddon Bridge Tuffs	Dunnerdale Group	Duddon Hall Tuffs		
Waberthwaite				Waberthwaite Tuffs		
Lingmell					Langdale Rhyolite	Airy's Bridge Group
Airy's Bridge	Paddy End Rhyolites		Kidson How Tuffs	Waberthwaite Tuffs and Worm Crag Group		
Whorneyside						
Birker Fell			Cockley Beck Group and Seathwaite Tarn Andesites	Lower Andesites	Mosedale Andesites	Birker Fell Andesite Group

* The Wrengill, Dow Crag and White Pike Andesites are interpreted entirely as syn-Borrowdale Volcanic Group sills in this memoir.

Shaded areas indicate that the formation is either not present in the area or was not differentiated.

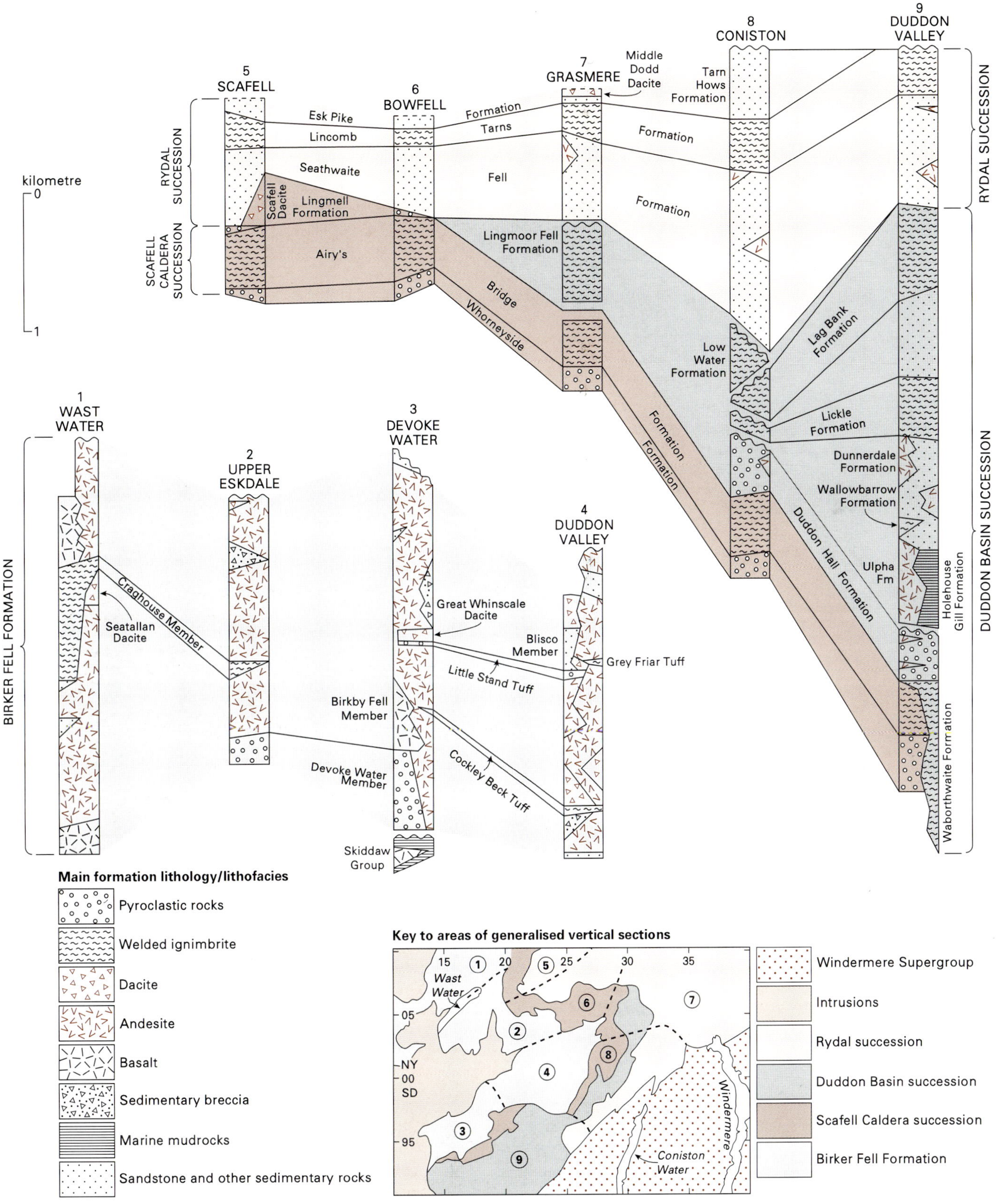

Figure 12 Lithostratigraphy of the Borrowdale Volcanic Group summarising the variations in the Birker Fell Formation and three volcaniclastic successions (Scafell Caldera, Duddon Basin and Rydal) across the district.

part of the district, are nine formations (Figure 12) comprising pyroclastic and volcaniclastic sedimentary rocks that are preserved only in that area, referred to geologically as the Duddon Basin. This succession contains the only marine strata in the group (Holehouse Gill Formation). The successions of both the Scafell Caldera and Duddon Basin are succeeded in the eastern part of the Ambleside district by the predominantly sedimentary Seathwaite Fell and Esk Pike formations separated by the major ignimbrite of the Lincomb Tarns Formation.

DEPOSITIONAL ENVIRONMENT

Ward (1876) maintained that the lowest part of the Borrowdale Volcanic Group had been deposited in a marine environment, but that much of the remaining part of the succession was subaerial. Evidence for marine conditions was cited by Geikie (1891) and Green (1919). Most subsequent accounts imply that subaerial and subaqueous environments co-existed within a chain of volcanic islands (Mitchell, 1956a; Fitton and Hughes, 1970; Millward et al., 1978; Turner and Wadge, 1979). However, lithofacies associations within the Ambleside district contain abundant evidence that supports the conclusion of Branney (1988a) that, with few exceptions, the Borrowdale Volcanic Group was emplaced subaerially. The general evidence for this is given below and examples are described under the lithostratigraphical details in Chapters 5 and 6.

Lithofacies which characterise a volcanic island environment generally consist of volcaniclastic rocks, interbedded with fossiliferous marine sedimentary rocks; hyaloclastite and pillow lava are also typical. The absence of a marine fauna has been long stated, and there are only two localities from which trace fossils have been recorded (Mitchell, 1956b; Suthren, in Branney, 1988a; Johnson et al., 1994). The Holehouse Gill Formation (Figure 1; Chapter 6) is unusual because it contains beds of non-volcanic sedimentary rocks of undoubted marine deposition. Numan (1974) recorded these dark grey mudstone beds in Holehouse Gill, near Ulpha, and their marine microflora was noted by Molyneux (1988). Overall there is no evidence to indicate that the Borrowdale Volcanic Group was deposited under widespread and long-lasting marine conditions.

Oliver (1954, 1961) favoured a subaerial environment because of the abundance of welded ignimbrites within the sequence, though it is now accepted that welding may occur subaqueously (summary by Cas and Wright, 1987). However, ash-fall and pyroclastic surge deposits, diagnostic of subaerial eruption and deposition, do occur in many parts of the sequence; examples within the Birker Fell Formation, and interbedded with ignimbrite in the Airy's Bridge and Lickle formations are given in Chapters 5 and 6 (Plate 4). Common at many stratigraphical levels within the succession are thin beds of eutaxitic lapilli-tuff, previously considered by Suthren and Furnes (1980), as indicative of a subaerial environment. However, eutaxitic-like textures may be produced by post-alteration or 'diagenetic' compaction (Branney and Sparks, 1990) and so cannot be taken as unequivocal evidence for subaerial eruption and deposition.

Some of the widespread and abundant laminated sandstone and siltstone sequences are indicative of subaqueous deposition, probably in ephemeral lakes. Also, sedimentary lithofacies indicating fluvial deposition occur at many levels within the Birker Fell Formation and comprise substantial parts of the later Dunnerdale, Caw and Seathwaite Fell formations. The top of some pyroclastic deposits (e.g. Duddon Hall Formation) were scoured and reworked locally by streams. Small rills eroded by surface runoff during the phreatomagmatic eruption of the Whorneyside Formation and draped by succeeding ash-fall layers indicate subaerial deposition (Branney, 1991). Interbedded sequences comprising reworked volcanic sediments occur between welded ignimbrite sheets throughout the upper part of the Borrowdale Volcanic Group, but rapid reworking of unconsolidated subaerial pyroclastic deposits into sedimentary basins may have resulted in over-representation of volcaniclastic sedimentary rocks in the sequence. Lacustrine environments were actively promoted by the extensional tectonic regime and volcanotectonic collapse.

Plate 4 Parallel laminated ash-fall basaltic tuff mantling an erosion surface. Throstle Garth Member, Scar Lathing, Eskdale [NY 2253 0490] (D 4016).

Unconformities are present at the base of the Borrowdale Volcanic Group and at the junction with the overlying Windermere Supergroup. Discordant surfaces are also common throughout the sequence, with features suggesting a wholly subaerial origin. There are no structures nor littoral facies rocks indicative of sea stacks, cliffs and shorelines. Many of the small-scale bedding discordances were caused by local reworking of deposits and some surfaces are mantled by parallel-bedded ash-fall tuff and lapilli-tuff (Plate 4). Other unconformities are widespread and characterised by angular relationships, for example at the base of the Whorneyside, Duddon Hall and Seathwaite Fell formations. However, unconformities within caldera successions cannot be taken to imply erosion or time gaps and internal angular discordances are a characteristic feature of piecemeal caldera collapse (Branney and Kokelaar, 1994).

MINERALOGY

Borrowdale Volcanic Group magmatic rocks typically contain abundant phenocrysts though, within the Birker Fell Formation, aphyric basalt characterises the Throstle Garth Member, and aphyric dacite the Great Whinscale Dacite. The phenocryst assemblage includes plagioclase, alkali feldspar, clinopyroxene, orthopyroxene, olivine, biotite, iron–titanium oxide and garnet (Plate 5). Accessory mineral phases include apatite and zircon. Hornblende phenocrysts, characteristic of many continental-margin andesite suites, occur only very sparsely suggesting that the Borrowdale Volcanic Group magmas were 'dry' and probably contained less than 3 per cent water (compare with Gill, 1981). Details of variations in the phenocryst assemblage within formations are contained in Chapters 5 and 6.

Olivine and orthopyroxene phenocrysts are never fresh, but the crystal habit of some chlorite pseudomorphs indicates that both were present, orthopyroxene generally in rocks ranging in composition from basaltic andesite to dacite, but olivine only in basalt lavas of the Throstle Garth Member. Clinopyroxene (diopside–augite), locally up to 10 mm across, occurs conspicuously in basalt and basaltic andesite. It is sporadically fresh, but mainly replaced by chlorite and in some rocks also by epidote; near the contacts with the granitic intrusions the replacement is actinolite–hornblende. Glomeroporphyritic aggregates of anhedral granular mafic minerals are common in the mafic and intermediate rocks. Altered biotite microphenocrysts occur in some dacite and rhyolite.

Laths of plagioclase, up to 3 mm, are the dominant phenocryst in most of the rocks (Plate 5b). Magmatic compositions of plagioclase are preserved locally, but albitisation is common and most phenocrysts are variably altered to sericite, epidote and locally, carbonate. The fresher crystals typically show a wide range of internal textures, including zones of melt inclusions, and discontinuous, normal, convolute and oscillatory zoning. Turbid phenocrysts of alkali feldspar occur in some dacite and rhyolite (e.g. Seatallan Dacite, Plate 5c; Little Stand Tuff and Bad Step Tuff).

Plate 5 Photomicrographs of lithologies from the Birker Fell Formation.

a. Porphyritic basaltic andesite from the Birkby Fell Member, containing clinopyroxene (epitaxially replaced by calcic amphibole) and plagioclase phenocrysts in a dark fine-grained groundmass containing tiny feldspar laths (registered no. E70009). Plane-polarised light; width of field 12 mm (PMS716).
b. Porphyritic andesite from the Birker Fell Formation. Phenocrysts of plagioclase typically showing cloudy (saussuritic) cores and unaltered, zoned margins. Subsidiary pyroxene phenocrysts, commonly as glomerophyric clusters, are pseudomorphed by expitaxial and/or fibrous amphibole (actinolite/actinolitic hornblende). The microcrystalline groundmass is mesocratic with local flow alignment of microlitic feldspar laths (E69693). Plane-polarised light; width of field 12 mm (PMS717).
c. Seatallan Dacite showing dominantly plagioclase phenocrysts in an originally glassy groundmass. The latter contains acicular quench microlites and crude perlitic cracks, and has recrystallised to a blebby, fine-grained felsic mosaic. Patchy epidote after plagioclase is common (E71122). Plane-polarised light; width of field 12 mm (PMS718).
d. Lapillistone from the Devoke Water Member composed of close-packed basaltic and paler crystalline volcanic clasts, together with minor crystal fragments. Juvenile basaltic clasts are non-vesicular, subangular and hyalopilitic indicating an origin by water–magma quench fragmentation (E69816). Plane-polarised light; width of field 12 mm (PMS719).
e. Welded lapilli-tuff from the Craghouse Member ignimbrite. A eutaxitic fabric is defined by paler, locally distorted fiamme in a darker vitric matrix. Broken plagioclase and chloritised pyroxene crystals together with small lithic fragments are also present (E71117). Plane-polarised light; width of field 12 mm (PMS720).
f. Cockley Beck Tuff displaying characteristic garnet (lower right) with corroded margin and alteration along fractures. Plagioclase crystals and lithic fragments are common, set in a fine grained tuff matrix (E69300). Plane-polarised light; width of field 12 mm (PM721).

Originally the groundmass of the lavas and sills varied from hypocrystalline to holocrystalline (Plate 5). Though all glass has long since devitrified, petrographical terms referring to the originally vitreous state of some rocks are used throughout this account. Hyalopilitic, trachytic and intergranular textures occur in basalt and basaltic andesite. In more felsic rocks glassy groundmass predominates, recrystallised to an altered, fine-grained devitrified mosaic of quartz, alkali feldspar, sericite and chlorite; some rocks have snowflake texture. Perlitic cracking is common and spherulites occur locally.

Welding textures in ignimbrite range from eutaxitic to parataxitic (Plates 5, 6), though a continuous foliation is present in some very high-grade tuffs, such as the Bad Step Tuff (Branney et al., 1992). Shards are preserved only locally.

Garnet

Phenocrysts of red almandine–pyrope garnet, up to 8 mm diameter are an unusual feature of the suite (Plates 5f,

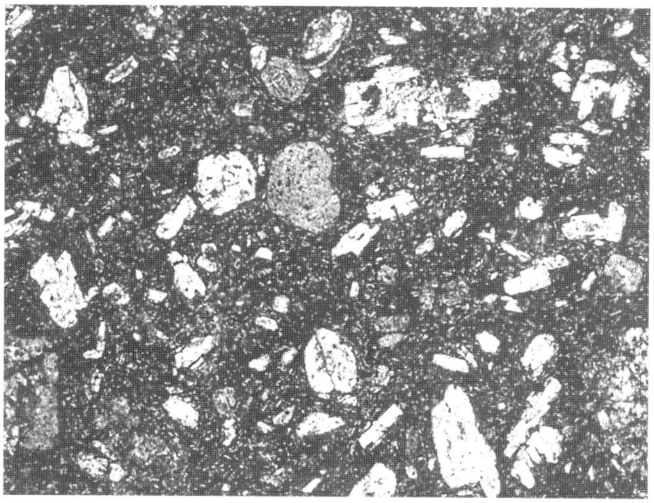

a.

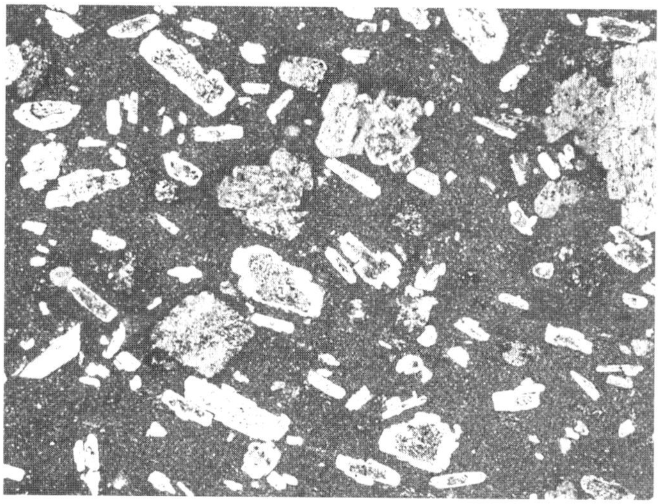

b.

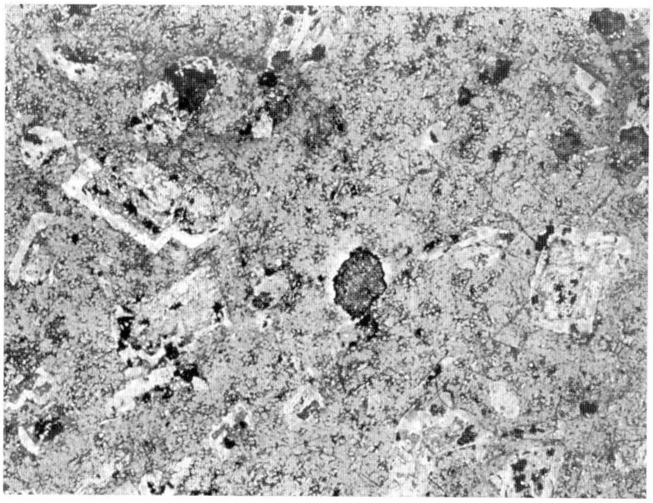

c.

d.

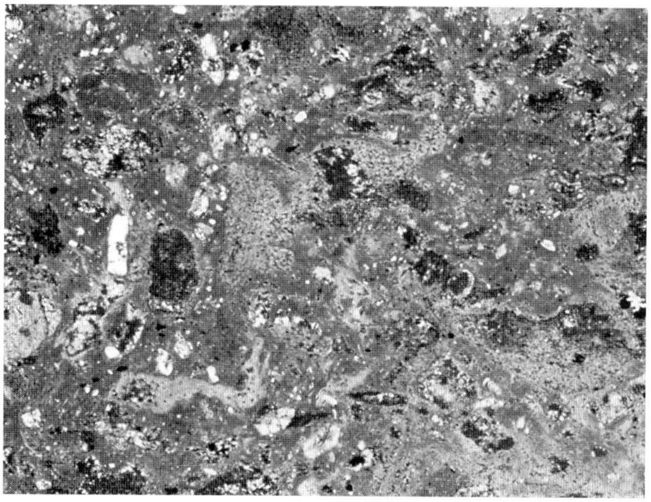

e.

f.

6d; Walker, 1904; Oliver, 1956a, b; Fitton, 1972; Fitton et al., 1982). They are common within some units in the andesite to rhyolite compositional range, though sparse examples have been found in basaltic andesite tuff in the Devoke Water Member. Garnet in the more basic rock compositions contains up to 31 mol per cent pyrope (Fitton, 1972). Most crystals are heavily corroded or resorbed and associated with alteration rims of chlorite and magnetite; some occur in aggregates with plagioclase. Garnet in dacite and rhyolite contains at least 9 mol per cent pyrope (Fitton, 1972). These crystals are generally much less corroded or altered and many are euhedral. Spessartite and grossular are minor components (Fitton, 1972). Inclusions of titanomagnetite and pyrite occur in garnet in andesite, whereas those in dacite and rhyolite contain zircon and apatite.

Fitton et al. (1982) indicated that garnet was mainly present in formations within the lower part of the Borrowdale Volcanic Group (i.e. within the Birker Fell Formation). But they are also conspicuous within the Airy's Bridge Formation, especially in the Long Top Member, the Bad Step Tuff, and throughout the Lingmell Formation. Garnet does not occur stratigraphically above the Lingmell Formation, though it is present in some minor intrusions (Beddoe-Stephens and Mason, 1991; McConnell et al., 1993).

GEOCHEMISTRY

The Borrowdale Volcanic Group suite shows a compositional range from basalt to rhyolite (Fitton and Hughes, 1970; Millward et al., 1978; Fitton et al., 1982; Moseley and Millward, 1982). The most common compositions occur at modes of 57 per cent and 67 per cent SiO_2. The rocks are mostly quartz and hypersthene normative and, in the silicic rocks, corundum normative. The intermediate members show no trend towards iron enrichment, with analyses plotting within the field occupied by calc-alkaline, medium- to high-K andesite and shoshonite on AFM (Millward et al., 1978) and total alkali–silica diagrams, though K concentrations are influenced by secondary element mobility. Discrimination diagrams, using immobile trace-elements such as Th–La (after Gill, 1981) and Ti–Zr (after Pearce, 1982), illustrate the general geochemical similarity of the Borrowdale Volcanic Group to orogenic andesite suites emplaced through continental crust, such as in Chile (Lopez-Escobar et al., 1977) and the western USA (Ewart, 1982). Selected analyses of rocks from the Ambleside district are given in Tables 7 to 12.

Geochemical effects of alteration

Alteration of the mineralogy of the Borrowdale Volcanic Group rocks is widespread and associated with modification of the rock chemistry; an assessment of the nature and extent of the chemical alteration will influence the interpretation of magma petrogenesis (O'Brien et al., 1985; Allen et al., 1987). Causes of alteration include post-emplacement hydrothermal alteration, burial metamorphism, contact metamorphism around the Eskdale and Ennerdale intrusions, cleavage development and the growth of carbonate-bearing assemblages (Chapter 10). Some major and trace elements (e.g. Na, K, Rb, Ba and Sr) are mobile during alteration and in the Borrowdale Volcanic Group are of limited use for discussing magma petrogenesis. Previous authors (e.g. Moseley and Millward, 1982; Fitton et al., 1982) have noted the high levels of K_2O, that K_2O is generally greater than Na_2O and K/Rb ratios are mainly low for andesites of this K_2O content. This may indicate that, in rocks with significant sericite, the K_2O content may not be a primary feature (Allen et al., 1987).

Since Fitton (1971) reported the first full major and trace element analyses of Borrowdale Volcanic Group rocks it has been assumed generally that the scatter shown by Harker-type diagrams (Figures 13, 14) was caused by variation in phenocryst content (reflecting the effect of crystal accumulation or depletion) and by post-emplacement chemical alteration. However, Beddoe-Stephens et al. (1995) showed that within the broad envelope of compositions within the Birker Fell Formation subgroups of lavas have coherent and tightly constrained trends, indicating that these rocks have retained much of their original magmatic signature.

Plate 6 Photomicrographs of rocks from the volcaniclastic successions of the upper part of the Borrowdale Volcanic Group.

a. Welded lapilli-tuff from the Long Top Member (Airy's Bridge Formation) with eutaxitic fabric defined by elongate darker fiamme containing plagioclase crystals and set in a pale, vitroclastic matrix. Lithic clasts comprise dark vesicular andesite and pale felsic fragments (E71102). Plane-polarised light; width of field 12 mm (PMS722).

b. Densely welded (vitrophyric) lapilli-tuff from the Crinkle Member (Airy's Bridge Formation). Vitric matrix shows a faint parataxitic foliation wrapping crystal and pale felsitic lithic clasts (lower left). A spaced fracture cleavage is orientated top to bottom (E67979). Plane-polarised light; width of field 12 mm (PMS723).

c. Welded, lithic-rich ignimbrite from the Lingmell Formation. Broken garnet (pale mineral, lower centre) is common in this unit (E67955). Plane-polarised light; width of field 12 mm (PMS724).

d. Scafell Dacite displaying characteristic euhedral garnet phenocryst. Smaller plagioclase phenocrysts show crude alignment within the very fine-grained flow-foliated (hyalopilitic) groundmass (E68106). Plane-polarised light; width of field 12 mm. (PMS725).

e. Lithic and crystal-rich lapilli-tuff from the Lincomb Tarns Formation. Typically contains dark andesitic and paler felsitic lithic clasts in a moderately welded vitroclastic matrix (E68254). Plane-polarised light; width of field 12 mm (PMS726).

f. Volcaniclastic sandstone from the Seathwaite Fell Formation. Comprises moderately sorted, angular to subrounded grains of partly sericitised plagioclase, fresh augitic pyroxene, volcanic rock fragments and chlorite pseudomorphs after pyroxene or glassy shards. Chlorite also locally infills primary intergranular porosity. Only minor very fine intergranular matrix is present (E686167). Plane-polarised light; width of field 12 mm (PMS727).

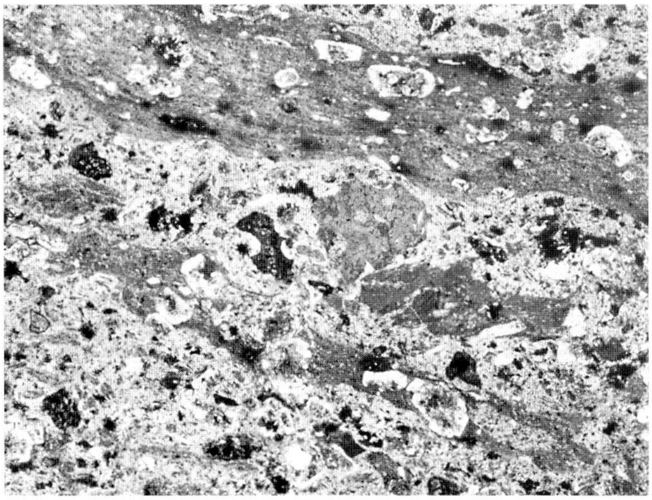

a.

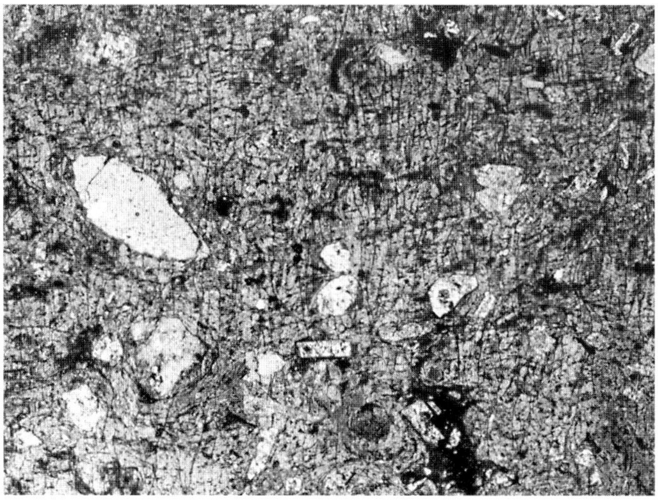

b.

c.

d.

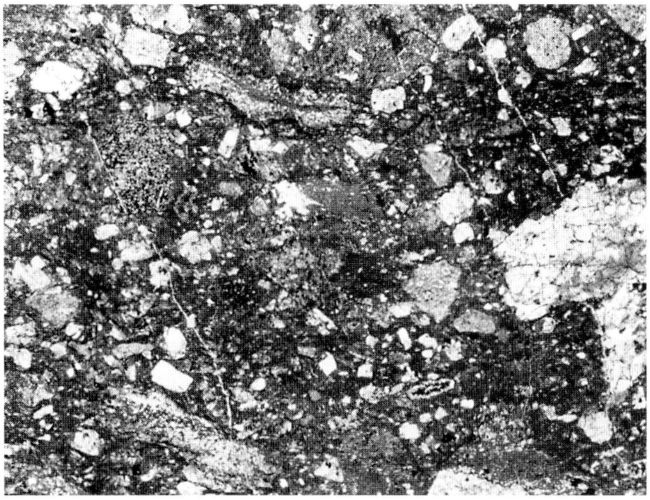

e.

f.

Table 7 Representative geochemical analyses of basalts from the Birker Fell Formation.

Sample no.	FJ1007	FJ1020	IY631	IY632	MG378	MG379	FJ1029	FJ1030	FJ1038	FJ1042	IY614
Grid ref.	SD 1513 9561	SD 1580 9503	NY 1076 0369	NY 1053 0351	NY 1395 1133	NY 1394 1128	NY 2274 0393	NY 2263 0409	NY 2278 0532	NY 2179 0421	NY 1994 0283
Section/ Member	Devoke Water	Devoke Water	Wrighthow	Wrighthow	Haycock	Haycock	Throstle Garth	Throstle Garth	Throstle Garth	Throstle Garth	Stony Tarn
Unit no.	17	31			7	8					1
wt%											
SiO_2	48.67	54.21	50.43	50.10	49.33	49.35	51.09	52.93	53.48	52.79	52.82
TiO_2	1.10	0.95	1.02	0.77	1.26	1.25	1.00	1.09	0.92	0.80	1.17
Al_2O_3	14.30	14.27	14.87	11.38	16.42	16.47	15.37	16.40	14.11	15.93	17.48
Fe_2O_3*	11.31	9.99	10.73	11.13	10.11	10.09	8.97	9.46	8.51	9.25	7.38
MnO	0.33	0.35	0.21	0.31	0.41	0.41	0.27	0.27	0.23	0.35	0.29
MgO	10.40	9.72	8.74	11.65	7.09	7.11	10.56	8.62	9.93	7.88	6.59
CaO	10.24	5.96	7.58	7.85	7.28	7.29	9.24	6.32	9.08	9.31	8.18
Na_2O	1.23	2.62	2.22	1.84	2.21	2.18	2.57	3.82	1.71	2.41	2.96
K_2O	2.27	2.12	0.46	1.22	2.87	2.85	0.74	1.23	1.78	1.16	1.20
P_2O_5	0.17	0.16	0.26	0.27	0.17	0.16	0.22	0.21	0.21	0.19	0.26
Total	100.02	100.35	99.37	99.32	99.97	100.81	100.03	100.35	99.96	100.07	99.60
LOI	1.89	2.73	2.85	2.80	2.82	3.65	2.25	2.57	2.32	1.49	1.27
ppm											
V	276	204	200	187	211	244	203	205	180	210	168
Cr	475	857	470	624	155	126	1083	872	943	554	189
Ni	144	296	146	150	51	46	340	239	298	88	51
Rb	117	70	19	30	99	26	30	72	43	36	41
Sr	248	173	270	256	184	208	252	222	236	327	226
Y	21	25	21	13	22	28	26	30	25	23	25
Zr	70	129	114	62	104	126	130	146	121	107	155
Nb	6	7	9	5	8	10	11	12	10	9	13
Ba	161	622	132	249	373	250	155	160	591	300	251
La	11	15	17	10	12	17	22	19	19	24	23
Ce	29	35	39	24	24	33	44	46	42	50	50
Th	3	5	5	2	4	3	7	8	6	8	7
Mg#	64.6	65.8	61.7	67.5	58.1	58.3	70	64.3	69.8	62.8	63.9
Lab.	RH	RH	BGS	BGS	BGS	BGS	RH	RH	RH	RH	BGS

Laboratories: BGS British Geological Survey, RH Royal Holloway, University of London. *Total Fe; $Mg^{\#}$ Mg number.
For BGS analyses totals include loss on ignition (LOI).

Table 8 Representative geochemical analyses of basaltic andesite and andesite from the Birker Fell Formation.

Sample no.	FJ998	FJ1022	FJ1009	MG374	MG396	CT 1485	CT1492	IY607	IY613
Grid ref.	1491 9586	1587 9492	1527 9551	1399 1149	1621 1049	2501 0154	2564 0086	2103 0439	1992 0300
Section	Devoke Water	Devoke Water	Devoke Water	Haycock	Haycock	Grey Friar	Grey Friar	Stony Tarn	Stony Tarn
Unit no.	8	33	19	3	25	1	9	8	2
wt. %									
SiO_2	54.17	60.04	57.59	55.51	58.27	56.19	59.37	58.39	60.27
TiO_2	1.60	1.00	1.26	0.95	0.87	0.93	0.75	0.86	0.96
Al_2O_3	16.80	16.25	17.81	17.97	18.89	17.14	18.01	19.18	16.98
Fe_2O_3*	9.58	7.84	7.78	9.61	5.95	7.58	5.32	6.28	6.71
MnO	0.20	0.24	0.20	0.33	0.23	0.20	0.33	0.14	0.23
MgO	5.13	3.78	2.93	3.05	2.52	4.05	3.10	1.97	2.88
CaO	7.00	6.09	7.27	4.80	4.38	6.04	2.34	6.03	4.76
Na_2O	3.35	2.53	2.64	3.00	3.58	2.53	3.92	2.29	2.29
K_2O	2.02	2.20	2.42	1.44	2.89	2.44	3.41	3.19	2.41
P_2O_5	0.29	0.21	0.29	0.28	0.21	0.23	0.19	0.21	0.23
Total	100.14	100.18	100.19	99.83	99.81	99.44	99.23	100.02	99.24
LOI	1.78	1.70	1.76	2.89	2.02	2.11	2.49	1.48	1.52
ppm									
V	271	179	207	74	90	121	82	98	106
Cr	173	145	96	6	35	71	37	38	69
Ni	62	24	33	3	12	17	9	10	20
Rb	40	65	67	47	92	69	93	129	101
Sr	364	288	431	250	220	302	223	257	200
Y	27	34	31	28	32	21	30	32	29
Zr	183	212	215	166	190	217	204	189	183
Nb	20	14	18	14	12	12	12	11	12
Ba	561	492	666	232	502	674	747	683	348
La	22	28	29	18	33	46	26	31	31
Ce	50	64	63	43	71	94	53	65	71
Th	5	9	9	4	12	12	9	9	15
Lab.	RH	RH	RH	BGS	BGS	BGS	BGS	BGS	BGS

* Total Fe

Laboratories: BGS British Geological Survey; RH Royal Holloway, University of London

BGS analyses totals include loss on ignition (LOI)

Birker Fell Formation

Compositions range from basalt (less than 52 per cent SiO_2) continuously through to dacite (63–69 per cent SiO_2), though most of the lavas lie within the range basaltic andesite to low-silica andesite (Figures 13, 14; Tables 7 to 9). The most common composition is basaltic andesite (53 to 57 per cent SiO_2). Volumetrically, most of the dacitic rocks are pyroclastic (Craghouse Member, Cockley Beck Tuff and Little Stand Tuff), though there are some lava flows. The presence of basalt with high MgO, Ni and Cr, particularly within the Throstle Garth Member, indicates relatively primitive, mantle-derived magmas.

Harker variation diagrams (Figures 13, 14) illustrate the calc-alkaline characteristic of these rocks with decreasing Fe_2O_3* (total iron as Fe_2O_3), MgO, TiO_2, CaO and V with increasing SiO_2. Al_2O_3 shows an inflected trend, increasing and then decreasing with differentiation (Figure 13b). The alkalis, and Rb, Ba and Sr, are scattered because of their post-emplacement mobility, but total alkalis (Na_2O + K_2O) have a significant positive correlation with SiO_2, suggesting that gross enrichment or depletion of total alkalis has not occurred. An exception to this may be some samples from the Scoat Tarn area [NY 160 104] which contain anomalously high K, possibly caused by the addition of K lost from the Ennerdale intrusion. The incompatible trace elements (Zr, Nb, Th, Y and rare earth elements (REE); Figure 14d, e, f) show well-defined positive correlations with SiO_2. Cr and Ni vary widely in the basic rocks, but there is a very abrupt, exponential decline towards more silicic compositions (Figure 14a, b).

The co-variation between immobile element concentrations is good throughout the suite. For example, Ce/Y ratios are restricted in range, increasing only slightly from mafic to felsic compositions (Figure 14g). An analogous La/Y plot was used by Fitton (1972) to rule out significant garnet control on the compositional spectrum of the Borrowdale Volcanic Group. The markedly concave Ni, Cr and MgO trends (Figures 13d; 14a, b) suggest that evolution of the suite was dominated by crystal–liquid fractionation of

Table 9 Representative geochemical analyses of ignimbrite and dacite from the Birker Fell Formation.

	Ignimbrite						Dacite		
Sample no.	MG392 Craghouse Member	IY609 Craghouse Member	IY636 Craghouse Member	IY637 Craghouse Member	CT1487 Cockley Beck Tuff	Little Stand Tuff**	MG389 Seatallan	CT1488	Great Whin-scale***
Grid ref. NY	1563 1035	1984 0392	1128 0292	1141 0260	2502 0148		1530 1961	2506 0143	
Section	Haycock	Stony Tarn			Grey Friar	Devoke Water	Haycock	Grey Friar	Devoke Water
Unit no.	21	6			3	30	18	4	31
wt. %									
SiO_2	60.39	64.01	61.02	65.24	61.79	67.74	67.55	63.50	63.02
TiO_2	0.71	0.57	0.76	0.51	0.83	0.43	0.47	0.79	0.94
Al_2O_3	18.89	17.75	18.37	17.53	18.14	16.16	16.36	16.98	16.20
Fe_2O_3*	4.73	4.42	5.38	3.15	5.92	3.30	3.71	6.19	5.87
MnO	0.17	0.17	0.15	0.15	0.21	0.07	0.18	0.16	0.14
MgO	1.02	0.70	1.53	0.67	1.46	1.15	0.75	1.18	1.57
CaO	2.67	3.25	4.66	2.77	2.64	2.57	1.55	1.88	3.55
Na_2O	3.08	2.99	2.73	3.66	1.36	3.53	2.73	2.87	4.53
K_2O	5.65	4.42	3.45	4.54	5.10	3.25	6.11	3.93	2.04
P_2O_5	0.25	0.22	0.26	0.14	0.20	0.19	0.19	0.30	0.32
LOI	1.77	1.38	1.27	0.93	1.71	–	1.32	2.05	1.48
Total	99.33	99.88	99.58	99.29	99.36	98.39	100.92	99.83	99.66
ppm									
V	39	34	59	23	60	33	25	58	112
Cr	6	13	17	10	30	17	7	30	1
Ni	3	3	5	1	14	8	2	14	9
Rb	146	156	110	118	198	133	153	142	57
Sr	290	257	277	249	214	285	215	156	268
Y	36	39	33	38	30	48	39	31	38
Zr	328	299	227	378	244	288	272	232	276
Nb	19	18	16	17	14	20	16	14	20
Ba	1886	810	696	1046	931	627	1468	684	592
La	40	42	32	39	48	–	46	47	33
Ce	91	90	72	85	101	–	86	91	69
Th	11	13	10	13	16	–	13	15	10

* Total Fe.
** Average analysis from K A Taylor — unpublished BSc dissertation, Univ. Sheffield.
*** Average analysis from Kanaris-Sotiriou et al., 1991.
All other analyses by BGS.

basaltic magma. Gross mixing between rhyolitic crustal melts and mantle-derived basaltic magmas, as invoked for the generation of some andesites (e.g. Eichelberger, 1975; Grove et al., 1982), does not appear to have been important. Least squares modelling of major oxides, and Rayleigh fractionation calculations for trace elements (Beddoe-Stephens et al., 1995) indicate that crystal fractionation of either olivine or orthopyroxene together with plagioclase, clinopyroxene, Fe–Ti oxide and apatite can account for nearly all of the variation within the suite. Some mixing between magmas lying along the compositional spectrum is not ruled out, and is a common feature of open-system magmatic evolution in andesitic provinces (e.g. Walker et al., 1993). This may be indicated by relatively high Ni and Cr contents in some intermediate rocks, and also by the presence in the succession of andesite–basaltic andesite composite sheets (Allen et al., 1987). Furthermore, a small amount of combined assimilation of pelitic crustal material is likely, as indicated by the occurrence of garnet (Fitton et al., 1982; Beddoe-Stephens et al., 1995).

Local successions within the Birker Fell Formation have subtly different, subparallel trends. In particular, rocks from the Haycock–Scoat Tarn area plots at lower MgO, Ni, Cr and higher Al_2O_3 for a given SiO_2 than elsewhere, whereas the Devoke Water rocks are slightly richer overall in V (Figures 13, 14). The data suggest that slightly different magma compositions were supplied in different parts of the region, but that similar differentiation processes generated the variation in composition of volcanic units within the localised areas.

Stratigraphical geochemistry

Geochemical profiles (Figures 15, 16) through the Birker Fell Formation in the Ambleside district provide evi-

Table 10 Representative geochemical analyses of rocks from the Scafell Caldera succession.

Sample no.	MG152	IY623	IY618	IY626	L16385	L16388	L16384	BST7	IY630	L16395	IY189	IY629	IY339
Grid ref.†	1510 0714	1742 0573	1731 0582	1730 0537	2046 9744	2056 9739	2035 9742	2487 0485	1172 0513	2097 9714	1693 0501	1665 0475	2057 0801
Unit	WSE	LTT	LTT	LTT	LTT	LTT	LTT	Bdp	Crk	Crk	LME	LME	SCD
wt %													
SiO_2	58.93	67.27	65.95	57.73	66.09	62.45	65.68	70.13	68.82	68.45	62.45	67.81	65.88
TiO_2	0.84	0.50	0.49	0.85	0.42	0.48	0.40	0.25	0.32	0.26	0.41	0.47	0.48
Al_2O_3	19.21	16.47	17.56	20.35	16.23	17.42	17.62	15.25	16.20	15.26	14.92	14.15	16.50
Fe_2O_3*	5.33	4.81	4.19	5.72	4.97	3.60	3.37	3.40	3.68	3.00	6.52	5.97	6.33
MnO	0.14	0.16	0.12	0.23	0.15	0.11	0.08	0.09	0.20	0.06	0.18	0.26	0.18
MgO	1.83	0.67	1.15	1.56	1.68	0.77	0.69	0.25	0.39	0.45	0.72	0.85	0.83
CaO	5.06	2.25	2.40	4.74	3.25	3.40	1.61	1.34	2.11	0.87	2.75	3.41	2.18
Na_2O	4.38	2.80	1.87	2.43	3.73	4.63	4.06	2.43	2.84	1.13	2.14	2.31	2.47
K_2O	1.36	4.89	3.67	3.96	1.98	4.30	6.21	5.24	5.01	8.02	3.43	2.15	4.81
P_2O_5	0.30	0.15	0.17	0.31	0.21	0.23	0.20	0.36	0.42	0.31	0.37	0.26	0.33
LOI	2.28	0.97	1.69	1.60	—	—	—	1.31	0.92	—	—	1.12	—
Total	99.56	100.94	99.26	99.48	98.71	97.16	99.92	100.05	100.91	97.81	93.89	98.76	99.99
ppm													
V	54	11	37	61	39	33	35	—	5	10	12	22	11
Cr	17	25	24	4	0	21	6	7	13	3	4	17	0
Co	—	—	—	—	20	19	22	—	—	11	—	—	—
Ni	4	0	6	2	10	11	15	6	0	18	2	2	2
Cu	—	—	—	—	5	5	7	6	—	6	—	—	—
Zn	—	—	—	—	20	10	7	39	—	6	—	—	—
Rb	43	169	140	143	65	116	189	202	165	229	162	86	188
Sr	426	203	192	293	414	387	252	92	134	175	167	152	156
Y	34	41	36	30	44	46	42	59	54	60	45	43	59
Zr	215	338	288	199	352	402	351	304	337	359	272	230	312
Nb	17	19	16	16	19	20	20	20	19	22	19	15	21
Ba	378	912	703	655	735	1180	1132	1074	1059	3196	658	423	928
La	29	39	40	27	34	42	43	67	51	68	48	47	50
Ce	76	86	86	69	74	89	86	160	133	131	103	103	111
Pb	—	—	—	—	—	—	—	10	—	—	—	—	—
Th	11	14	12	9	12	13	12	22	19	19	15	12	17
Lab.	BGS	BGS	BGS	BGS	Leic.	Leic.	Leic.	B	BGS	Leic.	BGS	BGS	BGS

* Total Fe. LOI Loss on ignition.
Laboratories: BGS British Geological Survey; Leic. Leicester University; B Branney et al. (1992).
Bdp Bad Step Tuff; Crk Crinkle Member; LME Lingmell Formation; LTT Long Top Member; SCD Scaffell Dacite; WSE Wet Side Edge Member.
† Sample numbers prefixed L are located in 100 km square SD; all others in NY.

Table 11 Representative geochemical analyses of rocks from the succession in the Duddon Basin and Coniston Fells.

Sample no.	L16396	L16401	E66804	E66818	E66808	E66824	E66810	E66814	E66837	E66816	E66835	E66834	E67318
Grid ref. †	1523 9015	1526 9101	2006 9065	2162 9694	2078 9158	2362 9782	2087 9162	2109 9162	2669 9754	2118 9158	2494 9368	2465 9374	2970 0499
Unit	Wb	Wb	ULP	WAL	KLN	KLN	KLN	PER	PER	STP	LBT	LBT	LRF
wt %													
SiO_2	54.09	58.05	59.04	61.09	62.34	67.31	63.43	70.56	70.78	70.80	63.82	65.02	64.43
TiO_2	0.97	0.85	0.80	1.05	0.80	0.81	0.76	0.27	0.32	0.24	0.81	0.68	0.75
Al_2O_3	15.87	17.88	15.89	16.59	11.89	15.18	14.84	14.54	13.63	14.18	16.03	14.34	15.66
Fe_2O_3*	7.47	6.05	6.21	7.82	5.88	4.52	4.91	3.33	4.63	2.77	5.29	5.48	5.28
MnO	0.19	0.16	0.17	0.21	0.17	0.13	0.13	0.12	0.25	0.09	0.15	0.18	0.13
MgO	3.59	2.37	2.53	2.50	2.11	1.58	1.41	0.91	0.96	0.61	1.56	2.22	1.55
CaO	5.94	5.38	4.65	2.21	6.27	2.45	4.00	0.19	0.15	1.41	2.30	2.55	1.77
Na_2O	2.87	2.70	1.32	3.05	2.61	4.28	3.56	2.46	1.68	2.61	3.40	2.99	3.21
K_2O	4.28	3.52	3.17	2.31	1.42	2.83	3.04	7.04	7.14	5.09	4.67	4.14	5.37
P_2O_5	0.21	0.36	0.22	0.34	0.19	0.24	0.21	0.08	0.10	0.07	0.23	0.20	0.21
LOI	–	–	6.32	2.69	6.70	1.39	4.55	1.17	1.14	2.37	2.43	2.78	1.40
Total	95.48	97.32	100.32	99.86	100.38	100.72	100.84	100.67	100.78	100.24	100.69	100.58	99.76
ppm													
V	176	89	89	140	90	51	56	4	8	5	75	61	56
Cr	44	23	29	2	108	17	4	4	11	5	38	35	22
Co	28	20	10	14	17	10	5	1	5	1	10	10	10
Ni	28	17	9	2	37	3	1	0	2	1	8	8	6
Cu	12	9	57	1	10	1	7	4	13	5	11	8	7
Zn	51	53	101	93	46	32	41	23	55	83	64	69	23
Rb	72	148	89	101	52	53	84	176	171	153	148	108	140
Sr	304	351	125	126	178	163	162	53	75	104	135	162	174
Y	34	37	32	28	21	31	30	36	40	44	34	32	36
Zr	214	230	238	167	143	233	224	404	399	330	238	222	271
Nb	15	18	14	12	8	13	12	21	20	20	15	13	17
Ba	1361	662	674	466	159	1276	614	1480	1941	868	716	720	1253
La	27	35	31	28	20	23	27	46	35	48	38	31	37
Ce	56	74	70	61	44	53	62	96	69	110	66	78	92
Pb	—	—	67	5	3	6	5	2	0	20	8	25	5
Th	10	10	11	7	5	9	10	17	17	18	13	10	13
Lab.	Leic.	Leic.	BGS	BGS	BGS	BGS	BGS	BGS	BGS	BGS	BGS	BGS	BGS

* Total Fe; LOI Loss on ignition.

† All grid references within 100 km square SD

Laboratories: BGS British Geological Survey; Leic. Leicester University;

KLN Kiln Bank Member; LBT Lag Bank Formation; LRF Lingmoor Fell Formation; PER Paddy End Member; STP Stickle Pike Member; ULP Ulpha Formation; WAL Wallowbarrow Formation; Wb Waberthwaite Formation.

Table 12 Representative geochemical analyses of Lincomb Tarns Formation, Glaramara Tuff and intrusive rocks.

Sample no.	B42	B43	B44	B45	B46	B47	B56	FJ971	FJ987	FJ990	FJ975	CT1586
Grid ref.†	2510 0970	2325 0755	2325 0755	2325 0755	2325 0755	2585 0815	2985 0505	2887 9910	3068 0045	3191 0198	3255 0252	2073 9166
Unit	LTa	LTa	LTa	LTa	LTa	GMT	GMT	INT	INT	INT	INT	INT
wt %												
SiO_2	63.31	65.23	65.61	63.45	62.93	57.26	74.45	54.21	57.99	60.93	72.05	77.07
TiO_2	0.53	0.51	0.53	0.57	0.63	1.30	0.46	1.38	1.07	1.57	0.36	0.07
Al_2O_3	16.52	16.01	15.70	16.80	16.77	16.39	12.40	17.61	16.84	18.27	14.31	12.96
Fe_2O_3*	5.50	4.95	5.19	5.13	6.17	7.88	3.52	9.15	8.39	6.51	3.21	0.88
MnO	0.21	0.18	0.19	0.18	0.24	0.24	0.12	0.27	0.27	0.09	0.05	0.02
MgO	0.85	1.12	0.99	1.05	1.16	2.79	1.77	4.49	3.99	2.54	1.44	0.38
CaO	1.95	4.18	3.16	3.71	3.18	5.46	0.32	5.17	5.96	3.53	0.28	0.05
Na_2O	2.85	4.14	4.13	4.50	2.97	3.99	1.05	2.32	2.53	2.61	2.35	1.69
K_2O	5.32	1.13	2.21	1.86	3.70	1.04	3.12	4.75	2.87	3.59	6.06	6.11
P_2O_5	0.30	0.28	0.27	0.32	0.32	0.21	0.08	0.29	0.24	0.37	0.08	0.01
Total	99.18	99.07	99.18	98.89	99.65	98.98	99.59	99.64	100.15	100.01	100.19	99.24
LOI	1.84	1.34	1.20	1.32	1.58	2.42	2.30	6.24	2.57	2.81	1.44	1.23
ppm												
V	21	31	34	32	34	173	53	279	224	216	12	2
Cr	9	32	31	19	17	36	20	116	58	53	12	5
Co	–	–	–	–	–	–	–	–	–	–	–	1
Ni	2	4	7	5	5	7	9	22	8	20	4	0
Cu	–	–	–	–	–	–	–	18	9	68	4	2
Zn	–	–	–	–	–	–	–	107	92	83	51	4
Rb	198	41	68	60	112	23	148	162	62	96	186	184
Sr	155	324	268	314	243	283	37	202	503	329	62	42
Y	51	51	48	57	47	29	23	32	26	31	52	38
Zr	303	279	263	321	295	185	204	147	146	221	304	179
Nb	17	16	16	17	17	13	12	15	12	23	21	19
Ba	968	322	867	630	700	144	438	885	841	734	762	347
La	50	46	46	55	47	26	17	29	21	31	75	28
Ce	110	100	104	117	97	50	56	62	48	67	100	71
Pb	–	–	–	–	–	–	–	7	10	8	2	5
Th	17	14	15	19	–	–	–	7	7	9	16	21
Lab.	BGS	BGS	BGS	BGS	BGS	BGS	BGS	RH	RH	RH	RH	RH

Laboratories: BGS British Geological Survey; RH Royal Holloway, University of London. * Total Fe.
For BGS analyses totals include loss on ignition (LOI).
GMT Glaramara Tuff, INT intrusive rocks, LTa Lincomb Tarns Formation.
† Grid references are within the 100 km square NY except for samples FJ971 and CT1586 which are from within SD.

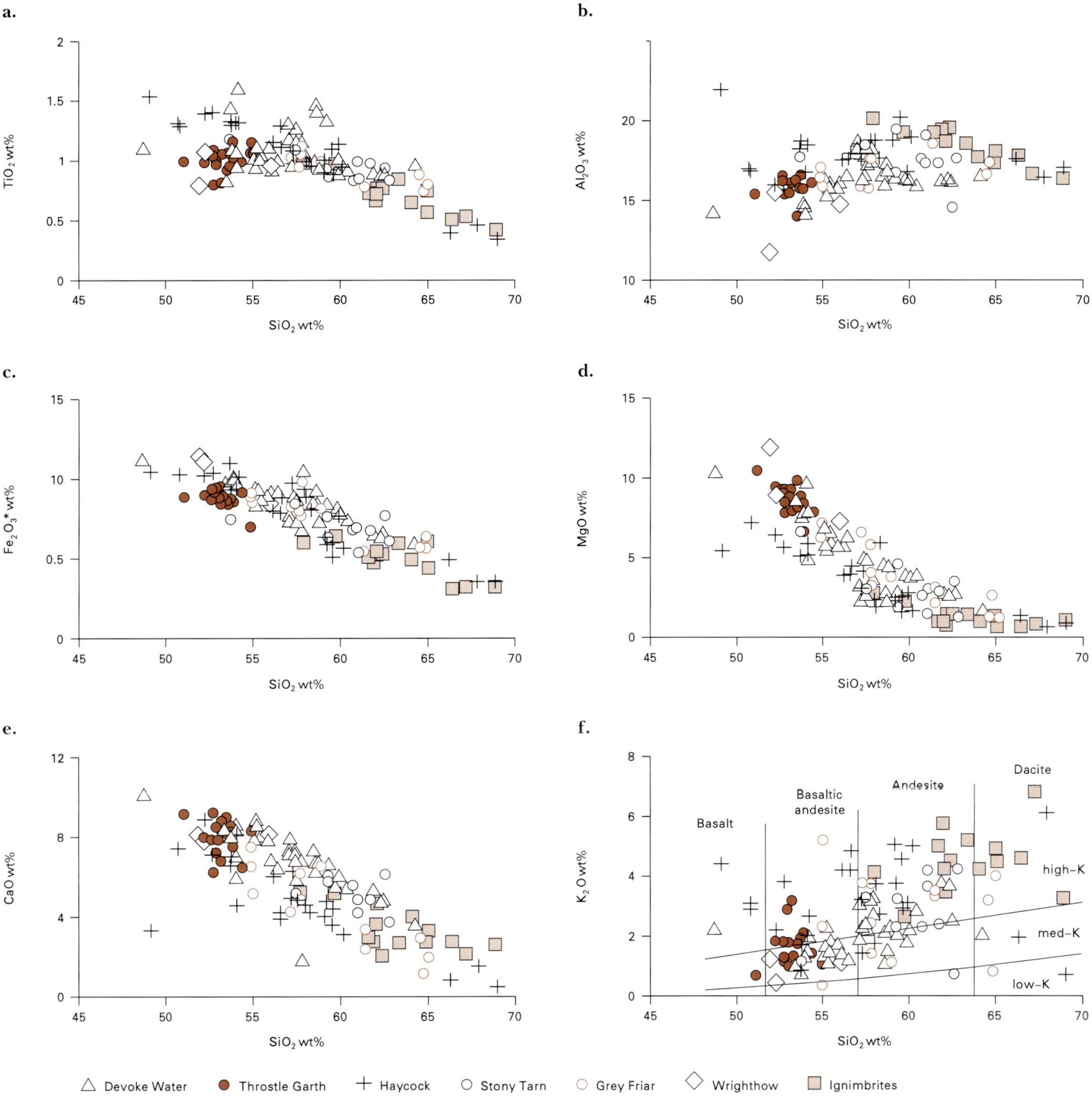

Figure 13 Major oxide Harker variation diagrams for the Birker Fell Formation. Includes analyses of ignimbrite from the Craghouse Member, Little Stand Tuff and Cockley Beck Tuff.
Fe_2O_3* total Fe as Fe_2O_3.

dence for systematic and cyclic variations in composition (Beddoe-Stephens et al., 1995). The section between Devoke Water and Stainton Pike is 2.6 km thick and completely traverses the Birker Fell Formation. In the northern part of the district, between Haycock and Red Pike, a 2.2 km section was sampled. Other profiles are between Cockley Beck and Grey Friar, and between Stony Tarn and Slight Side. Three main types of systematic geochemical variation are identified. Firstly, there is a trend to more mafic compositions with stratigraphical height, as seen in the lower part of the Devoke Water profile (Figure 15) and at the base and top of the Haycock profile (Figure 16). Secondly, a trend to more felsic compositions with stratigraphical height occurs, for example in the middle part of the Haycock profile (Figure 16). Thirdly, no overall trend with stratigraphical height is detected, as in the

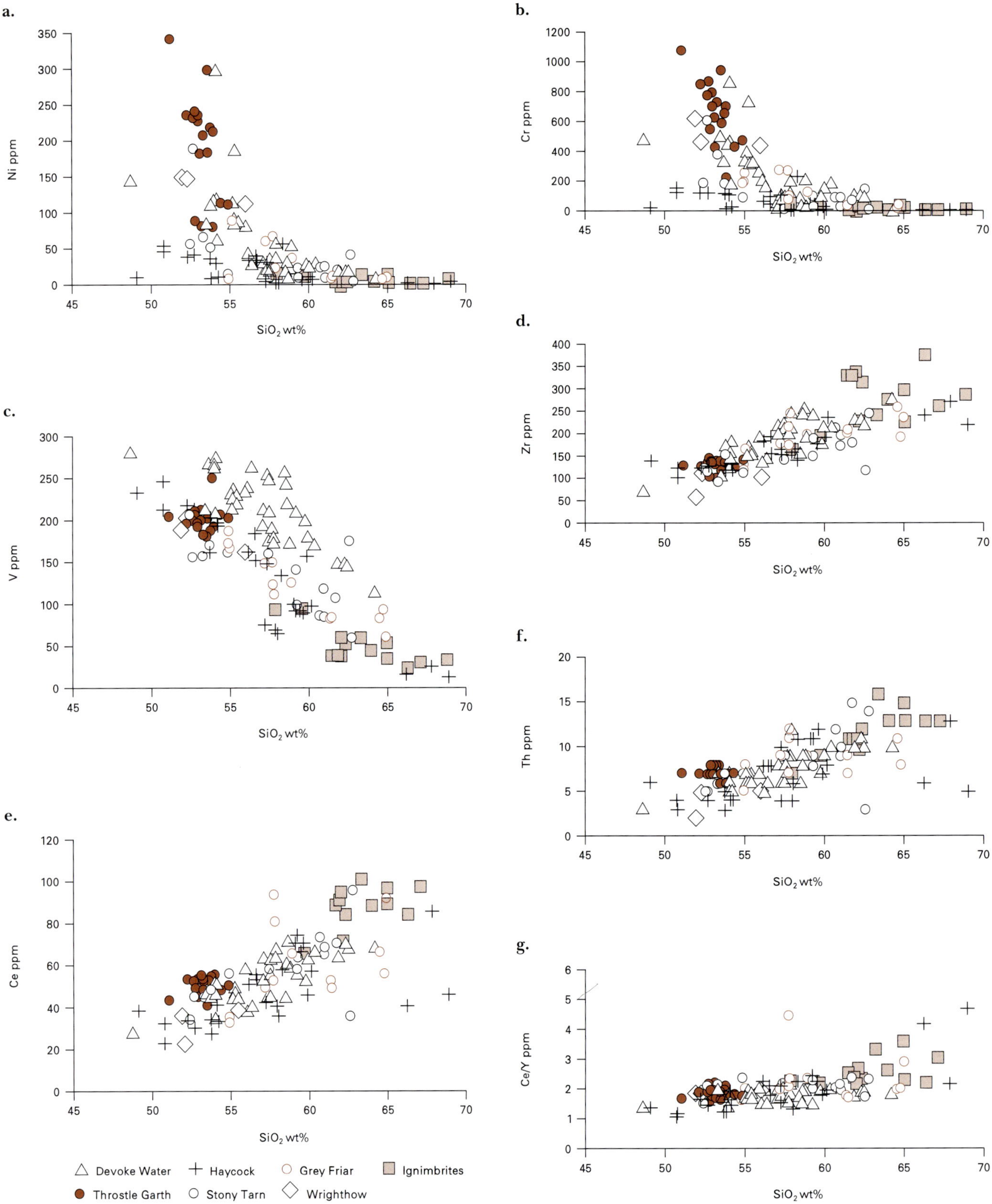

Figure 14 Trace element Harker variation diagrams for the Birker Fell Formation.

upper part of the Devoke Water profile (Figure 15), and through the Throstle Garth Member.

DEVOKE WATER PROFILE

Two groups are present: a lower one with an upward trend to more mafic composition and an upper one having higher Zr and Y values and a weak trend with stratigraphical height, for example in SiO_2 (Figure 15). The top of the lower group overlaps with the base of the upper group, suggesting that the lava pile at this locality was built from at least two centres erupting distinctive magmas. An anomalous basalt lava, (unit 17, Figure 15) was erupted immediately prior to the main upper group sequence.

The lower group comprises three parts (Figure 15a) each lying on a separate crystal fractionation trend (Figure 17), as the products of discrete fractionating batches of mantle-derived magma. Furthermore, within each part the most mafic (i.e. highest Ni, Cr) magmas were erupted last (compare Figures 15c, d). This pattern has been interpreted as the relatively rapid, periodic emptying of a zoned magma chamber caused by displacement created by the ascent of new batches of magma through the crust (Beddoe-Stephens et al., 1995). There is no clear pattern of evolution within the upper group of lavas (Figure 15), suggesting that eruption followed a steady-state pattern in which differentiation (crystal fractionation) processes were balanced by recharge, mixing and eruption.

The Great Whinscale Dacite and Little Stand Tuff lie off the trend displayed by upper group lavas for some elements (e.g. SiO_2, Zr) suggesting derivation from a different, more distant source. The Great Whinscale Dacite is certainly geochemically distinct from most of the other dacites in the Birker Fell Formation with its lower MgO, Ni and Cr, and higher MnO and Nb (Kanaris-Sotiriou et al., 1991).

HAYCOCK–SCOAT TARN PROFILE

Trends in the Haycock profile are systematic but different from those at Devoke Water (Figure 16). The basal part of the sequence trends upward to more mafic compositions. The subsequent trend to more felsic compositions culminated in eruption of the Seatallan Dacite and ignimbrites of the Craghouse Member. Overlying these is a basalt that lies off the upward trend to mafic compositions characterising the uppermost lavas (Figure 16). Beddoe-Stephens et al. (1995) interpreted the patterns to have arisen from the initial rapid eruption of a zoned magma chamber, producing progressively more mafic magmas, followed by a decline in the eruption rate that permitted differentiation processes to dominate and led to more evolved compositions. A hiatus in eruption of the sequence is represented by the thick volcaniclastic succession of the Eagle Crag Member (that crops out just north of the Ambleside district), and this corresponds to a compositional discontinuity between units 13 and 14 (Figure 16). Co-variation between Ce and Zr (Figure 18) suggests that the Seatallan Dacite, Craghouse Member and lavas above represent different magma batches to those below. The trend towards more mafic compositions in the upper part of the succession suggests emptying of a zoned magma body, the top of which must have evolved almost to rhyolitic compositions, and may have been sufficiently volatile rich to generate pyroclastic flows.

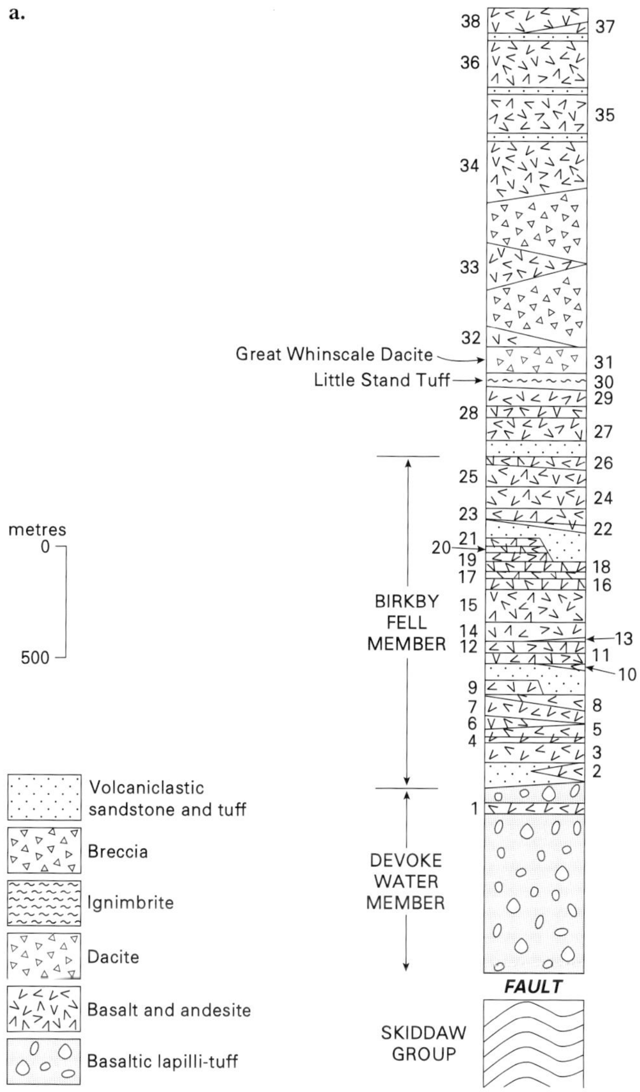

Figure 15 Stratigraphical geochemical variations through the Devoke Water section, Birker Fell Formation (after Beddoe-Stephens et al., 1995). The groups shown in **a** are discussed in the text.

Volcaniclastic successions

Pyroclastic rocks and the lavas and sills that constitute the upper part of the Borrowdale Volcanic Group range from basaltic andesite to rhyolite (Tables 10 to 12). The bulk composition of the many pyroclastic rocks may have been modified by the inclusion of variable volumes of lithic clasts and by eruption processes that resulted in the loss of fines. There is no compositional gap in the suite of rocks and the most common composition is dacite, emplaced mostly as pyroclastic deposits. The lavas of the Duddon

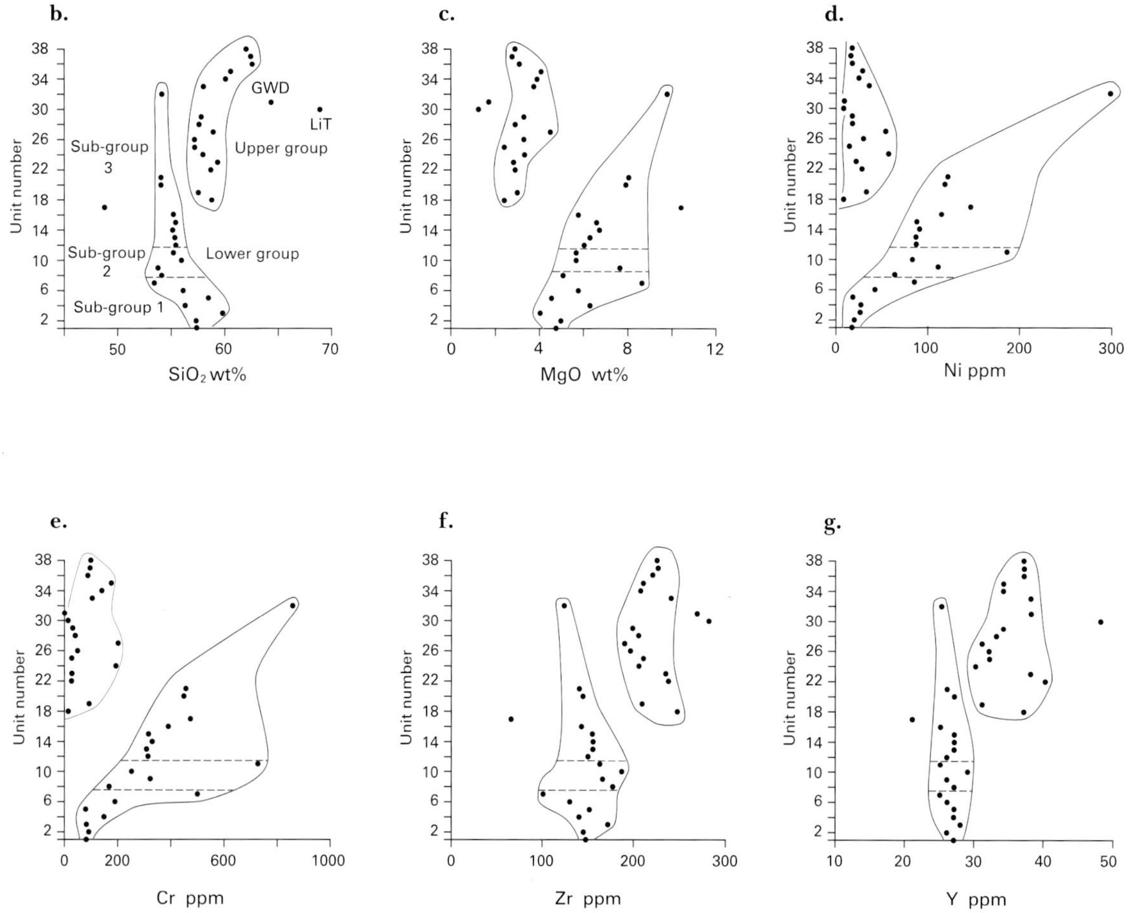

Figure 15 *continued.*
Abbreviations: GWD Great Whinscale Dacite; Lit Little Stand Tuff

Hall and Ulpha formations are basaltic andesite to low-silica andesite (53 to 60 per cent SiO_2). The Harker-type variation diagrams (Figures 19, 20) show that the volcaniclastic successions have co-magmatic trends with the Birker Fell Formation; clusters and trends within the suite are broadly related to stratigraphy. The Whorneyside, Airy's Bridge and Lingmell formations (Figure 12), form a coherent geochemical unit with simple trends showing decrease in TiO_2, Al_2O_3, Fe_2O_3, MgO, Nb, Zr, V and Ce with fractionation. This succession differs from a second group comprising formations within the Duddon Basin in having lower TiO_2, Fe_2O_3, MgO and V, and higher Al_2O_3 and Ce. Discrimination between the two groups is illustrated best by the TiO_2–Al_2O_3 plot (Figure 21). The more aluminous nature of rocks from the Scafell Caldera is reflected in the common occurrence of garnetiferous rocks within them, possibly related to higher degrees of contamination of parental magmas by pelitic crustal material.

The Waberthwaite and Wallowbarrow formations have parallel trends of similar bulk composition, but the former has consistently higher total alkalis, Zr, Th, Nb and Y, and lower TiO_2 and Fe_2O_3. The three ignimbrite members of the Lickle Formation (Figure 12) do not form a geochemical unit; the Kiln Bank Member has a trend parallel to the Waberthwaite Formation, but the Paddy End and Stickle Pike members, and an unnamed ignimbrite from the Side Pike Complex, consistently plot at the most evolved end of the field defined by rocks from the Scafell Caldera (e.g. Figure 19a).

The most evolved rocks occur in the Duddon Basin, but there is no coherent geochemical population. Significantly higher values of Zr have been found in rhyolitic rocks of the Duddon Basin compared with the Scafell Caldera. The trend of the latter (Figure 20a) shows Zr levelling off and beginning to decline at compositions greater than about 68 to 70 per cent SiO_2, indicative of precipitation and removal of zircon. By contrast, rhyolitic rocks of the Duddon Basin appear to have sustained a greater concentration of Zr into residual melts before significant zircon fractionation. Saturation levels of Zr in silicic melts is influenced by temperature and by compositional features, correlating positively with $(Na + K)/Al$ (Watson and Harrison, 1983). The more aluminous magmas within the Scafell Caldera may be reflected in the precipitation of zircon at lower Zr levels than the slightly less aluminous magmas of the Duddon Basin.

In the Scafell Caldera succession there are differences between the Long Top Member and the overlying Crinkle Member (including the Bad Step Tuff), Lingmell Forma-

tion and Scafell Dacite, though most oxides and trace elements lie on a common trend. The upper part is significantly enriched in Y, light REE (e.g. Ce), Th and P_2O_5 at a given level of differentiation as indicated by Zr or SiO_2 (Figures 20b, d, e; 22c, d). On Th–Zr and Y–Zr plots the lower and upper parts define subparallel trends. The transition from the Long Top Member to the overlying Crinkle Member is marked by an abrupt geochemical break, but there is no field evidence for a time gap suggesting perhaps a change in vent position that tapped a new batch of magma. These geochemical differences may have resulted from the strong partitioning of P, Th, Y and Zr into apatite and zircon. Beddoe-Stephens and Mason (1991) suggested that cumulate concentration of garnet into the upper units, particularly the Lingmell Formation accounted for enriched Y levels. Furthermore, the garnet invariably contains apatite inclusions which could account for elevated P_2O_5 and light REE contents.

A small number of analyses of rocks from the Lincomb Tarns Formation and Glaramara Tuff (Seathwaite Fell Formation) represent the only data available from the uppermost part of the Borrowdale Volcanic Group. There are geochemical similarities with the Airy's Bridge Formation; the Lincomb Tarns Formation generally plots along with dacitic parts of the Long Top Member (Figures 19–22), but with elevated Fe_2O_3, P and Ce, and Th and Y values comparable with the more silicic Crinkle Member and Lingmell Formation.

SYSTEMATIC GEOCHEMICAL VARIATION IN THE SCAFELL CALDERA

Geochemical variations in ignimbrite are partly the result of post-emplacement alteration and variation in the abundance of the components of pyroclastic flows. Systematic variations not attributable to these processes may arise from the geochemical structure of the magma chamber (e.g. Smith, 1979). Systematic variations in geochemistry are present in the Airy's Bridge and Lingmell formations in the Central Fells (Beddoe-Stephens and Mason, 1991, fig. 6) and in a profile through the Airy's Bridge Formation on White How [SD 204 974], west of Stonythwaite. The geochemical discontinuity between the Long Top Member and the units above is seen particularly in Mg number, MgO, Ce, Y, V, TiO_2 and Zr/TiO_2, but less so in P_2O_5, Th, Zr and MnO. That two magmatic cycles may be present is supported by plots of SiO_2–P_2O_5, Zr–Y and Zr–Th (Figures 19f, 22c, d). Higher MgO, Mg number, TiO_2 and V, and lower Th, Ce and Y indicate that the Long Top Member is overall less differentiated than the second cycle, which displays a trend from silicic to less silicic with stratigraphical height, characteristic of pyroclastic flow eruptions from zoned magma chambers (Smith, 1979). The Long Top Member cycle is more complex, though the least differentiated magmas were erupted last and there is a smooth zonation profile through the ignimbrite sequence.

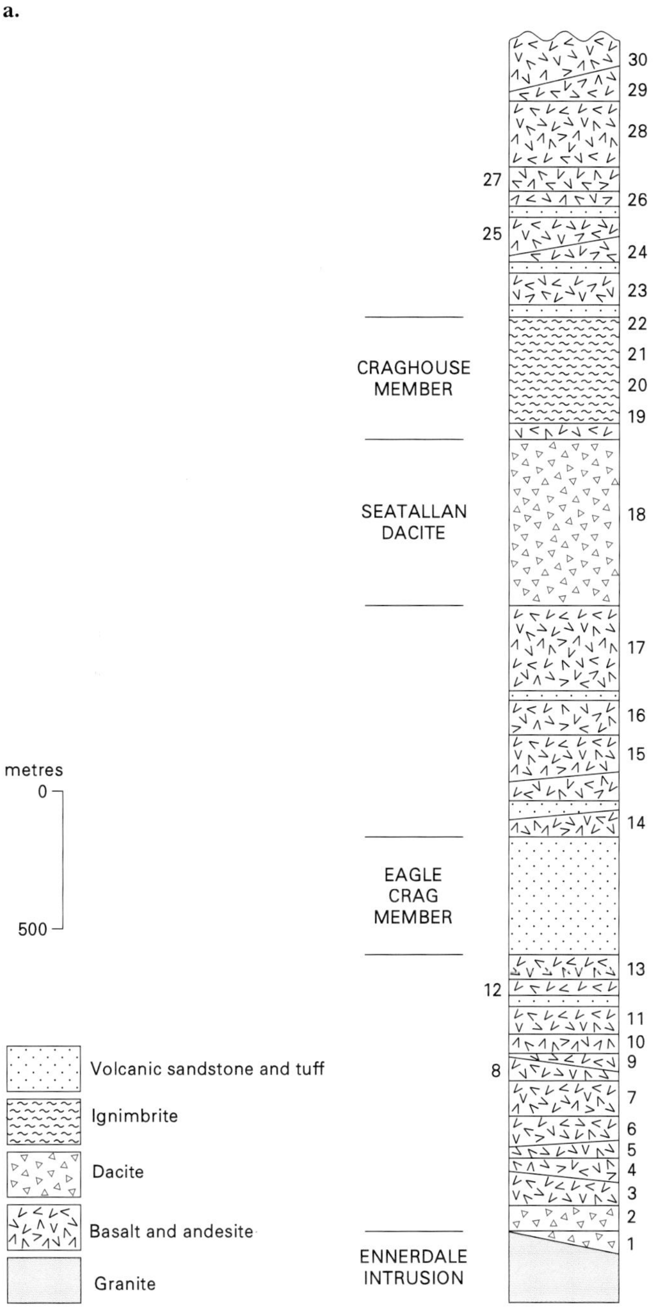

Figure 16 Stratigraphical geochemical variations through the Haycock section, Birker Fell Formation (after Beddoe-Stephens et al., 1995).

Radiogenic isotopes

Isotope data for the Borrowdale Volcanic Group are few, comprising analyses of only two andesites, a porphyritic dacite and the Great Whinscale Dacite from the Birker Fell Formation in the Ambleside district, and one andesite and three dacites from the Haweswater area (Thirlwall and Fitton, 1983; Kanaris-Sotiriou et al., 1991). Initial $^{87}Sr/^{86}Sr$ ratios from the Birker Fell Formation vary from 0.7063 to 0.7071, but the young Sr isotope age of these rocks indicates that they have probably lost some radiogenic Sr subsequently (Kanaris-Sotiriou et al., 1991).

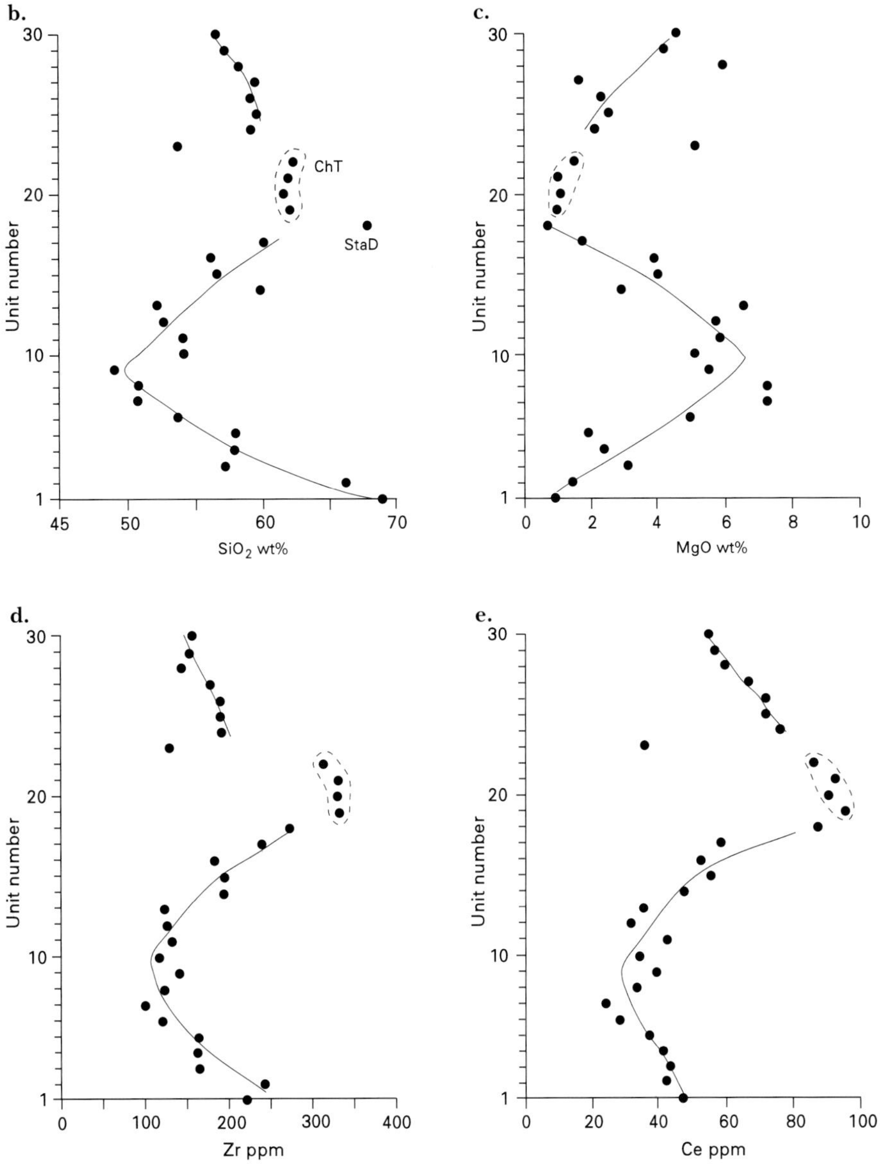

Figure 16 *continued.*
Abbreviations: ChT Craghouse Member; StaD Seatallen Dacite

$^{147}Sm/^{144}Nd$ values range from 0.1146 to 0.1391 and $^{143}Nd/^{144}Nd$ from 0.512105 to 0.512384. The initial Nd isotopic composition of the Great Whinscale Dacite was significantly more radiogenic than the others, implying derivation of that magma from a more primitive source (Kanaris-Sotiriou et al., 1991). The Nd values for all these rocks are comparable with those for volcanic rocks from the Central Volcanic Zone in the Andes where thick continental crust is known to be present beneath the volcanic pile (see Wilson, 1989 for summary of Andean data).

Magma source

The most mafic Birker Fell Formation compositions are consistent with near-primary, mantle-derived partial melts (Figure 23). The source characteristics of the basalts are illustrated by a MORB-normalised spiderdiagram of minor and trace elements (Figure 24; after Pearce, 1983). The pattern is typical of calc-alkaline basalt from a continental-margin volcanic province, and is characterised by highly enriched K, Rb, Ba and Th, and declining values from La to Y. Nb is enriched relative to mid-ocean ridge basalt (MORB), but forms a pronounced negative anomaly relative to Th and the light REE. The patterns produced by K, Rb and Ba are inconsistent in shape, almost certainly because of the mobility of these elements during alteration. However, elevated abundances of these elements together with the less mobile Th are typical of a subduction zone component (Pearce, 1983; Pearce and Parkinson, 1993) introduced by aqueous fluids derived from the subducted slab. The comparatively lower and declining enrichments of Nb, P and Zr over a baseine defined by Y and Ti (i.e. compared to MORB or oceanic island-arc basalts) are indicative of either a within-plate asthenospheric (OIB), sub-continental lithosphere or lower crustal component.

Hildreth and Moorbath (1988) have argued that in the Chilean Andes an OIB source or sub-continental lithospheric mantle is unlikely to be a significant contributor to the geochemical signature of erupted magmas. They considered that subduction-modified, MORB-like asthenospheric mantle and lower crustal material dominate the geochemical and isotopic characteristics of derived magmas. However, in the Andean segment studied by Hildreth and Moorbath (1988) primary basalts are not present. In the Throstle Garth Member, which contains basalt with near primary characteristics, it is more plausible that the pattern of increasing Y–Zr–Nb enrichment is attributable to the sub-crustal mantle wedge. The enriched levels of light REE relative to Nb and P (Figure 24) suggest that subduction-modified, enriched lithospheric mantle most likely acted as a source or interacted with ascending magmas (e.g. Davis and Hawkesworth, 1993; Davidson et al., 1988). The comparatively high SiO_2 content of the high Mg-number basalts (see Table 7) may reflect the melting of mantle modified by the influx of subduction zone hydrous fluids and mobile alkalis, as suggested by Gill (1981).

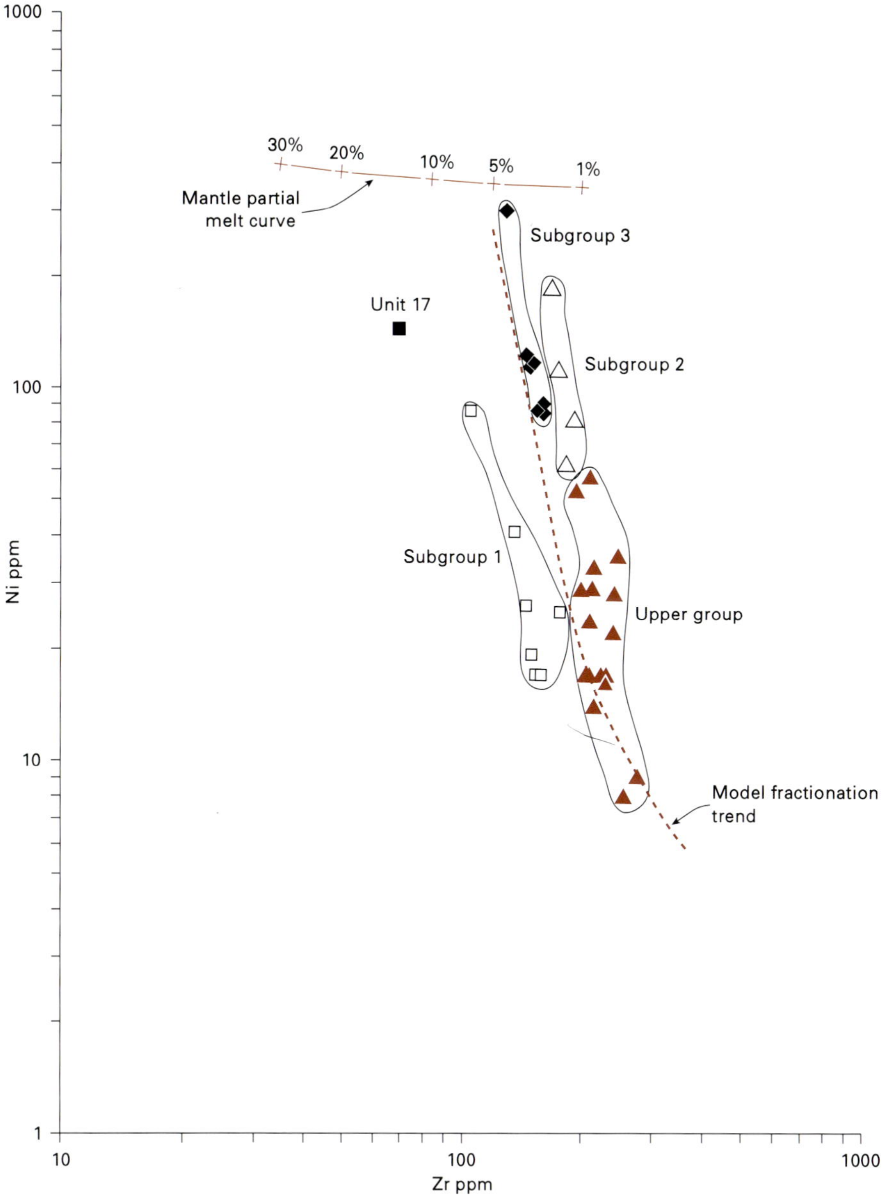

Figure 17 Log Ni–Zr covariation plot for the Devoke Water section showing subdivision of the lower group; subgroup 1, units 1–7; subgroup 2, units 8–11; subgroup 3, units 12 +.

Partial melt curve based on peridotite with 2400 ppm Ni and 11 ppm Zr. Fractionation curve based on Rayleigh modelling of olivine/orthopyroxene + clinopyroxene + plagioclase + oxide + apatite crystallisation assemblage.

FACIES MODELS

Low-profile andesite volcanoes of the Birker Fell Formation

Though the andesite succession of the Birker Fell Formation has been interpreted previously as clusters of stratovolcanoes (see Millward et al., 1978; Branney and Soper, 1988) there is little evidence to support this. Geikie (1891) proposed that it represented a multiple centred volcanic field. Green (1919, p.161) suggested that the volcanoes had low profiles such as those seen on Hawaii. Suthren (1977) stated that andesite lavas in the Derwent Water and Thirlmere areas in the adjacent Keswick district are laterally extensive and that there is no evidence for steep palaeoslopes. Petterson et al. (1992) set out evidence for the existence of many low-profile edifices. A low-relief topography at the upper boundary of the formation is shown by stratigraphical relationships with the overlying Whorneyside Formation. The top surface of this substantial and widespread deposit that buried the Birker Fell Formation shows little evidence of erosion prior to deposition of the succeeding Airy's Bridge Formation and may be taken as a datum; on this basis the pre-Whorneyside Formation topography had a relative elevation of no more than about 150 m (Branney, 1988b; 1991).

Petterson et al. (1992) considered that several features were critical to the interpretation of the morphology of the lava field within the Birker Fell Formation. The top and bottom of the lavas are generally remarkably regular and subparallel over some distance. Palaeoslopes are indicated by the geopetal structure of the laminated sedimentary rocks within small cavities in the tops of the lavas. If they had come to rest on the flanks of a stratocone then there would be systematic differences between dip and dip orientation of the sedimentary rocks in the cavities compared with the top of the lava flows. There is no discrepancy and the lavas must have been erupted on to an area of uniform low relief. This, and the occurrence of laterally persistent, conformable, bedded sedimentary rocks between lavas creates the impression of a dominantly horizontal, flat-lying lava pile built up by the interfingering of generally subparallel, conformable, tabular flows (Petterson et al., 1992, fig. 7). This interpretation is not consistent with facies models for continental stratovolcanoes which include a core facies dominated by pyroclastic breccia, stubby lavas, domes and avalanche-type flow deposits, and a distal facies dominated by mass wastage from the cone, including colluvial and alluvial deposits with grain size diminishing away from the main vent (Vessel and Davies, 1981).

Petterson et al. (1992) compared the Birker Fell Formation to the monogenetic volcano fields of Mexico and Central America (e.g. Bloomfield, 1975; Walker, 1981;

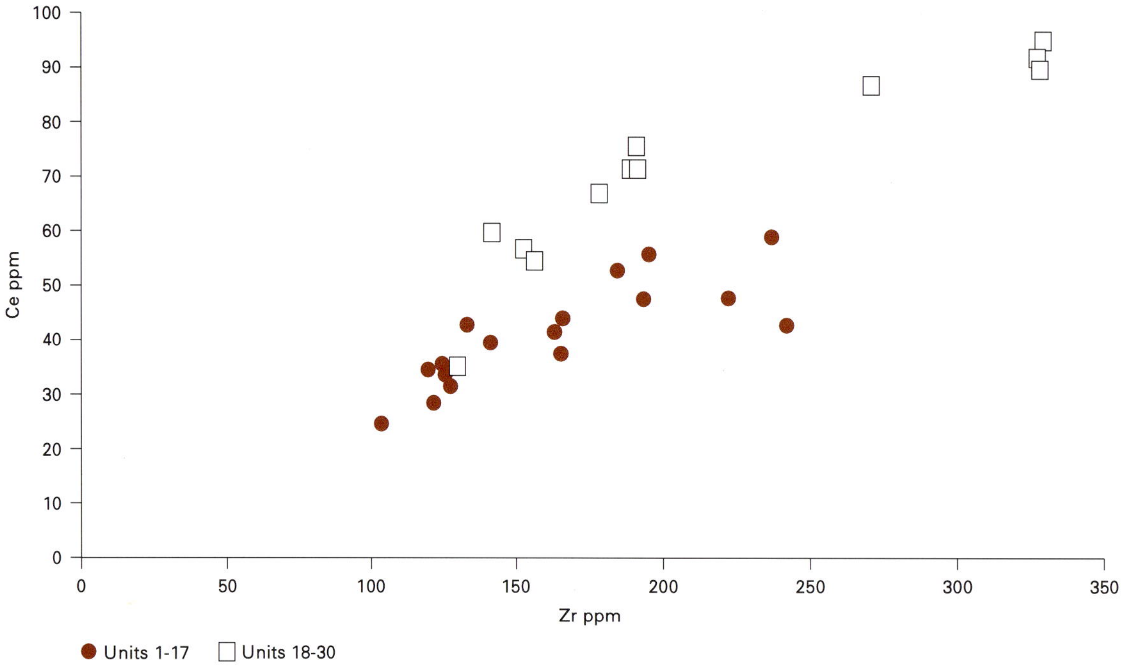

Figure 18 Ce–Zr covariation plot showing stratigraphical subdivision of the Haycock sequence into two magma types.

Martin-Del Pozo, 1982), and the low-profile plateau-andesite volcanoes of the Cascade province of the USA (Higgins, 1973; McBirney and White, 1982) and central Chile (Vergara and Munizaga, 1974; Hildreth et al., 1984). Syn-volcanic subsidence was considered an important aspect of the model (Petterson et al., 1992). Examples of low-profile andesite lava fields filling subsidence depressions resulting from earlier caldera collapse or tectonic extension are numerous, including for example, Santorini (Pichler and Kussmaul, 1980; Druitt et al., 1989). Topographically upstanding margins surrounding a Santorini-like model for the Birker Fell Formation could provide a source region for the clastic detritus within the interbedded volcaniclastic units (Branney, 1988b). Implicit in either model is that there were many vent sites. The andesite sheets are probably located over their vents which are largely concealed. However, potential vent locations have been identified for the Devoke Water Member in Eskdale, the Throstle Garth Member in the Lingcove area, the Cockley Beck Tuff around Cockley Beck and the composite andesites (Allen et al., 1987) in the upper part of the Duddon valley (Petterson et al., 1992, fig. 9).

The bias within the Birker Fell Formation towards basaltic andesite compositions, together with the eruption of primitive basalt, is a characteristic of plateau-like fields developed within extensional or transtensional environments and which promote ready flux of magma from mantle source to surface (Bacon, 1990). Polygenetic stratovolcanoes commonly show a long-term trend to more silicic compositions (Gill, 1981), whereas there is no evidence from the stratigraphical geochemical profiles from the Birker Fell Formation that this type of pattern dominated. Even though an individual section through the flanks of a stratocone would be unlikely to contain a complete eruptive sequence, it should still show evidence of the broad trend. Furthermore, sampled sections through a stratovolcano should show a broad correlation and co-genetic relationship between magmas from each section. In the Birker Fell Formation there is no chemical correlation between the Haycock and Devoke Water sequences, which are about 15 km apart. Though the units within each section form coherent co-genetic suites, the two sections exhibit separate differentiation trends and two discrete magmatic plumbing systems must have been operating. Sequences at Grey Friar and Stony Tarn are also quite different. The Wrighthow and Throstle Garth members also represent localised volcanic centres, erupting discrete, compositionally distinct magmas. Broadly, individual centres were spaced at distances of about 4 km.

The multiple centres model probably mimics regional variation in the mantle source and melt production. Figures 13d, and 14a, b show that the Haycock basalts are depleted in MgO, Ni and Cr for a given SiO_2 content compared with those at Devoke Water, and the two sequences lie on separate crystal fractionation trends. Thus, they cannot be related by, for example, polybaric olivine (± clinopyroxene) fractionation of a common parental melt. This difference may have been caused by the degree of melting locally (Figure 23), variable H_2O content, or heterogeneous mantle depletion/enrichment. However, the higher Al_2O_3 content and more evolved nature of basalt in the Haycock sequence indicates that magmas in this area were subject to a greater degree of high pressure (plagioclase-absent) fractionation prior to eruption. In the monogenetic volcanic field of the Cascade Range, Barnes (1992), Bacon (1990), and Prueher and McBirney (1988) similarly record the production of discrete batches of partial melt with different source characteristics feeding different volcanic centres. At Taal volcano Miklius et al. (1991) interpreted contemporaneous but different magmatic lineages as evi-

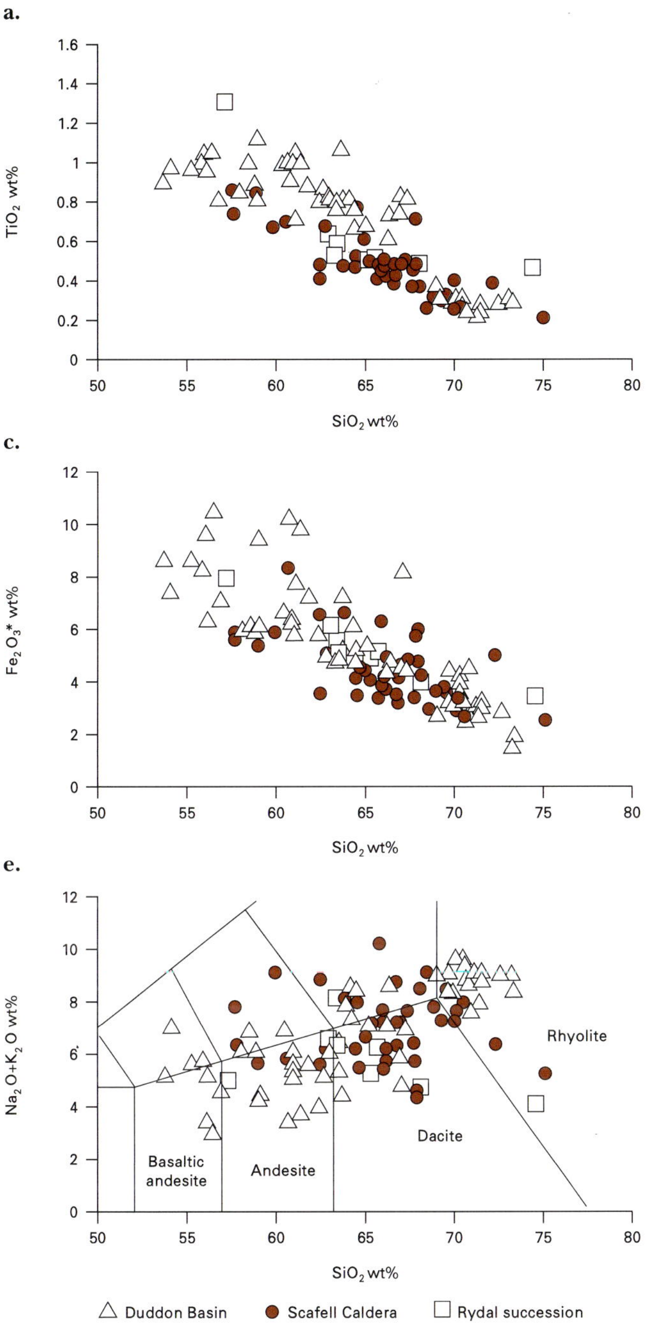

Figure 19 Major oxide Harker variation diagrams for selected rocks from the Scafell Caldera, Duddon Basin and Rydal successions.

Fe_2O_3* total Fe as Fe_2O_3.

dence for discrete magma batches processed independently through separately evolving plumbing systems.

Caldera-related magmatism

The broad-scale succession within the Borrowdale Volcanic Group of basic to intermediate lavas overlain by thick, ponded silicic ignimbrites, lava domes and lacustrine sedi mentary strata is typical of a 'caldera cycle' (Lipman, 1984). The diversity of pyroclastic facies within the volcaniclastic successions records rapid alternations in eruption style between magmatic and phreatomagmatic activity, changes in the location, geometry and number of vents, and rapid changes in highly irregular caldera-floor topography during subsidence. Individual voluminous eruptions were of a size almost always associated with calderas: for example, the Airy's Bridge Formation is up to 700 m thick and has been calculated to be greater than 120 km^3 (Branney and

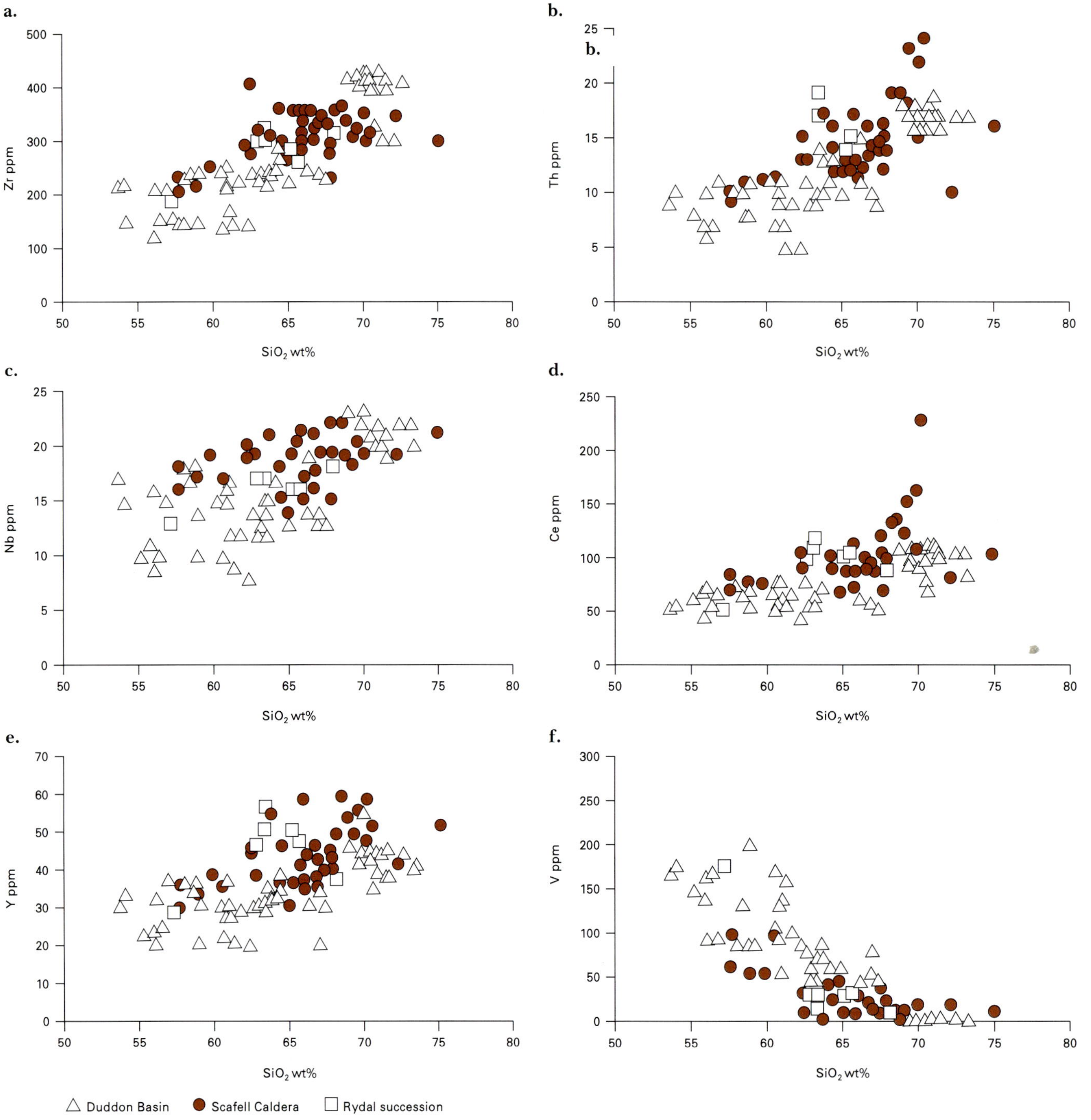

Figure 20 Trace element Harker variation diagrams for selected rocks from the Scafell Caldera, Duddon Basin and Rydal successions.

Soper, 1988). The intensely welded rocks, commonly with rheomorphic structures indicate high temperature emplacement and are typical of intra-caldera facies. Co-ignimbrite lag breccias suggest a location proximal to the eruption sites. Syn-volcanic faults with dip-slip displacements of hundreds of metres are abundant and closely spaced (locally less than 100 m apart; Chapter 9) and are associated with lenses of mesobreccia derived from fault scarps during eruption (Branney and Kokelaar, 1994). Intense soft-state deformation of tuffs is widespread and records syn-eruptive gravity sliding, spreading and disruption to form megabreccia (Branney and Kokelaar, 1994).

The Scafell Caldera, located broadly within the Scafell Syncline, was probably initiated by downsag during cal-

Figure 21 Al_2O_3–TiO_2 diagram for ignimbrites in the volcaniclastic successions.

Fields: solid line includes rocks from the Scafell Caldera succession

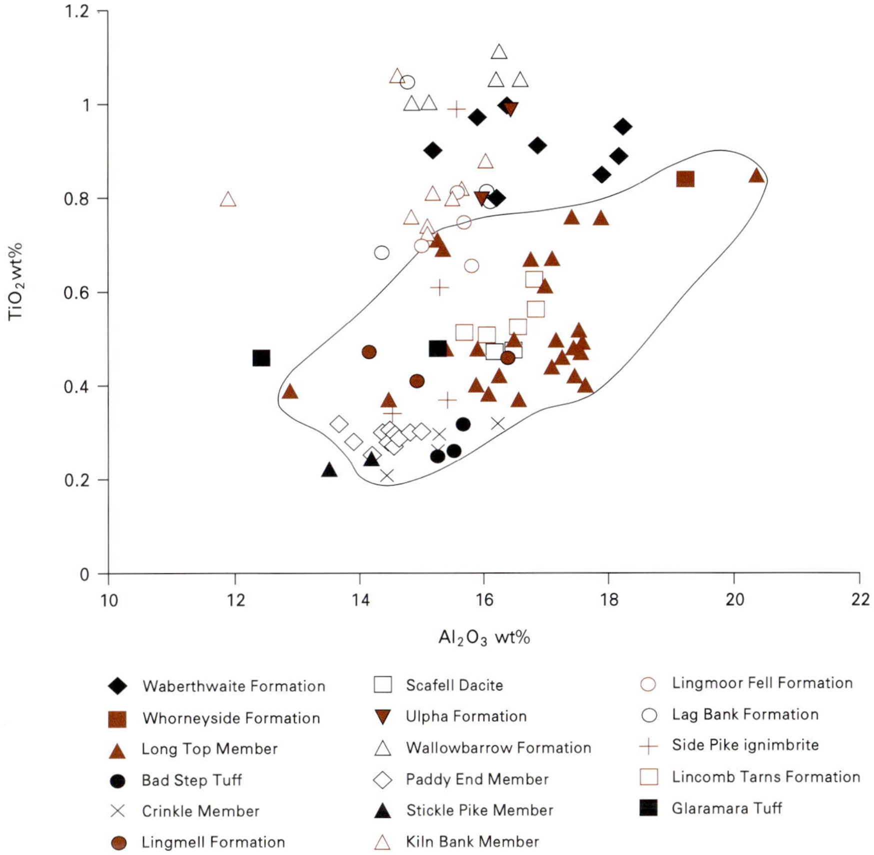

dera collapse (Chapter 9; Branney, 1988b; Branney and Soper, 1988; Davis, 1989; Channell and McCabe, 1992). Ignimbrites and succeeding caldera-lake sedimentary sequences thicken into it across contemporaneous volcanotectonic faults (Branney et al., 1993b; Branney and Kokelaar, 1994). The Ulpha Syncline is the only fold structure within the Borrowdale Volcanic Group to be truncated by the unconformity at the base of the Windermere Supergroup and is thus pre-Ashgill in age. The geometry of this fold and the absence of evidence for Ordovician compressional deformation suggest that a volcanotectonic origin is likely (Branney and Soper, 1988). Some ignimbrites and sedimentary formations thicken into the Ulpha Syncline and this feature, referred to in sedimentary terms as the Duddon Basin, probably developed through extensional tectonic subsidence.

Summary of the eruptive history of the Scafell Caldera

The sequence of events that led to the development of the Scafell Caldera have been described in detail by Branney and Kokelaar (1994). The fundamental change from the initial effusive phase of volcanism represented by the Birker Fell Formation occurred with the emplacement of ignimbrite that formed the lower part of the Whorneyside Formation. The centre of this eruption is believed to have been to the north-west of the Central Fells, in the adjoining Keswick district, and subsidence associated with the eruption of the pyroclastic flows allowed water to interact with the magma causing a change to phreatoplinian activity that deposited a widespread airfall ash blanket throughout the district (Branney, 1991). The creation of numerous low fault scarps during this eruption, probably accompanied by seismicity, caused widespread soft-sediment deformation of the ash deposit.

The overlying silicic pyroclastic flow deposits of the Airy's Bridge Formation were probably erupted immediately after the Whorneyside Formation because there is little evidence of erosion of the underlying deposit and rocks from both formations are intercalated on the north side of the caldera in the Keswick district. Eruption of these voluminous pyroclastic flows was accompanied by further subsidence that created increasingly large fault scarps, many of which were the sites of avalanches and landslides. The pattern of volcanotectonic faulting (Chapter 9) and thickness variations of the ignimbrites indicate that the Scafell Caldera developed as a piecemeal structure (Branney and Kokelaar, 1994, fig. 17). The eruption of large volumes of magma at high discharge rates preserved high temperatures within the ignimbrite sheets which then deformed in a ductile manner when disturbed by contemporaneous faulting. Subsidence intermittently allowed water to flow into the vents, producing short-lived phreatomagmatic air-fall deposits (Branney, 1988b; Davis, 1989). During the silicic phase of activity the caldera floor sagged, probably within major arcuate faults, where elongate fissures were partly filled with tuff (Branney and Kokelaar, 1994, fig. 16). The waning stage of this phase of volcanism is recorded within the Lingmell Formation and included the effusion of volatile-poor lava domes, such as the Scafell Dacite. The broad depression of the caldera was then filled by lacustrine deposition of ash and sediment (Kneller and McConnell, 1993; Branney and Kokelaar, 1994).

Duddon Basin

The succession in the Duddon Basin comprises a complex intercalation of pyroclastic, effusive and sedimentary facies that succeeded eruption of the Airy's Bridge Formation. Subsidence was much in evidence during aggra-

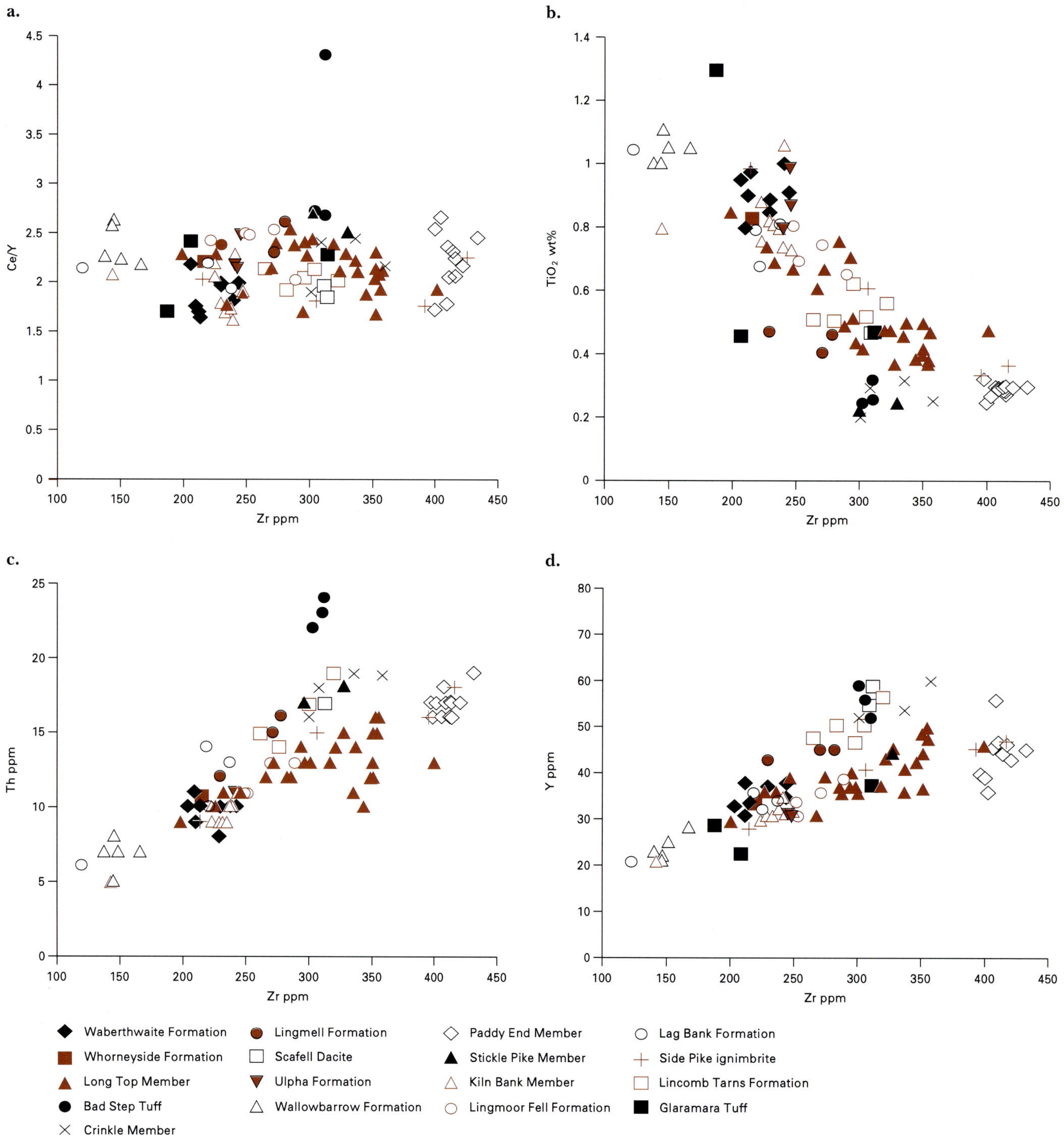

Figure 22 Trace element discrimination of ignimbrites using Zr as an index of fractionation for selected rocks from the Scafell Caldera, Duddon Basin and Rydal successions.

dation of the succession, demonstrated by the gradual thickening of pyroclastic and sedimentary formations towards depocentres within the Duddon Basin, and the presence of the marine Holehouse Gill Formation. Substantial thickening of ignimbrites occurred across the Grassguards, Stonythwaite and Baskill faults, but the abundant, closely spaced volcanotectonic faults affecting both the pre-caldera andesites and the intra-caldera rocks in the Scafell Caldera are absent. This suggests that the Duddon Basin may have been controlled largely by down-

Figure 23 Log Ni–Zr covariation plot of rocks from the Birker Fell Formation (excluding ignimbrites).

The partial melt and crystal fractionation curves are explained on Figure 17.

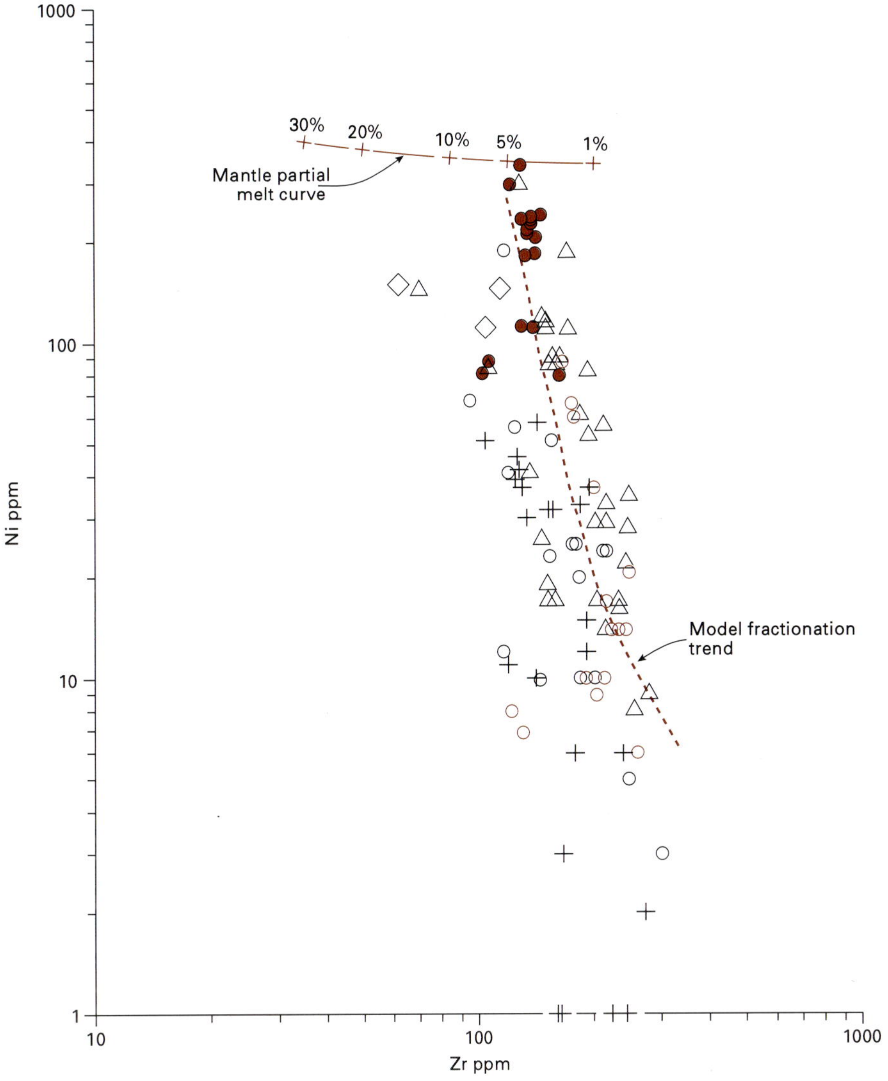

sagging and extension rather than a major period of caldera collapse. However, facies indicative of proximity to vents are present within several of the formations, implying that a number of eruption sites, including possible calderas, were associated with the basin.

Voluminous, high-grade andesitic rheomorphic ignimbrite with co-ignimbrite breccia was erupted to form the Waberthwaite Formation. The preservation of these probable intra-caldera facies rocks to the west of the Baskill Fault suggests that the south-western fells may have been the site of a caldera, but poor exposure of the rocks precludes a thorough appreciation of these events. A basin-wide change in eruption style took place with widespread air-fall andesitic ash and its associated lavas within the Duddon Hall Formation. The tuff has similar characteristics to the phreatoplinian tuff of the Whorneyside Formation. After the close of the phreatoplinian eruption subsidence resulted in a marine incursion into the volcanic field and deposition of mudstone of the Holehouse Gill Formation. Enhanced subsidence west of the Baskill Fault enabled a thick accumulation to be preserved there. The fine sediment must have been derived from exposed areas of basement rocks, possibly the Skiddaw Group, whereas the coarse sediment was reworked locally from the surrounding volcanic terrain. During this time andesite lavas of the Ulpha Formation interfingered with the marine sediments.

The remainder of the Duddon Basin succession comprises largely volcaniclastic sedimentation, alternating with pyroclastic flow eruptions. The Lickle Formation includes ignimbrites of increasingly evolved rhyolitic compositions, but the Lag Bank Formation represents a return to dacitic magma. The locations of the vents of these eruptions, probably calderas, are not known. The three ignimbrites within the Lickle Formation are relatively high grade; a very coarse mesobreccia is associated with the later Lag Bank Formation and is indicative of a location near to the vent.

The succession in the Duddon Basin probably represents a substantial time span because of the thicknesses of the sedimentary intercalations and the marine sequence. This is in contrast to events within the Scafell Caldera where volcanism was probably largely continuous. None of the major ignimbrite sheets nor sedimentary formations within the Duddon Basin succession have direct correlatives in the Scafell Caldera. However, because the base of the Seathwaite Fell Formation is highly diachronous, the Lingmell Formation and the lower part of the Seathwaite Fell Formation in the Central Fells must have been deposited penecontemporaneously with the succession in the Duddon Basin.

Succession in the Rydal and Ambleside area

The successions in the Scafell Caldera and the Duddon Basin are overlain by sedimentary strata intercalated with pyroclastic rocks. In the Scafell Caldera lacustrine sedimentation followed the Lingmell Formation volcanism without a break. Southwards into the Duddon Basin the base of the Seathwaite Fell Formation is highly diachronous and the sedimentary depositional phase began considerably later with the sedimentary rocks in the south probably correlated with the upper part of the formation in the Central Fells.

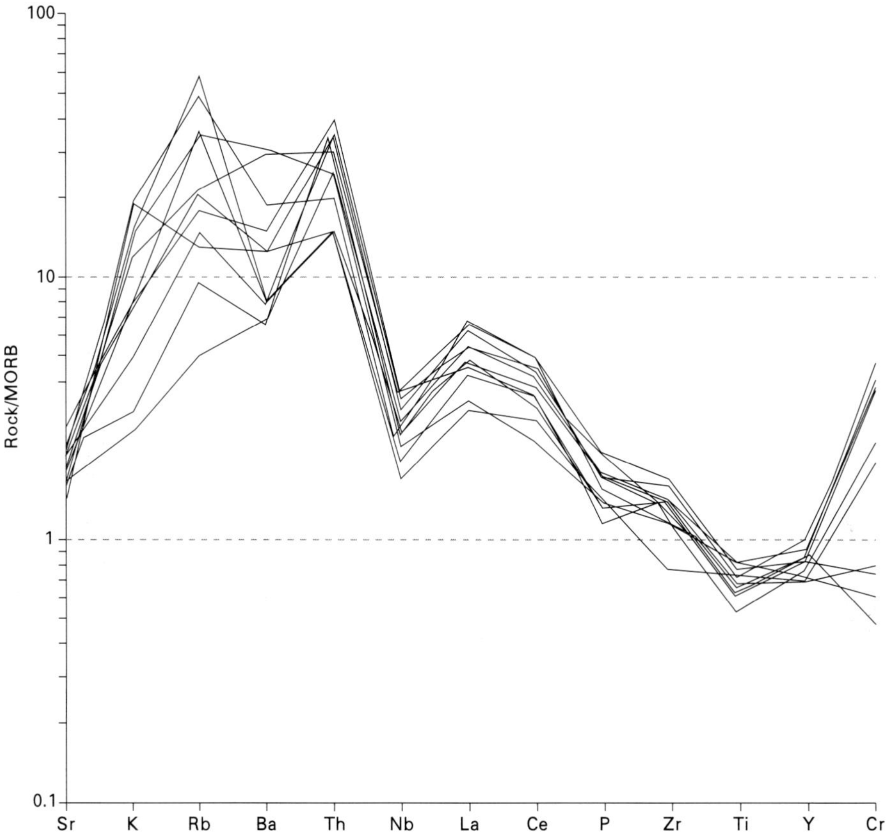

Figure 24 MORB-normalised spiderdiagram for representative Birker Fell Formation basalts. Normalising values from Pearce (1983).

Though the Seathwaite Fell Formation is largely sedimentary, there was some contemporaneous volcanic activity; gravity flows generated during an eruption deposited the Pavey Ark breccias, for example. Air-fall andesitic tuffs are preserved locally and very coarse-grained sandstone within the Seathwaite Fell Formation in the Coniston, Tilberthwaite and Kirkstone areas commonly contains pumice, probably resulting from contemporaneous eruptions. Abundant basaltic andesite and andesite sills were emplaced into the unconsolidated, water-bearing sediments of this formation and may have been contemporaneous with the eruptions. Some andesite sills in the Easedale–Rydal area may have broken surface to become extrusive locally.

Near the top of the Seathwaite Fell Formation widespread, thin silicic pyroclastic deposits of the Glaramara Tuff presaged eruption of further voluminous pyroclastic flows that formed the Lincomb Tarns Formation. This is a widespread, densely welded ignimbrite that may have resulted in caldera formation; though the location of the caldera is not known, it may coincide with the thickest preserved, and most complex facies which occurs in the Rydal and Ambleside areas, east of the Coniston Fault. The overlying Esk Pike Formation saw a return to subaqueous deposition including probable reworked ash from the Lincomb Tarns or contemporaneous eruptions (McConnell, 1993). These youngest formations within the Ambleside district have a fuller and more extensive development to the north and east of the district.

FIVE

Borrowdale Volcanic Group: Birker Fell Formation

The Birker Fell Formation (Petterson et al., 1992) comprises approximately 130 km^2 of the west of the Ambleside district adjacent to the Eskdale and Ennerdale intrusions, in Wasdale, Eskdale, and on Birker and Ulpha fells. Its thickness of 2200 to 2800 m comprises about one-third of the total Borrowdale Volcanic Group succession. The formation includes rocks that have been referred to variously as the Lower Andesites in the area south of Eskdale and around Devoke Water (Firman, 1957), as the Birker Fell Andesite Group in the upper Eskdale area (Oliver, 1961), as the Andesite and Dacite groups between Wasdale and the Ennerdale intrusion (Clark, 1964), and as the Cockley Beck Group and Seathwaite Tarn Andesites in the upper part of the Duddon valley and Seathwaite Fells (Mitchell, 1963) (Table 6).

In the district the base of the formation lies west of Devoke Water at Barnscar [SD 132 958], where basaltic lapilli-tuffs of the Devoke Water Member unconformably overlie mudstone and siltstone assigned to the Skiddaw Group. Close by, in Linbeck Gill on the west side of Devoke Water [SD 151 971], a 1.5 m-thick sliver of siltstone is exposed between granite and a hornfelsed basaltic andesite sill emplaced at the base of the volcanic sequence. Elsewhere within the district the base is cut out by the Eskdale and Ennerdale intrusions. The top of the Birker Fell Formation is defined by the base of the overlying Whorneyside Formation, an extensive andesitic ignimbrite and co-eruptive air-fall tuff (Branney, 1991) that marks a fundamental change from the lava-dominated to ignimbrite-dominated stratigraphy.

Andesite sheets, each up to 200 m thick, dominate the succession and on the published 1:50 000 scale geological map these are generally shown undivided (British Geological Survey, 1996). They include lavas and some sills, intercalated with thin units of volcaniclastic rocks. Where dips are shallow the weaker interbeds have weathered out to form prominent benches whereas the more resistant andesite typically forms steep crags. The resulting trap topography is locally very well developed (Millward et al., 1978; Moseley, 1983); good examples occur at Border End on Hard Knott [NY 225 018], between Crook and Green crags on Birker Fell [SD 196 985] (Plate 1), and around Scoat Tarn [NY 160 104].

In the subaerial, erosive environment that characterised the Borrowdale Volcanic Group episode the preservation potential of tephra and sediment was very much lower than that of the andesite sheets. Thus the volcaniclastic rocks are likely to significantly under-represent the explosive and sedimentary phases within the Birker Fell Formation. Intercalated pyroclastic rocks include air-fall tuff, pyroclastic surge deposits and ignimbrite ranging in composition from basalt to rhyolite. Sedimentary strata comprise siltstone, sandstone, conglomerate and breccia. These lithologies mostly form units 1 to 15 m thick, but 50 to 200 m are present locally. Distinctive facies associations are assigned member status on the 1:50 000 Series map (British Geological Survey, 1996) and are described below.

LITHOFACIES

Lavas and high-level sills

Green (1919) was the first to compare the brecciated andesite in the Borrowdale Volcanic Group with modern, autobrecciated lava. He considered many of the andesite sheets to be lavas, but was in no doubt that there were also many sills. Subsequently, it has been assumed that all sheets were lava (for example Firman, 1957; Oliver, 1961; Mitchell, 1963; Clark, 1964). The resurvey has identified some autobrecciated sills within the Birker Fell Formation on the basis of cross-cutting relationships and intrusion into overlying deposits. Chilled margins, increasing grain size towards the centre and contact metamorphosed country rock overlying the sheet are not present. Both lavas and sills may be expected in the andesite pile because recent volcanic successions commonly contain high-level sills, some of which breach their cover to become lavas. Branney and Suthren (1988) described peperitic intrusive features from the Borrowdale Volcanic Group and proposed that sills may constitute a significant proportion of the succession. However, it has proved difficult to establish from field evidence the proportion of sills within the Birker Fell Formation of the district.

The upper contact of the andesite sheets, commonly consisting of an open-network breccia with interstitial laminated sedimentary infill, is critical to the lava versus sill interpretation. Laminated internal sediment may be introduced by percolation of sediment down into cavities in the top of an autobrecciated lava, or may be relict host sediment following intruded magma into wet sediment. The lower contact is not diagnostic: lava flows can burrow into an unconsolidated substrate and/or thermally interact with a wet substrate. Hot magma intruded into wet sediment vaporises the sediment pore water and the resulting fluidisation causes significant sediment displacement (Kokelaar, 1982). Fluidisation may result in the following: brecciation of the intruding sheet; a structureless, thin, fluidised lamina between the intruded sheet and the host sediment; vesiculation of the host sediment; complex intermixing of host sediment and solidified magma resulting in a 'peppering' of the host sediment with numerous hydroclastically disrupted fragments of magma that are totally matrix supported (Kokelaar, 1982; Branney and Suthren, 1988). Thus, a spectrum of textures may be produced by the intermixing of sediment and magma, ranging from jigsaw-like

andesite breccias, to blocky breccias with sediment infill, to pods and pillows of andesite that are totally matrix supported. The term peperitic is used to describe these intrusive textures collectively.

Some peperitic-margined sills have been identified from the Birker Fell Formation during the resurveying of the district; for example on Middle Fell, Wasdale [NY 1530 0640]. However, the extrusive or intrusive origin of many sheets is equivocal. The preponderance of andesite sheets in the formation within this district exhibiting the following structures was taken by Petterson et al. (1992) to indicate that most sheets formed as lavas. However, opinions vary among workers as to the relative proportions of lavas or sills. The uppermost few metres of the open-network autobreccia contain finely laminated, medium- to fine-grained sandstone and siltstone. Sedimentary intercalations between the andesite sheets are mostly coarse sand and gravel grade. Furthermore, laminae within the cavities commonly reflect the form of the cavities with compaction, hour-glass bedding and cross-lamination suggesting infill by water percolation (Figure 25). These sedimentary characteristics are identical to those found within internal cavities in collapse breccias, hard grounds and palaeo-emersion surfaces in limestone successions (Aissaoui and Purser, 1983). Internal sediment is only present in the basal breccia of a small number of andesite sheets, but their original bedding structures have been destroyed or modified with trails and flames injected into the breccia. By contrast, in these same examples the internal sediment within the upper breccias retains a finely laminated structure. Eruption of lava into very shallow water or on to water-saturated sands is suspected in these cases.

An extrusive origin is clearly indicated in cases where bedded volcaniclastic rocks contain clasts identical to, and derived from, the underlying rock. Examples occur near Scoat Tarn [NY 162 104]. Also, clasts of the Great Whinscale Dacite occur reworked in adjacent sedimentary breccia on Little Stand [NY 2505 0310]. Sandstone containing fresh pyroxene crystals overlies an andesite sheet containing fresh pyroxene phenocrysts on Teighton How [NY 275 030] north of Wrynose Pass, within a succession otherwise lacking fresh pyroxene. Erosion and reworking of the top of the lava is the most likely explanation in these examples. Extrusion may also be indicated where a distinctive ash-fall tuff has a similar composition to the adjacent sheet; one example is the Seatallan Dacite which is overlain by a laminated silicic tuff.

Geochemical data from sections through the formation show regular systematic compositional variations with stratigraphical height (Figures 15, 16) compatible with sequential lava extrusion. The alternative explanation, that these were emplaced as sills and thus requiring a systematic intrusion sequence, is considered unlikely. However, the geochemical data may indicate the possible presence of some sills. For example, in the Haycock traverse (Figure 16), the basaltic andesite overlying the Craghouse Member does not fit into the regular geochemical pattern defined by the other sheets.

Draping of volcaniclastic rocks over the upper, irregular, blocky surface of the andesite results from differential compaction during burial in examples of lava and sill. However, mantling of air-fall tuff over irregularities in the upper surface of the sheet suggests an original microtopography and emplacement as a lava. Examples of this occur in the Throstle Garth Member in upper Eskdale [NY 220 040].

Simple and compound basaltic and basaltic andesite aa lavas

Basalt and basaltic andesite within the Birker Fell Formation are generally dark grey or greenish grey rocks with a grey or orange-brown weathered crust. They include lavas with extensive autobreccias comprising highly scoriaceous, clinkery or spinose, angular and irregular blocks and large slabs (Figure 26). Breccia comprises almost all of some lavas. In others there are ragged slabs of ropy, pahoehoe-like material. The inner part of the lava generally comprises dense, massive basalt or basaltic andesite becoming increasingly amygdaloidal and scoriaceous upwards. Most amygdales are elongate with variable orientation. These features are characteristic of modern aa lavas and are typical of simple lavas on White Pike [SD 150 958], that comprise most of the Birkby Fell Member. Lavas with similar characteristics are shown undivided on Sheet 38 Ambleside (British Geological Survey, 1996) between Tarn Crag and Brotherilkeld [SD 193 996–NY 216 013], in the Duddon valley near Black Hall [NY 240 012], around the shore of Wastwater [NY 155 058], and around Scoat Tarn [NY 160 104]. Most of these are within the lower part of the formation, but are not correlated with the Birkby Fell Member. Lava is mostly 10 to 50 m but ranges up to 100 m. Aa lava grades into block lava as the composition approaches the silica-rich end of the basaltic andesite compositional range.

Localised thick piles of autobreccia-dominated lava may represent compound lavas. Their structure is heterogeneous with irregular mounds of spinose, scoriaceous blocks containing abundant, apparently randomly orientated, lenticular tongues and channel fills of massive lava. Ragged, ropy, pahoehoe-like slabs are common. Channels are several metres thick by a few tens of metres long, with a sharp base, above which massive, non-amygdaloidal lava passes upwards into increasingly amygdaloidal material and a clinkery breccia top. The Throstle Garth and Wrighthow members in the upper part of the Birker Fell Formation contain compound lava fields, up to 240 m thick. Compound lavas in the former were ponded within a palaeotopographic depression (Petterson et al., 1992). Such lavas are commonly fissure-fed at slow effusion rates on basalt volcanoes (Walker, 1970; Pinkerton and Sparks, 1976). Relatively large volumes are erupted and many are ponded near to their vent system.

Porphyritic, micro-porphyritic and aphyric basalt varieties are present. Phenocrysts are euhedral to subhedral forming up to 30 per cent by volume. Clinopyroxene and plagioclase dominate the phenocryst assemblage (Plate 5a), with small amounts of an opaque mineral, and more rarely pseudomorphs after olivine; orthopyroxene becomes common in the more evolved rocks. Some lavas are pyroxene dominated, others are plagioclase rich. The

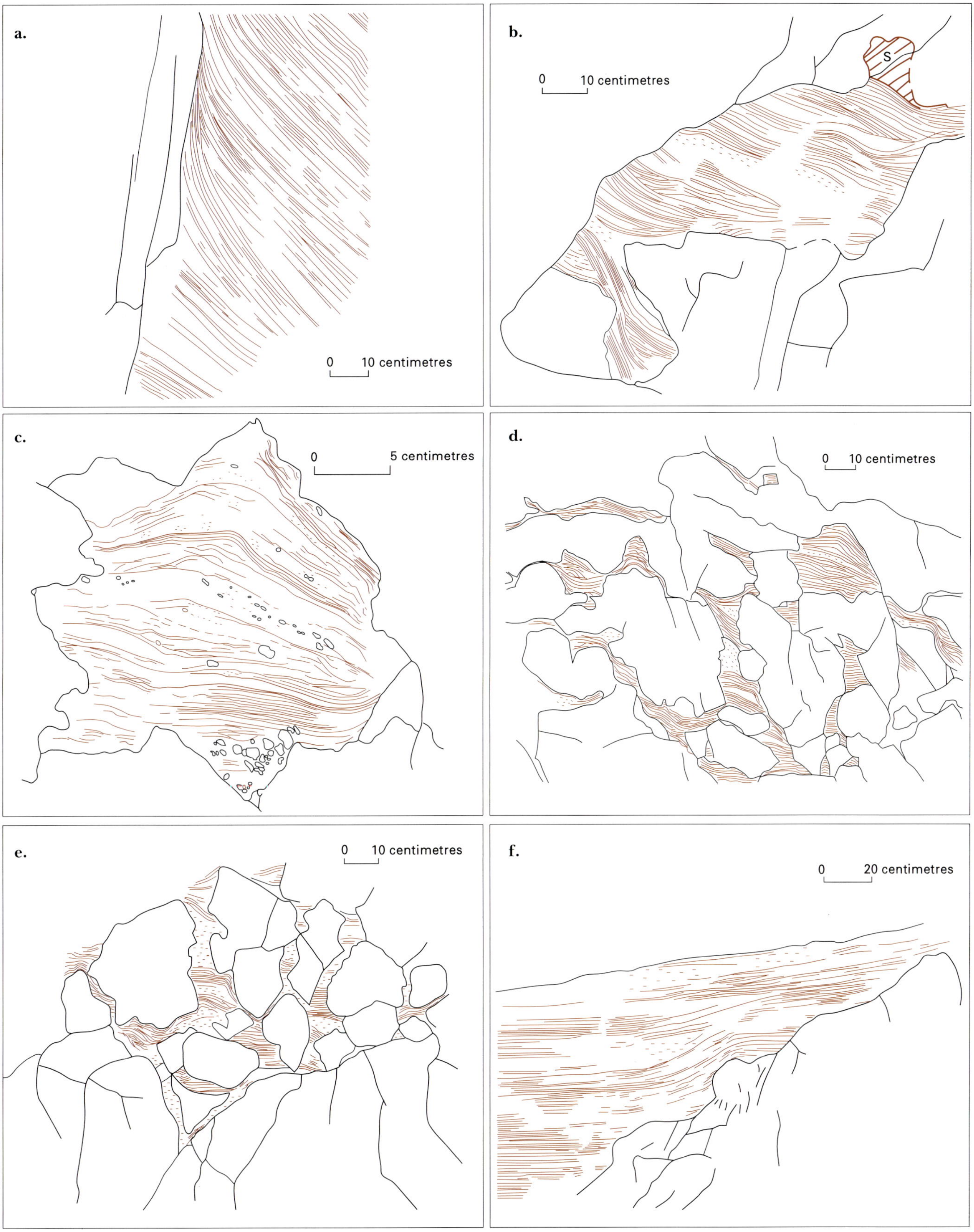
a.
0
10 centimetres
b.
0
10 centimetres
S
c.
0
5 centimetres
d.
0
10 centimetres
e.
0
10 centimetres
f.
0
20 centimetres

Figure 25 Field sketches illustrating the characteristics of internal sediment within autobreccia of andesite block lavas on Hard Knott and Harter Fell. Autobreccia clasts unornamented; lined lamination; dots massive siltstone and fine-grained sandstone.

a. Compaction and thinning of lamination adjacent to cavity wall.
b. Internal sediment within basal autobreccia: **s** — late infill with very fine-grained silica.
c. Hour-glass infill of cavity.
d and e. General perspective of internal sediment within autobreccia.
f. Overlap of lamination at top of andesite.

mafic phases are commonly altered to chlorite and epidote with prismatic or acicular amphibole mats locally developed, particularly within the contact metamorphic aureole of the Eskdale intrusion. Some fresh clinopyroxene is present in places but plagioclase is generally fresher than the pyroxene with only patchy replacement by epidote and sericite. The plagioclase is commonly zoned and internal corrosion textures are widely developed; chlorite inclusions are probably pseudomorphs after glassy relicts, a typical feature of rapidly formed volcanic plagioclase crystals.

The melanocratic to mesocratic groundmass varies from originally hypocrystalline to holocrystalline with textures including hyalopilitic, trachytic and intergranular. In most examples there is some devitrified glassy mesostasis. Clinopyroxene is fresh in the groundmass of some samples. The groundmass is commonly intensely altered to aggregates of quartz, feldspar, sericite, epidote, chlorite and opaque minerals, with amphibole and biotite developed near the Eskdale intrusion (Chapter 10).

Basaltic andesite, andesite and dacite block lavas

Sheets of basaltic andesite, andesite and dacite comprise most of the Birker Fell Formation. These are generally shown undivided on Sheet 38 Ambleside (British Geological Survey, 1996), although some thick, blocky dacite and rhyodacite sheets are depicted separately. Most of the sheets represent simple lavas (Figure 27), with massive, flow-banded and flow-jointed centres and autobreccia at the top, base, lateral margins, or locally throughout. The average thickness of the sheets is 150 to 200 m, but varies from 10 m to more than 200 m, locally with dacites attaining 250 to 300 m. The lateral extent of individual lavas varies from a few hundred metres to more than 5 km, but no general relationship between thickness and lateral extent has been established. Lava margins are rarely seen; they comprise steep, blocky breccia aprons.

The block lavas are asymmetrical in cross-section (Figure 27). Most basal autobreccia is up to 2 m thick, closely packed and rarely with sedimentary infill between the

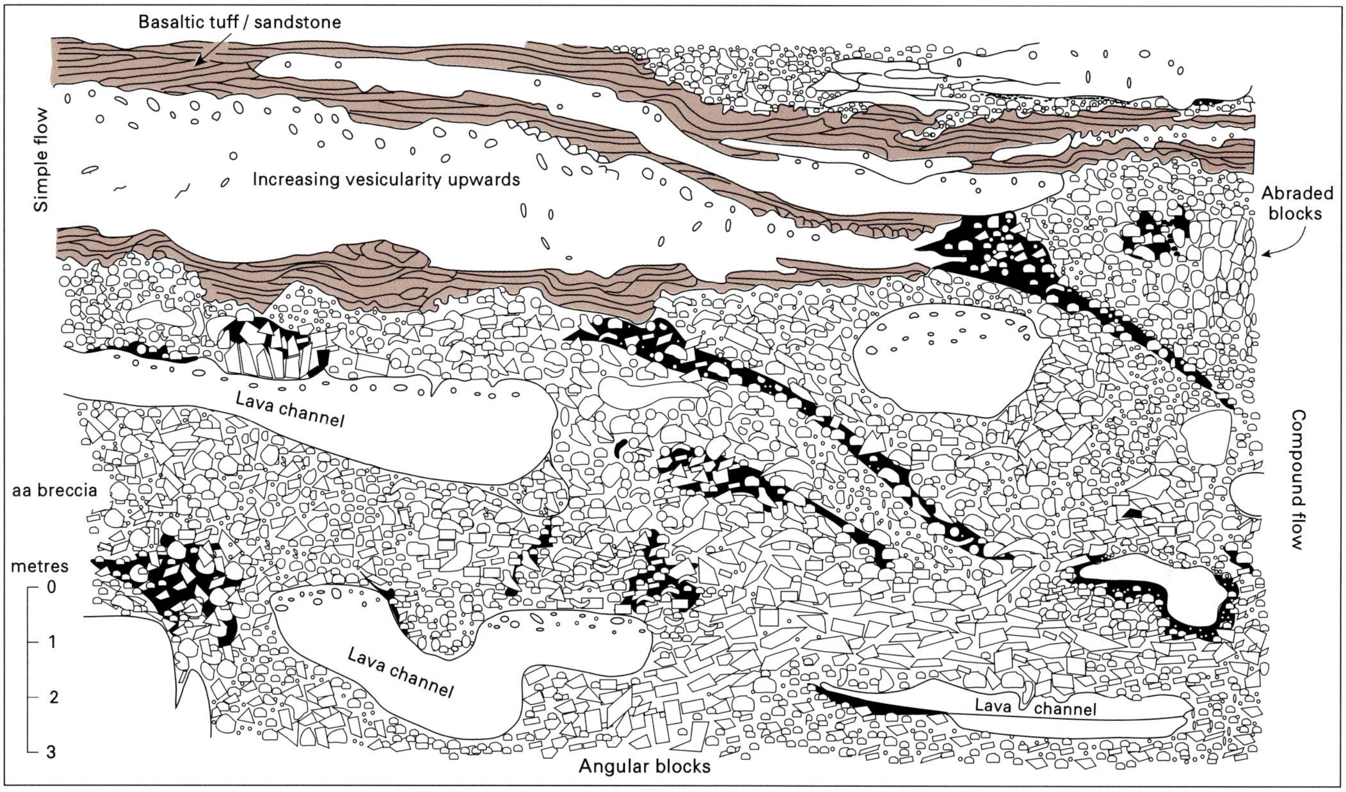

Figure 26 Morphological features of simple and compound aa basalt and basaltic andesite lavas within the Birker Fell Formation.

Figure 27 Schematic section through a blocky andesite lava, illustrating the main features.

Details of the internal sediment in the upper flow breccia are shown in Figure 25.

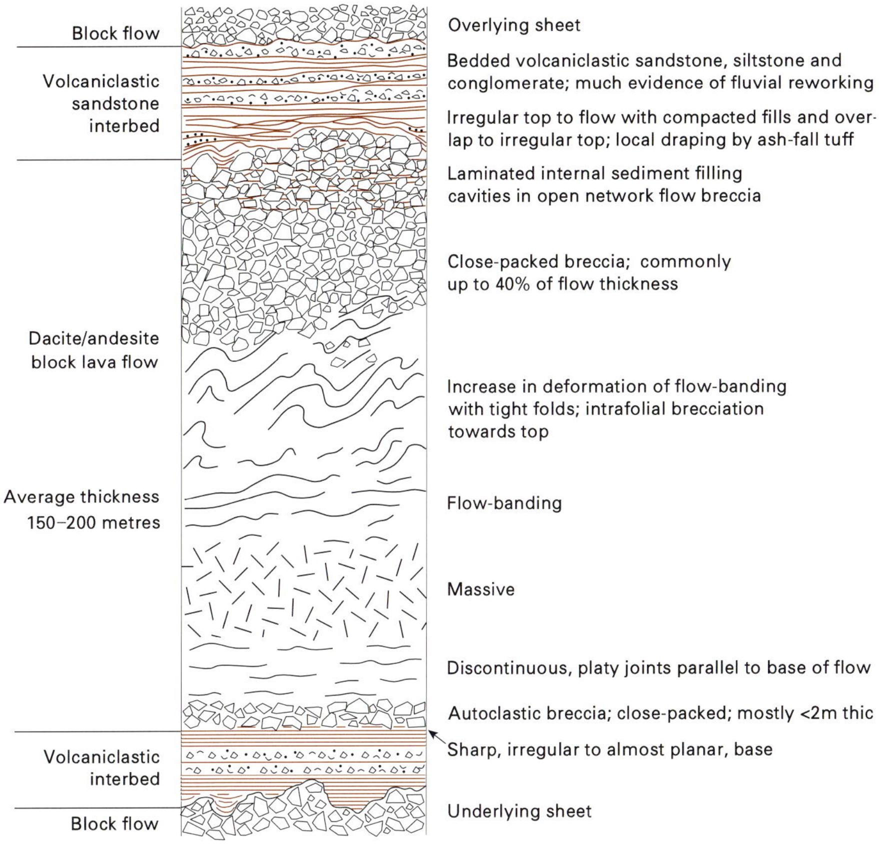

blocks, whereas the upper breccia comprises up to about 40 per cent of the sheet thickness. Blocks are mostly angular to subrounded and sub-equant, ranging from 0.15 to 1.5 m. The basal autobreccia passes up into coherent andesite with discontinuous platy jointing, parallel with the base of the sheet. Shears and brittle folds, including kink bands, affect these joints indicating that the joints were formed during late-stage flow. Above this, the central part of the lava is typically massive, passing gradationally upwards into flow-banded andesite that becomes increasingly flow-folded upwards and grades into the upper autobreccia.

Where underlying volcaniclastic deposits are of significant thickness, the base of the lava is parallel to bedding with little deformation of the subjacent strata; in the absence of volcaniclastic rocks between lavas, sheet bases are more irregular, following the top of the underlying lava. The extremely irregular upper surfaces may be mantled by pyroclastic rocks, but more commonly, the hollows contain laminated and thinly to medium-bedded volcaniclastic sandstone, breccia and conglomerate (Figure 25f). Cavities in the upper autobreccia are filled with laminated fine-grained sandstone and siltstone.

Most of the basaltic andesite, andesite and dacite block lavas are porphyritic. Where fresh, the euhedral to subhedral plagioclase feldspar laths, up to 3 mm in length, show oscillatory zoning, but the phenocrysts are variably replaced by sericite and epidote. Some plagioclase feldspar crystals contain rings of altered melt inclusions. The phenocrysts are commonly set in a fine- to medium-grained, mesocratic groundmass (Plate 5b) and comprise up to 40 per cent of the rock by volume. Other, generally subordinate, phenocryst phases include orthopyroxene, clinopyroxene, opaque minerals, and in acid andesite and dacite, rare garnet. Orthopyroxene is invariably altered to chlorite and epidote; by contrast, clinopyroxene is fresh in some rocks, altered to chlorite, epidote and quartz in others and, near to the Eskdale intrusion, mostly replaced by amphibole.

Typically, the groundmass is a fine-grained, interlocking, microcrystalline network of subhedral and euhedral feldspar commonly replaced by sericite and epidote, anhedral mafic phases mostly altered to chlorite, and anhedral opaque minerals. Accessory minerals include apatite and zircon; a brownish tinged apatite is characteristic of the Birker Fell Formation rocks. Groundmass crystals are usually randomly orientated, although in some cases a trachytic texture is present. In some rocks the groundmass is too fine grained to enable positive identification of individual mineral phases, and probably represents a devitrified, rapidly quenched liquid. Dacite mesostasis is more leucocratic than andesite mesostasis (a result of the higher modal abundance of felsic minerals) and commonly has a blotchy devitrified groundmass mosaic or perlitic cracking, evidence of an original glassy state.

Many of the andesites are amygdaloidal, indicating relatively high initial fluid partial pressures on eruption. Amygdale abundance and size increase towards the top of flows, though in some cases whole sheets are amygdaloidal. Amygdales are of chlorite, quartz, epidote and carbonate; amphibole or, rarely, prehnite may be present in places.

Garnetiferous dacite

On the west slopes of Border End [NY 220 015] and in Wrynose Bottom, upstream from Black Hall [NY 240 015 to 265 022], is a 150 to 200 m-thick, garnetiferous dacite, possibly a single block lava, that is distinguished from other lavas of this composition by its unusual phenocryst assemblage. It overlies the geochemically very similar Cockley Beck Tuff with which it may be co-genetic. Phenocrysts are of plagioclase (including some labradorite) pseudomorphs after pyroxene, garnet (up to 4 mm) and opaques. Xenoliths include microdiorite, gabbro, quartzite

and various garnet-bearing mineral aggregates that may be cognate. The groundmass is variably affected by flow-banding and perlitic cracking.

Composite andesite sheets

Composite lavas and sills constitute a volumetrically small, but volcanologically important component of the formation. They occur principally in the area around Border End [NY 229 017], on the summit of Harter Fell [SD 218 997], and between Little Stand [NY 245 033] and the Three Shires Stone area [NY 279 028]. Allen et al. (1987) described this previously unrecognised rock type and suggested an emplacement mechanism. The composite sheets comprise greenish grey basaltic andesite and pale brownish grey andesite. Andesite forms most of the volume of the sheets, ranging from 55 per cent for the unit on Little Stand to more than 70 per cent for that on Red How. Typically, the andesite is overlain by basaltic andesite and separated from it by a complexly interlayered zone (Figure 28). In some of the sheets there is additional basaltic andesite at the base but this is generally less than 4 m thick.

The andesite component is fairly uniform, massive or flow-jointed, non-amygdaloidal, and porphyritic. The basaltic andesite component is commonly highly amygdaloidal with a massive to flow-laminated and flow-folded lower facies passing upwards into autobreccia, similar to that present in the blocky andesite lavas previously described. The basal basaltic andesite layer, where present, is flow laminated and xenolithic, commonly with a thin, rubbly basal breccia.

The interlayered zone (Figure 28, Plate 7) comprises lenses, pods and irregular inclusions of andesite, generally up to 15 cm thick, within a matrix of basaltic andesite. Contacts between andesite and basaltic andesite are

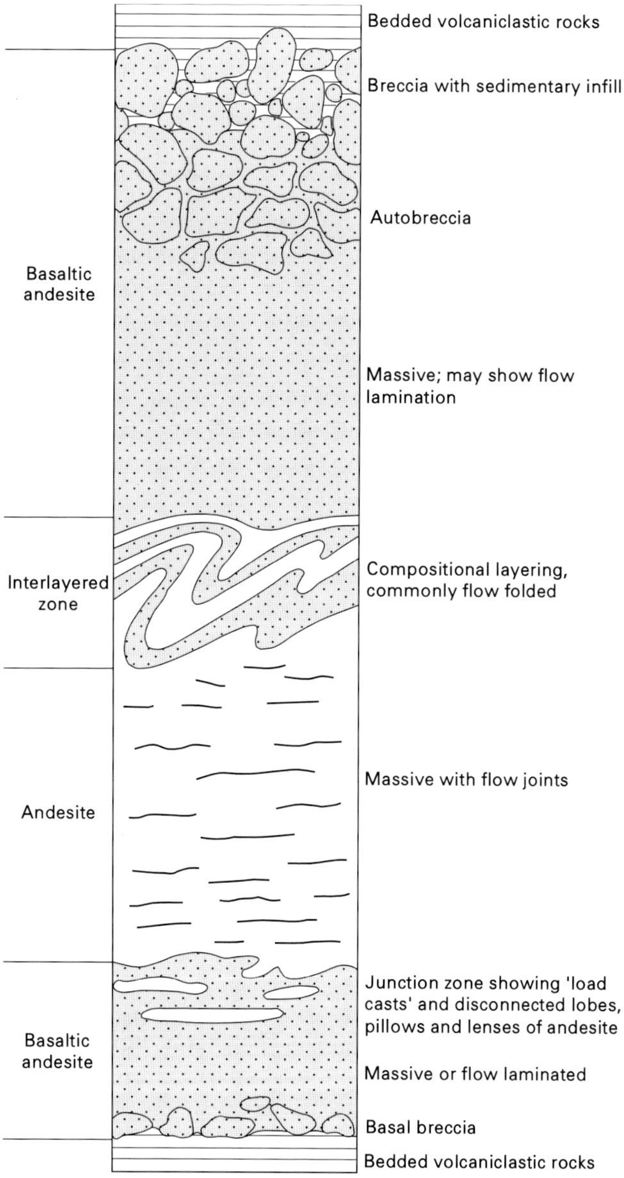

Figure 28 Idealised section through a three-part composite andesite, based on the Red How lava [NY 252 030] (after Allen et al., 1987).

Plate 7 Flow-folding in the upper part of a porphyritic, composite andesite lava.

The recessed layers are crystal rich and more mafic than the other layers. This mixed zone is typical of composite lavas in the Birker Fell Formation. Flow-folding on this scale is commonly also seen in blocky andesite. Border End, Hard Knott [NY 239 017] (D 4027). Hammer is 37 cm long.

irregular, sharp and sutured. The zone is commonly intensely flow-folded and the andesite layers may contain stretched amygdales. The junction between andesite and any lower basaltic andesite component is sharp and highly irregular with load casts, lobes and pillows of andesite penetrating the basaltic andesite below.

The thickness of these composite andesite sheets varies considerably. The sequence north of the river Duddon and east of Mosedale is 400 m thick and contains the thickest individual sheet, 100 to 250 m on Red How, where the outcrop extends over 2.5 km. Those units exposed on Border End, Harter Fell and Little Stand are only up to 40 m thick, but one of these has a lateral extent of at least 1.5 km. The uppermost composite andesite sheet on Little Stand has been interpreted as a sill by Branney and Suthren (1988) because of the presence of peperite at its upper contact.

Petrographically these rocks are similar to andesite and basaltic andesite elsewhere in the succession. Phenocryst concentration is 20 to 50 per cent by volume. In some sheets the concentration in the two components is broadly similar, though in the Red How lava the basaltic andesite contains up to 47 per cent and the andesite has only 20 per cent. Within the interlayered zone patches and blebs of andesite are totally surrounded by basaltic andesite. The andesite inclusions commonly have a rim of sphene and/or epidote, a silicified inner margin and zoned chlorite and quartz amygdales in the core.

Detailed whole-rock major and trace element analyses and mineral compositions of the composite lavas are discussed in Allen et al. (1987). The analyses confirm that the components of the composite flows are broadly similar to other basaltic andesite and andesite within the Birker Fell Formation. Systematic variations through the sheets on Little Stand and Red How were discussed by Allen et al. (1987) who related the chemical changes to varying phenocryst content, the assimilation of cognate xenoliths, magma mixing and fractional crystallisation. Samples taken from the zone of intermixing proved that geochemical as well as physical mixing of the two magmas had occurred.

Allen et al. (1987) considered that these sheets were produced by the simultaneous eruption through a single vent of two magmas of contrasting composition. Dry andesite magma in a high-level magma chamber was intruded by a wet, hotter, more basic magma that fractionated to produce a lower density magma able to rise and mix with the overlying andesite magma. These magmas were mixed only within their contact zones. The localised occurrence of such a specialised product lends support to the multiple vent model for the Birker Fell Formation (Petterson et al., 1992).

Pyroclastic rocks

Ignimbrite

Massive beds of very poorly sorted tuff and lapilli-tuff, interpreted as ignimbrite, provide the best stratigraphical markers within the formation. Most of the units are eutaxitic (best seen on weathered surfaces), and microscopic flattened and welded shards are seen in thin section. Some distinctive ignimbrites have member status, for example the garnetiferous, Cockley Beck Tuff, and the nodular, alkali-feldspar-bearing Little Stand Tuff. A group of andesitic and dacitic ignimbrite sheets has been defined formally as the Craghouse Member. Other ignimbrites are lithic rich, some containing abundant clasts of greywacke, sandstone, siltstone and silty mudstone, derived from the sub-volcanic Skiddaw Group. Many of these are preserved only locally, but the Grey Friar Tuff has an extensive outcrop. The thickest and most extensive of the ignimbrites are depicted on Sheet 38 Ambleside (British Geological Survey, 1996) and are described later in this chapter.

Pyroclastic fall deposits

Parallel-bedded, poorly, moderately and well-sorted tuff, lapilli-tuff and breccia, variously with either normal, reverse or symmetrical grading may be interpreted as pyroclastic fall-out deposits. These features are not uniquely diagnostic of ash fall-out and also may have been produced by secondary reworking of ash, for example by sheet wash or turbidity currents. However, good evidence for an air-fall is provided where beds of even thickness mantle either the irregular top surface of brecciated lava or a clear erosion surface (Plate 4). Examples are few, mainly because of the very poor preservation potential of the deposits, but one is preserved within the compound basalt lava succession of the Throstle Garth Member on Scar Lathing [NY 2253 0490], in upper Eskdale. Fall-out tuff is also associated with the Seatallan Dacite and Blisco Member, but no fall-out deposit has yet been traced extensively enough to constitute a useful stratigraphical marker. An example of a 1 m-thick, plinian pumice-fall deposit occurs on Long Scar [NY 2747 0363] overlying an autobrecciated andesite and 3 m below the Little Stand Tuff. Accretionary lapilli, characteristic of both air-fall and pyroclastic surge deposits, are sparse within the Birker Fell Formation. Examples of beds containing small accretionary lapilli occur within the Throstle Garth Member in upper Eskdale on the western end of Long Crag [NY 2274 0522] and adjacent to Lingcove Beck [NY 2380 0507], and in the Blisco Member at Long Scar [NY 2747 0363].

Pyroclastic surge deposits

No pyroclastic surge deposits of significant lateral continuity have been identified in the Birker Fell Formation of the district. Lenses of low-angle cross-laminated tuff, interpreted to have been deposited from pyroclastic surges occur mostly associated with ignimbrite, for example within the Cockley Beck Tuff.

Rocks of hydroclastic origin

Eruptions produced by the interaction of water and magma are common in contemporary andesite volcanoes, though most of the resulting fragmental deposits have low preservation potential. The contribution of phreatomagmatic and phreatic eruptions to the Birker Fell Formation may therefore not be reflected fully by the few hydroclastic products preserved. However, a significant hydrovolcanic

phase is thought to be recorded by the Devoke Water Member. This type of eruptive mechanism commonly incorporates fragments of rocks from underlying formations and so the widespread presence within the Devoke Water Member of Skiddaw Group clasts supports the interpretation of a hydroclastic origin.

On a smaller scale, east of Harter Fell on Dropping Crag [SD 2250 9945], probable hydroclastic deposits are interbedded with highly porphyritic andesite lavas over a strike length of about 1 km. These volcaniclastic rocks comprise heterogeneous, very poorly sorted tuff-breccia which contains angular, non-vesicular clasts up to 30 cm in diameter. Bedding is typically wavy and lenticular with many low-angle truncations and dune-like structures. Scattered ballistic blocks are present and some block or bomb sags are unoccupied.

Volcaniclastic sedimentary rocks

Many of the volcaniclastic units within the Birker Fell Formation are composite, containing both pyroclastic and sedimentary rocks; some are ambiguous in interpretation. The commonest facies present is a siltstone to coarse-grained sandstone with parallel lamination or thin bedding and normal grading. An air-fall origin is possible, either onto level terrain or into water, but equally the deposit may have been produced by sheet wash or turbidity currents. There are also many deposits that are undoubtedly sedimentary. The more substantial volcaniclastic units are shown on Sheet 38 Ambleside (British Geological Survey, 1996), but some, including the Dale Head Member, are defined formally. Sedimentary rocks also comprise significant or minor facies within other members, for example the Blisco and Throstle Garth members respectively.

Siltstone and sandstone commonly form parallel-bedded or laminated, well-sorted units grading up to coarse-grained and pebbly sandstone. Some facies are dominated by cross-bedded, coarse-grained sandstone with steep-sided channels containing basal pebble, cobble and boulder lag gravels; imbricated rip-up clast conglomerates occur at a number of localities (Plate 8). In alternating fine- and coarse-grained deposits soft-sediment deformation and small-scale synsedimentary faulting is common.

The rocks represent immature sediments with varying proportions of heterolithic, glassy, dominantly angular to subangular clasts, crystal fragments, scoria and pumice. Rock fragments are mainly non-vesicular or weakly amygdaloidal andesite and dacite, texturally similar to the andesites of the Birker Fell Formation. Crystal fragments form a highly variable proportion with plagioclase and pyroxene dominant. Andesite scoria and silicic pumice, commonly porphyritic and devitrified, are generally subordinate components, and may be flattened and deformed, giving beds rich in these clasts a eutaxitic fabric. Flattening of the clasts probably resulted from loading compaction subsequent to their alteration to clay minerals (see Branney and Sparks, 1990), rather than from welding.

Coarse sedimentary breccia and conglomerate are generally of limited extent, typically filling channels or valleys, but some extensive breccia units are shown on the published 1:50 000 scale geological map. One of these, the Dale Head Member, is a valley-confined massive debris-flow deposit in the central part of the district; other examples are on Ulpha Fell, south-east of Woodend [SD 170 957], and in upper Eskdale. On Ulpha Fell, up to

Plate 8 Intraclast conglomerate within a succession of volcaniclastic rocks.

A channel up to 40 cm deep is filled with conglomerate containing platy clasts of silty sandstone up to 10 cm long. The clasts are similar to the material from which the channel is cut. High Crag, east of Dale Head, Duddon valley [NY 2473 0077] (L 3166)]. Hammer is 37 cm long.

480 m of very poorly exposed breccias crop out locally; a lower, weakly bedded, very poorly sorted breccia with angular to subangular clasts typically 2 to 3 cm across is overlain by massive beds containing clasts up to 20 cm diameter. The poorly sorted, clast-supported breccia in upper Eskdale [NY 213 046], underlying the Throstle Garth Member, contains flattened pumice and blocks of dacite and andesite, and impersistent lenses of bedded volcaniclastic sandstone and gravel. They may be eruption-induced mass-flow deposits generated by, for example, floods or the collapse of a lava dome. Coarse debris-flow breccias with angular blocks up to 1.5 m occur on Pike of Blisco [NY 2706 0428] near the top of the Blisco Member.

STRATIGRAPHY

The stratigraphy of the Birker Fell Formation is summarised in Figures 29 and 30. Despite the lateral impersistence of many units, in the area to the south-east of the Eskdale intrusion through to Hard Knott [NY 230 020] and Stonesty Pike [NY 247 037], a coherent internal stratigraphy is recognised using the more extensive members described below. By contrast, to the north in upper Eskdale and Wasdale, the internal stratigraphy of the formation is difficult to establish because of the absence of good marker beds. Though the dominantly andesite successions in these areas are similar, stratigraphical links between them are poorly known beyond the common presence of the Devoke Water Member at the base and the Whorneyside Formation that overlies the formation throughout the district. A substantial thickness of felsic volcanic rocks occurs in the central parts of the succession: the Seatallan Dacite and Craghouse Member in the northern area, and the Little Stand Tuff and Great Whinscale Dacite in the south. These events may have been broadly synchronous.

In the northern area, most of the Birker Fell Formation consists of porphyritic andesite block lavas interbedded with sills. The thickness ranges from 1850 m in upper Eskdale to about 2200 m in Wasdale and 2400 m in the west of the district around Nether Wasdale (Figure 29). The Eskdale and Ennerdale granites have removed the base of the Birker Fell Formation there, but in upper Eskdale the uppermost part of the Devoke Water Member, which forms the basal unit of the formation in the southern area, is present. Pyroxene and plagioclase-phyric basaltic andesite lavas form the lowest rocks north-west of Wast Water. They are mostly simple aa lavas but are not formally named because of their local distribution. The Wrighthow and Throstle Garth members occur in the upper part of the formation and comprise compound basalt lavas. Intercalations of volcaniclastic sandstone and siltstone, in units up to 100 m thick, are present locally within the succession but none constitute stratigraphical markers. The main stratigraphical correlation in the northern area is provided by the Seatallan Dacite and the overlying Craghouse Member, the most evolved members present.

In the southern part of the district, the Birker Fell Formation south of Devoke Water [SD 155 966] is 2600 m thick. The dominant, uniform andesite succession is divided by a group of widespread and distinctive stratigraphical marker units occurring about 1200 m above the base of the formation (Figure 30). These include the nodular Little Stand Tuff, the aphyric Great Whinscale Dacite (Kanaris-Sotiriou et al., 1991), the lithic-rich Grey Friar Tuff, and the Blisco Member, a mixed pyroclastic and sedimentary succession. The Little Stand Tuff and the Great Whinscale Dacite were grouped previously as the Tongue House Member (British Geological Survey, 1991; Petterson et al., 1992). In Branney (1988b) and Branney and Soper (1988) the Little Stand Tuff, Great Whinscale Dacite and the Blisco Member were referred to jointly as the Blisco Formation. Lower in the formation, between 600 and 800 m above the base, is the garnetiferous Cockley Beck Tuff (Figure 30), which crops out from Foxbield Moss, just west of Crook Crag [at SD 191 986] to Cockley Beck [NY 251 017]. It is overlain by a thick and probably coeval, garnet-bearing dacite lava.

At the base of the Birker Fell Formation south-west of Devoke Water is a thick sequence of basic lapilli-tuffs, the Devoke Water Member, and the overlying basaltic lavas of the Birkby Fell Member (Figure 30). Farther east, an unnamed succession of plagioclase and pyroxene-phyric basaltic aa lavas occurs above and below the Cockley Beck Tuff between Tarn Crag and Brotherilkeld [NY 215 012]. These may correlate with basalts underlying the Dale Head Member breccias near Cockley Beck (Figure 30). Composite lavas (Allen et al., 1987) and sills are largely restricted to Hard Knott, Harter Fell and north of the River Duddon between Mosedale and Wrynose Pass and occur near the middle of the succession above and below the Little Stand Tuff–Great Whinscale Dacite–Blisco Member markers.

The uppermost part of the Birker Fell Formation, above the Little Stand Tuff and associated markers, is 700 to 750 m thick on Cockley Beck Fell and near Seathwaite Tarn, and 900 to 1200 m thick on Ulpha Fell. Compared with the lowest part of the succession volcaniclastic rocks form a more substantial part in these areas. North-east from Grey Friar, on the south side of the Duddon valley, sandstone and breccia units between andesite sheets thicken eastwards (Figure 30, columns 10–14). On Ulpha Fell there are also significant thicknesses of breccia, sandstone and lapilli-tuff (Figure 30, columns 3–6).

Seatallan Dacite

In the north-western part of the district, the Seatallan Dacite represents the earliest significant silicic eruptive event within the Birker Fell Formation. The dacite crops out from Cat Bields [NY 130 070], through to the type area of Seatallan [NY 140 085], and on the slopes between Haycock and Scoat Tarn [NY 152 105]. It also extends westwards and northwards into the Gosforth and Keswick districts, respectively. A lithologically and geochemically similar dacite crops out on Latterbarrow Crag [NY 126 027] 1.5 km south of Nether Wasdale and 5 km south-south-west of Seatallan. It is uncertain whether this is the same lava or a separate co-eruptive effusion. The overall strike length is about 9 km, which is unusually large for a

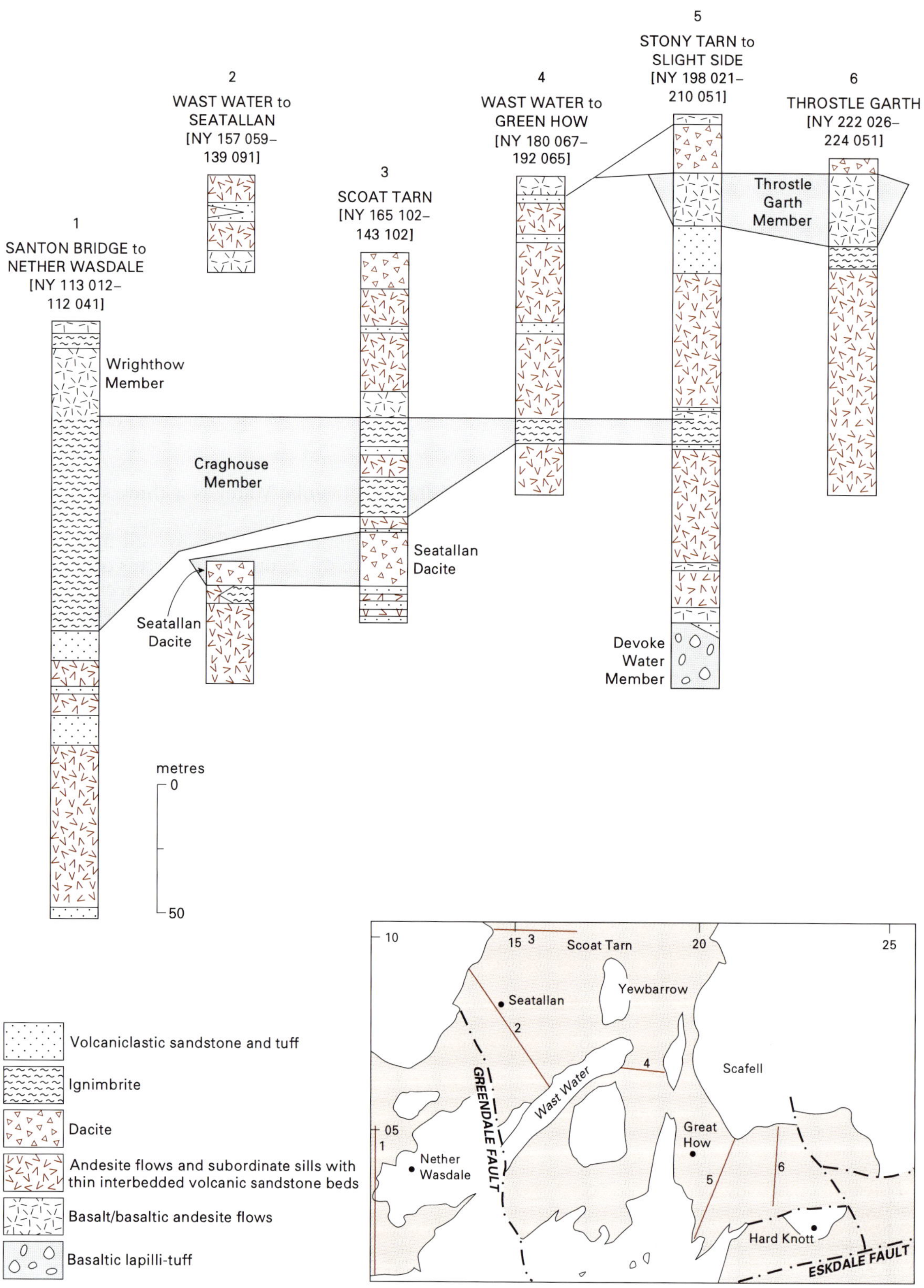

Figure 29 Correlation of selected generalised vertical sections through the Birker Fell Formation in the north-western part of the district.

dacite lava. In the district the top of the dacite is only seen near Scoat Tarn. The thickness is 150 to 200 m, but because the base and top are not exposed in a single fault block the full thickness is not ascertained.

The Seatallan Dacite is a lava and, with a silica content of about 67 per cent, is one of the most acidic rocks in the Birker Fell Formation (Table 9). Well-developed flow-banding and locally flow-folding is displayed clearly on well-etched exposures. Exposures have a blocky appearance with three orthogonal joint sets. Marginal autobreccias are thin, for example west of Scoat Tarn, comprising only the uppermost 5 m. The dacite is typically

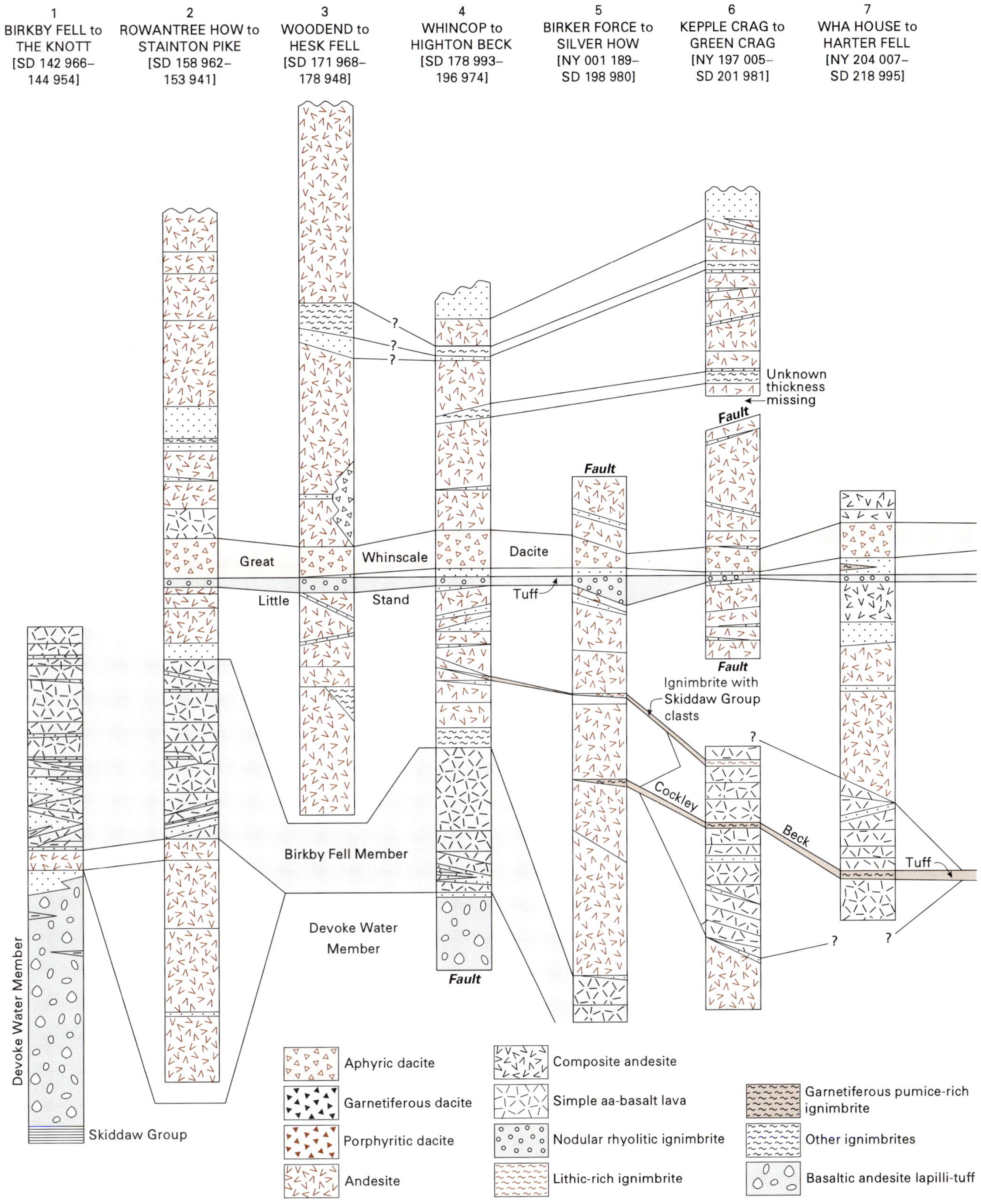

Figure 30 Correlation of selected generalised vertical sections through the Birker Fell Formation in the south-western part of the district. Inset map shows location of sections.

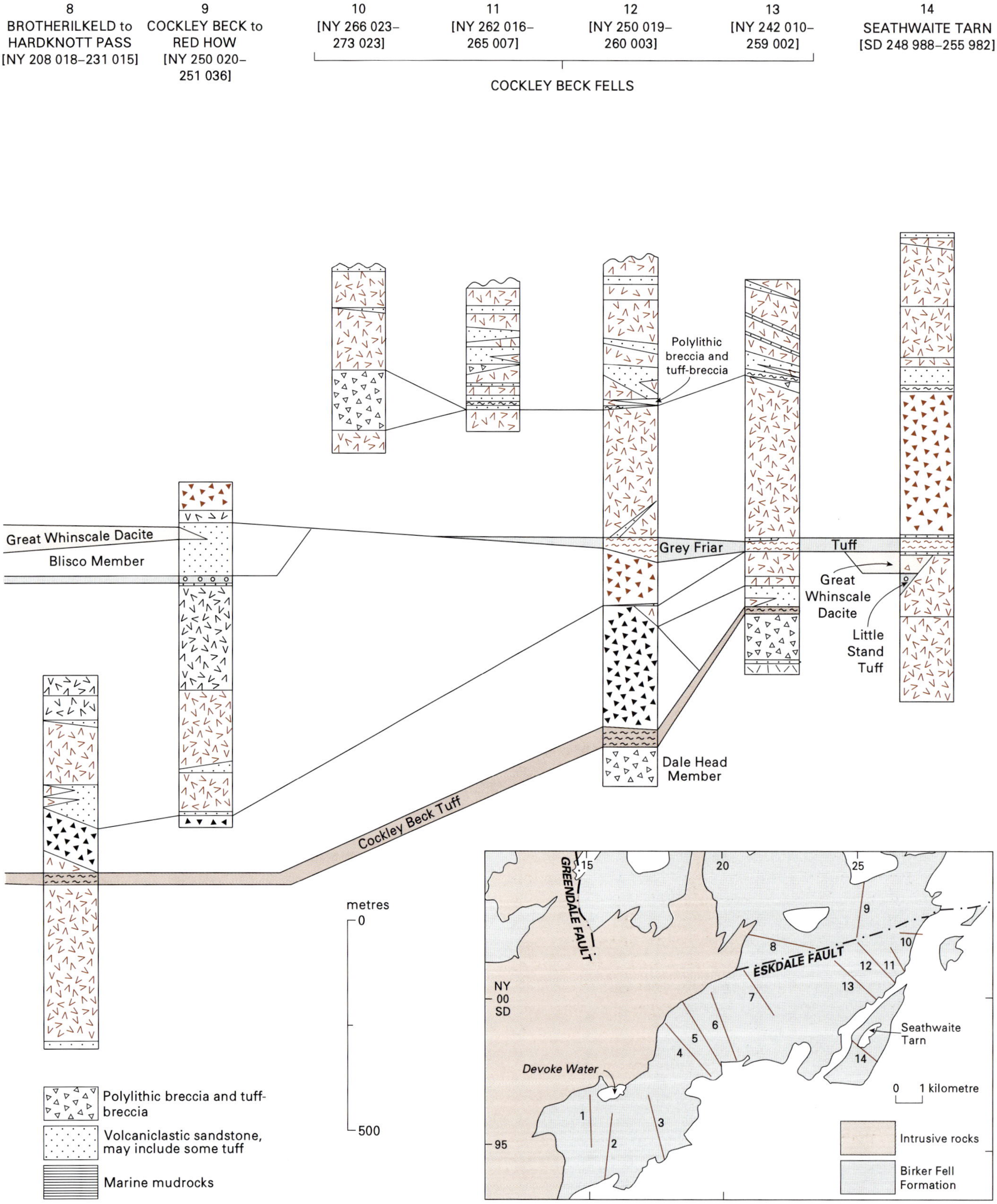

Figure 30 *continued.*

porphyritic (Plate 5c), comprising plagioclase and subordinate alkali feldspar, opaque minerals and sparse, thin, platy chlorite pseudomorphs, possibly after biotite. Phenocrysts are commonly 3 to 5 mm across, though larger crystals occur locally, and constitute up to 30 per cent of the rock by volume. The weathered surface is characteristically rough because the feldspar phenocrysts are more resistant than the fine-grained groundmass. The leucocratic groundmass consists of fine- to medium-grained anhedral feldspar and quartz devitrification mosaic. Epidote commonly replaces feldspar.

An underlying ignimbrite and a thin cover of air-fall tuff are considered to be co-eruptive with the Seatallan Dacite. Beneath the lava on the north side of Seatallan [NY 138 085] for a strike length of about 1 km, is a 30 to 50 m-thick, eutaxitic, dacitic ignimbrite. Overlying the Seatallan Dacite on Scoat Fell [NY 153 106], just to the north of the district, is 0.5 to 1 m of thinly bedded to laminated fine-grained dacitic air-fall tuff.

Craghouse Member

The Craghouse Member comprises welded andesitic and dacitic ignimbrite, probably erupted during the earliest caldera-forming activity recorded in the Borrowdale Volcanic Group. Parts of the member crop out in the north-west of the district, south-west of Nether Wasdale [NY 113 027] its type locality, around Scoat Tarn [NY 158 104], at Wasdale Head [NY 180 062] and east of the Eskdale Granite from Great How to Heron Crag [NY 195 045 to 216 030]; correlatives occur to the north and west in the adjacent Keswick and Gosforth districts. The thickness varies abruptly from 30 m east of the Dorehead Fault to 800 m north of Santon Bridge; data from west of the district prove a thickness in excess of 1100 m. These variations are considered to have been fault controlled. The Craghouse Member rests on undifferentiated andesite at Wasdale Head [NY 180 062] and Great How [NY 198 039]. North of Santon Bridge, at the type locality, ignimbrite rests on a thick succession that includes volcaniclastic rocks and andesite. At Scoat Tarn [NY 158 104] an intensely brecciated and amygdaloidal basaltic andesite sill has been intruded between the Seatallan Dacite and the ignimbrite sequence.

A number of characteristics are associated with individual localities. West of the Ennerdale intrusion [NY 113 027] the Craghouse Member comprises a single ignimbrite cooling unit, 800 m thick, of eutaxitic lapilli-tuff with a pale pink-grey weathered crust. There is a well-developed planar fabric but, for example, north of Santon Bridge, the welding fabric is locally deformed by tight folds. Chloritised fiamme typically measure 40 × 5 mm, but in places range up to 150 × 10 mm. The tuff is crystal rich and contains sparse andesitic and felsitic lithic clasts. Locally, at the base are 1 to 2 m of thinly bedded air-fall, pumice lapilli-tuff.

At Scoat Tarn [NY 158 104] a number of distinct units within the ignimbrite can be mapped approximately 1 km north–south along strike. The lowest unit, 100 m thick, comprises homogeneous, massive, white to pale pink crystal-rich tuff with abundant broken and euhedral feldspars, sporadic rounded fiamme and sparse lapilli-grade andesite lithic clasts. The tuff passes upwards into eutaxitic lapilli-tuff, 150 m thick, and composed of at least three flow units, each forming a prominent topographic feature. The flow units show inverse coarse-tail grading of fiamme so that in general, a lower crystal-rich, relatively fiamme-poor zone becomes fiamme rich in higher parts with fiamme typically in the size range 30 × 13 mm to 150 × 10 mm. Brecciated and amygdaloidal basaltic andesite and andesite sills have been intruded into the tuff sequence.

At Wasdale Head [NY 180 062] and around Great How [NY 198 039] most of the member comprises massive, homogeneous, eutaxitic, crystal-rich lapilli-tuff ranging in thickness from 30 to 100 m. On Great How ignimbrite overlies a 2 to 3 m-thick bed of laminated to thinly bedded sandstone with a sharp, irregular contact. A 15 cm-thick, crudely layered, basal zone to the ignimbrite grades up into 0.5 m of eutaxitic lapilli-tuff and massive lapilli-tuff. The ignimbrite is overlain by a very crystal-rich, 1 m-thick tuff with punch-and-swell bedding structures and interpreted possibly as a co-ignimbrite ash-cloud surge deposit. At Wasdale Head, the massive Craghouse Member has a relict streaky but recrystallised vitroclastic matrix and overlies a 50 cm-thick bed of volcaniclastic sandstone.

The Craghouse Member rocks are petrographically variable. The crystal-rich tuff contains 30 to 40 per cent plagioclase, some opaque minerals and subordinate chlorite pseudomorphs after mafic phases. Crystals are rarely euhedral, and most are rounded to subangular fragments. Elongate crystals are preferentially orientated subparallel to the welding fabric. The elongate and ragged, chloritised fiamme within the coarser rocks are commonly plastically deformed around rigid crystals and lithic clasts. Lithic clasts, mainly rounded to subrounded fragments of andesite, dacite, and parataxitic dacitic tuff, are also present. The groundmass is a very fine-grained mosaic after devitrified glass, commonly displaying a streaky texture that is deflected around the rigid crystals and lithic clasts (Plate 5e). The matrix also locally contains very fine-grained, broken feldspar and opaque crystals.

Wrighthow Member

The Wrighthow Member crops out in a small fault-bounded block on the western margin of the district, 1 km west of Nether Wasdale [NY 108 037]. The type area and most of its limited outcrop is west of the district, where it probably overlies the Craghouse Member. The type area is Wrighthow Crags, 500 m north of Hollins Farm [NY 1070 0370]. The succession of basalt lavas comprises up to 200 m of scoriaceous clinopyroxene and clinopyroxene-plagioclase-phyric basalt. Lenses of massive basalt are interdigitated with much scoriaceous, clinkery breccia forming an assemblage characteristic of a compound aa lava field. In the porphyritic basalt clinopyroxene, replaced by actinolite and/or chlorite, occurs as subhedral to euhedral phenocrysts or glomeroporphyritic clusters up to 8 mm across. Plagioclase occurs as microphenocrysts in some rocks and is altered to sericite and epidote. The melanocratic groundmass is generally

very fine grained and largely replaced by a fine mat of actinolitic amphibole microlites.

Throstle Garth Member

A sequence, up to 310 m thick, of grey-green aphyric to locally sparsely olivine-phyric basalt lavas crops out in upper Eskdale, almost at the top of the Birker Fell Formation (Figure 29). They are interpreted mainly as compound lavas and fill a palaeotopographical depression west of Scafell to Lingcove Beck, Adam-a-Cove and Hard Knott [NY 191 068 to 243 043 to 232 019]. The original steep sides to this feature are preserved in the west on High Scarth Crags [NY 215 043] and in the east on Adam-a-Cove. Individual compound lava fields are up to 240 m thick. Thin, simple lavas of pyroxene-phyric basalt are restricted to the top of the sequence west and south of Scafell [NY 198 068 and 210 049]. The basalt lavas overlie breccia and eutaxitic lapilli-tuff in the north-west and andesite elsewhere. They were first recorded on the Geological Survey old series six-inch sheets as 'slaggy grey greenish lava, vesicular green lava and slaggy breccia'. Oliver (1961) referred to them as the Throstle Garth Aphanitic Andesites and Branney (1988b), Branney and Soper (1988) and Branney and Kokelaar (1994) included them in their Lingcove Formation. The type area is Throstle Garth [NY 226 040].

The few metres of open-network breccia at the top of lavas is commonly filled with laminated, fine-grained sandstone and siltstone. Lavas are separated by up to 10 m of thinly bedded and laminated medium- to coarse-grained basaltic volcaniclastic rocks. Reworked deposits with channels, cross-bedding, scours and lag gravels are dominant, but locally there are beds of basaltic tuff mantling subaerial erosion surfaces (Plate 4). Accretionary lapilli beds occur locally.

The Throstle Garth Member mostly lies within the aureole of the Eskdale granite and the rocks are generally pervasively altered. However, contact metamorphism is nowhere complete and original textures are preserved. The rocks are aphyric to sparsely porphyritic with a holocrystalline groundmass containing a fine-grained interstitial texture or a trachytic texture with abundant plagioclase microlites, prismatic amphibole (possibly after pyroxene), opaque minerals and secondary sphene. Relict fresh clinopyroxene is preserved in a few samples. Patchy areas of unresolvable interstitial mesostasis are common. Phenocrysts constitute no more than 8 per cent by volume and most abundant are small aggregates, up to 2 mm across, of quartz, chlorite and acicular amphibole as pseudomorphs with the characteristic shape and fracture pattern of olivine (E69390, E69463). Sporadic plagioclase and/or clinopyroxene phenocrysts are also present.

Devoke Water Member

The Devoke Water Member comprises 550 to 620 m of weakly bedded basaltic andesite lapilli-tuff mainly exposed in the type area on Birkby Fell [SD 146 965 to 155 960] south-west of Devoke Water. Thinner, much faulted successions crop out around Stony Tarn [NY 200 025] and north-east of Whincop [SD 179 992]. Small outcrops survive in the roof of the Eskdale granite and cap the hills north-east of Boot on Great Barrow [NY 181 015 and 185 016]. These isolated remnants suggest that in its original extent the lapilli-tuff covered an area with a radius of at least 4.5 km. These strata form the lowest part of the Birker Fell Formation and rest unconformably on the Skiddaw Group at the western end of Birkby Fell at Barnscar [SD 132 959].

Most of the sequence is a poorly to moderately sorted, close-packed, heterolithic lapilli-tuff and tuff. The maximum general grain size is 20 mm but sporadic angular blocks up to 100 mm are present. Most of the clasts are non-amygdaloidal, devitrified or holocrystalline basaltic andesite and are angular to subrounded. The principal clast type is a turbid, very fine-grained, aphyric to slightly pyroxene-phyric, non-scoriaceous basaltic andesite with a weak hyalopilitic texture (E69816; Plate 5d). Locally, a pyroxene-phyric basaltic andesite with a pronounced trachytic groundmass is dominant, and at one locality on Birkby Fell [at SD 1477 9611] a plagioclase-phyric basaltic andesite with a dark, microcrystalline groundmass and relict hyalopilitic texture is an important component (E69800). The pyroxene-phyric clasts have a similar texture and mineralogy to the nearby (and probably related) basaltic andesite sheets intruded within the upper few metres of the Skiddaw Group south-west of Devoke Water. Accretionary lapilli occur sporadically throughout the sequence, but individual beds containing these rarely exceed 3 to 4 cm in thickness.

Crystals comprise a very small proportion of the tuff and most are euhedral amphibole pseudomorphs after clinopyroxene. They are similar in shape and size to those in the dominant lithic clasts and a few have a thin layer of fine-grained basaltic andesite attached to part of the crystal margin. Locally, fresh plagioclase occurs and is of a similar composition to phenocrysts within the lapilli. A few garnets have been recorded from Birkby Fell (E69797) occurring as free crystals and phenocrysts in the clasts. The reddish, almandine garnets are intensely fractured, with secondary development of biotite, amphibole and quartz. Though these rocks lie within the contact metamorphic aureole of the Eskdale granite, recrystallisation is rarely complete and even at the granite contact, relict clastic textures are preserved within the lapilli-tuff. Metamorphic minerals are coarsely developed in the original pore spaces of the lapilli-tuff.

Subspherical to tabular silty mudstone, siltstone and fine-grained sandstone clasts are an significant constituent. The clasts generally comprise less than 10 per cent of the lapilli-tuff (e.g. E69797), but in some exposures [e.g. SD 1474 9612] laminae are composed almost entirely of platy, locally imbricated mudstone and silty mudstone clasts up to 5 mm (E69833). These sedimentary clasts were probably derived from the Skiddaw Group basement.

A weak parallel-bedding fabric is commonplace, though some units are internally massive. Weakly bedded parts of other units may pass gradationally into a well-defined thinly bedded upper part with low-angle cross-stratification, sporadically cut by small channels. In some areas the changing orientation of very large-scale cross-

bedding in relation to the structural dips gives an appearance of chaotic bedding. Some internally cross-stratified bed sets are 60 to 70 m thick.

The Devoke Water Member is interpreted to have been deposited by the proximal reworking of contemporaneous vents. Weak bed-forms, accretionary lapilli, absence of a fine-grained matrix and the presence of a dominant clast lithology indicate a pyroclastic origin. The angular, non-vesicular clasts and common inclusion of probable Skiddaw Group lithic clasts further suggest that the main process was hydrovolcanic, either involving a surface body of water or an aquifer. The distribution of the rocks over a 4.5 km radius, together with the large-scale cross-bedding and localised reworking suggests that they form part of a proximal, fan deposit produced by the local erosion of a tuff-ring field. Parts of original cones may be present in places. The low preservation potential of the original structures may explain the poor lateral stratigraphical continuity of the Devoke Water Member.

Birkby Fell Member

A sequence of thin basic aa lavas interbedded with volcaniclastic rocks overlies the Devoke Water Member in two outcrops: the larger, designated as the type area, to the south-west of Devoke Water on Woodend Height, White Pike and The Knott [SD 143 952], and the smaller about 5 km farther north-east around Whincop [SD 180 990]. The Whincop outcrop laterally interdigitates with pale coloured, blocky andesite lavas farther east. The member is 430 to 480 m thick south-west of Devoke Water and up to 350 m at Whincop. It comprises at least 23 sheets of dark grey to greenish grey, clinkery, autobrecciated, porphyritic, basaltic andesite and subordinate andesite, most of which are interpreted as simple aa lavas, but possibly a small number are sills. Most of the sheets are lenticular and less than 50 m thick, but some range up to 100 m. South-west of Devoke Water the lowest lava is a distinctively flow-banded andesite with a clinkery autobrecciated top. North of White Pike and the Knott the member rests, apparently conformably, on a reworked, well-bedded, fluvial sandstone in the uppermost part of the Devoke Water Member. Across a substantial fault, on Woodend Height [SD 156 956] the lavas overlie acid andesite.

Interbedded volcaniclastic rocks locally make up a significant proportion of the sequence. Units up to 50 m thick are present in the lower part of the member, but in the upper part only a few metres are generally preserved and no volcaniclastic sedimentary interbeds are present between some lavas. Thinly parallel-bedded to laminated, medium- and coarse-grained sandstone is predominant, grading up to a thin capping of siltstone at the top of some beds. Channelling and cross-bedding occur in places. The lowest beds in each volcaniclastic unit commonly fill irregularities in the top surfaces of the underlying lava and in some cases mantling is also seen.

Groups of three or four successive lavas generally have very similar petrographic characteristics. Primary igneous textures are mostly well preserved (Plate 5a) though there is variable overprinting by contact metamorphism near the Eskdale intrusion, where acicular amphibole is most conspicuous, commonly occurring in apparently random orientations throughout the rock and in some samples (e.g. E70015) obscuring the original groundmass texture; biotite is very common in rocks from the Whincop outcrop. The primary mineralogy comprised plagioclase, pyroxene and a fine, granular opaque mineral all of which are variably preserved. Plagioclase in some rocks is relatively fresh, with phenocrysts typically displaying multiple zoning and internal resorption textures (e.g. E70011), whereas in other specimens a cloudy, sericitic pseudomorph is all that remains. Fresh clinopyroxene is sporadically preserved (e.g. E70011), but in many thin sections relict twinning typical of clinopyroxene is seen faintly within the amphibole pseudomorphs.

The Birkby Fell lavas are porphyritic throughout, containing 7 to 35 per cent by volume phenocrysts. In most examples, plagioclase phenocrysts are equal to, or only slightly more abundant than, those of pyroxene. The exceptions are the uppermost four flows where pyroxene comprises less than 3 per cent by volume compared with 12 to 30 per cent plagioclase. One basalt (E70016) contains only very small amounts of phenocrystic feldspar, with 12 per cent fresh euhedral clinopyroxene and 14 per cent amphibole pseudomorphs after another mafic phase, possibly olivine. It is geochemically distinct from subjacent and suprajacent lavas (unit 17 in Figures 15, 17; Table 7).

Groundmass textures are variable but most of the rocks are very fine to fine grained and generally holocrystalline; the amount of devitrified glass present is never substantial. Trachytic, seriate, intergranular and intersertal textures are common, the last two occurring together in many examples.

Dale Head Member

Volcaniclastic breccia and sandstone comprising the Dale Head Member crop out from west of Black Hall to Cockley Beck and Hinning House Fell [NY 230 010 to 253 007] in the Duddon valley and on the north-west slopes of Border End, Hard Knott [NY 225 022]. The type area is Dale Head [NY 241 006] and Rowantree How [NY 249 006]. The member rests on andesite and tuff within the Birker Fell Formation. There is little angular discordance at the base except west of Black Hall where the breccias abut a steep palaeoslope. A maximum thickness of 220 m occurs east of the river Duddon, between Dale Head and Rowantree How [NY 2493 0061] but the member thins abruptly, or is faulted out, along strike to the north-east, west and south-west (Figure 30).

The base of the member is exposed only in the Dale Head Close area [NY 245 007], south of the Cockley Beck Fault, where a basal unit comprises 3 m of eutaxitic lapilli-tuff. West of the River Duddon, about 500 m west of Black Hall [NY 2340 0110], a similar lithology of unknown thickness fills a steep-sided cleft in the underlying andesites. Most of the member comprises massive sheets of breccia, locally separated by interbeds of volcaniclastic sandstone. The breccia is mainly clast supported with little discernible matrix. The angular to subangular clasts

of varieties of andesite, along with rare pink rhyolite, are generally less than 10 cm across, but exceed 1 m in places. There is no apparent pyroclastic component in the breccia.

Sandstone interbeds increase in thickness and abundance upwards through the member. They are mainly parallel laminated with bedding well defined by the grain-size distribution. Scours and small channels are commonly filled with cross-bedded and graded sandstone and load casts and flame structures occur where breccia overlies volcaniclastic sandstone. Dish and pillar structures occur in some of the laminated beds. There are many local facies variations in the member. For example, on Rowantree How, east of Dale Head, it comprises a single graded breccia–sandstone unit up to about 30 m thick. Around Little Garter How [NY 249 015] east of Cockley Beck, there are two units of breccia, separated by a 1 m-thick bed of laminated volcaniclastic sandstone which, in places, infills irregularities in the upper surface of the lower breccia unit. North of Little Garter How, the sandstone is missing and the upper breccia overlies an irregular erosional surface on the lower one.

Dykes produced by dewatering of the sediments occur at High Crag [NY 2464 0090 and 2466 0079] about 300 m south-west of The Brow. One of the dykes, about 1 m wide and trending 050°, cross-cuts crudely stratified breccia and contains clasts of laminated volcaniclastic sandstone with dish and pillar structure. The dyke is truncated by an erosion surface at the base of the overlying bedded sandstone unit; these relationships indicate dewatering and erosion before deposition of the overlying sandstone. Another dyke, up to 1.4 m wide and at least 20 m in lateral extent, contains blocks of breccia.

The Dale Head Member was interpreted previously as vent agglomerate (Mitchell, 1963). However, the lithofacies, contact relationships and changes in geometry suggest deposition by sedimentary processes within a palaeovalley. There is no pyroclastic component and the breccia units are reinterpreted as mass-flow deposits. The high concentration of clasts and small amounts of matrix suggest that the transporting medium was water which effectively winnowed-out any original fine-grained component. The sandstone interbeds, some of which form graded tops to the mass-flow deposits, indicate much reduced flow regimes.

Cockley Beck Tuff

The Cockley Beck Tuff is a garnet-bearing, crystal-rich, dacitic lapilli-tuff, generally with a well-developed eutaxitic texture (Plate 5f). It crops out in the central part of the district, extending for some 7 km from just west of Crook Crag [SD 191 986], to the western slopes of Hard Knott [NY 223 023], and Cockley Beck [NY 250 018]. It was identified originally in the Cockley Beck area by Mitchell (1963), but was not formally named. The type area is defined between Cockley Beck and Dale Head Close [NY 247 004], where the tuff overlies the Dale Head Member; elsewhere it overlies basaltic andesite or andesite sheets between 400 and 800 m above the base of the Birker Fell Formation (Figure 30). In the Hard Knott and Cockley Beck areas, the tuff is overlain by a probably co-genetic garnetiferous dacite lava; in other areas it is variously overlain by volcaniclastic rocks, andesite or basaltic andesite sheets. At Cockley Beck, the tuff is a compound ignimbrite comprising massive, eutaxitic lapilli-tuff, interbedded with thin units of laminated and cross-laminated, garnet-bearing coarse tuff, interpreted as a pyroclastic surge deposits. The tops of some tuff units are locally channelled and reworked into gravel-grade lag deposits, some containing rip-up clasts of fluvially reworked tuff. The thickness of the member varies from less than 5 m in the south-western part of its outcrop to 25 m in the type area around Cockley Beck; the thicker parts of the sequence are also the coarsest grained. This thickness and grain-size pattern indicates that the source eruptions probably occurred in the vicinity of Cockley Beck.

At Cockley Beck, fiamme are abundant, up to 40 cm long, and normally coarse-tail graded. Systematic fining south-westwards to a maximum fiamme size of 15 mm is accompanied by a decrease in their abundance. Subrounded volcanic lithic clasts, up to 2 mm in size, occur in places. Crystals form up to 30 per cent by volume, and include plagioclase (up to 16 per cent), garnet (8 per cent), mafic phases (up to 5 per cent), and a little biotite and an opaque mineral. Euhedral, subhedral and broken crystals of plagioclase are variably sericitised, with zoning preserved in the less altered examples. Prismatic pseudomorphs after pyroxene, and opaque-riddled, chloritic, ragged platy pseudomorphs after phenocrystic biotite occur sporadically. Garnets are mostly subhedral, up to 2 mm in diameter, and commonly altered to fine-grained mats of pale green-brown biotite with granular quartz along irregular cracks (E69633). Fresher garnet crystals contain inclusions of opaques, zircon and apatite; more rarely the garnets poikilitically enclose early formed, small, corroded, plagioclase crystals. Some small, highly corroded, garnet fragments occur as cores to plagioclase glomerocrysts. In most specimens the matrix is a pale coloured, turbid, fine-grained felsic devitrification mosaic. A vitroclastic texture has been recorded in the matrix locally, comprising aligned, flattened and welded shards (E68535).

Unnamed ignimbrite markers

West of Hard Knott [NY 2237 0162], in Gaitscale Gill [NY 2573 0222], north-east of Kepple Crag [NY 2050 0008] and east of Birkerthwaite [SD 186 985 and SD 192 984] pockets of a thin, impersistent, pale weathered, massive, lithic-rich, eutaxitic lapilli-tuff, extensively recrystallised with much fine-grained amphibole and opaque oxide, lie between the Cockley Beck Tuff and the Little Stand Tuff (Figure 30). Remnants are preserved in hollows within an otherwise uniform andesite sequence, but it is not certain that these outcrops are of a single, correlated unit. Elongate fiamme are common. Lithic clasts are abundant, up to 2 mm, and mostly of silty mudstone and feldspathic sandstone probably derived from the underlying Skiddaw Group; scattered angular quartz grains are also present.

Little Stand Tuff

The distinctively white-weathered, nodular, crystal-rich rhyodacitic Little Stand Tuff crops out for about 16 km from Yoadcastle [SD 155 950] in the south-west of the district, north-eastwards to Pike of Blisco [NY 274 045]. It is a very significant stratigraphical marker within the andesite sequence, occurring about 1200 m above the base of the formation (Figure 30). Firman (1957) included the outcrop of this unit between Yoadcastle and Harter Fell [SD 220 998] as part of his 'Great Whinscale Rhyolite' (see Great Whinscale Dacite). The type area is from Little Stand to Pike of Blisco.

The Little Stand Tuff generally overlies andesite sheets but on Little Stand and Pike of Blisco it is separated from them by up to 4 m of impersistent, laminated volcaniclastic siltstone, cross-stratified sandstone and a subtly bedded lapilli-tuff. It is overlain on Yoadcastle by the Great Whinscale Dacite, but north-eastwards these are separated by an increasing thickness of volcaniclastic rocks. On Pike of Blisco, the Little Stand Tuff is overlain by bedded sandstone comprising material reworked from the tuff. The top and base of the Little Stand Tuff are typically sharp and regular with only minor undulations, but in the north-east of the outcrop the top surface is hummocky and more irregular. The thickness of the Little Stand Tuff throughout the district is mostly between 7 and 20 m ranging up to 27 m in the type area. Localised and abrupt increases in thickness occur south-east of Woodend [SD 169 960] where 35 m are preserved, and north of Smallstone Beck [SD 195 982] where 60 m are present, probably as a result of ponding in substrate undulations.

Plate 9 Nodules (lithophysae) in the Little Stand Tuff, Harter Fell, 350 m north-west of summit [SD 2150 9984] (D 4029). Hammer head is 17 cm long.

The lowest 1 to 1.5 m are very fine grained, flinty and homogeneous but in places there is diffuse bedding with low-angle cross-laminae and discontinuous lenticles rich in crystals, indicating deposition from an initially turbulent pyroclastic flow. In the main body of the tuff, a primary welding foliation is seen locally with scattered fiamme and a weak remnant eutaxitic fabric. In the type locality, the massive central part passes up into a crystal-rich, weakly stratified facies.

A distinctive feature is the abundance, in diffuse zones, of subspherical siliceous lithophysae up to about 6 cm in diameter (Plate 9). Their distribution is variable; south of Yoadcastle and south-east of Woodend there are high concentrations in the top 2 to 4 m but few lithophysae occur below this level. On Harter Fell, Little Stand and Pike of Blisco lithophysae occur in the central part of the tuff and are particularly abundant at the base and top of this zone; on Harter Fell the nodules at the base of the central zone are flattened parallel to bedding. Where the Little Stand Tuff thins out against andesite topography on Little Stand [NY 2460 0336] and at Black Wars [NY 2710 0448] on Pike of Blisco lithophysae are absent. Where the lithosphysae are most abundant their spacing is no greater than 10 cm and their margins are sharp and regular. The outer part of each lithophysa is texturally similar to the matrix; the fine foliation within the lithophysae is sub-concentric with its margin. The central, drusy quartz-filled cavity is variably arcuate to cigar-shaped and in some cases cuts the margin of the lithophysa and the adjacent rock matrix.

The tuff is generally crystal-rich, containing up to 50 per cent euhedral to anhedral crystals, the latter mainly as angular fragments. The crystal component is almost entirely feldspar, much of which is plagioclase, commonly much altered to sericite and fine, fibrous aggregates of amphibole. Alkali-feldspar crystals, with a typically turbid brownish staining, are up to 3 mm; some crystals appear to be weakly perthitic and may enclose small anhedral quartz grains (E69722). There are some rare chlorite pseudomorphs after a mafic mineral and accessory minerals include an opaque phase and apatite. Granophyric and trachytic lithic clasts up to 4 mm in size occur sporadically throughout. The matrix is a fine-grained, devitrified felsic mosaic, preserving traces of a primary foliation and faint relics of a smeared-out vitroclastic fabric.

Great Whinscale Dacite

The Great Whinscale Dacite is a remarkably uniform, aphyric dacite lava within the central part of the Birker Fell Formation and is exposed over a distance of about 13 km, from west of Yoadcastle [SD 155 950] in the south-west of the district to Little Stand [NY 248 032]; fault-

bounded remnants are also present at Seathwaite Tarn [SD 250 987] and at Long Crag in the Duddon valley [SD 227 977]. The dacite was described by Firman (1957) as the Great Whinscale Rhyolite, and by Kanaris-Sotiriou et al. (1991). The type locality is Great Whinscale [SD 196 989], on Crook Crag.

In the south-west of the district, the Great Whinscale Dacite directly overlies the Little Stand Tuff, but farther north-east these two members are separated by up to 10 m of volcaniclastic rocks (Figure 30). Generally the dacite is overlain by andesite, but south of Wood End [SD 169 958] it is overlain by breccias, on Little Stand by a variable sequence of volcaniclastic rocks and at Seathwaite Tarn by the Grey Friar Tuff described below. A consistent thickness of 75 to 85 m is maintained throughout most of the outcrop ranging down locally to about 25 m. The outcrop on Little Stand has a dome-shaped cross-section with steep sides mainly of autobreccia. A poorly exposed steep-sided termination to the outcrop is also mapped on the north-west shores of Seathwaite Tarn a few metres north of Tarn Beck [SD 250 988]; this may be an original margin to the lava or a erosion surface.

The Great Whinscale Dacite is homogeneous, medium to dark grey, splintery, white-weathered, and aphyric with a well developed flow-foliation parallel to the underlying strata. The base is generally planar and the lowest 3 m comprise autobreccia. Locally the underlying bedded volcaniclastic rocks are truncated, with flames of sediment within the autobreccia. The upper surface of the lava is regular with only small-scale undulations and 15 to 25 m of autobreccia are commonly present in the thickest development at Crook Crag and Maiden Castle [SD 222 993] on the south-east side of Harter Fell. In both autobreccia units the clasts commonly range up to 0.3 m diameter, and are more rarely up to 1 m across. They are angular to subangular, close-packed, have clear internal foliation and in some localities there is visible perlitic cracking indicating that the dacite originally was an obsidian lava. The lower autobreccia contains many clasts with contorted flow foliation. Minor internal sediment is only present in the upper autobreccia.

Above the basal autobreccia some 4 to 6 m of the dacite have a strong platy flow-foliation that is widely folded on a small scale, but overall lies parallel to the base of the dacite and to bedding in the underlying strata. The central part of the dacite is dominated by discontinuous joints with 1 to 4 cm spacings that are co-planar with the foliation beneath; there is generally a narrow transition zone in which continuous foliation, and discontinuous fine jointing are both present. The principal exceptions to the simple internal structure described above occur on Great Whinscale where most of the central part of the dacite is foliated and flow jointed with large-scale folding probably related to ramping during flow. Vesiculation has been recorded in the upper part of the dacite at two localities on Harter Fell [SD 217 997 and SD 222 989] where the uppermost 6 m contain a variable concentration of chlorite and quartz amygdales in what must have been a highly pumiceous part of the lava. The amygdale-rich layers alternate with foliated dacite and merge upwards into autobreccia.

The Great Whinscale Dacite is aphyric with perlitic cracks indicating an originally glassy groundmass (Kanaris-Sotiriou et al., 1991, fig. 5). Locally, lath-shaped plagioclase microlites (around 0.05 to 1.00 mm × 0.01 mm) are aligned in a trachytic texture which becomes pilotaxitic in the central parts of the lava; there the plagioclase laths tend to be larger and in places dominate the matrix. Compositional zoning, expressed mainly in the proportion of chlorite present, occurs in some marginal parts of the lava. Rare microphenocrysts of plagioclase occur and these may be included in glomerophenocrystic aggregates with epidote pseudomorphs after mafic microphenocrysts (probably orthopyroxene) (E69657). No evidence of vitroclastic texture has been found in any of the samples examined. The groundmass, which forms the bulk of the rock, now consists of fine-grained felsitic material (E69654, E69950). Detailed X-ray diffraction analyses of samples from Harter Fell establish an overall mineralogy that is in remarkably close agreement with the norm for these rocks, the main minerals present being orthoclase, sodic plagioclase, quartz and chlorite (Kanaris-Sotiriou et al., 1991). The hollow form of the plagioclase laths in the groundmass, with the centres filled by fine-grained chloritic material, is strongly indicative of rapid crystallisation.

Grey Friar Tuff

The Grey Friar Tuff is a poorly sorted, eutaxitic, dacitic, very coarse lapilli-tuff and tuff-breccia, characterised by its content of lapilli and small lithic blocks. It is up to 75 m thick and crops out from the Seathwaite Fells [SD 240 995] to the Cockley Beck Fells [NY 260 014]. The type area lies around Grey Friar and Seathwaite Tarn [NY 250 000 to SD 240 980]. At Seathwaite Tarn, it directly overlies the Great Whinscale Dacite and thence becomes a key stratigraphical marker within the Birker Fell Formation on Grey Friar where the Little Stand Tuff and Great Whinscale Dacite are absent (Figure 30). The base of the tuff is irregular and on Cockley Beck Fell, alongside Troughton Gill [NY 2580 0156], the tuff fills an irregular palaeotopography. It was previously mapped as a 'streaky rhyolitic tuff' by Mitchell (1963).

The clast population is heterogeneous, dominated by much Skiddaw Group mudstone, pink, fine-grained rhyolite and various types of andesite. Fiamme, 5 to 300 mm long, comprise 25 to 40 per cent of the rock with the larger ones having a ragged, fibrous texture (Petterson et al., 1992, figure 3c). The matrix comprises a medium- to coarse-grained tuff that includes scattered plagioclase crystals. Plastic deformation of fiamme around the more rigid lithic clasts suggests that the deposit is welded but shard textures have not been observed in the matrix to confirm this.

Blisco Member

Pyroclastic and sedimentary facies comprise the Blisco Member, a distinctive assemblage up to 120 m thick that crops out between Little Stand, Cold Pike and the Pike of Blisco [NY 247 034 to 282 047]. The base of the Blisco

Member (British Geological Survey, 1996) is defined at the base of a prominent sandstone, informally termed the Long Scar sandstone that mostly overlies the Little Stand Tuff, though at the northern end of Little Stand the sandstone rests on, and is sharply truncated against, an andesite body. The Blisco Member is overlain by andesite sheets, one of which has been interpreted as an intrusion by Branney and Suthren (1988). The top surfaces of the remaining sheets are not exposed in this area and so it cannot be determined whether they are intrusive or extrusive. If they are intrusive then the top of the Blisco Member is not exposed; if they are extrusive, the interpretation preferred here, they belong to the overlying parts of the Birker Fell Formation. Branney and Soper (1988) included the volcaniclastic rocks, together with the Little Stand Tuff and the Great Whinscale Dacite, within their Blisco Formation, which represents an important explosive volcanic phase within the dominantly effusive Birker Fell Formation.

The Long Scar sandstone (not separately identified on Sheet 38 Ambleside), is a well-stratified, fine- to very coarse-grained sandstone made up of mainly andesitic grains. It contains abundant siltstone intraclasts, erosion surfaces, cross-stratification and ripple cross-lamination, numerous small sandstone- and conglomerate-filled channels, and is intensely epidotised. Parallel and ripple-laminated, normally graded siltstone and fine-grained sandstone intervals up to 12 cm thick, with erosional bases, are interbedded with thicker lenses of trough cross-bedded coarse- and very coarse-grained sandstone and pebbly sandstone. Accretionary lapilli have not been observed. Semi-brittle soft-sediment deformation structures are common. The Long Scar sandstone varies in thickness from 10.5 m on Pike of Blisco [NY 2747 0363] to 20.9 m on Red How [NY 2530 0295] and is interpreted as recording ephemeral sheet-flood sedimentation and deposition from shallow-water turbidity currents (Branney, 1988b).

Five main lithofacies have been recognised above the Long Scar sandstone (Branney, 1988b). They have variable relationships and extent, but most can be traced for some distance on Cold Pike and Red How. The first comprises two units massive eutaxitic, andesitic lapilli-tuff, ponded against near-vertical palaeocliffs [NY 2693 0431 and 2719 0427]. These rocks are very poorly sorted with inverse coarse-tail grading of fiamme and thin inverse-graded basal layers. The lower one is 21.5 m thick and the higher one is 6 m thick. They are interpreted as ignimbrite (Branney, 1988b). A second lithofacies consists of massive to thinly bedded, poorly sorted, coarse breccia and conglomerate that may be ungraded, normal or reverse graded, or coarse-tail graded. Many of these beds have a bimodal grain-size distribution with large blocks in a poorly sorted sand, silt or mud matrix in which clast and matrix-supported fabrics occur. This lithofacies was probably deposited from cohesive debris flows and other high-concentration sediment-gravity flows. Thirdly, thinly stratified eutaxitic tuff and lapilli-tuff contain sporadic large, matrix-supported lithic blocks. Units in this facies thicken laterally from 0.1 to 1.5 m in palaeotopographical hollows, for example on Black Crag [NY 2687 0426]. The beds of tuff are locally interstratified with low-angle, cross-laminated fine- to medium-grained sandstone. A fourth lithofacies comprises massive and cross-laminated siltstone similar to some parts of the Long Scar sandstone, and probably of a similar origin. The fifth lithofacies is an impersistent thinly bedded fine accretionary lapilli-tuff unit that occurs on Pike of Blisco and is distinctively white-weathered.

In all of these lithofacies, most lithic clasts are of porphyritic acid andesite, similar to the grains within the andesitic Long Scar sandstone and suggesting a common source. Pumice is chloritic and epidotic; its generally flattened shape probably formed during diagenesis and burial rather than by welding compaction. Many units within the Blisco Member are cut by scours and vertical-sided erosional gullies up to 10 m deep, which suggest a subaerial depositional environment. Pervasive epidotisation is a widespread feature with nodular epidote abundant locally.

SIX
Borrowdale Volcanic Group: volcaniclastic successions

The Borrowdale Volcanic Group above the Birker Fell Formation consists predominantly of volcaniclastic rocks reflecting a fundamental change from eruptions dominated by lava effusion to explosive activity. In this account the lithostratigraphy of the upper part of the Borrowdale Volcanic Group is described in three sections based on eruptive and depositional centres in the Central Fells, the Duddon Basin and in the Rydal area (Table 1, Figure 12). Penecontemporaneous intrusions have been emplaced into these volcanic successions and, because of their probable close genetic relationship with the extrusive rocks, they are described at the end of this Chapter rather than in Chapter 8.

SCAFELL CALDERA SUCCESSION

The eruptive centres in the Central Fells produced dominantly pyroclastic rocks constituting the Whorneyside, Airy's Bridge and Lingmell formations (Figure 31). These rocks crop out principally within the Scafell Syncline in the northern part of the Ambleside district and in the adjacent Keswick district. Parts of the succession extend to the Coniston and Ulpha fells.

Whorneyside Formation

The Whorneyside Formation consists of andesitic tuff and lapilli-tuff forming two distinct lithofacies. The lower one comprises massive, eutaxitic lapilli-tuff and is interpreted as an ignimbrite; the upper one consists of laminated and thinly parallel-bedded, coarse-grained and fine-grained tuff considered to be a phreatoplinian ash-fall deposit (Branney, 1991) (Figure 32). The type area is on the south side of Crinkle Crags where the formation is 120 m thick and is named from Whorneyside Force [NY 261 054] (Branney, 1991). The outcrop extends from the type area, south-westwards to the Coniston Fells and the Duddon valley as far as Stainton Pike [SD 153 940], and northwards to Borrowdale in the adjoining Keswick district. Oliver (1961) included parts of the Whorneyside Formation in the basal division (unit 4d) of his Airy's Bridge Group. The name Whorneyside Ignimbrite was applied to the lower lithofacies by Branney (1991) and Branney and Kokelaar (1994), but designated as the Wet Side Edge Member by the British Geological Survey (1991); in this memoir Wet Side Edge Member is retained. The formation also incorporates parts of Firman's (1957) Worm Crag Group and Waberthwaite Tuffs in the Duddon valley, and part of Mitchell's (1963) Kidson How Tuffs in the Coniston Fells.

The base of the formation is sharp and rests unconformably on the Birker Fell Formation (Branney, 1991; Petterson et al., 1992). The main thickness variations in the formation (Figure 31) record the original dispersal of the tuff, but this is modified by major slumping and minor penecontemporaneous erosion. In the Central Fells the maximum relief on topography beneath the Whorneyside Formation was probably no more than 150 m, as indicated by thickness variations in the Wet Side Edge Member, and the andesite field was probably completely buried by the ignimbrite (Branney, 1991). The thickest development of the Wet Side Edge Member is north-west of the Duddon valley, where more than 400 m are present around Great Worm Crag [SD 193 969] (British Geological Survey, 1991).

The Wet Side Edge Member consists of brown-weathered, massive, poorly sorted, pumiceous andesitic lapilli-tuff. It is locally stratified at the base, where pyroclastic flow deposits up to 1 m thick, are interbedded with pyroclastic surge and ash-fall deposits (Branney, 1988b). The ignimbrite comprises several massive layers, some individually more than 30 m thick (Figure 32), with subtle internal variations in fiamme and lithic-lapilli contents, differences in the length to thickness (aspect) ratio of fiamme, and laterally impersistent stratification. Fiamme, crystals and abundant lithic lapilli are set in a recrystallised matrix in which vitroclastic textures are sporadically preserved. Fiamme form up to 20 per cent of the rock and have aspect ratios between 5:1 and 10:1. In places the largest fiamme define vertical inverse grading, possibly the result of the buoyancy of pumice relative to the lower parts of the flow during emplacement (Sparks, 1976), or because of unsteady flow conditions (Branney and Kokelaar, 1992).

Plagioclase crystals up to 3 mm long generally comprise 12 to 40 per cent of the lapilli-tuff. Epidote, chlorite and amphibole pseudomorphs after mafic phenocrysts (5 per cent) and apatite crystals (less than 1 per cent) are also present. Most of the largest crystals are euhedral and tabular, or glomerocrystic, but many are either abraded or fragmented and aligned subparallel to the eutaxitic foliation. Lithic lapilli are ubiquitous, 2 to 10 mm in size, and on average constitute 5 per cent of the rock. Most are non-vesiculated andesite, possibly derived from conduit walls during eruption. The matrix is recrystallised, consisting of a microcrystalline aggregate of quartz, sericite, chlorite and opaques; some unaltered fine-grained plagioclase crystal detritus occurs throughout. Locally, blocky and cuspate shards up to 2 mm are preserved.

At Whorneyside Force the top of the Wet Side Edge Member passes gradationally upwards, over about 10 m, into the overlying bedded tuff (Figure 32; Branney, 1991). The bedded tuff is about 30 m thick in the type section, but laterally it varies abruptly from 0 to 100 m, largely caused by intraformational slumping and sliding induced

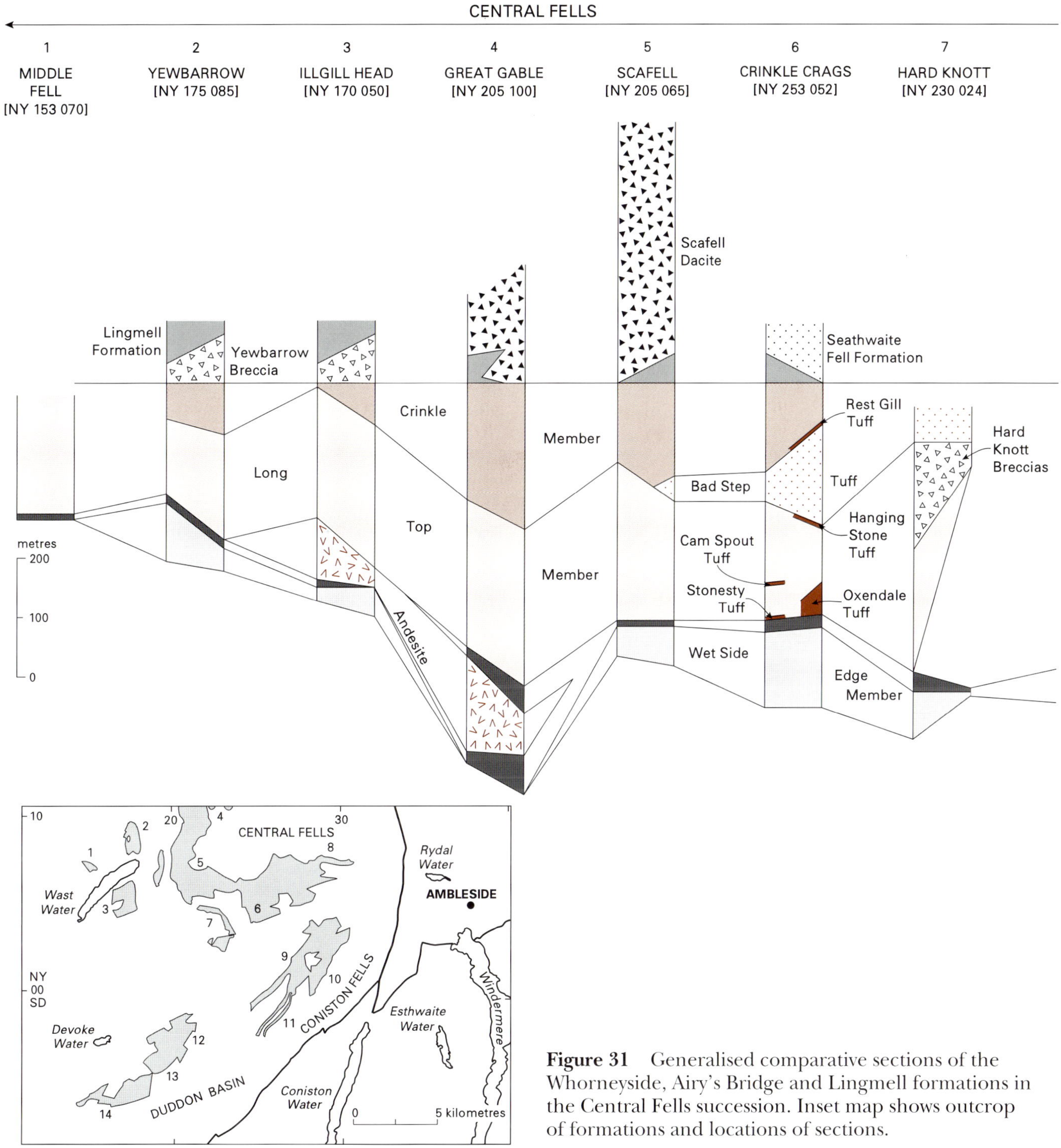

Figure 31 Generalised comparative sections of the Whorneyside, Airy's Bridge and Lingmell formations in the Central Fells succession. Inset map shows outcrop of formations and locations of sections.

by caldera subsidence (Branney and Kokelaar, 1994). Individual beds and laminae are traceable for tens of metres showing no significant variation in either thickness or grain size. Exposures in the Central Fells show that the thickest beds are consistently the coarsest grained and that beds thicker than 10 cm show diffuse internal stratification, caused by subtle variations in grain size and sorting (Branney, 1991). Bed contacts are commonly gradational, and both distribution and coarse-tail vertical grading may be either normal or reverse, though neither type is dominant. There is no cyclicity at any scale and beds are not grouped naturally into sets on grain-size or grading characteristics. This random stratification style has been interpreted to record changes in deposition caused by ash showers resulting from convective instabilities within the base of an umbrella cloud (Branney, 1991).

The upper, laminated tuff facies varies from fine to coarse tuff with lithic and crystal fragments up to 1.5 mm.

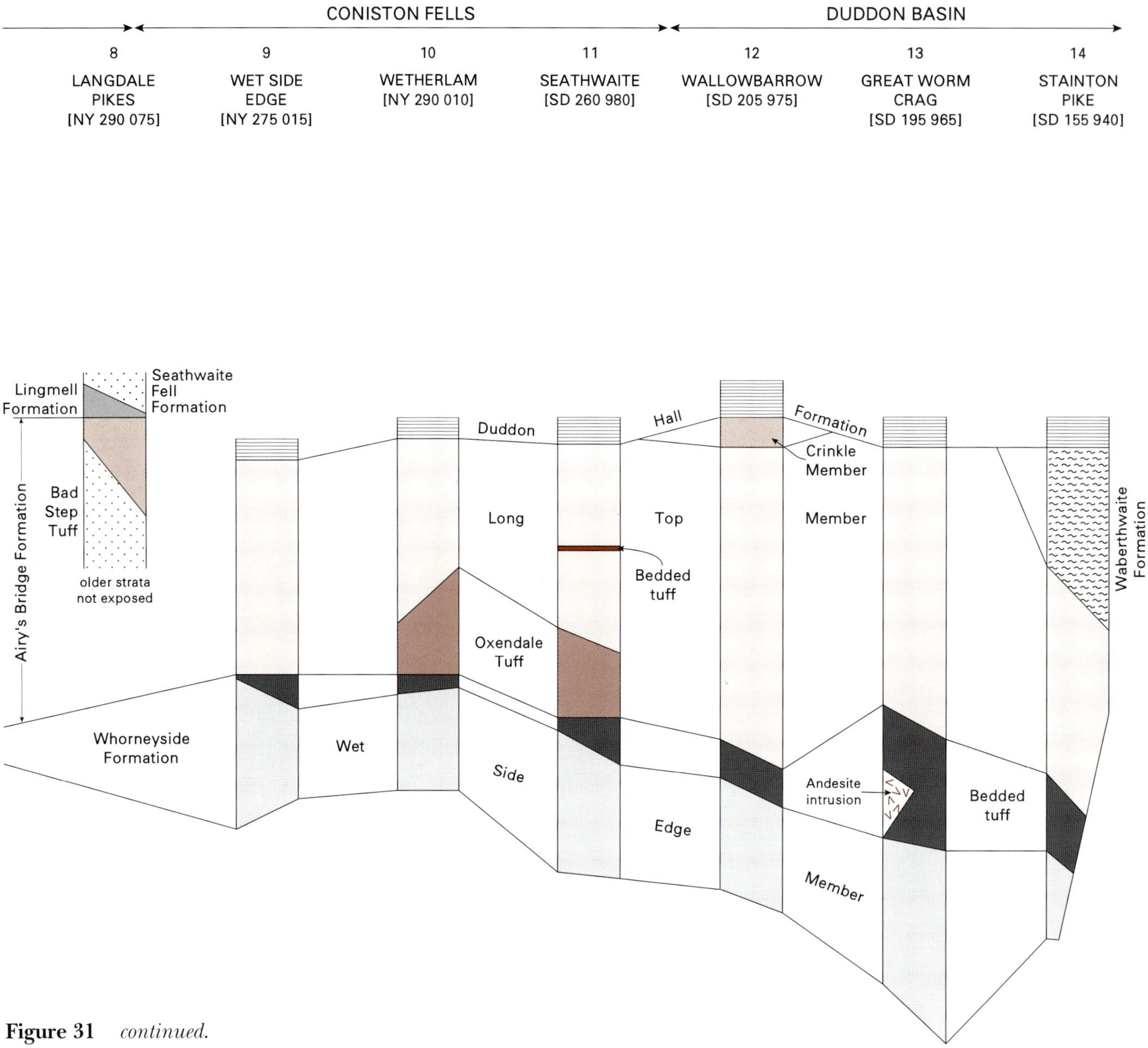

Figure 31 *continued.*

Sorting is extremely variable but generally poor. Crystal abundance varies widely between laminae; it averages about 40 per cent and rises to greater than 50 per cent in crystal-tuff layers. The crystal and lithic clast suites appear to be identical to those in the underlying ignimbrite. Locally, ash pellets and accretionary lapilli, up to 6 mm in diameter, are abundant forming laterally persistent framework-supported beds 2 to 7 cm thick. Most of these lapilli have a single rim and show no evidence of abrasion. Together with block impact sags and unabraded, angular crystal fragments they attest to the pyroclastic origin of the tuff. Most of the lithic clasts are smaller than 1 mm diameter and have angular splintery shapes; many glassy types are variably perlitic.

Intraformational erosion surfaces and syndepositional soft-state deformation structures are present in the bedded tuff producing local bedding discordances. Erosion surfaces are mostly seen towards the top of the tuff where rills up to about 0.5 m across and deep are present. Most of the rills are filled with ash-fall tuff that mantles the erosion surface and any slumped or rotated blocks that are present at the rill margins. Sparse rills up to 1.5 m deep have massive or crudely stratified, poorly sorted fills interpreted to result from sporadic ash slurries. Intraformational bedding discordances resulting from syndepositional sliding and slumping are abundant. Ash-fall tuff drapes the back scars and the deformed strata in the gravity-induced slide sheets and slumps.

The syndepositional deformation of the bedded tuff commonly results in local but marked angular discordances at an otherwise conformable contact with the overlying Airy's Bridge Formation. However, near Sourmilk Gill [NY 230 122] at the head of Borrowdale, in the adjacent Keswick district, and on White How [SD 204

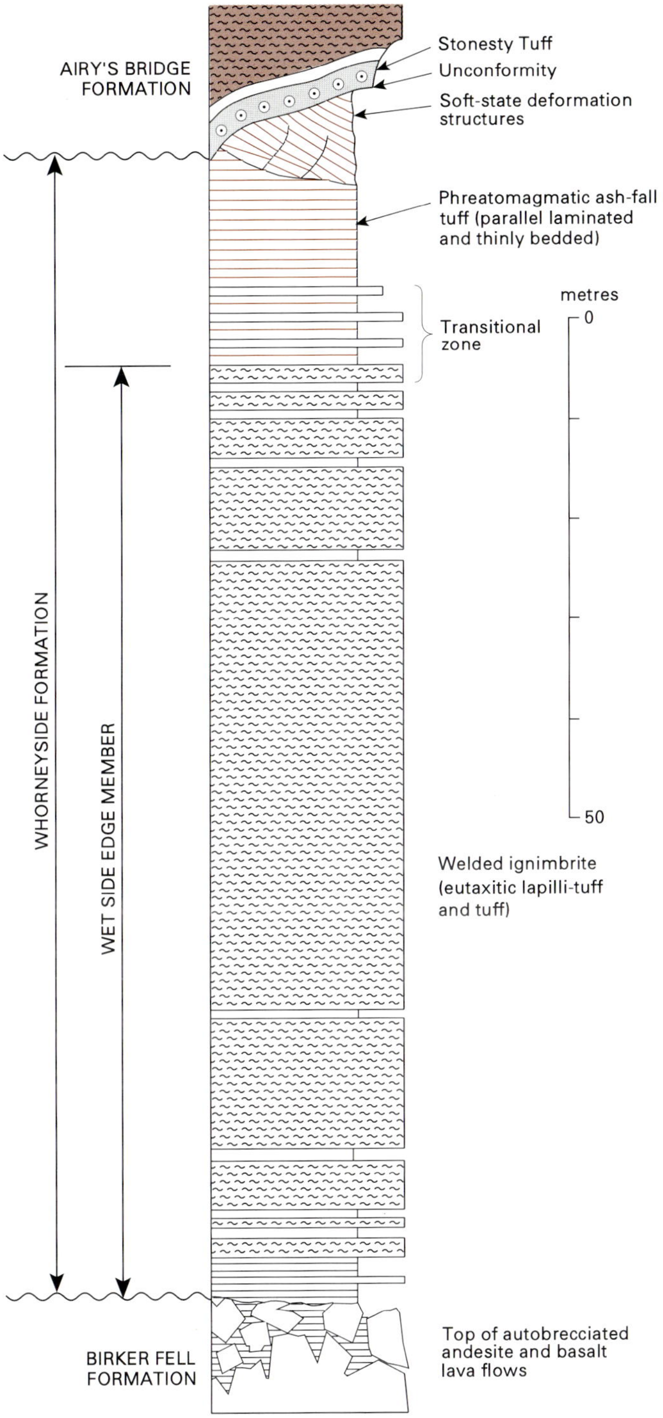

Figure 32 Type section of the Whorneyside Formation, near Whorneyside Force [NY 247 039 to 259 051] (modified from Branney, 1991).

974] in the Duddon valley, there are several metres of bedded tuff intercalated with the basal tuffs of the Airy's Bridge Formation. Thus, the eruption and deposition of the two formations overlapped briefly and the interval of erosion and slumping must have been contemporaneous with that overlap.

Branney (1991) discussed the mechanism of eruption of the Whorneyside Formation. In his model the vent was considered to have been located on the south-west side of Buttermere [NY 174 147] in the Keswick district. The most significant evidence for this is the increase northwards in the maximum size of lithic fall-out fragments from the Central Fells towards Borrowdale, where the bedded tuff was deposited subaqueously and locally redistributed in a shallow basin or in ephemeral lakes. In Borrowdale, 50 cm-diameter ballistic lithic clasts with associated impact structures occur and the proximal ash and lacustrine facies are cut by penecontemporaneous andesitic intrusions thought to record the volcanic centre. Water within the basin was considered to have interacted with magma causing the eruption to become phreatoplinian.

On Great Worm Crag [SD 193 969] the Wet Side Edge Member comprises a tuff-breccia with blocks up to about 25 cm across. Most of the formational thickness increase occurs within the tuff-breccia which may have been ponded there, but the overlying bedded tuff member is also as thick as 200 m in places. The bedded tuff succession fines upwards and there is an accompanying decrease in both bed thickness and grain size; coarse tuff and lapilli-tuff at the base are weakly stratified, becoming well bedded in the uppermost 50 m with a predominance of parallel, thin beds. The base of the formation at Great Worm Crag is markedly unconformable on the Birker Fell Formation (Figure 33). The outcrop preserves a vertical section through a V-shaped depression approximately 1.5 km wide and 1 km deep. The third dimension of the section is unknown, but the feature may represent the remnants of a vent fill, or possibly buried topography.

Airy's Bridge Formation

The Airy's Bridge Formation is a dacitic to rhyolitic pyroclastic sequence, up to 1 km thick, dominated by intensely welded ignimbrite comprising mainly tuff and lapilli-tuff, locally with intercalations of pyroclastic breccia and mesobreccia. Some of the ignimbrite has a lava-like appearance (Branney et al., 1992). Thin layers of bedded tuff separating ignimbrite sheets have unique lithofacies characteristics that make them ideal stratigraphical markers. The main outcrop extends south from the type area on Scafell to the Coniston Fells and thence south-westwards to the Duddon Basin as far as Rowantree Force [SD 143 935] on Waberthwaite Fell, where it is overstepped by the Waberthwaite Formation (Figure 31). Outliers occur in graben around Wast Water and on Hard Knott. The northern extent of the outcrop is outside of the district.

The formation was described first in the Central Fells as part of the Airy's Bridge Group by Oliver (1954; 1961). This included Green's (1915) 'streaky rocks' and correlated in part with Marr's (1916) Sty Head Garnetiferous Group. Oliver (1961) recognised four subdivisions of the group; a basal unit (4d) of andesitic tuff and lapilli-tuff (now part of the Whorneyside Formation) and the overlying sheets of dacitic to rhyolitic rocks (units 4c to 4a) which now constitute the Airy's Bridge Formation (Branney and Soper, 1988; Branney and Kokelaar, 1994). Some rhyodacitic pyroclastic rocks in the Coniston Fells

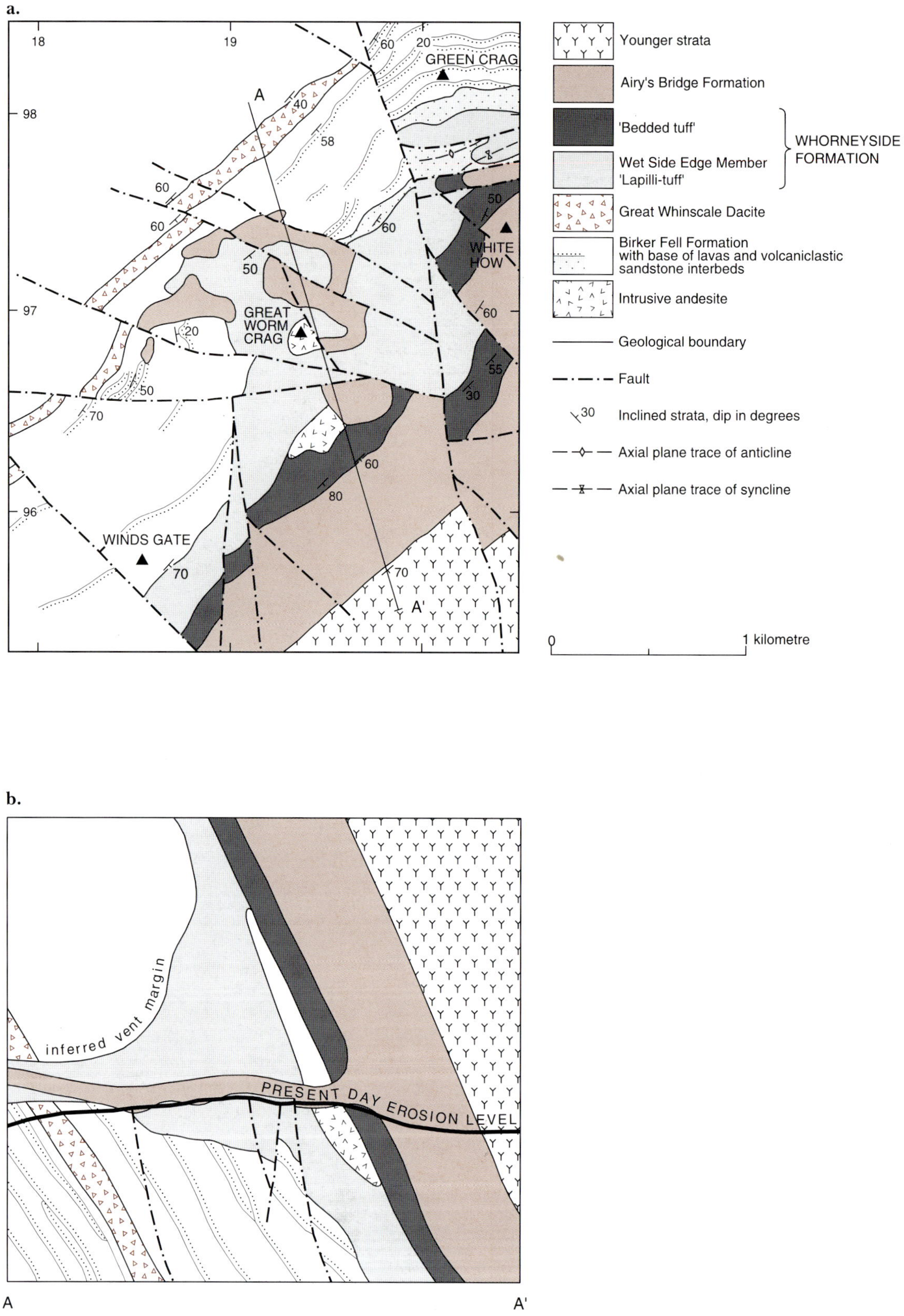

Figure 33 Diagram showing the discordant relationships between the Whorneyside Formation, Airy's Bridge Formation and the Birker Fell Formation around Great Worm Crag.

a. Simplified geological map.
b. Diagrammatic cross-section showing vent interpretation.

and Duddon valley previously assigned to other stratigraphical units are included within the Airy's Bridge Formation on lithological, geochemical and stratigraphical grounds. These include part of the Paddy End Rhyolites and most of the Kidson How Tuffs (Mitchell, 1940; 1963), and parts of Firman's (1957) Waberthwaite Tuffs and Worm Crag Group.

The type area for the Airy's Bridge Formation lies in the Scafell area. Its name is taken from Airy's Bridge [NY 2237 1035] on Styhead Gill, in the adjoining Keswick district. The locality, though not geologically significant, is shown on previous editions of the Ordnance Survey map of that area, but not on the current series. The formation is divided into an older sequence of massive and bedded eutaxitic ignimbrites referred to as the Long Top Member, and a younger suite of higher grade, rheomorphic ignimbrites and mesobreccias, the Crinkle Member (Figure 34; Long Top and Crinkle tuffs of Branney, 1988b and Davis, 1989). Four distinctive thin marker tuffs are widespread; the Stonesty Tuff generally defines the base of the formation. The lowest ignimbrite of the Crinkle Member has distinctive lava-like characteristics and is called the Bad Step Tuff (Branney et al., 1992; 1993a). A similar lava-like rhyolitic ignimbrite, the Oxendale Tuff, occurs locally in the Long Top Member.

The base of the formation is generally sharp and usually rests on a high-relief surface caused by severe slumping of the Whorneyside Formation prior to deposition of the Stonesty Tuff (Branney, 1988b; 1991). The stratigraphical relationships and thickness variations (Figures 31, 35) result mainly from irregular contemporaneous subsidence during piecemeal caldera collapse that accompanied eruption of the formation, particularly the Crinkle Member (Branney and Soper, 1988; Davis, 1989; Branney et al., 1992; Branney and Kokelaar, 1993).

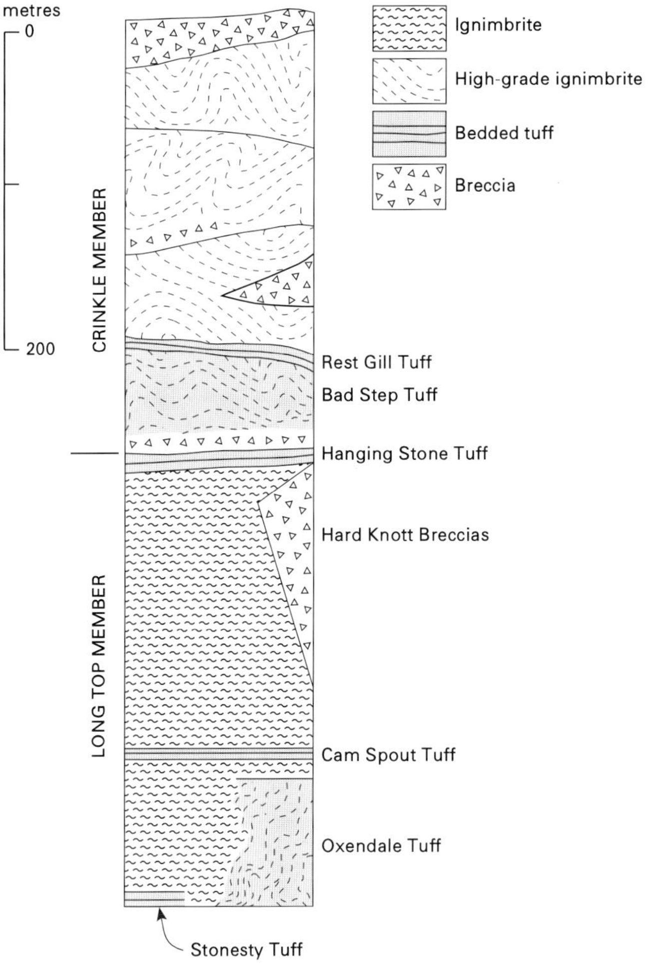

Figure 34 Stratigraphy of the Airy's Bridge Formation.

Long Top Member

The Long Top Member comprises massive and bedded eutaxitic dacitic tuff and lapilli-tuff and is up to 310 m thick. A basal unit of rhyolitic, welded tuff and two widespread, but thin, intercalations of bedded phreatomagmatic tuff and accretionary lapilli-tuff form important stratigraphical markers. At the top of the Long Top Member on Hard Knott [NY 231 023] is a thick succession of breccia (Hard Knott Breccias). The type area is at Long Top [NY 245 047] on Crinkle Crags and throughout the Scafell area, from the north of Pike o' Blisco, to Crinkle Crags, Scafell and Great Gable (Figure 35a). Outliers occur on Yewbarrow [NY 176 085], Illgill Head [NY 167 045] and Middle Fell [NY 153 066]. In the Coniston Fells [NY 306 029 to SD 255 973] the member comprises all of the Airy's Bridge Formation (Figure 31). The Long Top Member also crops out in the Duddon valley, south-west of the Grassguards Fault [SD 214 980 to 144 935] and within the Side Pike Complex [NY 282 036] (see Chapter 9).

The Long Top Member consists of welded, eutaxitic tuff and lapilli-tuff in thin to thick beds and massive units up to 30 m thick. Grading profiles are preserved in the massive units of eutaxitic tuff. 'Domain breccia' (Branney and Kokelaar, 1994) is a fused rheomorphic breccia in which, though the clast outlines are obscured, the clasts are recognisable from differing or rotated welding fabrics in adjacent domains. It occurs associated with rheomorphic folds and intracaldera faults. Reverse-graded lithic clasts and fiamme occur at the base of some massive units and others have reverse-graded fiamme towards their tops. Normal size- and concentration-grading of lithic clasts and normal concentration-grading of crystals are also widely developed. Thinly and medium-bedded, and low-angle cross-bedded, eutaxitic tuffs form a subordinate part of the succession. Cross-sets may be inclined at 10 to 15° to the principal bedding surfaces.

In the Coniston Fells, on Birk Fell, Wetherlam and Red Dell [SD 293 018 to 285 000] the Long Top Member comprises several distinctive ignimbrite sheets. Above the Oxendale Tuff in this area (Figure 31) about 5 m of bedded tuffs are locally developed [for example SD 2878 0077 to 2873 0060]. These are medium to thickly bedded, commonly wavy, parallel beds of tuff and lapilli-tuff which in places contain small accretionary lapilli, up to 5 mm in diameter, many of which are broken. Sequentially above are homogeneous, fine lapilli-tuff, a pink-weathered, lithic-rich, eutaxitic lapilli-tuff characterised by surface pock marks, and a white-weathered, platy, strongly foliated fine lapilli-tuff. The last of these is cut out south-westwards by a

Figure 35 Distribution and thickness variation of members in the Airy's Bridge Formation.

a. Long Top Member.
b. Crinkle Member.

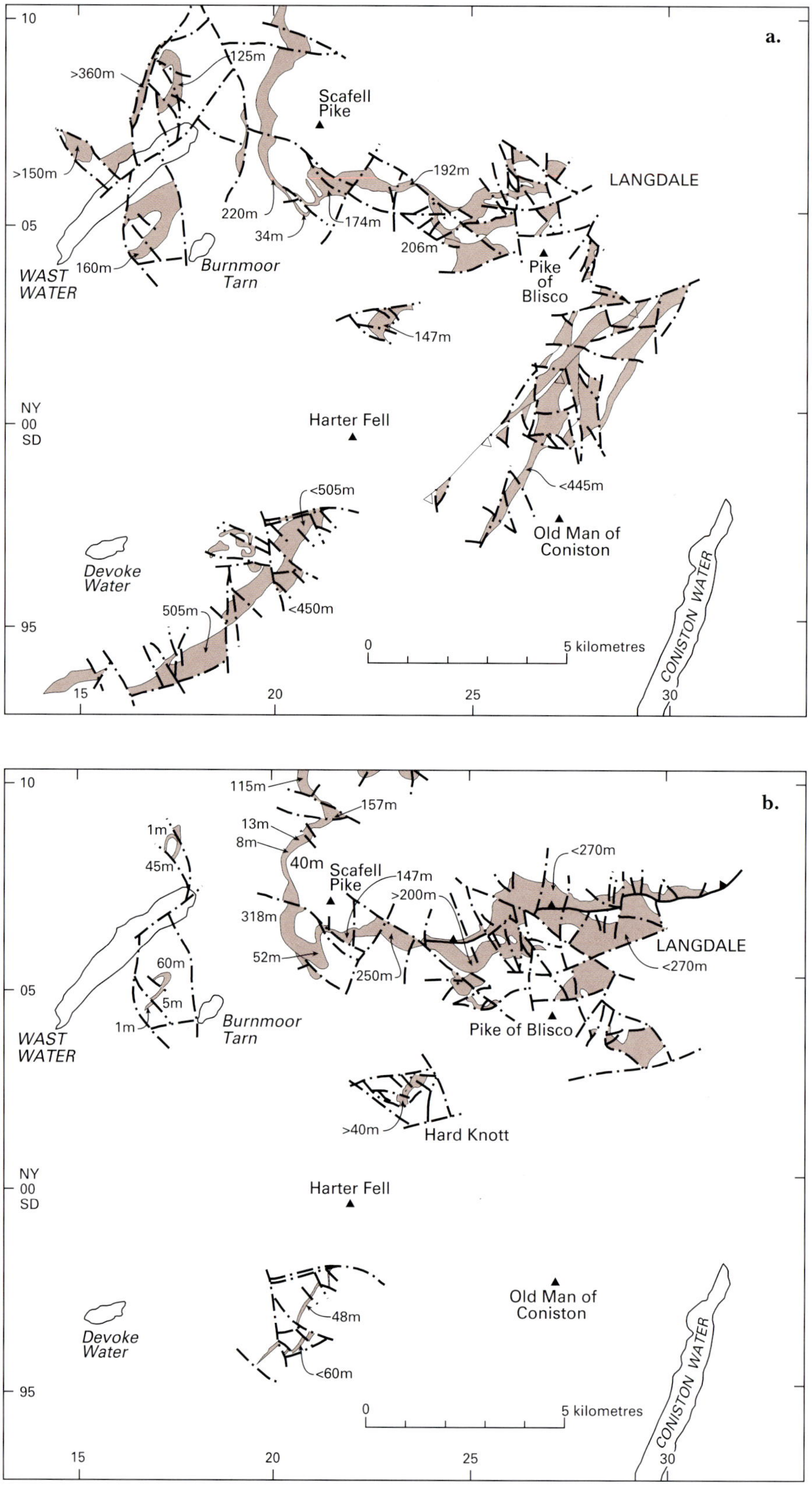

pumice-rich eutaxitic lapilli-tuff. Detailed correlation of these units with those in the Central Fells has not been made.

Typically, Long Top Member rocks are poorly sorted, containing a mixture of crystals, fiamme and lithic lapilli in a fine tuff matrix (Plate 6a). Most fiamme are less than 5 cm long but some range up to 15 cm. Fiamme generally have an aspect ratio of about 15 to 1, but aspect ratios of greater than 40 to 1 occur on Crinkle Crags. Undeformed pumice is locally present, for example on Bell Crags [NY 298 148]. Lithic clasts generally comprise less than 10 per cent of the rock, though there is some variation between individual units, and most are dacitic and angular with either aphyric or porphyritic textures. The largest clasts occur within two eutaxitic units at the top of the member in the Crinkle Crags area where welded tuff fragments up to 15 cm in diameter comprise up to 15 per cent of the rock. The crystal content ranges up to 55 per cent by volume. Plagioclase is dominant and occurs as phenocrysts in some fiamme. Apatite, zircon and opaque minerals occur throughout in small amounts. Garnet is present in some beds and locally, for example in one bed [NY 231 119] on Base Brown just to the north of the district, it comprises half of the crystal content. In Adam-a-Cove [NY 248 044] on Crinkle Crags, a tuff layer contains abundant euhedral garnets (Branney, 1988b; Davis, 1989). The original matrix of fine tuff has been replaced by a cryptocrystalline quartz–feldspar mosaic in which vitroclastic textures are only rarely preserved.

The Long Top Member thins northwards (Figure 35a), from about 500 m on Hesk Fell, 400 m on the Coniston Fells, 180 to 205 m around Crinkle Crags and Scafell, to 75 to 105 m on the northern limb of the Scafell Syncline near Base Brown [NY 225 124] just to the north of the district. Major rheomorphic slumps have caused local thickness variations; for example, thinning the member to the south of the summit of Scafell [NY 2100 0518] and thickening it at Black Crag, Scafell [NY 202 070] and in Crinkle Gill [NY 255 048] (Branney and Kokelaar, 1994).

Oxendale Tuff

The Oxendale Tuff is a 60 to 100 m-thick, pink-weathered, lava-like, rhyolitic welded tuff at the base of the Long Top Member in Oxendale [NY 260 052], at the head of Great Langdale. It also crops out on Whorney Side [NY 2600 0512], Green Tongue [NY 2704 0491] and near Horse Crags [NY 2780 0330 and NY 2845 0380] within the Side Pike Complex. Fault-bounded outcrops occur on Wetherlam [NY 288 011] and south-east of Seathwaite Tarn [SD 267 996 to 253 971]. On the south side of The Band [NY 2670 0546] and south-east of Seathwaite Tarn the member rests unconformably on slumped bedded tuff of the Whorneyside Formation. The contact is sharp and irregular, with a basal breccia that has a matrix of fine tuff derived by fluidisation from the underlying bedded tuff, indicating that the Oxendale Tuff was emplaced before the Whorneyside tuff had dried out (Branney, 1988b). The Oxendale Tuff is distinguished from the rest of the Long Top Member by its lava-like flow-banded fabrics and its rheomorphic folds and autoclastic breccias. The presence of fiamme and possible gradation upwards into the lower-grade part of the Long Top Member, north of The Band [NY 2645 0642], support its tentative interpretation as an ignimbrite.

Bedded stratigraphical marker tuffs

Three distinctive marker tuffs occur within the Long Top Member in the Central Fells, the **Stonesty, Cam Spout and Hanging Stone tuffs** (Figure 34; Branney, 1988b; Davis, 1989). Thin, parallel bedforms and lamination predominate, with sand-wave bedforms, pinch-and-swell structures, and other low-angle cross-stratification. Though the tuffs are now altered to a microcrystalline quartz–feldspar aggregate, closely spaced fine laminae indicate that the original grain size was mostly fine to coarse tuff (c. 1/16 mm). Most pumice fragments retain their original angularity. Lithic fragments are rare (less than 1 per cent) and very small, comprising mainly aphyric felsite, commonly with perlitic cracks, and rare porphyritic felsite. The marker tuffs are generally crystal poor, but sporadic crystal-rich beds occur in the Cam Spout and Hanging Stone tuffs; most crystals are zoned plagioclase with some prismatic apatite and zircon.

Accretionary lapilli are abundant and concentrated in discrete beds. Common reverse grading of the accretionary lapilli coincides with normal, coarse-tail grading of the tuff. The fine grain size and abundance of accretionary lapilli are typical features of phreatomagmatic eruptions and Branney et al. (1993a) considered that these marker tuffs were emplaced from phreatomagmatic pyroclastic surges and by fallout of phreatomagmatic ash.

The white-weathered and flinty **Stonesty Tuff**, generally less than 1 m thick, occurs at the base of the formation. It is laminated to thinly bedded, with low-angle cross-stratification. Accretionary lapilli occur in several beds and, locally near the top of the tuff, a bed also contains flattened pumice clasts. The Stonesty Tuff is apparently absent in the Great Langdale area where the Oxendale Tuff occurs at the base of the formation, and consequently the relative age of these two units is not known.

The **Cam Spout Tuff** is a cream- and brown-weathered fine tuff. It is laminated and thinly bedded, locally with low-angle cross-stratification and accretionary lapilli. The tuff is remarkably extensive even though it is generally less than 2 m thick. The maximum thickness is 6 m on the northern side of the Scafell Syncline, near Sourmilk Gill [NY 2275 1203] and 2.1 m are present on Cam Spout Crag. Stratigraphically, the Cam Spout Tuff is 75 m above the base of the Long Top Member on Cam Spout Crag, approximately 25 m above the Stonesty Tuff at the head of Langdale, and 5 to 15 m above the top of the Oxendale Tuff where that is present. The **Hanging Stone Tuff** is lithologically similar to the Cam Spout Tuff and generally less than 1 m thick.

Hard Knott Breccias

Around the summit area of Hard Knott [NY 231 024] mesobreccia is interstratified with eutaxitic tuff, lapilli-tuff and laminated tuff. The succession is up to 120 m thick and occurs at the top of the Long Top Member in this area. West of Hard Knott summit three breccia units occur, but to the east only the middle unit is present, resting directly on the Whorneyside Formation (Davis, 1989). The Hard Knott Breccias are a local development and probably accumulated adjacent to scarps.

The lowest unit, 22 to 50 m thick, consists of two white-weathered parataxitic lapilli-tuffs and an intervening massive bed of breccia. The tuffs contain more than 20 per cent lithic clasts, generally less than 2 cm in size, but with some up to 40 cm. Most of the clasts are of Long Top Member lithologies, along with rare amygaloidal basalt and porphyritic andesite. Fiamme up to 10 cm long form up to 25 per cent of the rock. Plagioclase crystals occur in concentrations up to 15 per cent at the bed bases. Near the top of each tuff layer are thin, discontinuous laminae of crystal-poor fine tuff which includes sporadic accretionary lapilli. The unit is interpreted as ignimbrite with intercalated co-ignimbrite ash-fall tuffs (Davis, 1989).

The middle unit comprises up to 59 m of massive, clast-supported breccia with some thin interstratified tuff (Plate 10). The breccia consists predominantly of lithic blocks derived from the underlying ignimbrites. Other clast types present include eutaxitic andesitic lapilli-tuff, amygaloidal basalt and porphyritic andesite. No juvenile component has been recorded. Clast size varies from less than 1 cm to over 4 m and blocks greater than 30 cm commonly form more than 50 per cent of beds. The sand to pebble-grade matrix forms up to 25 per cent of the rock and contains less than 2 per cent broken and angular plagioclase crystals with no identifiable pumice. The basal contacts of breccia beds are sharp and discordant, overlain by inversely graded layers. The uppermost 1 m has normal coarse-tail grading. Lateral variations in thickness are irregular but overall the unit thins towards the north. These beds are interpreted as non-cohesive debris-flow and rock-fall avalanche deposits (Davis, 1989).

The uppermost unit consists of 18 to 45 m of breccia with interbedded tuff and volcaniclastic sedimentary rocks. Compared with the underlying beds it contains substantial juvenile material, including pumice (generally less than 20 per cent) and euhedral garnets, and a greater propor-

Plate 10 Hard Knott breccias.
Clast-supported breccia beds with interstratified tuff and lapilli-tuff. Blocks within the breccia were derived from fault scarps that exposed ignimbrite within the underlying Long Top Member (Airy's Bridge Formation). North side of Hard Knott [NY 2316 0253] (D 4033). Observer is 1.7 m tall.

tion of interbedded tuffs. Eutaxitic breccia and lapilli-tuff beds are 1 to 10 m thick, lithic rich and have proportions of clast types similar to the massive to poorly sorted breccia beds. Lithic clasts up to 2 m occur but the average size is less than 5 cm and sorting is poor. The matrix contains up to 30 per cent crystals, mostly plagioclase with subordinate garnet. Most beds are massive; reverse grading of lithic clasts and crystals occurs near the base of some beds and in others crystals and lithic clasts are normally coarse-tail graded. The interbedded tuff and lapilli-tuff are parallel- and cross-bedded with abundant crystals and pumice fragments; thin, crystal-poor tuff beds cap some of the breccia layers. The uppermost breccia unit comprises lithic-rich ignimbrites, locally with underlying ground-surge and overlying co-ignimbrite ash-fall deposits.

Bad Step Tuff within the Crinkle Member

At the base of the Crinkle Member in the southern part of the Scafell Syncline is an extremely high-grade, lava-like, rhyolitic ignimbrite (Plate 11; Branney et al., 1992). Crudely bedded, heterolithic, pyroclastic breccia with a eutaxitic matrix at the base of the sheet grades up into a thick, flow-laminated, locally amygaloidal, lava-like centre. Rheomorphic structures become increasingly abundant in the upper part and a rheomorphic autobreccia is present at the top (Figure 36). Locally, for example on Crinkle Crags [NY 254 055], the Bad Step Tuff is brecciated throughout. The gradational passage from basal breccia into the flow-laminated centre suggests that it is a single ignimbrite. This interpretation is reinforced by the presence of an 8 m-thick lithophysal zone spanning the gradational contact between breccia and flow-laminated lava-like rock. The field appearance of this ignimbrite had resulted in previous interpretations as intrusive rhyolite (Hartley, 1932) and lava-flow (Oliver, 1961; Moseley, 1983).

The type section is at Bad Step [NY 2487 0485], a small cliff near the summit of Crinkle Crags. The outcrop is from near Pike de Bield Moss [NY 233 062] in upper Eskdale to Whitegill Crag [NY 298 070] in Great Langdale; an outlier occurs on Hard Knott. The tuff is recognised only in the southern limb of the Scafell Syncline; elsewhere it is either absent or its correlatives misidentified.

Plate 11 Foliated lava-like tuff within the Bad Step Tuff (Airy's Bridge Formation). North side of Hard Knott [NY 2323 0246] (D 4032). Hammer head is 17 cm long.

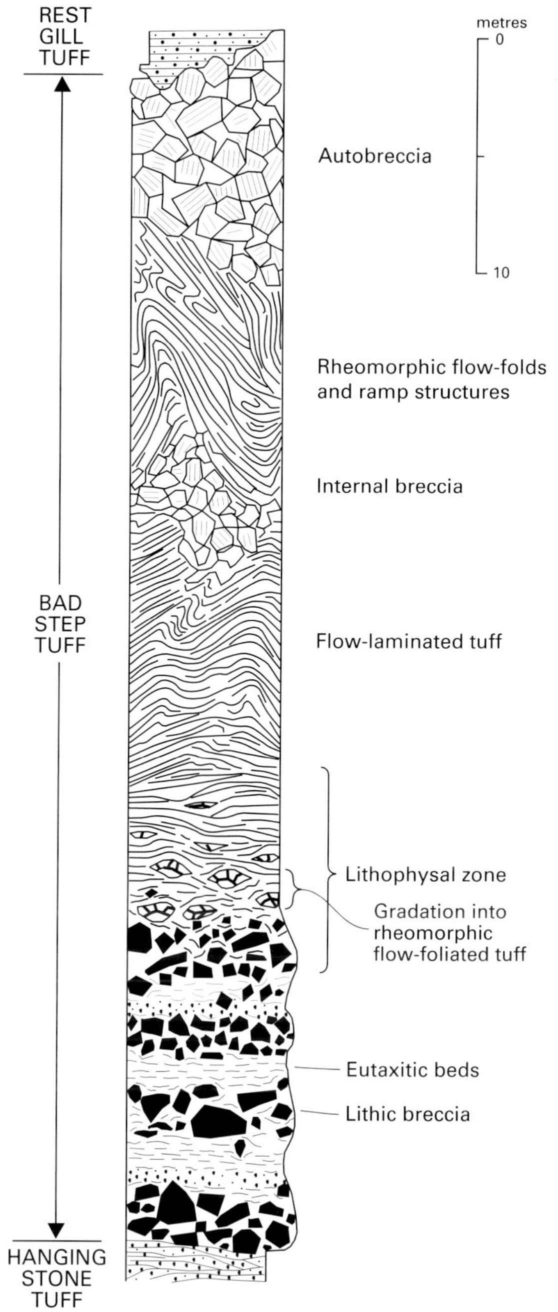

Figure 36 Schematic vertical section through the Bad Step Tuff showing lithological variations in the high-grade ignimbrite (after Branney et al., 1992).

Originally, the Bad Step Tuff covered an area of at least 35 km^2 and its magmatic volume was more than 5 km^3 (Branney et al., 1992). The thickness of the Bad Step Tuff varies from 40 m at Rest Gill [NY 246 054] to more than 400 m in Great Langdale [NY 280 063], where the base is not exposed. Many of the thickness changes occur abruptly across faults which Branney et al. (1992) interpreted as evidence for ponding of the tuff within a very irregular palaeotopography created by faulting during eruption-related, piecemeal volcanotectonic subsidence.

On Bad Step and Shelter Crags [NY 2459 0542 and 2530 0553] the basal, co-ignimbrite lithic breccia is up to 15 m thick. Some blocks in the breccia have altered rinds or radial joints and they occur in a eutaxitic matrix with lithophysae and patchy silicification. Branney et al. (1992) considered these features as evidence of hot emplacement. There is a crude stratification parallel to the basal contact defined by a combination of the variations in size and relative abundance of lithic blocks and lapilli, and in the variations in grain-size of the matrix. The fiamme have frayed terminations and show ductile compaction where they are in contact with lithic blocks. Most of the blocks and lapilli are welded tuff and lapilli-tuff derived from the Long Top Member. Rare, well-rounded clasts of andesite are also present in the breccia and Branney et al. (1992) argued that these may have been either stoped from the conduit wall during eruption of the tuff or derived from contemporaneous fault scarps. If the latter interpretation is true then, given that andesites are at least 140 m stratigraphically below the Bad Step Tuff, any fault scarps may have been at least that height.

Volumetrically, the flow-laminated centre comprises most of the Bad Step Tuff. Rheomorphism and recrystallisation have destroyed vitroclastic textures though perlitic cracking and pseudomorphs after spherulites indicate the former presence of glass. Flow laminae up to 2 mm thick are defined by mineralogical and textural contrasts and many are accompanied by discrete fractures or partings that can be traced laterally for up to 1 m. Quartz and chlorite amygdales up to 1 cm across occur in swarms; some are distorted or flattened. The original vesicles probably record welding compaction following vesiculation and indicate that volatiles released after welding caused brittle failure of the surrounding rhyolite. Branney et al. (1992) suggested that the central part of the Bad Step Tuff was probably impervious and trapped gas to form the lithophysae.

The principal phenocrysts in the foliated tuff are zoned, euhedral or slightly resorbed plagioclase (oligoclase–andesine) crystals; they are up to 3 mm across and comprise 7 to 12 per cent of the tuff. Also present are fresh garnets less than 4 mm across with resorbed edges and feldspar rims (up to 1 per cent), alkali feldspar, and chlorite and opaque pseudomorphs after mafic crystals; apatite and zircon occur as accessory minerals. Lithic clasts in the foliated tuff are less than 5 mm and sparse.

The flow-laminated zone passes gradationally upwards across a very irregular boundary into autoclastic breccia, from 5 to 40 m thick. At its base the breccia is clast supported, clasts having angular margins and jig-saw fits with adjacent blocks. The breccia becomes more of an open framework upwards and in the uppermost 5 m, interstices are filled with fine tuff that was probably washed in by percolating water. Some of this fine tuff is silicic and may represent the co-ignimbrite ash associated with the Bad Step Tuff eruption. However, andesitic tuff was probably washed in from the succeeding Rest Gill Tuff eruption.

Crinkle Member above the Bad Step Tuff

The rest of the Crinkle Member comprises eutaxitic, parataxitic and foliated lapilli-tuff with interbeds of mesobreccia,

and thin units of bedded tuff. The maximum thickness developed beyond the northern margin of the Ambleside district. In most places a distinctive marker unit, the **Rest Gill Tuff**, either forms the base of the member or overlies the Bad Step Tuff, filling irregularities and interstices in its autobrecciated top. The Rest Gill Tuff comprises up to 3.5 m of laminated and thinly bedded tuff, with a distinctive brown and turquoise-coloured weathered surface. Scour and fill structures, and sandstone intraclasts indicate aqueous reworking. Accretionary lapilli are absent. The tuff is named after the type locality near Rest Gill [NY 2446 0558 to 2500 0536] on Crinkle Crags (Branney, 1988b), but has a widespread outcrop within the Scafell Syncline. The maximum thickness occurs near Hanging Stone [NY 2270 1192]. The Rest Gill Tuff is interpreted tentatively as an aqueously reworked, phreatomagmatic, ash-fall tuff (Branney et al., 1993a).

The type area of the Crinkle Member is on Crinkle Crags, between Long Top [NY 2485 0475] and Shelter Crags [NY 2495 0580]. The main outcrop extends around the Scafell Syncline from Great Gable [NY 209 102] to near Raw Pike [NY 309 072] and Kettle Crag [NY 280 049] in Great Langdale. Outliers are seen on the north-west side of the Duddon valley, between the Grassguards and Baskill faults [SD 215 985 to 190 951], on Yewbarrow [NY 176 085] and Illgill Head [NY 173 051] (Figure 35b). The member is absent from the Coniston and Seathwaite fells.

In the Central Fells the thickness of the member exceeds 200 m (Figure 35b). Abundant rheomorphic folds and fabrics are characteristic of the Crinkle Member and folds have wavelengths varying from less than 1 mm to more than 100 m. They are commonest in the middle of the sequence and are particularly well developed on the northern limb of the Scafell Syncline where up to half the thickness of the member is deformed (Davis, 1989). Clast-supported rheomorphic domain breccias are extensive (Branney and Kokelaar, 1994). Towards the top of some component ignimbrites rheomorphic folds pass upwards into an overlying autobreccia. Mesobreccia is intercalated with the ignimbrite succession on Shelter Crags [NY 249 055], The Band [NY 258 059] and along the north side of Great Langdale. Massive, densely welded and rheomorphic, rhyolitic tuff and lapilli-tuff with vitric, crystal and lithic components comprise more than 90 per cent of the sequence. Eutaxitic and parataxitic fabrics are usually present. Locally, up to seven separate ignimbrites have been identified within the succession (Davis, 1989). Most of these are massive though the basal part of some is thinly bedded. The bedding is defined by changes in the size and abundance of fiamme and, rarely, by variations in crystal and lithic concentrations. The ignimbrites (including the Bad Step Tuff), are generally of higher grade than those in the Long Top Member (Plate 6b).

On the north-west side of the Duddon valley, on the west side of Iron Crag [SD 209 971] the Crinkle Member is 35 to 40 m thick. The base is sharp and overlies a slightly irregular erosion surface. The basal 2 m of the member comprise laminated to thickly bedded tuff, of very variable grain size, that may be laterally equivalent to the Rest Gill Tuff. Above this are two high-grade ignimbrites separated by up to 1 m of bedded tuff. The lower ignimbrite is 20 m thick, lithologically uniform, crystal poor and devoid of lithic clasts. It has a parataxitic fabric and weak internal stratification. Lithophysae occur near the base and are elongated within the eutaxitic foliation. A prominent bedding plane defines the base of the overlying thinly bedded tuff. The upper ignimbrite is 15 to 17 m thick, lithologically similar to the lower one, though the presence of lithic lapilli, a eutaxitic foliation and the absence of lithophysae make it distinctive. About 1 km south of Iron Crag up to 35 m of mesobreccia locally overlie lapilli-tuff and the member attains its maximum thickness of 75 m.

Throughout the Crinkle Member fiamme are generally extremely flattened and stretched with aspect ratios varying from 5:1 to 200:1. Trails of steeply inclined and vertical fiamme are common, particularly adjacent to faults where they provide evidence for fault movement during emplacement (Branney, 1988b; Branney and Kokelaar, 1994). By comparison with the Long Top Member the Crinkle Member is crystal-poor, from less than 8 per cent in the lowest parts of the sequence to 15 per cent towards the top. The dominant crystal phase is tabular plagioclase (oligoclase) less than 2 mm in length; sparse mafic phenocrysts are replaced by chlorite and epidote. Accessory crystal phases include euhedral garnet up to 4 mm, opaques, apatite and zircon. The lithic clast component gradually increases upwards, from less than 5 per cent near the base to a maximum of approximately 20 per cent; clasts are mainly fine-grained felsite, with sparse welded tuff. The matrix is a microcrystalline quartz–feldspar mosaic with small amounts of sericite and chlorite; its original glassy nature is indicated by common perlitic cracks.

Lingmell Formation

In the Central Fells a diverse sequence of garnet-bearing effusive, pyroclastic and intercalated sedimentary rocks from various sources occurs stratigraphically between the Airy's Bridge and Seathwaite Fell formations; they constitute the Lingmell Formation (Figures 31, 37; Kneller et al., 1993a). The formation includes local breccia accumulations on Shelter Crags and in outliers on Yewbarrow [NY 174 085] and Illgill Head [NY 166 048]. A distinctive lava flow, the Scafell Dacite, occurs within the top of this sequence on Great Gable [NY 215 100], Lingmell [NY 206 080] and Scafell [NY 207 065]. The type area is Lingmell to Great Gable [NY 206 064 to 214 100].

The base of the formation varies from unconformable to conformable, and is gradational locally, particularly in Langdale, from underlying Crinkle Member ignimbrites. Both upper and lower boundaries are liable to be diachronous. The upper boundary is generally abrupt, but in the south-east of the outcrop it is locally gradational into the overlying Seathwaite Fell Formation where the upper boundary is taken at the top of the uppermost richly garnetiferous breccia or lapilli-tuff. Excluding the Scafell Dacite, the formation is generally less than 60 m thick, though on Illgill Head about 80 m are present. The

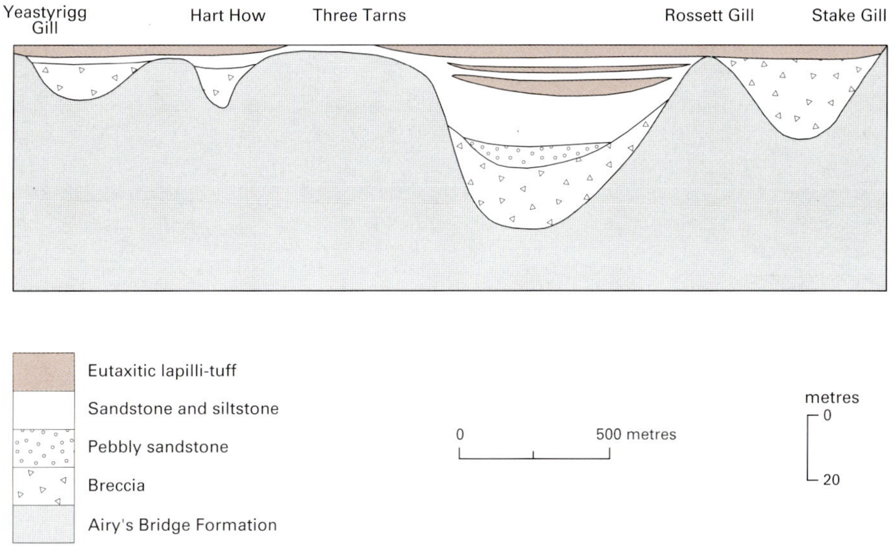

Figure 37 Diagram illustrating topographical control on thickness and facies variations in the Lingmell Formation on the southern limb of the Scafell Syncline.

main, pyroclastic part of the formation has been referred to as the Naples Needle Member (Kneller et al., 1993a) after the Napes Needle Tuffs and Breccias of Davis (1989).

North of Scafell, the thickest sequence occurs in hollows in the irregular top of the Airy's Bridge Formation (Figure 37). Up to 1.5 m of laminated and cross-laminated tuff and siltstone containing sparse felsitic lithic clasts, plagioclase crystals and single-rimmed accretionary lapilli are present locally at the base. Soft-state deformation and slumps are abundant. The basal fine tuff is overlain by bedded, garnetiferous eutaxitic breccia, lapilli-tuff and coarse tuff (Plate 12). Beds of eutaxitic breccia and lapilli-tuff, 0.25 m to 1.5 m thick, contain lithic clasts mainly of Airy's Bridge Formation lithologies, as well as porphyritic andesite, amygaloidal basalt and rare, tabular blocks of laminated tuff and siltstone that have been plastically deformed. Fiamme have low aspect ratios. Crystals of garnet (Plate 6c), zircon and apatite occur throughout, but are concentrated in crystal-rich laminae and thin beds with low-angle cross-sets (Davis, 1989). A bed of fine tuff caps the pyroclastic succession.

Eastwards from Scafell, sandstone and siltstone are interbedded with pyroclastic rocks in a succession that ranges in thickness from 2 to 60 m (Figure 37). The sedimentary rocks contain loading and syndepositional deformation structures. Breccias form the base of the thickest parts of the succession, occupying hollows in the irregular top of the underlying Airy's Bridge Formation. The youngest beds are more laterally persistent, but only one thinly bedded lapilli-tuff, high in the succession, is widespread.

On Illgill Head and Yewbarrow, Airy's Bridge Formation ignimbrites are overlain by the massive, coarse **Yewbarrow breccia** that locally makes up most of the Lingmell Formation. The breccia is up to 50 m thick, poorly sorted, predominantly clast supported and contains blocks up to 2 m across, almost entirely of welded lapilli-tuffs from the underlying formation. The upper part of the breccia becomes finer grained and crudely stratified and is overlain by thinly bedded and laminated garnetiferous tuff, accretionary lapilli-tuff and garnetiferous eutaxitic ignimbrite.

The facies within the formation have diverse origins, including rock-fall avalanche, high- and low-concentration pyroclastic flow and pyroclastic fall-out deposits.

Plate 12 Lingmell Formation.
Beds of welded lapilli-tuff with variable fiamme and lithic clast contents; thinly bedded unit in upper centre of picture contains accretionary lapilli and was derived either from pyroclastic surges or phreatomagmatic fall-out deposits. West side of Lingmell, Wasdale [NY 2050 0835] (D 4350). Hammer is 37 cm long.

Kneller et al. (1993a) suggested, from the distribution and thickness variations, that one of many vents, now occupied by rhyolite, was on Rosthwaite Fell [NY 260 120] to the north of the district. In the Scafell area, subaerial deposition seems probable, based on the absence of unambiguous evidence for subaqueous deposition (Kneller et al., 1993a). By contrast, eastwards in Langdale the locally intercalated sedimentary rocks with cross-bedding indicate limited erosion and aqueous or subaqueous reworking of the pyroclastic rocks.

SCAFELL DACITE

The Scafell Dacite is a garnetiferous lava dome with peripheral lava flows that crops out in the axial region of the Scafell Syncline, from Little Narrow Cove [NY 2250 0675] on the south side of Scafell to Aaron Crags [NY 2293 1038] on Seathwaite Fell, just to the north of the district. The dacite straddles, and attains its maximum thickness at, the Little Narrow Cove fault which had major contemporary displacements and was possibly the eruption site (Kneller et al., 1993a).

The dacite mainly overlies volcaniclastic rocks of the Lingmell Formation, but in places rests directly on the Airy's Bridge Formation. The contact with the underlying strata is irregular and locally intrusive into the underlying fine-grained tuff (Davis, 1989). There is considerable relief on the upper surface (Figure 38), with a central dome and a peripheral flow (Kneller et al., 1993a). The flow is widespread, 100 to 125 m thick, whereas the dome has a maximum exposed thickness of 425 m. On Criscliffe Knotts [NY 215 085] the original slope of the dome surface was greater than 25° and at Aaron Crags [NY 2293 1038] ranged from 15 to 25°.

The margins of the dacite are autobrecciated, and locally clasts have been re-incorporated into the flow-banded central parts. Interstices in the autobreccia, particularly near the termination of the flow, are commonly filled with volcaniclastic siltstone (Davis, 1989). Poorly sorted, matrix-supported breccia is present as aprons adjacent to the steep margins of the dome and as wedges into the overlying Seathwaite Fell Formation (Figure 38). Some of the breccias were deposited during eruption of the Scafell Dacite, but others represent its local reworking.

The dacite is moderately porphyritic (Plate 6d) with zoned plagioclase dominant. Epidote commonly replaces the core of the plagioclase crystals, particularly near the vesiculated margins of the lava flow. A few pyroxene phenocrysts are replaced by chlorite and epidote. Euhedral garnet phenocrysts are a minor, but conspicuous, phase. Single garnet crystals are up to 7 mm across and glomerophyric aggregates of 2 or 3 crystals are also present. They are mainly fresh, with only minor epidotic alteration along fractures and small apatite and opaque oxide inclusions are common. The matrix is very fine grained and hyalopilitic, but is variably recrystallised to chlorite and sericite. Perlitic cracking and banded perlite occur

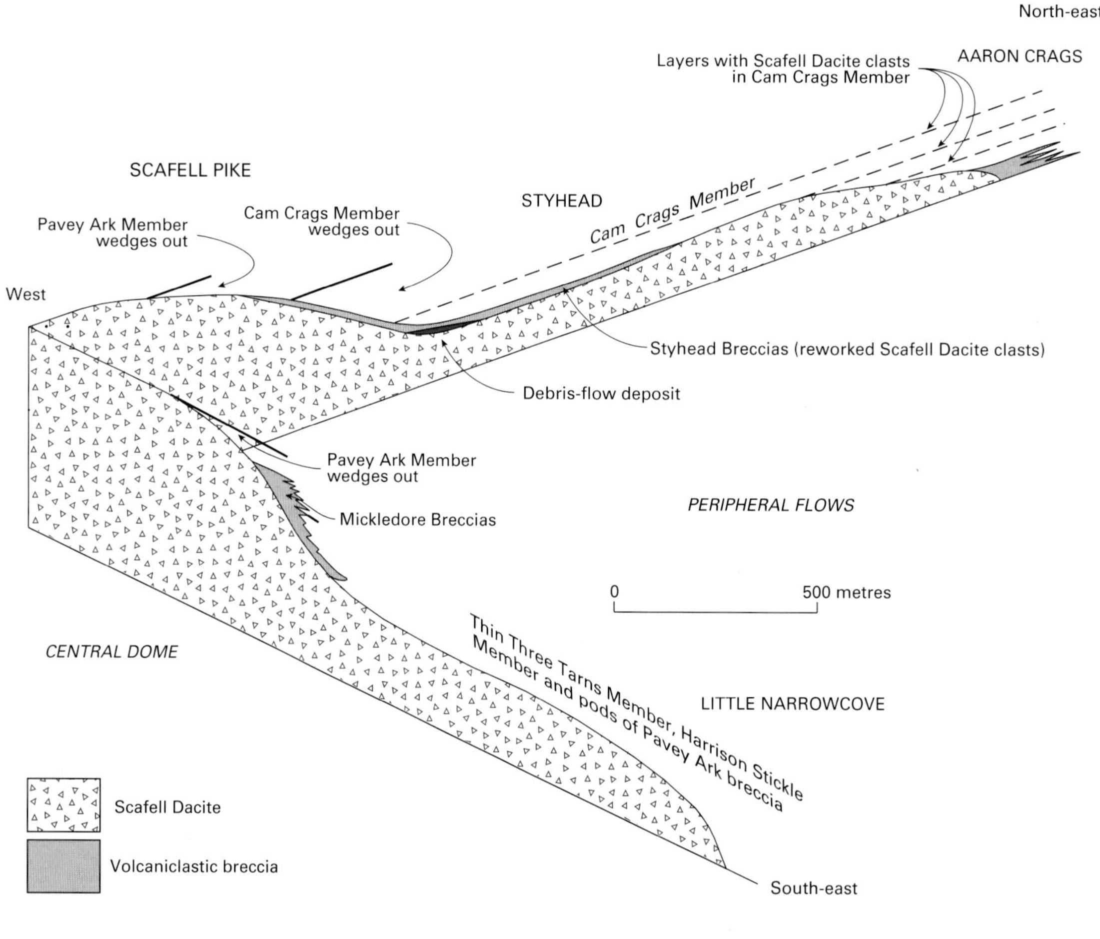

Figure 38 Morphology of the Scafell Dacite showing the marginal breccias, distribution of clasts in, and members of, the overlying Seathwaite Fell Formation.

extensively. Much of the dacite is flow banded, the spacing varying from 0.2 to 30 cm. Flow-folds are present locally with some, such as those on the east side of Lingmell [NY 2100 0822], having amplitudes of over 25 m and lateral continuity for up to 100 m.

SUCCESSION IN THE DUDDON BASIN

The succession in the Duddon Basin comprises approximately 3 km of pyroclastic and sedimentary rocks that overlie the Airy's Bridge Formation in the south-west of the district. Nine formations are present that are not recognised elsewhere within the district (Figure 39). One of these, the Holehouse Gill Formation, contains the only known marine strata in the Borrowdale Volcanic Group. The succession thins abruptly north-eastwards across the Grassguards Fault, which defines the margin of the Duddon Basin, on to the Coniston Fells where the overall thickness is approximately 1.5 km and some formations are absent. The thinning continues farther north-east until the division probably dies out in Great Langdale beneath the unconformity at the base of the Seathwaite Fell Formation. The generally eastward dip and younging of these strata means that the oldest formations crop out in the west of the district.

Waberthwaite Formation

The Waberthwaite Formation unconformably overlies the Airy's Bridge, Whorneyside and Birker Fell formations, and comprises andesitic, lithic-rich welded lapilli-tuff that crops out in the core of the Ulpha Syncline on Whitfell and Corney Fell [SD 155 925] extending southwards into the adjacent Ulverston district; the type area is Corney Fell [SD 151 902 to 161 917]. Included are the Waberthwaite Tuffs of Firman (1957). The base of the formation is taken at the change from generally lithic-free dacitic rocks of the Airy's Bridge Formation to andesitic, lithic-rich lapilli-tuff. These rocks were included previously within the Airy's Bridge Formation (British Geological Survey, 1991), but recently geochemical analysis (Figures 19 to 22) shows the Waberthwaite Formation to be distinct from the Airy's Bridge Formation.

The succession comprises poorly sorted, and largely unbedded, variably lithic-rich lapilli-tuff and mesobreccia and a unit, up to 20 m thick, of parallel-laminated and thinly bedded crystal-rich tuff. Porphyritic andesite sheets are present locally in the succession, though it is uncertain whether these are contemporaneous lavas or subsequently intruded sills. The formation is at least 1300 m thick.

The lapilli-tuff contains variable proportions of fiamme, crystals and lithic clasts in a fine-grained quartz–feldspar matrix, commonly with a fine devitrification mosaic. Perlitic cracking and vitroclastic textures are preserved locally. Fiamme, in places enclosing plagioclase crystals, are sparse to abundant, typically ragged, irregular and eutaxitic to parataxitic. The crystal component, 5 to 40 per cent by volume, is mainly turbid, altered plagioclase, with conspicuous chlorite, epidote, and/or biotite and actinolite pseudomorphs after pyroxene and an opaque mineral commonly altered to leucoxene. Lithic clasts are mostly small, angular to subrounded fragments of basalt, andesite, tuff and welded tuff; rare very fine-grained sandstone and siltstone fragments occur in some units, probably derived from the underlying Skiddaw Group. The lowest 100 m of the formation are crystal-rich and more densely welded than the rest of the sequence with a distinctive parataxitic texture locally passing into a platy foliation with small-scale syndepositional folds and shears. In the western part of the outcrop, within the contact metamorphic aureole of the Eskdale intrusion, the lapilli-tuffs have a brownish hue because of overprinting with biotite and actinolite. Fiamme are generally replaced by fine-grained chlorite and epidote or, within the aureole, by biotite and/or actinolite. Outside the aureole sericite, epidote, chlorite and quartz dominate the alteration mineralogy.

Mesobreccia occurs south of the Whillan Beck Fault, between Burn Moor [SD 147 921] and Littlecell Bottom [SD 146 918] in units having a maximum thickness of 175 m and passing gradationally up into the welded lapilli-tuff. The breccia is very poorly sorted, unbedded, ungraded, clast-supported and contains angular, nonspherical to crudely spherical blocks and large lapilli of white-weathered welded tuff and foliated felsite up to 0.4 m across. Some originally softer, rounded basaltic or andesitic blocks and elongate fragments of intermediate to acid spatter are locally present.

Duddon Hall Formation

The Duddon Hall Formation comprises predominantly bedded, andesitic ash-fall tuff and lapilli-tuff. It crops out throughout the Duddon Basin and Coniston Fells and extends southwards into the adjoining Ulverston district. Outliers occur on Whitfell [SD 163 930] and Burn Moor [SD 145 923]. The thickness of the formation varies from 150 to 450 m. The type area lies between the Grassguards and Stonythwaite faults, between Wallowbarrow Heald [SD 210 970] and Hollin House Tongue [SD 227 965]. The name is modified from Firman's (1957) Duddon Hall Tuffs and applies to the same rocks. Younger, bedded parts of the Duddon Bridge Tuffs (Mitchell, 1956b) are equivalent; Mitchell's Barrow and Ivanscar andesites are now interpreted as intrusions. On the Coniston Fells the Duddon Hall Formation was included previously as part of the Lower Tilberthwaite Tuffs of Mitchell (1940).

The sequence comprises parallel-laminated to thickly bedded, medium- to coarse-grained tuff with subordinate lapilli-tuff, breccia, and fine porcellaneous tuff (Plate 13). Many thin beds are laterally persistent and a mixture of normally, inversely and multiply graded units is characteristic. Bedding mantles intraformational erosion surfaces and volcanic bombs with bomb sags occur locally, indicating a subaerial pyroclastic-fall emplacement. On crags west of Levers Water [SD 2745 9914], beds of accretionary lapilli and amygdaloidal tuff are present locally suggesting a pyroclastic surge component to the eruptions.

Within the type area the thickness of the formation increases from 150 m adjacent to the Grassguards Fault

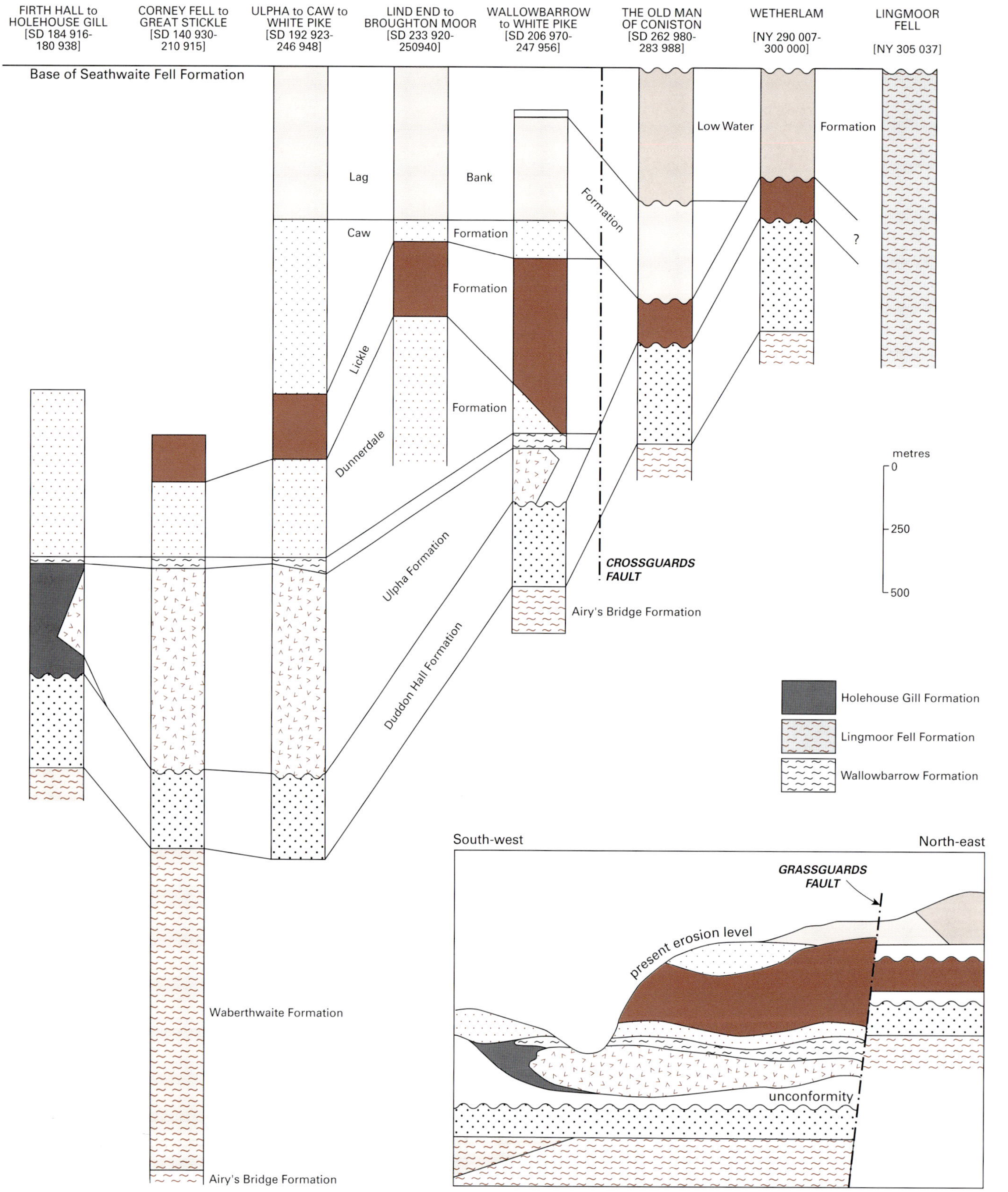

Figure 39 Comparative generalised vertical sections through the stratigraphy of the Duddon Basin and Coniston Fells.

Plate 13 Parallel, thinly bedded ash-fall andesitic tuff, Duddon Hall Formation.

The abrupt change in grain size from bed to bed, along with the varied bed grading profiles, are typical of the widespread phreatoplinian deposit. Hollin House Tongue [SD 2273 9663] (D 4130). Hammer is 37 cm long.

to 350 m near the Stonythwaite Fault (Figure 40, Column 4). In the Coniston Fells up to 450 m are present. From Erin Crag [SD 282 995], the formation thickens north-east on the flanks of Wetherlam and south-west between Levers Water and Goat's Hawse. About 250 m are present on Dow Crag and Buck Pike excluding the andesite sills. Thickness variations in the formation are attributed to the original tephra distribution, to the presence of ignimbrites and to the inclusion of lavas and sills.

Generally the base of the Duddon Hall Formation is marked by a sharp change from welded tuff to the overlying bedded andesitic tuff. Locally, andesite intervenes and is included within the Duddon Hall Formation. In the Coniston Fells there is local overstep by breccia. In the south-west of the district, south of Ulpha Park [SD 190 910], there is a minor unconformity with andesite lavas overlain by tuffs (Figure 40).

The basal 1 to 2 m of the formation in the type area (Column 4, Figure 40) comprise parallel-laminated, thinly and medium-bedded, coarse tuff followed by thickly and very thickly bedded coarse lapilli-tuff that passes up into laminated to thickly bedded coarse tuff and lapilli-tuff. Overall, the succession fines upwards and there is a corresponding decrease in bed thickness. Individual beds show various grading profiles, with reverse grading the most common. Abundant soft-state, syndepositional deformation structures, including disrupted bedding, slumps, slides and intraformational breccia beds are associated with contemporary gravity-induced mass movement.

Ignimbrites, mostly less than 5 m thick, are locally common. South of the Grassguards Fault these are mainly near the base; but in the type area an ignimbrite with abundant lithic clasts and fiamme, and 10 to 30 m thick, occurs at the top of the formation. In contrast, sheets of massive and very poorly sorted, ungraded, matrix-supported lapilli-tuff and tuff-breccia, 50 to 140 m thick, dominate the sequence in the Coniston Fells. Five are seen in the Wetherlam area (Figure 40, Column 7), four sheets on Goat's Hawse (Column 5) and three north-east of Levers Water (Column 6). These rocks are scoria and lithic rich with angular, ragged and fiamme-like juvenile fragments. Some of the larger clasts have an amygdaloidal centre but dense outer part, possibly indicating post-emplacement compaction as a result of alteration, a conclusion supported by the presence of fiamme-like and equant clasts in the same bed. Angular to subrounded, non-vesicular andesite and dacite clasts comprise significant proportions of these deposits. The distinctive, subspherical surface hollows of the lowest-but-one sheet above Dow Crag and north-west of The Old Man of Coniston are weathered carbonate replacement aggregates of unknown origin. The thin uppermost sheet, south-east of Wetherlam, is a splintery, white-weathered welded tuff. A dense, porphyritic dacite clast, in places plastically deformed, dominates the lithic clast population in one of the sheets suggesting a block-and-ash flow origin, perhaps generated from the collapse of a lava dome. The other units are probably ignimbrites.

Locally, fluvial scours and channels occur at the top of the bedded tuff; for example on Hollin House Tongue [SD 2283 9681] and north-west of Peel Crag [SD 2290 9659 and 2298 9667]. The channel on Hollin House Tongue is at least 100 m wide and up to 20 m deep with locally steep margins. The basal channel-fill consists mostly of homogeneous unbedded, coarse-grained sandstone, probably deposited from a high-concentration particle flow, with some interbeds of clast-supported, channel-margin collapse breccia. This is overlain by up to 3 m of white-weathered, parallel bedded, eutaxitic tuff and accretionary lapilli-tuff that is confined to the channel. Reworking of the tephra generally increases to the north-east, particularly on the Coniston Fells.

Holehouse Gill Formation

The Holehouse Gill Formation comprises a sequence of dark grey silty mudstone, sandstone and granule con-

Figure 40 Comparative generalised vertical sections through the Duddon Hall Formation showing thickness and facies variations.

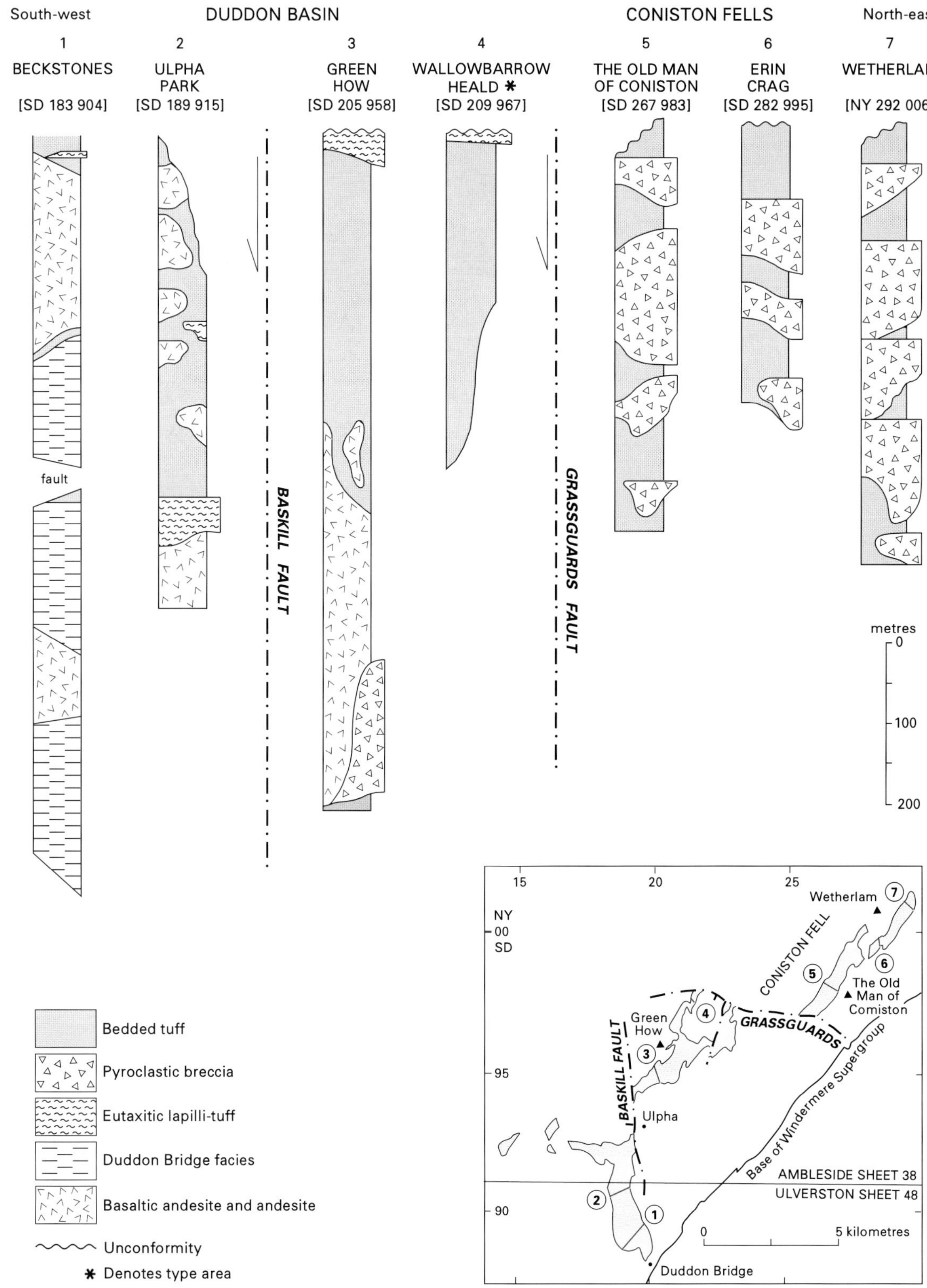

glomerate (Figure 41). The base is unconformable on the Duddon Hall Formation, the low-angle discordance indicating south-easterly onlap of the marine strata. The formation is 450 m thick in the type section at Holehouse Gill. North-east of the type section the strata interdigitate with the Ulpha, Wallowbarrow and Dunnerdale formations (Figure 41), and south-westwards the Holehouse Gill Formation pinches out laterally and passes gradationally upwards into volcaniclastic sandstone of the Dunnerdale Formation. The northward dipping succession in Holehouse Gill and on The Pike [SD 186 934] may thicken northwards down dip, towards the Bigertmire Pasture

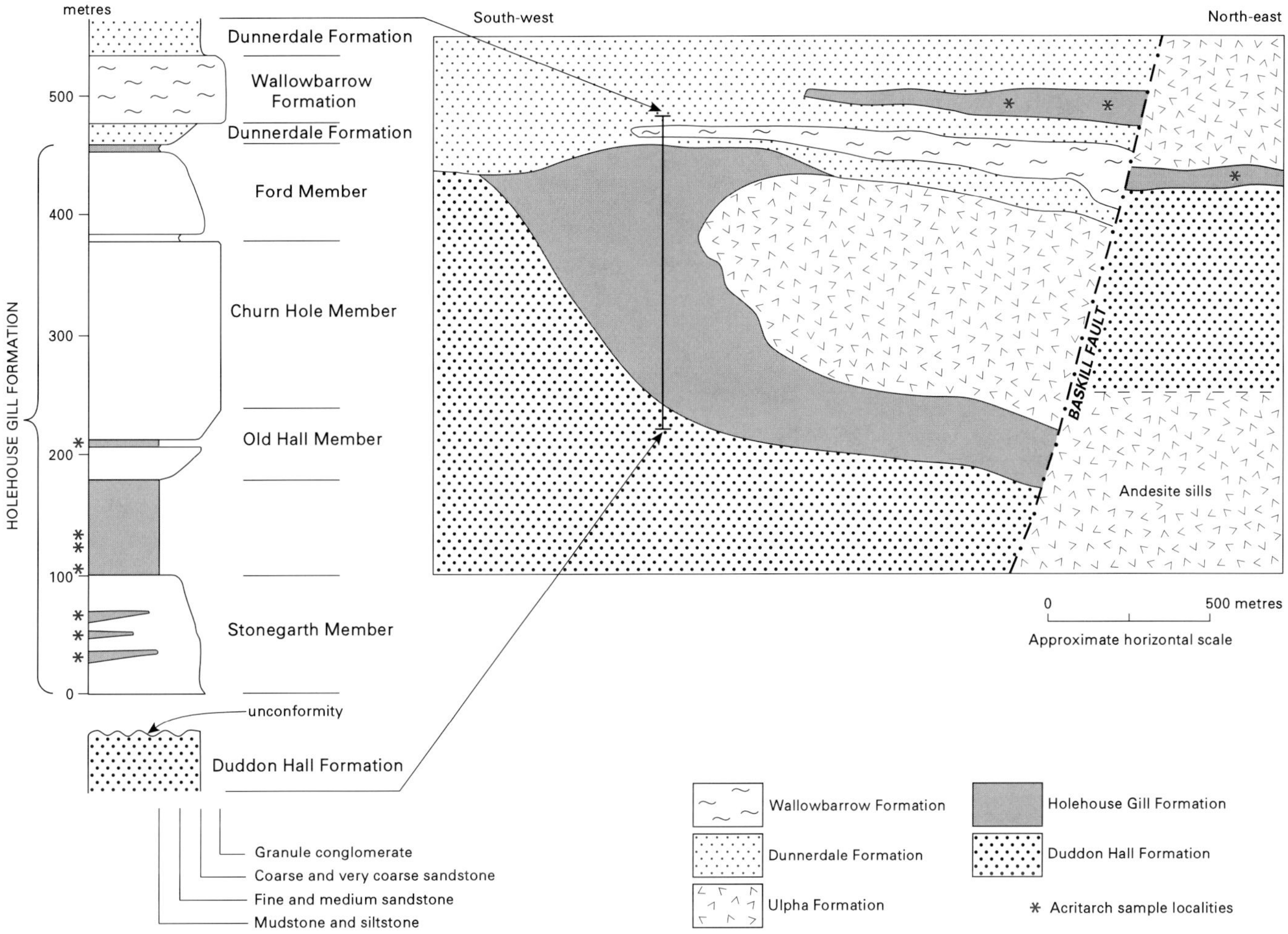

Figure 41 Holehouse Gill Formation type section and schematic horizontal section showing stratigraphical relationships with adjacent formations.

The complex interfingering with younger formations probably results mainly from syndepositional displacement on the Baskill Fault and the Bigert Mire Pasture Fault which trends parallel to, and north of, the horizontal section. Andesite lavas of the Ulpha Formation flowed into the depression from the south-east.

Fault. Contemporary displacements probably occurred on this fault which is considered to be the primary control on sedimentation of the Holehouse Gill Formation.

The base of the formation is exposed 300 m east of Old Hall Farm [SD 185 294], where the uppermost part of the underlying Duddon Hall Formation has been reworked to form the lowest 10 to 15 m of sandstone in the **Stonegarth Member** (Figure 41). The sandstone forming the upper part of the Stonegarth Member and the succeeding **Old Hall** and **Ford members** is lithologically fairly uniform, medium to very coarse grained, poorly sorted and texturally immature. Plagioclase crystals and various crystalline volcanic rock fragments are the dominant grain types, but quartz crystals, pumice fragments and siltstone clasts are also present. The silty mudstone matrix is a turbid, microcrystalline aggregate containing much sericite and carbonate, together with disseminated opaque minerals. In the Stonegarth Member, the proportion of mudstone matrix increases gradually upwards towards the overlying silty mudstone unit (Figure 41). Several discrete interbeds of mudstone also occur.

The **Churn Hole Member** comprises massive, clast-supported granule to pebble conglomerate and is approximately 150 m thick. Interbeds of laminated siltstone and sandstone are present in the basal 10 m of the member and some sandstone also occurs in the uppermost 20 m. The conglomerate consists mainly of igneous rock granules and plagioclase crystals. Most of the granules are less than 10 mm across, but range up to 25 mm; they are angular and subangular and mostly of intermediate to basic composition with aphyric, porphyritic and trachytic textures. Some mudstone, siltstone and sandstone fragments are also present. Most of the plagioclase crystals are euhedral and unabraded, but some are fragmental. The lithological uniformity and structureless nature of the conglomerate may indicate deposition as a voluminous mass-flow in

which the largest particles were concentrated and from which fines were elutriated.

The sedimentary rocks were noted first during the primary six-inch geological survey and a century later, Numan (1974) described them as marine pelites stratigraphically located at the top of Firman's (1957) Duddon Hall Tuffs. Poorly preserved acritarchs were recorded by Numan from the pelites but no taxa were determined. However, their discovery provided confirmatory evidence for marine deposition within the Borrowdale Volcanic Group and subsequent resampling yielded a probable early Caradoc flora.

The early Caradoc age for the formation is based on the occurrence of *Veryhachium irroratum* and *Frankea sartbernardensis*. *V. irroratum* most probably indicates a Caradoc or younger age, although with some limitations, as discussed by Molyneux (1988). *F. sartbernardensis* appears in the Arenig and ranges into the upper Ordovician, its highest documented occurrence in the Anglo-Welsh area being in the lower Caradoc (Servais, 1993). The joint occurrence of these species therefore favours a lower Caradoc age. The acritarch assemblage also contains long-ranging taxa that indicate a general Ordovician age and some that may be reworked, possibly from the Skiddaw Group.

Ulpha Formation

The Ulpha Formation comprises block lavas of plagioclase-phyric andesite, locally interbedded with thin units of volcaniclastic rocks. A maximum thickness of 800 m occurs on the east side of the Duddon valley near Ulpha [SD 198 935] (Figure 39); the type area is south-east of Ulpha [SD 190 920 to 200 920]. The formation thins southwards to approximately 400 m on Bleansley Bank in the Lickle valley [SD 210 890] in the adjoining Ulverston district. North-eastwards, the outcrop extends north of Hollin House Tongue [SD 227 963] where it wedges out near to the Grassguards Fault. Across the Duddon valley, the andesites thin abruptly and wedge out east of Holehouse Gill [SD 182 927] and west-south-west of Stonythwaite [SD 219 968]. The lensoidal form of the outcrop suggests that these rocks probably formed a broad, low-relief lava shield.

Mitchell (1956b) described the lavas south-east of the River Duddon as the Ulpha Andesites and divided them into a lower, Ivanscar group and an upper, Bleamsley Bank group, separated by a persistent band of lithic tuff. Firman (1957) extended the usage of Ulpha Andesites to lavas north-west of the River Duddon, but included only Mitchell's Bleamsley Bank Andesites, considering the pyroxene-phyric Ivanscar Andesites to be interbedded with the underlying Duddon Hall Tuffs. Firman's scheme is adopted in the present account and the lava sequence is formalised as the Ulpha Formation. The base of the formation is taken at the sharp and concordant base of the lowest plagioclase-phyric andesite in the succession. The lavas mostly overlie the Duddon Hall Formation but near Ulpha, on the north-west side of the Duddon valley, they rest on, and are intercalated with, the Holehouse Gill Formation.

The block lavas have massive to flow-jointed centres and marginal autobreccias. Abundant plagioclase phenocrysts and sporadic chlorite pseudomorphs after pyroxene are present, along with common mafic xenoliths up to 10 cm across. South of Ulpha, at least four lavas, each 100 to 150 m thick, are indicated by interbedded units of volcaniclastic rocks. The youngest lava terminates abruptly on the north side of Yew Pike [SD 203 925], the termination probably representing the original, steep flow margin. North of Ulpha intraformational autobreccias are the only indication of lava margins. Throughout most of the district a few metres of volcaniclastic sandstone occur at the top of the formation, thickening southwards from Yew Pike where the sediment was disrupted by loading during emplacement of the overlying Wallowbarrow Formation.

Wallowbarrow Formation

The Wallowbarrow Formation is an heterolithic, andesitic lapilli-tuff with a locally developed basal pyroclastic breccia. It has a maximum thickness of 70 m between Wallowbarrow Crag [SD 222 967] and the Grassguards Fault, the designated type area. The formation thins southwards from Wallowbarrow into the adjacent Ulverston district. An outlier occurs on Long Crag [SD 225 980] just north of the Grassguards Fault. The formation is equivalent to the 'coarse, unbedded tuffs' within Firman's (1957, p.51) Wallowbarrow Crag Group and on the south-east side of the River Duddon, was described by Mitchell (1956b) as a thin, flow-brecciated andesite and included by him in his Ulpha Andesites as the youngest flow.

The base of the formation is well defined by the marked lithological contrast with the underlying Ulpha Formation. South of Yew Pike [SD 203 925] the base is very irregular through loading of the pyroclastic rocks on to unconsolidated sediments at the top of the Ulpha Formation. Near Wallowbarrow Crag [SD 212 968] and north of Hollin House Tongue [SD 230 972], the formation oversteps the Ulpha Formation to rest directly on the Duddon Hall Formation, and north-west of Holehouse Gill to overlie the Holehouse Gill Formation. In the Long Crag outlier the formation either rests unconformably on, or is faulted against, the Birker Fell Formation.

In the type area around Wallowbarrow Crag, the formation comprises pyroclastic breccia overlain by eutaxitic lapilli-tuff. The matrix- to clast-supported breccia contains a population of various angular, intermediate and some acid lithic clasts up to 30 cm across. The coarse tuff matrix is locally eutaxitic and contains abundant plagioclase crystals. The breccia is generally massive, but in some places it is crudely stratified and at one locality rheomorphic folds are present. The overlying lithologically and texturally uniform eutaxitic lapilli-tuff has a sharp base. There is a gradual reduction in the size and abundance of clasts southwards from the type area, such that the division of the formation into breccia and eutaxitic tuff is less evident. The formation is considered to be an ignimbrite and the separation into pyroclastic breccia and lapilli-tuff at Wallowbarrow is attributed to density grading. Because the grade and thickness of the ignimbrite and

the size and abundance of lithic clasts decrease south-westwards, the vent might have been located to the north-east of the outcrop area, perhaps adjacent to the Grass-guards Fault.

Dunnerdale Formation

The Dunnerdale Formation is a coarsening-upwards succession of volcaniclastic sandstone. The formation thins north-eastwards along strike from about 250 m in the type area in the Dunnerdale Fells [SD 210 920], towards Wallowbarrow. In the faulted outlier north of Pike Side [SD 185 936], up to 400 m of strata are preserved, but the upper part of the succession there comprises an atypically coarse breccia. These rocks were previously the 'lower group' within the Dunnerdale Tuffs of Mitchell (1956b); part of Firman's (1957) Duddon Hall Formation is included within the new division.

The base of the formation is taken at the base of the lowest bed of volcaniclastic sandstone overlying andesite of the Ulpha Formation or bedded tuff of the Duddon Hall Formation. Near Bigert Mire [SD 178 925], in the outlier north-west of the River Duddon, silty mudstone of the Holehouse Gill Formation passes gradationally upwards into volcaniclastic siltstone and fine-grained sandstone of the Dunnerdale Formation, implying uninterrupted sedimentation there, whilst elsewhere the Wallowbarrow and Ulpha formations were being deposited.

In the type area, from near Yew Pike [SD 203 925] southwards to Croglinhurst [SD 212 898] in the adjacent Ulverston district, the lowest third of the formation consists of thinly bedded, fine-grained sandstone with siltstone interbeds and sedimentary characteristics of turbidite deposits. The grain-size and bed thickness increase upwards and the remainder of the formation comprises thickly and very thickly bedded, coarse- and very coarse-grained sandstone. The upward change in bedding characteristics may indicate a gradual change in depositional style from low- to high-density turbidity currents. Syndepositional, soft-state deformation and water-escape structures are common throughout the formation, particularly in its uppermost strata. Abundant dish and pillar structures, and sedimentary dykes indicate pore-water escape during deposition. Scours and cross-bedding are present locally towards the top of the succession, indicating fluvial deposition.

Trace fossils have been recorded at one locality in the Dunnerdale Formation, one of only two known occurrences in the Borrowdale Volcanic Group. They were first noted by Mitchell (1956b) who recorded 'tracks' on a sandstone bedding plane at Lum Pot [SD 2174 9045] in the banks of the River Lickle. These have been identified subsequently as nonmarine arthropod trails, probably made by a myriapod-like organism, and have been assigned to the ichnogenera *Diplichnites* and *Diplopodichnus* (Johnson et al., 1994). Variation in the morphology of the trails is thought to have been caused by drying-out of the substrate. These trace fossils record some of the earliest freshwater arthropods in the geological record (Johnson et al., 1994).

Lickle Formation

The Lickle Formation comprises three members, mainly of rhyolitic and dacitic pyroclastic rock, separated by volcaniclastic sandstone, bedded tuff, andesite and basaltic andesite (Figure 42). The formation comprises Mitchell's (1956b) Lickle Rhyolites and the 'more rhyolitic upper division' of his underlying Dunnerdale Tuffs, which forms the Kiln Bank Member in this memoir. The overlying Paddy End Member is taken from the Paddy End Rhyolites of Mitchell (1940) in the Coniston Fells. The Stickle Pike Member is newly defined. The three members crop out in the Duddon Basin, but only the Paddy End Member in the Coniston Fells. The type area for the Lickle Formation is the well-exposed ground to the south-west of Caw [SD 230 944]. A thickness of 350 to 400 m is generally seen, excluding the intercalated andesite and basaltic andesite sheets; locally up to about 600 m are present. The base of the formation is sharp and well defined by the contrast in lithology between the pale weathered pyroclastic rocks and the underlying Dunnerdale Formation. In the Duddon Basin, the base of the formation is apparently conformable on the Dunnerdale Formation. On the Coniston Fells, only the Paddy End Member is present and this overlaps the Duddon Hall Formation north-eastwards to rest on the Airy's Bridge Formation.

Kiln Bank Member

The Kiln Bank Member is a dacitic, crystal-rich coarse tuff, typically eutaxitic and containing sparse non-vesicular lithic lapilli. It is interpreted as an ignimbrite and, throughout most of the outcrop, it is apparently a single unit, although on the south side of Tarn Hill [SD 209 916] there is probably a second, thin unit near the top. The ignimbrite is welded except for the uppermost part which contains equant fiamme and crude stratification. The type area is the south-east side of the Duddon valley around Kiln Bank [SD 213 940], where the member is 140 to 180 m thick with a sharp, planar base on the Dunnerdale Formation. South of Tarn Hill near Old Hutton [SD 2070 9192 to 2076 9090], thin beds of tuff with accretionary lapilli locally occur at the base and are interpreted as pyroclastic surge deposits. On Wallowbarrow Heald, an outlier of dacitic to rhyolitic, densely welded ignimbrite, containing abundant lithic lapilli is included within the member, though it is probably a separate, locally preserved unit.

A thin succession of volcaniclastic sandstone, bedded tuff, pyroclastic breccia, and basaltic andesite overlying the ignimbrite is included within the Kiln Bank Member on Sheet 38 Ambleside. The lower part of this succession in the type area comprises 20 to 50 m of thinly bedded sandstone, locally with interbedded siltstone, breccia and tuff (Figure 42). The sandstone becomes fine grained upwards, though beds of markedly different grain size are superposed throughout. Sedimentary structures include cross-stratification, bedding discordances and scours. Overlying these beds are andesitic tuff-breccia, bedded rhyolitic tuff and andesite in a variable association that commonly fills irregularities at the top of the sandstone.

Figure 42 Comparative generalised vertical sections through the Lickle Formation, Caw Formation and parts of Lag Bank and Low Water formations showing thickness and facies variations.

PADDY END MEMBER

The Paddy End Member is a high-grade, rhyolitic ignimbrite varying from homogeneous, vitric-crystal tuff to a lava-like tuff. It is the only representative of the Lickle Formation in the Coniston Fells (Figure 42). The type area for the member is south-west of Levers Water [SD 278 989] in the Coniston Fells. Throughout this area the Paddy End Member comprises a single sheet of homogeneous tuff 150 to 170 m thick. It rests unconformably on the Duddon Hall Formation and the base is defined by the sharp change from bedded andesitic tuff to splintery, pink-weathered rhyolitic rock. Autobreccias occur at the base, and in the uppermost 40 m the member locally passes up into a eutaxitic lapilli-tuff containing sporadic elongate fiamme. The original glassy state is indicated by the presence of perlitic cracks. In thin section the texturally inhomogeneous, finely devitrified groundmass is commonly turbid, obliterating primary textures. However, in places relict vitroclastic textures indicate that the rock is a densely welded tuff. Lithic clasts are sparse, locally comprising subrounded felsite, less than 1 mm across. The crystal content is 2 to 5 per cent by volume, comprising plagioclase feldspar and very sparse chlorite with epidote pseudomorphs after a mafic mineral. The oligoclase is euhedral to subhedral and fragmented; most crystals are unzoned, but some show patchy disequilibrium reactions with the replacement of lamellar-twinned oligoclase by unzoned albite. Alteration of the feldspar includes minor saussuritisation to total replacement by sericite and epidote; small patches and veinlets of fine granular quartz are commonly present.

The thickness throughout much of the Duddon Basin is 90 to 100 m, thinning in the south to only 30 m. The base is sharp, resting on an erosion surface cut in the underlying succession, and locally the Paddy End Member oversteps the Kiln Bank Member to rest on the underlying Dunnerdale Formation. The rheomorphic characteristics of the Paddy End Member are gradually lost south-westwards from the Stonythwaite Fault. There, towards the top there is a gradational reduction in the fiamme aspect ratio and the uppermost 10 to 15 m are only partly welded. Locally, the uppermost part of the member is stratified, for example up to 7 m of bedded ash-fall tuff crop out on Great Stickle [SD 2117 9159].

Up to 10 m of bedded rhyolitic tuff with interbeds of accretionary lapilli-tuff, occur locally beneath the ignimbrite and probably represent precursor ash-fall deposits. Apparent welding, rheomorphic structures, lithophysae and secondary silicification suggest that these beds have been sintered by the overlying high-grade ignimbrite. South of the Stonythwaite Fault, the Paddy End Member is overlain by 10 to 30 m of volcaniclastic sandstone, typically with thin parallel bedding but with some medium and thick interbeds. The grain size is mostly medium to fine, though some thick beds are coarse or very coarse grained and contain scattered pebbles. Scours and cross-bedding occur on The Knott [SD 2433 9320] and syndepositional, soft-state deformation structures are present south of Stickle Tarn [SD 2139 9270].

Lapilli-tuff, up to 240 m thick, overlies the Paddy End Member between the Grassguards and Stonythwaite faults (Column 5, Figure 42) and in Throng Close [SD 2410 9815]. It is massive, welded, lithic rich, poorly sorted and locally has a eutaxitic texture. The dominant lithic clast is an angular, pink-weathered rhyolite probably derived from the underlying ignimbrite. At the base a clast-supported breccia consists almost entirely of interlocking dacite clasts up to 60 cm across. The breccia passes gradationally upwards across complex, steep contacts into heterolithic matrix-supported lapilli-tuff which also becomes finer grained upwards.

STICKLE PIKE MEMBER

The Stickle Pike Member is a pale-weathered rhyolitic eutaxitic to parataxitic lapilli-tuff, containing non-vesiculated, juvenile lithic fragments, and interpreted as a welded ignimbrite. The welding fabric is locally enhanced by silicification and vitroclastic textures are preserved commonly in the groundmass. The ignimbrite is remarkably uniform throughout the Duddon Basin. Columnar cooling joints are present locally, for example on Stephenson Ground Crag (Plate 14) [SD 2339 9346]. The uppermost 10 m of the member are weakly welded. In the type area on Stickle Pike [SD 2120 9276] the thickness is 90 to 100 m, but towards the south-east margin of the Duddon Basin it thins to 20 m. The sharp base is broadly conformable on underlying strata, but locally the member fills channels, for example on Great Stickle [SD 2117 9159] and south of Stainton Ground Quarries [SD 2222 9322].

Caw Formation

The Caw Formation consists of a coarsening-upwards sequence of volcaniclastic sandstone, breccia and conglomerate previously known as the Caw Tuffs which formed the basal unit of the Tilberthwaite Tuffs of Mitchell (1956b). The diachronous base of the formation is well defined in locations where parallel-laminated sandstone overlies rhyolitic lapilli-tuff of the Stickle Pike Member; but where conglomerate and breccia form the basal deposits the base is uncertain. The sandstone lithofacies (Plate 15) is conformable on the underlying Lickle Formation, except for low-angle discordances locally. The poorly sorted, medium- to coarse-grained sandstone is mostly thinly parallel bedded or laminated. Cross-bedding, ripples and convolute laminae occur in places and some scours up to 1 m wide in the basal part of the succession contain coarse, gravel-lag layers. Tuff interbeds up to 2 m thick are common in the basal part and many contain accretionary lapilli up to 1 cm (Plate 16) and sporadic volcanic bombs. Load structures and disrupted bedding occur locally in sandstone beneath some of the thicker tuff beds. The sandstone passes gradationally over a few tens of metres into the overlying heterolithic breccia and conglomerate. This lithofacies contains abundant pyroclasts, including pumice, up to 0.15 m across. The breccia is mostly massive, though weak and laterally impersistent bedding planes and some sandstone interbeds are present.

Plate 14 Fanned columnar joints in welded ignimbrite from the Lickle Formation. Stephenson Ground Crag, near Broughton Mills [SD 2339 9346] (D 4135). Observer is 1.8 m tall.

The formation is 710 m in the type area on Caw [SD 230 944] (Figure 39). Southwards, the formation thins to less than 10 m over a distance of 2.5 km and to the north it wedges out in about 2 km. The 10 m-thick sequence in the south consists entirely of sandstone. This thickens progressively northwards and is overlain by an increasing amount of conglomerate and breccia, such that on Stephenson Ground Crag [SD 236 934], about half-way towards the type section, 80 m of sandstone are overlain by 50 to 150 m of conglomerate and breccia. On Caw, the sandstone is 160 m and the breccia 550 m thick. North-eastwards from Caw the reverse trend is apparent. The marked marginal thinning of the formation and its components indicates deposition within a sedimentary basin no more than about 5 km wide.

Lag Bank Formation

The Lag Bank Formation, formerly the Lag Bank Tuffs of Mitchell (1956b), comprises dacitic lapilli-tuff interpreted as an ignimbrite. It crops out throughout the Duddon Basin, extending into the Coniston Fells (Figure 39). In the type area, near Lag Bank [SD 247 943], the ignimbrite is about 600 m thick and at least 400 m are present on The Old Man of Coniston [SD 272 978]. The base is sharp and well defined where the Lag Bank Formation overlies the Caw Formation, by the change from sedimentary breccia to welded lapilli-tuff. On the Coniston Fells the Lag Bank Formation overlies the Paddy End Member.

Most of the formation comprises lithologically uniform lapilli-tuff, typically containing abundant white-weathered juvenile plagioclase-phyric dacite lapilli. Fiamme are only locally abundant enough to give a eutaxitic texture. The crystal component is mainly plagioclase. Relict vitroclastic textures in unaltered parts of the matrix show that the lapilli-tuff is moderately to densely welded, but that the welding intensity varies, generally lessening towards the top of the formation. Columnar jointing is notable in the lowest and middle parts of the formation on Lag Bank. The uppermost 10 m of the formation are crudely stratified in places, with interbeds of tuff-breccia and laminated tuff up to 15 cm thick containing accretionary lapilli. The latter probably represent ash-fall or pyroclastic surge deposits.

Breccias are present locally at, or near, the base of the formation. South-west of The Old Man of Coniston, from north of Blind Tarn [SD 2620 9689] to Goat Crag [SD 2682 9749], a wedge of very poorly sorted, clast-supported breccia up to 50 m thick contains angular blocks up to 10 m across, of eutaxitic lapilli-tuff, bedded accretionary lapilli-tuff (similar to that of the Duddon Hall Formation) and foliated rhyolitic tuff (Paddy End Member). Up to 90 m of breccia of similar appearance occur adjacent to the Stonythwaite Fault on High Moss Close [SD 242 960]. These probable mesobreccias pass gradationally upwards, over a few metres, into lapilli-tuff. East of Stephenson Ground Scale [SD 237 940] a basal breccia contains non-vesiculated clasts, up to 0.8 m across, in a flow-foliated, folded and brecciated lapilli-tuff matrix. The breccia appears to have formed in situ through the interaction between the hot ignimbrite and the underlying wet sediments. The soft-sediment deformation structures in the underlying sandstone probably resulted from loading by the Lag Bank Formation.

Low Water Formation

More than half of the maximum thickness of 600 m of the Low Water Formation comprises two sheets of homogeneous welded dacitic lapilli-tuff, interpreted as ignimbrite. Along most of the narrow outcrop from Low Water [SD 275 983] to Greenburn Beck [NY 310 030] the formation overlies the Paddy End Member, but at the southern end it rests on the Lag Bank Formation (Figure 39). The type area is at Lad Stones, Above Beck Fells [NY 2935 0010 to SD 2940 9930]. The Low Water Formation was included previously in Mitchell's (1940) Upper Tilberthwaite Tuffs. A well-developed eutaxitic texture is present throughout, with fiamme up to 12 × 2 cm comprising up to 20 per cent of the rock. The lapilli-tuff is generally lithic rich, with pale weathered, angular, non-vesicular clasts mostly less than 3 cm across and rarely up to 8 cm. A clast-supported, very poorly sorted lithic-rich

Plate 15 Ripple-laminated sandstone within thinly bedded and laminated succession of the Caw Formation.
Note the low-angle bedding discordances. Broadslack Beck, Stephenson Ground, near Broughton Mills [SD 2342 9365] (D 4137). Hammer head is 17 cm long.

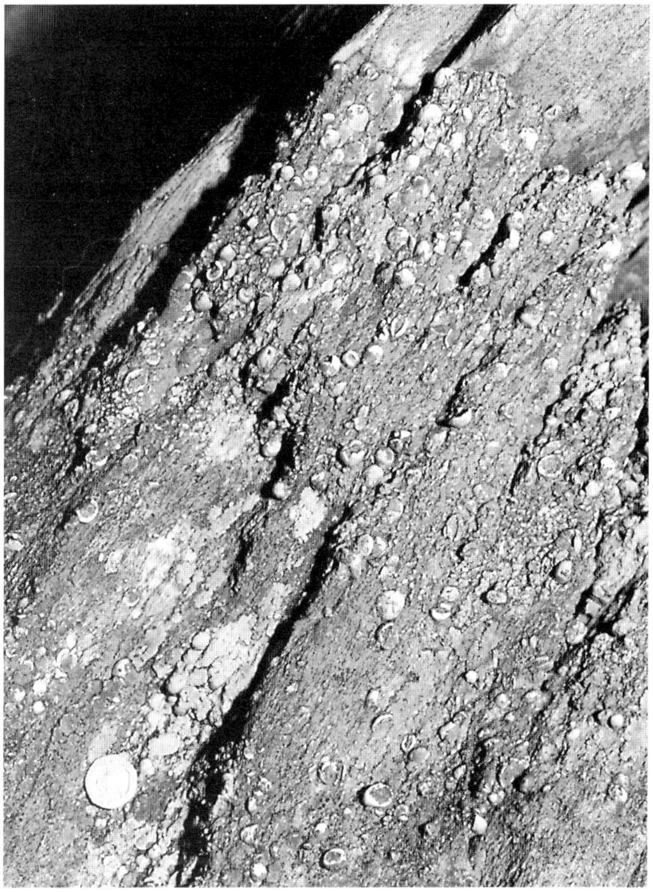

Plate 16 Accretionary lapilli-tuff.
These are characteristic of air-fall and phreatomagmatic deposits at many localities throughout the Borrowdale Volcanic Group. This example is from the Caw Formation, exposed at Ravens Crag [SD 2236 9250], east Dunnerdale, near Broughton Mills (D 4141). Coin is 21 mm across.

zone up to 10 m thick occurs at the base containing blocks up to 70 cm across and is probably a lag breccia.

The ignimbrite sheets are separated by up to 35 m of laminated tuff and lapilli-tuff, locally containing beds of accretionary lapilli. In some parts compound beds, up to 4 m, thick contain alternations of lithic-rich and fiamme-rich lapilli-tuff separated by parallel-laminated to thinly bedded dacitic tuff and lapilli-tuff. On Above Beck Fells [SD 292 998] and in the surrounding area, up to 220 m of thinly to thickly bedded sandstone and pebbly sandstone are interbedded with coarse tuff and fine lapilli-tuff at the base of the formation. Locally, the sandstone is interbedded with dark grey silty laminae and is probably at least partly water laid.

Lingmoor Fell Formation

The Lingmoor Fell Formation comprises dacitic welded lapilli-tuff, interpreted as ignimbrite, and crops out on Lingmoor Fell [NY 303 044] between Great and Little Langdale; this constitutes the type area [NY 290 040 to 300 030]. Three ignimbrite sheets, separated by intercalated bedded tuff or breccia, constitute the sequence on the west side of Lingmoor Fell. The base is not seen and the thickness of the formation exceeds 1400 m. Fault-bounded outcrops of dacitic lapilli-tuff that occur in a similar stratigraphical position to the east of Ambleside are assigned provisionally to the Lingmoor Fell Formation on Sheet 38 Ambleside. The formation may include, or correlate with, the Low Water Formation in the Coniston Fells. Formerly, the strata were part of the intrusive Langdale Rhyolite of Hartley (1932), but were included by Moseley (1990; 1993) in the Airy's Bridge Formation.

The lowest ignimbrite, exposed on the north-east side of Bleamoss Beck [NY 301 032 to 298 036], has a strong near-vertical rheomorphic foliation in places, including folds of generally less than 1 m amplitude and wavelength, and patches of domain breccia. Shallow irregularities in the top of this ignimbrite are filled with glassy, fine crystal (plagioclase) tuff. The overlying 50 m-thick, weakly parallel-laminated tuff is coarse grained and contains scattered small lithic lapilli as well as poorly formed,

single-rimmed accretionary lapilli. In places there is poorly developed cross-stratification.

The middle unit, 100 m thick, is well exposed in Gill Grains [NY 302 037] and consists of massive, eutaxitic lapilli-tuff, succeeded by 20 m of heterolithic breccia containing eutaxitic and parataxitic tuff clasts set in a eutaxitic lapilli-tuff matrix. The clasts vary from subangular to rounded, up to 0.6 m across, and may have been derived from the Bad Step or Crinkle members of the Airy's Bridge Formation.

The uppermost unit comprises more than 600 m of dacitic lapilli-tuff, interpreted as ignimbrite, along with subordinate pyroclastic surge and ash-fall deposits (Branney, 1988b). Minor variations in the proportions of lithic and pumice lapilli and in the development of eutaxitic foliation are evident. Abundant plagioclase crystals are present throughout the matrix. In general, the eutaxitic foliation dips gently northward, broadly coincident with dips in the overlying Seathwaite Fell Formation. Also, parts of the formation are preserved in fault blocks in the Side Pike Complex [NY 290 052].

SUCCESSION IN THE RYDAL AREA

A widespread, thick and mainly sedimentary sequence, comprising the Seathwaite Fell Formation, overlies the Lingmell Formation in the Central Fells and overlaps a number of formations in the Duddon Basin. The Seathwaite Fell Formation is overlain by the Lincomb Tarns Formation, an extensive ignimbrite cropping out widely in the north-eastern part of the district. The Esk Pike and Tarn Hows formations and the Middle Dodd Dacite occur as outliers on the highest fells and beneath the unconformity with the overlying Windermere Supergroup. Stratigraphical relationships are summarised in Figure 43.

Seathwaite Fell Formation

The Seathwaite Fell Formation consists mainly of subaqueously deposited volcaniclastic sandstone with intercalations of pyroclastic lithofacies and penecontemporaneous sills. The type area is on Seathwaite Fell, in Borrowdale [NY 230 100] but the outcrop is extensive. The name is modified from Oliver's (1961) Seathwaite Fell Tuffs which was defined originally when stratigraphical correlations between the Central and Coniston fells had not been established. The newly defined formation in the Ambleside district includes the Upper Tilberthwaite and Yewdale Bedded tuffs of Mitchell (1940), the Walna Scar Quarries/Quarry Tuffs of Mitchell (1956b; 1963) and the Bedded Tuffs in Langdale and Grasmere of Hartley (1925; 1932). Though Oliver's designation does not have precedence, Seathwaite Fell Formation is used because of the location of the well-exposed type area. In the Coniston, Tilberthwaite, Grasmere and Ambleside areas the sequence is much faulted.

The base of the formation is sharp and diachronous. In north Langdale and on Bowfell, the Seathwaite Fell Formation is apparently conformable on the Lingmell Formation, but farther west it onlaps the Scafell Dacite. On Lingmoor Fell and in the Coniston Fells there is a marked angular discordance. This is exposed at a number of localities but best illustrated south-west of The Old Man of Coniston, near Cove Quarries [SD 271 973], where laminated sandstone and an interbedded dacitic lapilli-tuff that is used as a datum, onlap an erosion surface with at least 100 m of relief cut across the near-vertical Lag Bank Formation (Plate 17).

The thickness of the Seathwaite Fell Formation in the Central Fells is 300 to 540 m, but on the west side of Lingmoor Fell it is only 30 m thick. In the Coniston Fells the formation is 580 to 1100 m thick, excluding the sills and discounting thinning caused by the imposition of a strong cleavage in steeply dipping rocks. East of the Coniston Fault, in the Rydal area, there are at least 400 m; the base of the formation is exposed rarely. South of Snarker Pike [NY 390 077] the thickness is more than 800 m.

Many lithofacies make up the formation (Plates 18 to 21) and in the Scafell Syncline distinctive lithofacies have been combined to form members (Kneller and McConnell, 1993; Figures 43, 44). Elsewhere, only the more substantial, interbedded pyroclastic units have been recognised informally as members. Overall, the succession coarsens upwards indicating increasing rates of sedimentation through its depositional history. Synsedimentary, soft-state deformation and water-escape structures occur throughout the formation and attest to its rapid deposition (Plate 18; Parker, 1966). Interbedded pyroclastic rocks vary from thin beds of fine tuff to breccia units more than 100 m thick. The formation is also host to andesite and basaltic andesite intrusions, some of which were considered previously to be lavas and named as part of the volcanic succession, for example the Wrengill Andesites of Mitchell (1940) (Table 6). The inter-relationships of the sedimentary and pyroclastic components with older strata and faults reveal contemporaneous tectonic activity throughout the history of the depositional basin in which the rocks accumulated.

Lithofacies variation in the Seathwaite Fell Formation of the type area in the Scafell Syncline (Figure 44) is described below, followed by the principal variations and correlations eastwards across the Coniston Fault and south-westwards into the Coniston Fells. The members present in the Central Fells are described in detail by Kneller and McConnell (1993) and their distribution shown on the component 1:10 000 scale British Geological Survey maps for the Ambleside district. Of the named subdivisions only the Cam Crags and Pavey Ark members are shown on Sheet 38 Ambleside.

Three Tarns Member

The basal member in the Central Fells consists predominantly of white-weathered laminated and thinly bedded, fine-grained sandstone and siltstone. The base is sharp where the member overlies pyroclastic rocks in the Lingmell Formation and locally where it rests on the Airy's Bridge Formation and the Scafell Dacite, but gradational and arbitrary where the upper part of the Lingmell Formation has been reworked subaqueously. The thickness is mostly 50 to 60 m, but the member wedges out against the Scafell Dacite.

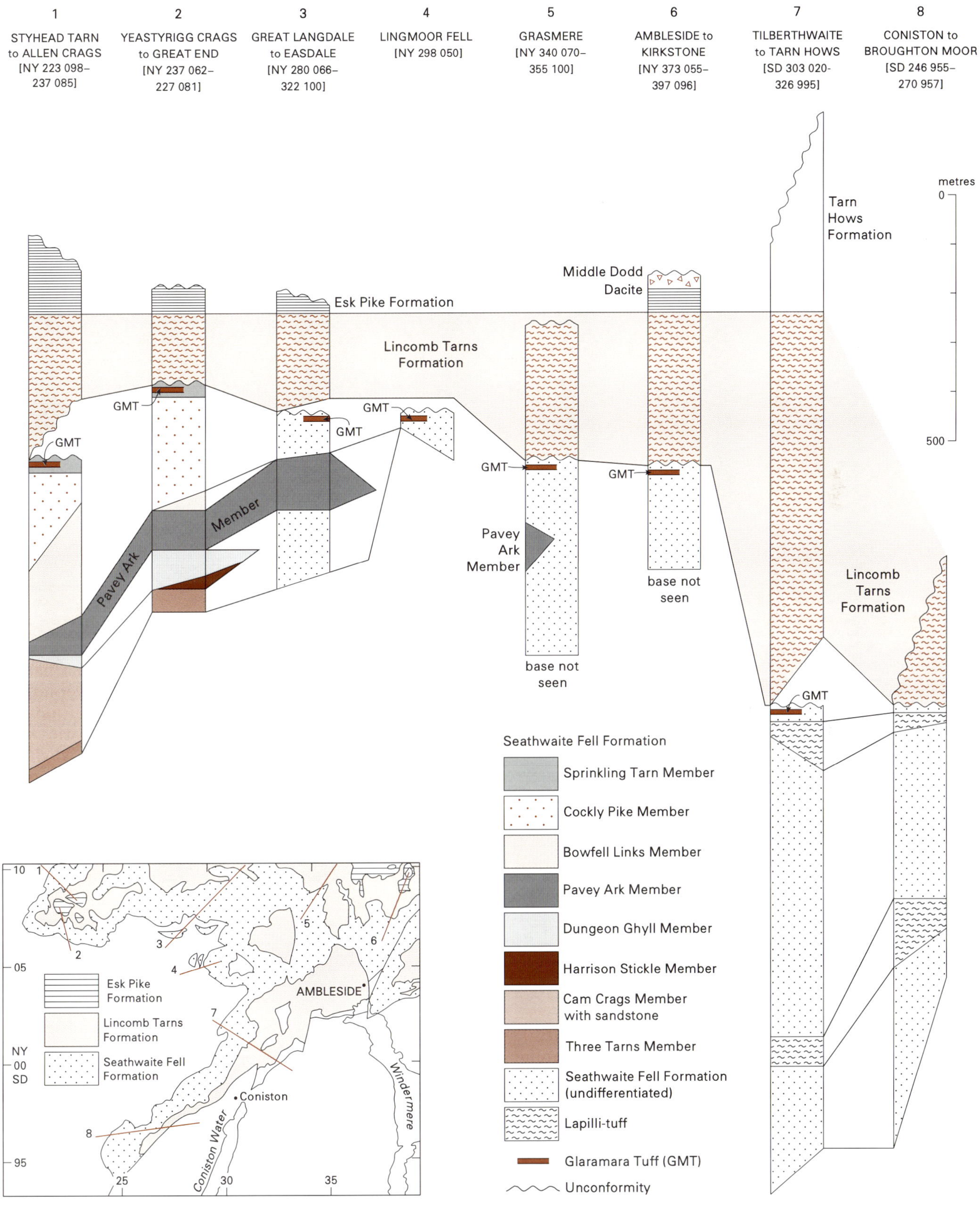

Figure 43 Comparative generalised vertical sections through the volcanic succession from the Seathwaite Fell Formation to the unconformity with the overlying Windermere Supergroup.

Plate 17 Pre-Seathwaite Fell Formation collapse as seen in the south slopes of The Old Man of Coniston [SD 268 964].

The basal sandstone beds in the Seathwaite Fell Formation overstep the underlying massive ignimbrite of the Lag Bank Formation and just beyond the centre skyline the ignimbrite within the Seathwaite Fell Formation rests directly on the Lag Bank Formation. The upper and lower contacts of the Paddy End Member dip very steeply (dip and strike shown) indicating that it had been rotated prior to onset of the clastic sedimentation of the Seathwaite Fell Formation (D 4037).

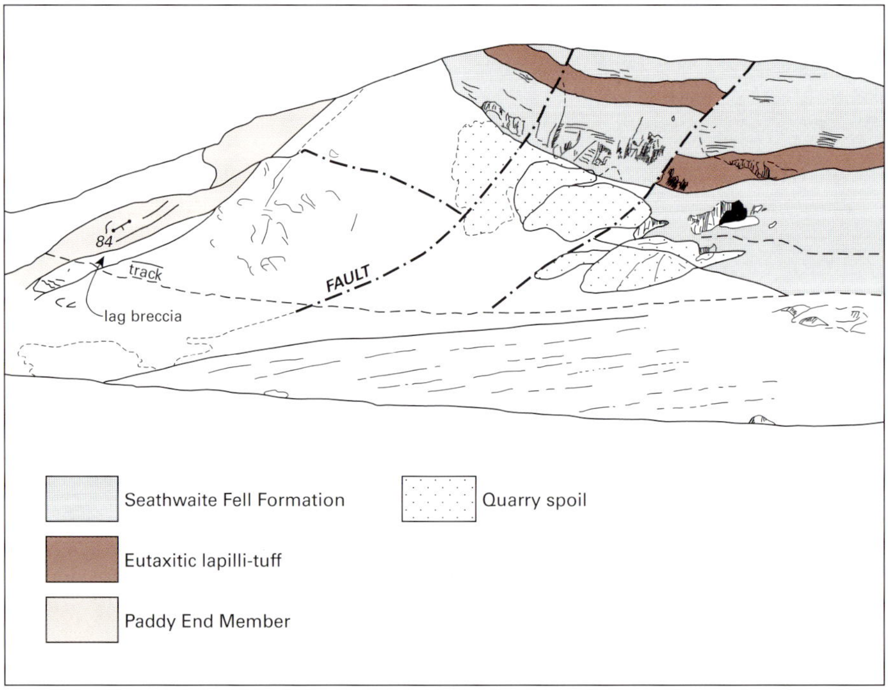

Kneller and McConnell (1993) interpret the dominantly thinly bedded, mostly normally graded sandstone as either a turbidite, or possibly subaqueous ash-fall, basin-floor deposits, although evidence of wave reworking is apparently widespread. Fiamme-like pumice clasts, commonly concentrated towards the top of some thick graded sandstone beds are probably the result of diagenetic flattening. The inclusion of pumice indicates either a pyroclastic origin or incorporation of locally reworked tephra. The thick beds of fiamme-bearing sandstone overlie strata with convolute lamination, the disturbance of which might have been caused by seismicity accompanying the pumice-generating eruptions and also responsible for triggering turbidite flow; alternatively, it may have resulted simply from rapid loading of the unconsolidated sediments. The uppermost parts of the Three Tarns Member have been disturbed by loading from the overlying Harrison Stickle Member breccia where present.

HARRISON STICKLE MEMBER

This member is a lithologically variable succession of pebble-grade breccia, lapilli-tuff and pebbly sandstone that overlies the Three Tarns Member on the southern side of the Scafell Syncline. The maximum thickness of 30 m occurs in the type area at Harrison Stickle [NY 282 073] where poorly sorted, clast-supported pebble- to cobble-grade breccia forms two distinctive lithofacies. The lower has a reverse-graded base, a variable matrix content and is 2 to 6 m thick. The upper is coarser, clast supported and contains three symmetrical, normal to reverse-graded breccia units with thin, central siltstone interbeds.

The clast population comprises various types of andesite and basaltic andesite, and larger (but rarer) blocks of dacite and rhyolite. The apparent welding of some breccia units in the type area led Kneller and McConnell (1993) to suggest deposition from pyroclastic flows into water. Several such ignimbrites are present around Pike of Stickle [NY 274 074] and some of the lapilli-tuff has a eutaxitic texture. The localisation of welding fabrics in the

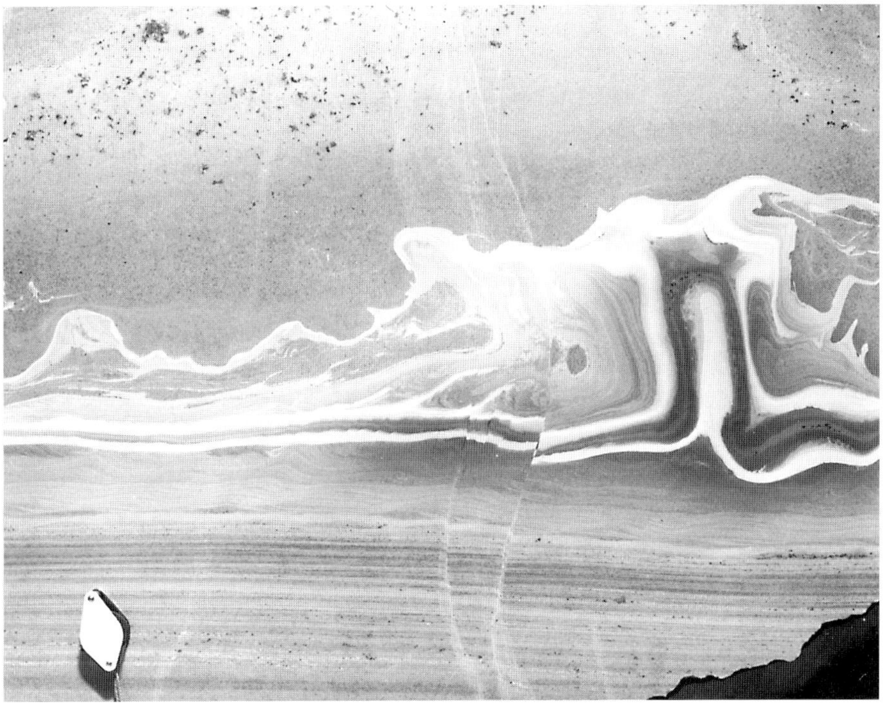

Plate 18 Parallel-laminated fine-grained sandstone and ripple cross-laminated sandstone in the lower part of the picture are overlain by mudstone that has been deformed by rapid loading from the overlying sandstone to produce the diapir and flame structures, Seathwaite Fell Formation. Cut slab from Broughton Moor [SD 253 945] (D 4149).

Plate 19 Massive coarse-grained sandstone, graded beds of pebble conglomerate and channel cross-bedded sandstone, Seathwaite Fell Formation. Low Long Crag, Torver High Common [SD 266 962] (D 4041). Hammer is 37 cm long.

eastern part of the outcrop and the distribution of thicker and coarse-grained flow-units, suggest a source in the east or south-east.

CAM CRAGS MEMBER

Up to about 230 m of pebble-grade breccia and pebbly, coarse-grained sandstone with rhyolitic and dacitic clasts constitute the Cam Crags Member. It overlies either the thinly bedded Three Tarns Member or the Scafell Dacite and is present in the north of the district. The type area is Thorneythwaite Fell to Glaramara [NY 240 120 to 247 105], in the Keswick district. The pebble breccias are both clast supported (gravelly) and matrix supported, with a locally well-defined parallel stratification and some very low-angle cross-bedding. At the northern extremity of the district the sequence is dominated by sandstone, but on Seathwaite Fell [NY 223 095] the size and proportion of dacite cobbles and the coarseness of the clast-supported dacitic breccia interbeds increase towards the margin of the Scafell Dacite. The member is wedge-shaped and has residual dips of about 15° to the south and south-east which are interpreted as indicative of the original depositional slope. It may be part of a system of coalesced, steep-fronted gravel deltas supplied from the north-north-west (Kneller and McConnell, 1993).

DUNGEON GHYLL MEMBER

The Dungeon Ghyll Member varies from siltstone to breccia and is characterised by widespread syndepositional deformation and slumping so that bedding is rarely apparent, especially in the finer-grained lithologies. The outcrop is discontinuous and the member locally rests on the Three Tarns, Cam Crags or Harrison Stickle members. West of the River Esk, the Dungeon Ghyll Member interdigitates with the Mickledore Breccia on the margin of the Scafell Dacite (Figure 44).

The type area is near Dungeon Ghyll in the Langdale Pikes [NY 285 066] where the member consists of up to 60 m of bedded and massive, fine- to coarse-grained sandstone. Elsewhere, the lower part is finer

Plate 20 Cross-bedded coarse-grained sandstone and pebbly sandstone. Pebbles are mostly of volcaniclastic siltstone rip-up clasts, Seathwaite Formation. Low Long Crag, Torver High Common [SD 2668 9605] (D 4044). Hammer head is 17 cm long.

Plate 21 Channel fill of poorly sorted and cleaved volcaniclastic conglomerate. Seathwaite Fell Formation, Low Long Crag, Torver High Common [SD 266 961] (D 4040). Hammer head is 17 cm long.

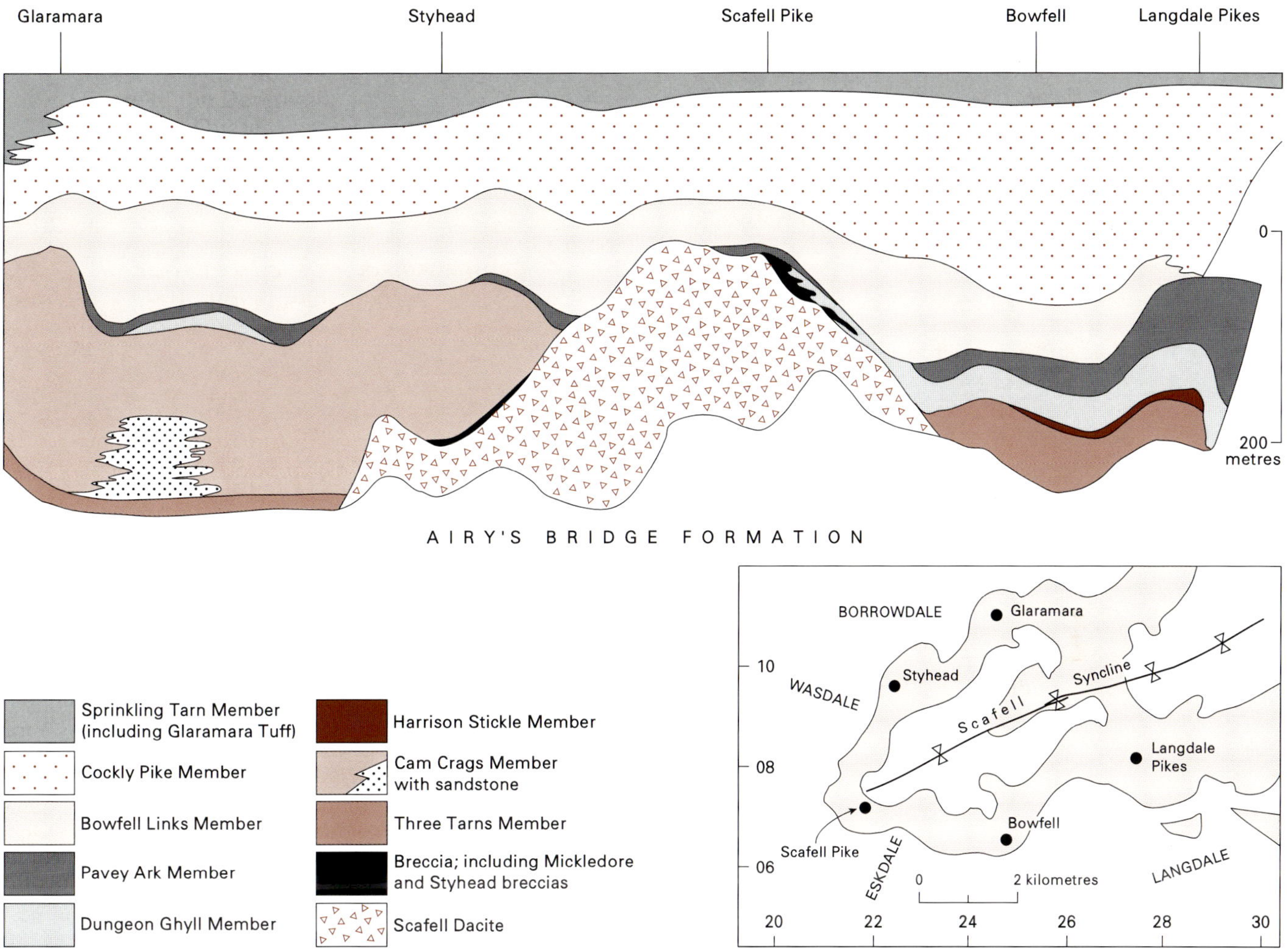

Figure 44 Diagram showing the thickness and lithofacies variations of the Seathwaite Fell Formation in the Scafell Syncline.

The thickness of the Scafell Dacite has been reduced by 39% to allow for compaction of the volcaniclastic sedimentary rocks in order to show best its original geometry (modified from Kneller and McConnell, 1993). Inset shows outcrop of the Seathwaite Fell Formation in the Scafell Syncline.

grained, largely siltstone with interbeds of lapilli-tuff, whereas the upper part commonly contains pebbly sandstone and breccia. The lithologies of the upper part are interpreted as gravity current deposits (Kneller and McConnell, 1993); they are commonly graded, with inverse grading at the base in places, and accumulated in palaeotopographical hollows. Widespread vague and chaotic relict bedding, but local preservation of well-bedded sections, particularly near the base of the succession, probably indicate that original bedding structures have been disrupted subsequently, probably by liquefaction, loading and mass-flow (Kneller and McConnell, 1993). Locally, these sediments may have been emplaced as a single, relatively low-viscosity, dilute debris flow, but generally a more complex, multiphase origin seems likely. Epidotic concretions, up to 80 cm in diameter, occur as a poorly understood feature of the upper part of some thick beds of sandstone.

Pavey Ark Member

The Pavey Ark Member consists of brown-weathered, massive breccia and weakly bedded, pebbly sandstone generally overlain by a variable sandstone to siltstone unit. In the Central Fells the thickness commonly ranges up to 80 m and exceptionally may reach 200 m, but to the north of Grasmere, just east of the Coniston Fault, more than 500 m are probably present. This wide range in thickness reflects the style of deposition of the member, filling hollows in an irregular palaeotopography; the breccia rests on the Cam Crags or Dungeon Ghyll members and wedges out against the Scafell Dacite (Figure 44). Locally, for example at White Gill [NY 300 068] in Great Langdale, the breccia is associated with a rotated fault block and is up to 200 m thick against the contemporaneous fault scarp.

At Pavey Ark [NY 285 078], the type area, the main lithofacies is a massive and generally poorly sorted, clast-

to matrix-supported, heterolithic coarse breccia. A pervasive, bedding-parallel foliation is commonly present. Locally, flattened, ragged and embayed vesicular andesite spatter is concentrated in two layers, about 30 and 50 m above the base of the member. The originally highly fluid pyroclasts are typically less than 40 cm long but range up to 1.7 m. The breccia becomes finer grained and stratified at the top, containing metre-scale channels and trough cross-bedding before grading abruptly upwards into the overlying sandstone and siltstone.

The breccias are interpreted to result from eruption-related subaqueous gravity flows (Kneller and McConnell, 1993). The presence of the andesite spatter indicates close association with an eruption and their general decrease in size westwards suggests a source farther east. Slurry breccias, pebbly mudstones and the incorporation of chaotic sediments are indicative of debris flows. The westward increase in the development of bedding, coarse-tail grading and sorting indicate transition to a more dilute, turbulent regime, again indicative of an easterly source, probably to the north of Grasmere where the member is thickest and the igneous component most dominant. There the lithofacies contains many andesite blocks probably derived from a part intrusive, part extrusive, magma body emplaced at the sediment–water interface. The westward progress of the debris flows was impeded by the barrier formed by the Scafell Dacite.

Bowfell Links Member

Thinly bedded, parallel and ripple cross-laminated siltstone and fine-grained sandstone are the dominant lithofacies of the Bowfell Links Member with some coarse-grained sandstone beds, locally containing 'fiamme'. Many individual beds are normally graded and some have ripple-laminated tops. The succession is generally between 30 and 50 m thick but in the type area, north-east of Scafell Pike, up to 120 m are present. The base is usually sharp above the coarser underlying members and is taken where uniform thin bedding first appears. The Bowfell Links Member usually overlies the Pavey Ark Member but locally oversteps it to rest on the Cam Crags Member, the Dungeon Ghyll Member or the Scafell Dacite. The Bowfell Links Member thins over the dacite and is only a few metres thick above the thick development of the Pavey Ark Member on the west side of the Langdale Pikes (Figure 44).

Soft-state deformation occurs locally and shows a close, spatial relationship with faults across which thickness and lithofacies variations indicate syndepositional movements. Examples are seen adjacent to the Woof Gill Fault in Langstrath and on the south side of Bowfell summit where there is a narrow zone of chaotic and disturbed bedding adjacent to a fault zone in which the sandstone is brecciated (Kneller and McConnell, 1993).

The dominantly fine-grained and thinly bedded sequence is cut by a few channels filled with thickly bedded or massive pebbly sandstone. Overall, palaeocurrent indicators show sediment dispersal towards the north-east, though in Great Langdale there is evidence for local distribution towards the north-west. The strata are interpreted as low- and high-density turbidite deposits, predominantly from surge-type flows (Kneller and McConnell, 1993). The presence of wave ripples suggests maximum water depths of a few tens of metres.

Cockly Pike Member

Massive or weakly stratified, coarse- to very coarse-grained sandstone is the dominant lithology with subordinate interbedded siltstone, finer-grained sandstone and breccia. The base is taken where the succession changes, mostly very abruptly, from the siltstone and fine-grained sandstone that form the underlying Bowfell Links Member to massive, very coarse-grained sandstone. Usually the base is well defined but there is some interdigitation locally with the Bowfell Links Member. The thickness varies from 55 to 250 m, increasing southwards across a wide outcrop within the Scafell Syncline. Contemporaneous faulting occurred during deposition of the member, on structures that were also active when the underlying Bowfell Links Member was laid down.

Three distinctive lithofacies are recognised. The lowest unit consists of very thickly bedded to massive, coarse- to very coarse-grained sandstone 10 to 100 m thick. The basal few metres of the sandstone are commonly stratified but, generally, bedding within the unit is only defined by laterally impersistent amalgamation surfaces and unsystematically graded layers. Interbedded debris-flow deposits are present locally. In the central lithofacies stratification is better developed and there are interbeds of conglomerate with angular eutaxitic lapilli-tuff cobbles, fiamme-bearing sandstone beds and possibly bedded tuff layers. The uppermost lithofacies consists of either well-bedded parallel to cross-stratified gravelly sandstone, or poorly bedded coarse-grained sandstone with interbeds of siltstone and fine-grained sandstone, and containing many steep-sided scours.

The massive sandstone beds in the lowest part of the Cockley Pike Member were interpreted by Branney et al. (1990) as a disordered turbidite facies deposited from high-density flows. Channelling is fairly widespread, particularly in the south of the Scafell Syncline, which accords with the limited palaeocurrent data for sediment dispersal to the north and north-west. The better-developed stratification in the overlying middle lithofacies indicates decreasing rates of sedimentation with breccio-conglomerate filling transient channels. Fiamme-bearing sandstone, the presence of pyroclastic fall deposits and local beds of fine tuff record contemporaneous episodes of volcanism during sedimentation. In the uppermost lithofacies, the proportion of siltstone and fine-grained sandstone increases eastwards which suggests that water depths and the distance from the sediment source increased in this direction. Developments of cross-stratified gravelly facies within the uppermost lithofacies apparently arise from reworking of the local basal facies of the overlying Sprinkling Tarn Member. These interdigitating stratigraphical relationships indicate penecontemporaneous erosion and redistribution of the uppermost member and establish that its base is diachronous.

Sprinkling Tarn Member

The youngest member of the Seathwaite Fell Formation in the Central Fells comprises interbedded coarse-grained

pebbly sandstone and pyroclastic rocks. The facies association distinguishes the member from the underlying Cockly Pike Member although sandstone in both members is lithologically similar and contact relationships are complex and diachronous. The gradational base of the member is taken where pyroclastic rocks first appear in the sandstone succession. Two distinctive pyroclastic facies are present: banded andesitic ash-fall deposits and a succession of rhyodacitic flow, fall-out and surge deposits known as the Glaramara Tuff. The Sprinkling Tarn Member is present throughout the Central Fells, with a maximum thickness of 70 m in the type-area near Sprinkling Tarn [NY 228 091].

Planar bed forms and cross-bedding are moderately to well defined in the sandstone, with trough cross-stratification locally well developed in scour and channel fills. Abrupt lateral facies variations result from either contemporaneous erosion or fault-controlled sedimentation. Bed forms and sedimentary structures include low-angle ripple cross-lamination, some low-angle cross-bedding and channels, slump folds and brecciation, shallow scours that are draped by younger fall-out deposits, high-angle soft-state faults and low-angle slump detachment surfaces.

Banded pyroclastic fall-out deposits occur in units of parallel-laminated and thinly bedded, fine and coarse crystal-rich tuff up to 6 m thick. This assemblage occurs below, above and interstratified with, the Glaramara Tuff. Bedding is defined by abrupt, vertical variations in grainsize; most beds are ungraded but a few have fine tuff laminae at the top. The characteristics suggest that these are phreatomagmatic fall-out deposits (Kneller and McConnell, 1993).

Glaramara Tuff

The Glaramara Tuff is a widespread unit within the uppermost 20 to 30 m of the Sprinkling Tarn Member. It comprises rhyodacitic pyroclastic flow, fall-out and surge deposits and, because of its distinctive appearance and chemistry compared with the overall andesitic nature of the Seathwaite Fell Formation, is an important stratigraphical marker. It may have been a precursor to the major ignimbrite eruptions of the ensuing Lincomb Tarns Formation (Kneller and McConnell, 1993). The base of the tuff is sharp and rests on a low-relief, erosion surface. It is about 20 m thick in its type area on the summit of Glaramara [NY 2460 1050].

In the type area, four lithofacies are present, but some may be absent elsewhere. At the base, an unwelded ignimbrite up to 5 m thick comprises lithic-rich, eutaxitic, coarse lapilli-tuff containing fiamme up to 20 mm. The lapilli-tuff fills channels and depressions in the underlying topography. The lower part is massive and the upper part lithic rich. The ignimbrite is overlain by 1.2 m of stratified, gravelly breccia, with low-angle and repose-angle cross-bedding.

The second lithofacies consists of about 6 m of bedded coarse tuff with an upward decrease in grain size and bed thickness so that the weakly stratified basal part passes up into a distinct centimetre- to decimetre-scale bedded upper part. Planar, low-angle cross-sets and dune bed-forms are present and locally there are very fine tuff laminae. Sparse, single-rimmed, accretionary lapilli up to 7 mm occur throughout the unit, but are concentrated in finer-grained beds in the centre and towards the top. The bedded coarse tuff is overlain by 3 to 6 m of nodular, eutaxitic lapilli-tuff that locally contains a bed of laminated tuff 2 m above the base. The lower part is massive and ungraded, whereas the upper part has inversely graded fiamme and a concentration of nodules at the top. A massive tuff, up to 20 cm thick and containing accretionary lapilli, forms the top of the subdivision.

The uppermost unit consists of up to 8 m of massive to bedded pebbly sandstone and breccia overlain by thinly bedded sandstone. These beds rest on an erosion surface and fill gullies in the underlying lapilli-tuff, but are locally absent. Scours, rip-up clasts, winnowed accumulations of broken accretionary lapilli and rare ripple marks are present in the sandstone.

Regional variations

Outside of the Central Fells, the Seathwaite Fell Formation has not been divided formally into members though similar facies are recognised. On Blea Rigg [NY 290 090] and to the east as far as the Coniston Fault and northwards into the Keswick district, the formation crops out in fault blocks with variable bedding dips. The members identified in the Central Fells, particularly the older ones, become progressively less well differentiated eastwards. The heterolithic breccia of the Pavey Ark Member is similar to that in the type area, but the thickness increases eastwards to a maximum of 120 m. East of the Coniston Fault, the formation has an extensive outcrop, but lithofacies variation within the coarsening-upwards sandstone succession in this area is poor. It is uncertain whether this represents a lateral facies change across the Coniston Fault or an increase in thickness of the Cockly Pike Member facies. The presence of breccia to the north-east of Grasmere that is probably part of the Pavey Ark Member, and the Glaramara Tuff near the top of the succession throughout the north-east of the district, both favour the latter interpretation.

In the Coniston Fells the basal part of the Seathwaite Fell Formation comprises thinly bedded and laminated siltstone and fine-grained sandstone, similar to the Bowfell Links Member of the Central Fells. A massive dacitic lapilli-tuff separates this from coarse-grained sandstone, locally with beds of coarse tuff, which comprise most of the formation. The sandstone and tuff facies may be equivalent to the Cockly Pike and Sprinkling Tarn members of the Central Fells. Pyroclastic rocks locally near the top of the succession are correlatable with the Glaramara Tuff.

The basal lithofacies in the Coniston Fells succession comprises up to 150 m of thinly bedded and laminated fine- and medium-grained sandstone and siltstone with sparse interbeds of pebbly coarse-grained sandstone; the pebbles are mostly siltstone intraclasts. Laminae and thin beds of fine tuff occur throughout the succession and some small-scale mantled erosion scars clearly demonstrate their origin as pyroclastic fall-out deposits. Bedding and lamination is predominantly planar and commonly poorly defined, but wavy bed forms and ripple cross-lamination, including climbing ripples, also occur in places.

Normal grading is common but beds are generally poorly sorted. Soft-state deformation structures, particularly convolute bedding, occur throughout the succession; load casts, microfaults and slumps are present locally (Plate 18).

Dark grey and brownish grey laminated siltstone and fine-grained sandstone are present locally near the base of the succession, the colour distinguishing them from the usual green-grey of the overlying strata. Micropalaeontological samples from Brandy Crag Quarry [SD 2840 9838], a slate trial 300 m east of Blind Tarn [SD 2649 9676] and from the Walna Scar Quarries [SD 248 958] yielded rare and poorly preserved acritarchs, including *Veryhachium?* sp., *Timofeevia?* sp., *Uncinispaera?* sp. and *Michystridium?* sp. and acanthomorph acritarchs. Unfortunately, none of these genera is demonstrably indigenous or diagnostic of age, but the presence of *Timofeevia?* sp. suggests that some may have been derived from older strata. The acritarch assemblages either indicate a paralic marine environment, additional to the Holehouse Gill Formation, or that the rocks that contain them have been derived from older marine strata. If the latter is true the Skiddaw Group is the most obvious source of material within the known age range of *Timofeevia?* sp.

Dacitic lapilli-tuff overlies the fine-grained sandstone lithofacies association throughout most of the Coniston Fells and is approximately 100 m above the base of the formation. However, farther south-west in the Duddon Basin, and to the north-east of High Tilberthwaite, it lies up to 370 m above the base of the formation and within the coarse-grained lithofacies association. Thus the sedimentary lithofacies boundary is diachronous. The lapilli-tuff is poorly sorted and contains abundant white-weathered lithic lapilli. It is generally 60 to 135 m thick, but it is locally absent.

The coarse lithofacies association consists mostly of massive and thickly bedded, coarse-grained and pebbly sandstone along with well-bedded and laminated, medium- to coarse-grained sandstone with some siltstone interbeds (Plate 19). On the Coniston Fells this succession maintains a fairly uniform thickness of approximately 580 m, but in the Duddon basin it thickens to 850 m and in the Tilberthwaite to Little Langdale area it is estimated to vary from 750 to 1100 m. Bedding is generally poorly defined with planar bed forms being predominant. Locally, cross-bedding is common including trough (Plate 20) and ripple forms. Sedimentary structures that are associated with rapid rates of deposition occur throughout the succession and include convolute laminae, load-induced sediment flames, ball and pillow structures and dewatering pipes; soft-state syndepositional deformation and microfaults are also common. Towards the top of the succession scours and channels, some containing pebble and cobble conglomerate, indicate fluvial deposition. Steeply incised channels up to 10 m deep and filled with chaotic, polymict breccia occur locally, for example at High Tilberthwaite [NY 3040 0192] and on Torver High Common [SD 2594 9522] (Plate 21). Some beds of andesitic coarse tuff and fine lapilli-tuff up to 30 cm thick are present, particularly north of High Tilberthwaite [NY 3037 0171]. These are parallel bedded and well sorted with normal, reverse and multiple grading characteristics indicative of an ash-fall origin.

A tuff-breccia, 25 to 80 m thick, within the uppermost 150 m of the Seathwaite Fell Formation is interpreted as a block-and-ash-flow tuff. It contains abundant lithic clasts, has a eutaxitic texture and characteristically has pocked surfaces where pumice has been removed by weathering. The lithic blocks are concentrated in the basal 10 m and occur in sufficient abundance around Church Beck [SD 294 980] and Little Arrow Moor [SD 275 967] to form a breccia that passes gradationally upwards into overlying massive, lithologically uniform lapilli-tuff. In places where the breccia is absent, a fine-grained basal layer is usually present. The base is sharp and the member rests on a low-relief erosion surface. The uppermost 2 m of the lapilli-tuff are stratified. Bedding in the top of the lapilli-tuff is most clearly defined to the south-west of the Walna Scar Road [SD 267 961]. East of the Coniston Fault, south-west of Arnside, the tuff-breccia is cut out by substantial erosion at the base of the Lincomb Tarns Formation.

Above the block-and-ash-flow deposit the Glaramara Tuff has been recorded, for example on Holme Fell [NY 314 008], Long Crag [SD 2988 9850] and south-west of The Bell [SD 283 974]. There, white-weathered, splintery, bedded fine tuff, locally with accretionary lapilli is interbedded with thin sequences of eutaxitic lapilli-tuff. Up to 30 m of parallel-laminated and cross-laminated, fine-, medium- and coarse-grained sandstone separates the tuff from the overlying Lincomb Tarns Formation.

Lincomb Tarns Formation

The Lincomb Tarns Formation comprises predominantly eutaxitic, dacitic lapilli-tuff, the product of large-volume pyroclastic flow eruptions. Associated ash-fall and surge deposits are represented by interbeds of tuff-breccia and bedded tuff. Thin intercalations of sandstone present locally indicate periods of fluvial deposition between eruptions. Previously, the formation has been referred to as 'felsitic and basic tuffs' in the eastern part of the Ambleside district (Hartley, 1925), Yewdale Breccia in the Coniston Fells (Mitchell, 1940), and the Lincomb Tarns Formation in the Central Fells (Oliver, 1954; 1961). The type area is near Lincomb Tarns [NY 241 093], on the ridge between Glaramara [NY 246 103] and Allen Crags [NY 238 088], where the formation is 290 m thick (McConnell, 1993). Elsewhere the formation varies considerably in thickness (Figure 43). On Broad Crag [NY 218 077] up to 170 m are present and farther east, in Easedale, 150 to 200 m are present, thickening to 200 to 250 m in the Rydal area in the north-east of the district. In the Coniston Fells area, more than 800 m occur on Black Fell [NY 340 021] and Tom Heights [NY 326 002].

The base of the formation is sharp and rests on a low-relief erosion surface cut in the Seathwaite Fell Formation. The basal lithofacies of the ignimbrite indicate that deposition occurred both subaerially and subaqueously on a westerly inclined palaeoslope (McConnell, 1993). The relief on the erosion surface is illustrated at Tarn Crag [NY 3015 0946] where a mound, approximately 20 m long and 5 m high, is preserved. This mound influenced deposition of the overlying pyroclastic deposits: a crystal-rich pyroclastic surge deposit normally present at the base of the forma-

tion is absent above the mound, a lag breccia is preserved on its upstream side and the eutaxitic foliation in the lapilli-tuff mantles the top and lee slope. These relationships suggest that the mound formed a solid, lithified obstacle, but elsewhere flames of sandstone have been injected into the base of the lapilli-tuff, indicating that the underlying sediments had not been lithified before emplacement of the pyroclastic rocks. Commonly, at the base of the formation there is a parallel- and cross-bedded crystal-rich tuff interpreted as a pyroclastic surge deposit (McConnell, 1993). This is 1.5 m thick on Rossett Pike [NY 251 075] and is overlain by a series of thin (less than 1 m) pyroclastic flow units with well-developed reverse concentration-grading and coarse-tail grading of fiamme. The basal sequence coarsens eastwards towards Codale Head [NY 290 089] where it consists of cross-bedded tuff-breccia and massive to stratified coarse tuff. The base of the formation between Broad Crag [NY 216 077] and the north side of Great End [NY 225 088] has been brecciated in situ and reworked to produce a variable succession of breccia, sandstone and tuff, up to 10 m thick.

Most of the formation comprises eutaxitic lapilli-tuff containing unevenly distributed fiamme, typically having aspect ratios of 3:1 to 10:1 and ragged flame-like ends characteristic of flattened pumice. Fiamme contain 20 to 30 per cent feldspar crystals, perlitic cracks and epidote and calcite amygdales. The lithic lapilli are mostly juvenile in origin, comprise porphyritic dacite, mostly less than 4 cm across, and form 15 to 20 per cent of the rock by volume; commonly, they are normally graded within flow units. The crystal component is dominated by feldspar, with subordinate mafic minerals replaced by chlorite. Accessory amounts of apatite, iron oxides and, rarely, garnet are also present. On Greenup Edge [NY 286 105] and Low White Stones [NY 280 102] there is a coarse unit that contains up to 30 per cent by volume white-weathered felsic blocks up to 12 cm across. East of the Coniston Fault, on the northern slopes of Red Screes [NY 398 090] intraformational lenses of breccia have a welded matrix and grade into eutaxitic lapilli-tuff. South-west of the Brathay Fault, the rock contains a polymict clast population dominated by non-vesicular pink-weathered rhyolite and dark green-grey, fine-grained basaltic andesite, the latter having cumulose margins (Plate 6e). The abundance of the two clast types, both of probable juvenile origin, suggests that these ignimbrites may have originated from mixed magma eruptions.

Internal grading characteristics of the eutaxitic lapilli-tuff comprising the main body of the formation indicate deposition from complex pyroclastic flow regimes. Individual flow units are defined by the reverse size and abundance grading of fiamme, locally accompanied by normal grading of lithic clasts. The basal flow units in much of the Central Fells are 1 to 5 m thick. Higher in the formation, and, typically towards the north-east, the flow units become thicker and less well defined. In some of these the eutaxitic texture is only seen in the uppermost 0.5 m. Bedded tuff occurs locally within the massive lapilli-tuff, for example a 2 m-thick unit comprising interbedded accretionary lapilli-tuff and lapilli-tuff crops out on Great End [NY 2266 0853] and Ill Crag [NY 2240 0738]. This, and other examples, probably represent ash fall-out and pyroclastic surge deposits.

A particular feature of the formation in the Coniston Fells area is columnar cooling joints which are spectacularly displayed, for example in Torver Beck [SD 273 963], on Little Arrow Moor [around SD 283 973], on Foul Scrow [SD 293 978] on the south side of Church Beck and at Long Crag [SD 299 983] north of Coniston. Mitchell (1940) interpreted the columnar-jointed outcrops as separate rhyolite flows, but the ubiquitous eutaxitic texture indicates that these are zones within a multiple-flow, compound cooling unit of ignimbrite (Millward, 1980). Complexity in the emplacement and cooling history of this unit is illustrated by the variations in cooling joint size. At Long Crag, two columnar zones have average column diameters of 11 cm and 5.5 cm, but in Torver Beck there are three zones with columns averaging 9, 20 and 23 cm across (Millward, 1980). The columnar zones have sharp boundaries and the column diameters remain constant along their length. Locally, for example at Long Crag, there is small-scale variation in both the plunge of the columns and the angle between the plunge of the column and the dip of the welding foliation.

Esk Pike Formation

The Esk Pike Formation consists of commonly silicified volcaniclastic mudstone and siltstone with interbeds of sandstone and lapilli-tuff. Originally named as the Esk Pike Hornstone (Oliver, 1961), it crops out in the north of the district as outliers on the fell tops of Great End and Allen Crags [NY 226 084 to 236 086], north of Easedale [NY 325 098] and around Scandale Head [NY 385 100]. The thickness ranges from 40 to 100 m, but up to 170 m are present in the type area on Esk Pike [NY 236 075]. The Esk Pike Formation may be coeval with the Tarn Hows Formation (Figure 43).

On Esk Pike, the base of the formation is taken at the erosion surface above which reworked lapilli-tuff of the Lincomb Tarns Formation passes gradationally into an interbedded sandstone and siltstone succession (McConnell, 1993). The transition is well exposed on Esk Pike [NY 2357 0736], where the erosion surface is cut in stratified, eutaxitic lapilli-tuff. Two metres of clast-supported breccia, composed of reworked lapilli-tuff, lie above the erosion surface. These are succeeded by 30 cm of crudely stratified pebble breccia above which thin and medium beds of sandstone alternate with siliceous mudstone. A similar transitional sequence, 0.5 to 15 m thick, is present on Allen Crags [NY 2367 0882].

Above the basal transition the succession consists of massive or bedded to laminated siliceous mudstone, light greenish grey on fresh surfaces and distinctively white weathered. The rock has a subconchoidal fracture and locally contains siliceous nodules, several centimetres in diameter, as well as epidote-rich nodules, 1 to 2 mm in diameter. Syndepositional deformation is widespread and locally was sufficiently intense to destroy bedding and form massive units of mudstone, some of which are now isolated pods in disturbed sandstone beds. Lithologically the siliceous mudstone is similar to the fine tuff

interbeds in the underlying Lincomb Tarns Formation. A few euhedral plagioclase crystals, chlorite pseudomorphs (up to 0.3 mm and possibly after amphibole) and rare grains of quartz are present in very fine-grained matrix. The presence of such crystals suggests a pyroclastic origin, and Oliver (1961) further noted that the feldspar laths showed a preferred orientation possibly caused by settling. McConnell (1993) considered that the mudstone may be redeposited ash either from contemporaneous eruptions or from the Lincomb Tarns Formation.

Some coarse-grained interbeds are normally graded from clast-supported breccia to sandstone. In the area around Scandale Head, the formation consists of alternations of thick pebbly sandstone beds, commonly graded, with finer-grained thinly bedded and laminated sandstone and siliceous, white-weathered mudstone. Ripple and trough cross-bedded sets are present with common synsedimentary slumping, brecciation and ramp structures. Erosional channels, commonly with box-shaped cross-sections [NY 3840 0944], contain coarser gravel lags.

Pyroclastic rocks occur throughout the succession. On Allen Crags, pumice-rich tuff and lapilli-tuff occur in beds up to 2 m thick [NY 2333 0862 and 2350 0872]. Erosional scours at the top of these beds are commonly infilled with sandstone. On Esk Pike, beds of pumice-rich lapilli-tuff up to 1 m thick are present near the base of the succession. All of these examples contain flattened pumice, but the thinness of the beds and their subaqueous depositional environment suggest the collapse of the pumice was produced by diagenetic flattening.

Tarn Hows Formation

The succession of tuff, lapilli-tuff and interbedded volcaniclastic sedimentary rocks that overlies the Lincomb Tarns Formation between the Coniston Fault and Windermere makes up the Tarn Hows Formation (Moseley, 1993). The type area lies between Tarn Hows and Iron Keld Plantation [NY 330 000 to 338 008]. The lithologies are much affected by a strong, penetrative cleavage and some of the sequence was recorded previously as 'andesite and rhyolite' by Hartley (1925). The formation is not present elsewhere within the district, but these rocks may be a facies variation of the upper part of the Lincomb Tarns Formation or coeval with the Esk Pike Formation. The base of the formation is poorly exposed, but is taken at the change from dacitic (Lincomb Tarns) to andesitic tuffs; the lithological change is accentuated by a significant increase in cleavage intensity. The sequence is unconformably overlain by Windermere Supergroup rocks and the maximum thickness of 420 m is seen between the Coniston and Brathay faults. East of the Brathay Fault, the formation is up to 100 m thick and contains volcaniclastic sandstone and siltstone interbeds.

Most of the sequence comprises coarse tuff and fine lapilli-tuff with a characteristic rough weathered surface with conspicuous surface holes caused by the preferential erosion of pumice or diagenetic carbonate aggregates. The rocks are generally unbedded or very weakly bedded, poorly sorted and ungraded. The constituents include a variable, but generally high proportion of lithic clasts with subordinate pumice and crystal debris; there is generally much fine-grained matrix. The crystal component is dominated by euhedral and broken feldspars with subordinate pseudomorphs after mafic minerals. A dominantly bimodal lithic clast population is made up of angular and subrounded, non-vesicular, dark, andesitic and pale acid fragments, mostly 0.5 to 6 mm across, but in places ranging up to 20 cm; welded tuff clasts are abundant in parts. The proportion of wispy streaks, after pumice, varies substantially but locally pumice is undeformed and not abraded so that arcuate and irregular, broken edges are preserved. Because there is no evidence for welding in the matrix, the local flattening of the pumice clasts may be a diagenetic effect.

In the area between Tarn Hows and the Brathay Fault, two important facies variations are present. The first of these is a splintery, acid, eutaxitic lapilli-tuff, up to 220 m thick. The welding foliation is shown by changes in devitrification grain sizes, with the fiamme altered to a coarse, snowflake texture. Subhedral oligoclase crystals and fragments, commonly corroded, comprise the crystal component but lithic clasts are absent. The uppermost 40 m of the formation in this area comprise grey, flinty, very fine-grained volcaniclastic sandstone, containing scattered small feldspar crystals and locally a small number of altered pumice fragments.

Middle Dodd Dacite

The Middle Dodd Dacite is a white-weathered, feldspar-phyric dacite with a rough surface and hackly fracture. It is probably the youngest formation preserved in the district forming outliers capping Middle Dodd [NY 398 097], where up to 75 m are present, and Little Hart Crag [NY 388 100]. The top of the unit is not preserved. The base is broadly conformable with the underlying Esk Pike Formation but some cross cutting, between the two outliers, seems likely because the thickness of the Esk Pike Formation is just 40 m on Middle Dodd compared with 100 m on Little Hart Crag. An intrusive contact is considered unlikely as there is no interaction with the underlying sedimentary rocks and a more likely explanation is that the dacite flow is preserved in hollows on an erosion surface.

The dacite is flow banded and locally brecciated, particularly near its base. It contains abundant feldspar phenocrysts and glomerocrysts and has a flow-banded, fine-grained groundmass. A lithologically similar dacite forms a discordant intrusion about 500 m south of Middle Dodd at Red Screes [NY 396 084]. Though its phenocryst content is lower, the intrusion may have been the feeder to the Middle Dodd Dacite.

PENECONTEMPORANEOUS INTRUSIVE ROCKS

Many penecontemporaneous minor intrusions are present within the volcaniclastic rocks of the upper part of the Borrowdale Volcanic Group. Some may have been feeders to the extrusive rocks and there is evidence that many were intruded with little overburden.

Basaltic andesite

Large volumes of basaltic andesite were intruded into the Dunnerdale and Lickle Formations on the south-east margin of the Duddon Basin. The intrusions are largely coincident with the South Borrowdales Lineament (Lee, 1989) south-westwards from Broughton Moor [SD 245 935], and their emplacement may have been controlled by extension on a basement fracture. The intrusions are irregular sheets, parallel or subparallel to bedding, and commonly follow stratigraphical contacts of ignimbrites with volcaniclastic sedimentary rocks, the boundary between the Dunnerdale and Lickle Formations being favoured particularly. Typically, peperitic breccias are present adjacent to the amygdaloidal tops of the sheets; basaltic andesite clasts within these breccias have chilled margins. The overlying sandstone is baked for several metres away from the contact and contains amygdales formed from boiling of the pore water. Massive, chilled basaltic andesite at the lower contact is generally in direct contact with sandstone. The central parts of the sheets typically have irregular, blocky cooling joints. The largest intrusion, the Hawk Andesites of Mitchell (1956b), attains a maximum thickness of 500 m at its type locality, The Hawk [SD 241 927]. A few curtains and rafts of sandstone and zones of peperitic breccia are present within the otherwise compositionally uniform mass, suggesting that emplacement occurred in several pulses. The basaltic andesite is pyroxene-phyric with scattered glomerocrysts, subhedral and euhedral phenocrysts of clinopyroxene up to 7 mm across in a seriate-textured, plagioclase groundmass. Much of the rock remains fresh but, where altered, chlorite replaces pyroxene and sericite replaces plagioclase.

In the Seathwaite Fell Formation on the Coniston Fells, basaltic andesite is the dominant intrusive rock, but in the Duddon Basin and north-east of Tilberthwaite it occurs together with a porphyritic andesite suite. Some basaltic andesite sills also occur to the east of the Coniston Fault, but are difficult to distinguish from the locally predominant andesite suite. The basaltic andesite is dark greenish grey, splintery, highly amygdaloidal, and pyroxene- and plagioclase-phyric. The intrusions are 15 to 200 m thick, most have outcrop lengths of 1 to 4.5 km and some are gently cross-cutting. Typically, peperitic breccias occur in the upper parts, and the adjacent sedimentary rocks have been fluidised, homogenised, locally vesiculated and metamorphosed. Examples of these high-level sills have been described from the Church Beck area [SD 295 988] (Branney and Suthren, 1988). Complete brecciation of the basaltic andesite occurs locally and one example on Colt Crag [SD 280 980] was interpreted, erroneously, as vent agglomerate (Watson, 1984). Phenocrysts and microphenocrysts of pyroxene and plagioclase comprise up to 15 per cent of the rock by volume though there is much variation. Pyroxene is mostly replaced by chlorite, quartz and carbonate minerals, and plagioclase by sericite with epidote and carbonate minerals. Pilotaxitic and trachytic textures in the fine-grained groundmass are ghosted by intense alteration to sericite and carbonate. Most of the sills are cut by a network of hydrothermal veinlets.

Andesite

Plagioclase-phyric andesite intrusions are particularly abundant within the Whorneyside Formation in the south-west of the district. All of these are probably concordant though contacts with the host rocks are rarely seen. Andesite on the north-western margin of the Duddon Basin and north-east of High Kiln Bank [SD 215 942] was intruded into the Lickle Formation to form an irregular body with discordant contacts, some of which are coincident with faults. The andesite is plagioclase-phyric and petrographically similar to an adjacent andesite lava within the Lickle Formation, with which it was probably associated.

Brownish grey, highly feldspar-phyric andesite sills occur widely throughout the district in the Seathwaite Fell Formation. Some of these were previously interpreted as lava flows and designated as the Wrengill Andesites by Hartley (1925; 1932) and Mitchell (1940; 1956b; 1963). They are concentrated at the top of the formation, possibly because the overlying Lincomb Tarns Formation acted as a barrier to the ascending magma. These sills may therefore postdate deposition of the Lincomb Tarns Formation and possibly represent one of the last magmatic episodes preserved within the Borrowdale Volcanic Group in the district. However, the large outcrop of andesite around Easedale Tarn [NY 314 088 to 316 079] may have been in part extrusive. It is typically massive and sparsely porphyritic, locally with conspicuous nodules, 0.10 to 1 m in diameter. The uppermost part is rubbly and a diffuse flow-foliation is present; it is overlain by breccia of the Pavey Ark Member. Similar heterogeneous massive to brecciated andesitic rocks occur adjacent to the Coniston Fault around White Moss Common [NY 345 066].

Generally, the sills have massive centres with irregular sets of cooling joints and brecciated margins; some are flow-jointed. Jointing patterns and internal breccia zones in some of the thicker sills suggest that they were emplaced by multiple injection. In the example capping Lingmoor Fell, extending on to the north side of Great Langdale around Silver Howe, the lower part with an irregular, blocky set of cooling joints is separated from the upper closely spaced flow-jointed part by breccia. Many andesite sills exceed 200 m, but this sill is probably the thickest within the district at 250 to 320 m.

The marginal breccias at the base and top of the sills are mostly peperitic. Particularly good examples occur at the base of a sill on Lingmoor Fell and the north side of Great Langdale where the basal breccia is approximately 10 m thick. Locally, north of Chapel Stile [NY 318 056], the breccia matrix consists of homogenised Glaramara Tuff, that forms flames extending up to 5 m into the sill. This sill also contains some rafts of deformed Glaramara Tuff. Usually, the sills are only sparsely amygdaloidal towards their tops and there is rarely any evidence of vesiculation in the overlying host rock.

Microporphyritic varieties also occur, for example on Blea Rigg [NY 296 081] where the intrusion has an elongate, oval-shaped outcrop 500 by 70 m, orientated approximately north–south (McConnell et al., 1993). The intrusion is brecciated in places at its margin and the Seath-

waite Fell Formation wallrock is also brecciated. This may be a result of hydrothermal activity following the intrusion of the andesite: the interstices are filled by a network of quartz veinlets that contain finely disseminated galena. The andesite is much altered with plagioclase, which forms 20 per cent of the rock, replaced by calcite, epidote and chlorite, and ferromagnesian minerals replaced by chlorite and epidote. Poorly preserved perlitic cracks are present in the microcrystalline groundmass.

Mixed-magma porphyritic andesite suite

A suite of dykes and irregular minor intrusions characterised by composite intrusion and magma-mixing are present in the Scafell to Bowfell area of the Central Fells (McConnell et al., 1993). They intrude the Airy's Bridge, Seathwaite Fell and Esk Pike formations. Contact relationships, such as peperitic breccias and intrusion-related soft-state deformation, indicate that the host sedimentary rocks were not lithified fully when the intrusions were emplaced.

The dykes range up to 30 m wide; typically, they have basaltic margins and andesitic centres that contain basaltic enclaves up to 1 m in diameter. The transition from the basaltic margin to the andesitic centre is variably sharp to gradational and the basic margins may contain multiple chilled contacts indicating polyphase intrusive pulses; up to 15 are present in the 1 m-wide margin of the dyke at Calfcove Bield [NY 224 080]. Basic enclaves in the andesitic zones are rounded and lobate, and since neither enclave nor host is chilled at their contact there was little temperature difference between them. The form of the intrusions varies and is apparently related to the lithology of the host country rock or at least to its degree of lithification at the time of intrusion. This association is illustrated on Pike de Bield [NY 236 069]; there, in the Lincomb Tarns and Esk Pike formations, two straight, parallel north-east-trending dykes merge into a complex, irregular branching network in the underlying Seathwaite Fell Formation. The location of the intrusion at Pike de Bield also appears to be controlled by a volcanotectonic fault that was active during the deposition of the Seathwaite Fell Formation. The intrusion follows the fault plane but branches off into the sandstone country rock which was apparently unlithified at the time.

Another possible component of the mixed-magma suite in the same area is the intrusive breccia at Ore Gap, an elongate body a little over 1 km in length and up to 250 m wide, trending north-east from Pike de Bield. The breccia is clast supported, and consists mainly of porphyritic andesite clasts up to 1 m in diameter with sporadic clasts of rhyolite, basalt and volcaniclastic sandstone, set in a matrix mainly of smaller andesitic fragments and accessory garnet. The breccia is crudely stratified adjacent to the near-vertical wall of the intrusion. The breccia is cut by irregular bodies of mixed-magma porphyritic andesite, including a central, near-vertical sheet that extends along the entire length of the breccia. This assemblage is interpreted as a vent agglomerate (McConnell et al., 1993).

Composite andesite–rhyolite

Irregular, composite sheets intrude the Caw Formation outlier on the interfluve between Dunnerdale and Long Mire becks [SD 225 925]. They are discordant and attain a maximum thickness slightly in excess of 30 m. The two components exploited the same fissure at different times and, with all internal contacts sharp, there is no evidence for magma mixing. The rhyolite appears to be earlier, and had solidified before the plagioclase-phyric andesite was injected into and alongside it; however, there is no direct evidence, such as xenoliths or apophyses, to confirm the temporal relationships. The sheets are thickest at their northern extremities where they comprise mostly rhyolite. The proportion of andesite increases southwards with a flow foliation developed locally, as the sheets thin and wedge out in that direction.

Garnetiferous dacite

A suite of garnetiferous dacite dykes in the Eskdale area was investigated by Beddoe-Stephens and Mason (1991) and, on mineralogical and geochemical criteria, was shown to be co-magmatic with the Scafell Dacite. Principal occurrences of these dykes are at Hard Knott [NY 230 015], Little Stand [NY 250 027], Great How [NY 195 036] and Hardrigg Gill [NY 195 060]. Typically, they are feldspar-phyric with euhedral garnet (almandine) phenocrysts, up to 4 mm, set in a very fine-grained, flow-aligned (hyalopilitic) matrix. Close to the margin of the Eskdale granite, hornfelsic textures are developed and garnet is recrystallised.

Rhyodacitic suite

Porphyritic rhyodacitic plugs on Wansfell [NY 393 040], east of Ambleside, within the Seathwaite Fell Formation caused extensive physical and chemical alteration of the host sedimentary rocks which must have been unlithified at the time of intrusion. The overlying Lincomb Tarns Formation has been deformed by their emplacement. The chilled margins of the intrusions commonly have fiamme, eutaxitic textures and lithic lapilli, characteristics that are normally associated with extrusive pyroclastic rocks. Peperitic breccias occur particularly in the Jenkin Crag [NY 383 028] plug. At the highest levels of the largest plug the intrusion margin is mostly gradational and poorly defined. At lower levels of this example and in the smaller plugs the contacts are sharp and well defined. Interaction between the magma and groundwater has extensively altered the country rocks such that fine-grained sandstone and siltstone may be altered for more than 1 km away from the observed contact with the plug; the affected rocks are baked and indurated, and may be either homogenised or hydraulically brecciated to contain patches of dense, metasomatically recrystallised rock cut by abundant quartz veins. The largest plug has produced a north-east-trending and plunging pericline in the Lincomb Tarns Formation. The hinge of this fold extends for at least 3 km suggesting that the plug is an

elongate body with only its south-western part exposed. Doming of the Lincomb Tarns Formation and disruption within the Seathwaite Fell Formation to accommodate the intrusion have produced a highly irregular contact.

Rhyolitic suite

On the southern limb of the Scafell Syncline, a suite of mainly concordant rhyolite sheets is intruded into bedded tuff of the Whorneyside Formation. Locally, these intrusions are discordant and extend upwards into the Airy's Bridge Formation, but they do not penetrate younger stratigraphical units and may have been associated with eruption of the Airy's Bridge Formation (Kneller et al., 1993a; McConnell et al., 1993). These intrusions vary from aphyric and microcrystalline to flow-banded and garnetiferous.

Rhyolitic intrusions in Great Langdale are contemporaneous with the Seathwaite Fell Formation. For example north of Whitegill Crag [NY 297 074], a sill of pink, aphyric, rhyolitic breccia up to 5 m thick cuts fine-grained sedimentary rocks in the Three Tarns Member. The rhyolitic sill is brecciated throughout its exposed length of 800 m; it consists of angular clasts up to 25 cm across in a matrix of homogenised and baked siltstone and fine-grained sandstone. The breccia has a very distinctive appearance because the clasts have been removed preferentially by weathering, leaving the matrix protruding. Isolated rhyolite clasts occur up to 1 m below the sill, and are presumed to have sunk through the sedimentary rocks which must have been unlithified at the time the sill was emplaced. High-level intrusion is likely, soon after deposition of the host strata.

The youngest suite of concordant and subconcordant rhyolitic intrusions intrudes the Seathwaite Fell and Lincomb Tarns formations in the north-east of the district, to the south and east of Little Langdale. The intrusions are adjacent to the Coniston fault over a distance of about 5 km, from Oxen Fell [NY 323 017] to Rydal [NY 356 052] but the intense faulting in this ground precludes determination of the number of sheets present. The maximum thickness of at least 280 m occurs on the north side of Oxen Fell. The rhyolite is pink-weathered and splintery. It contains up to 10 per cent by volume of plagioclase phenocrysts and glomerocrysts up to 3 mm across. The plagioclase is moderately fresh and some crystals are zoned. The groundmass has a mottled devitrification texture containing sericitised spherulites up to 4 mm in diameter. The contact between the rhyolite and the host country rock is sharp and well defined with only local marginal irregularities. Between Skelwith Bridge [NY 344 033] and Loughrigg Tarn [NY 344 043], a crude flow-foliation is present within and the contact of a rhyolite sheet and in a quarry [NY 341 037] north-west of Skelwith Bridge, the top of a rhyolite sheet contains fiamme and sparse lithic clasts suggesting that local vesiculation and fragmentation occurred.

A few rhyolite dykes and sills are also present as part of a suite that occurs more widely in older parts of the Borrowdale Volcanic Group succession on the north-western side of the Duddon Basin. The largest dyke follows a north-west-trending fault alongside Hollow Moss Beck [SD 206 931] and has an outcrop width of 70 m. Two high-silica rhyolite sills are present close to the top of the Dunnerdale Formation between Tarn Hill [SD 208 919] and Out Moss [SD 206 907]. The one emplaced lower in the formation is up to 40 m thick. The upper sill is less than 15 m thick and on Tarn Hill transgresses northwards and pinches out in the basal part of the Lickle Formation.

SEVEN

Ordovician–Silurian: Windermere Supergroup

The Windermere Supergroup was defined originally as the Windermere Group by Moseley (1984). It crops out in the Ambleside district as a sequence of folded and cleaved, predominantly marine sedimentary rocks, which unconformably overlies the Borrowdale Volcanic Group, and elsewhere is itself unconformably overlain by Carboniferous strata. The depositional history spans the period between the late Ordovician cessation of volcanism and deformation of the Acadian orogeny, the latter probably of Emsian age (Soper et al., 1987). Within the district, the Windermere Supergroup ranges from Cautleyan (mid-Ashgill) to Ludfordian (late Ludlow) in age (Table 13); it is in excess of 8 km thick. Elsewhere in north-west England the range extends from Longvillian (mid-Caradoc) to Přídolí (Ingham and McNamara, 1978; Rickards, 1978). Except for parts of the thin and variably developed basal facies the sequence is exclusively marine. The lithostratigraphical nomenclature employed here is based on a revised scheme, developed during the recent resurvey and applied throughout the Windermere Supergroup outcrop, by Kneller et al. (1994). The graptolite biostratigraphy has been reported by Rickards (1964a; 1967; 1969; 1970), Hutt (1974; 1975) and White (in Lawrence et al., 1986); zonal determinations cited in this memoir are taken from these works.

The basal part of the Windermere Supergroup, the Dent Group (Table 13; Figure 45) is of late Ordovician age and comprises a thin sequence of predominantly sedimentary rocks of relatively shallow-water origin with a number of non-sequences and minor unconformities. These rocks contain shelly faunas (Plate 22), and commonly show evidence of extensive bioturbation.

Though deposition was apparently continuous through the Ordovician–Silurian boundary, a change in lithofacies and biofacies occurs near the top of the Hirnantian, with the first indications of an anoxic environment. Thereafter, the Silurian is represented by an apparently continuous sequence deposited in relatively deep water. Possible local control of deposition by synsedimentary fault movement is evident from sedimentary variations across some major structures.

The lower part of the Silurian succession consists largely of graptolite-bearing hemipelagite with rare shelly fossils (the Stockdale Group; about 100 m thick), but progressively more turbiditic sand, silt and mud was introduced through the late Wenlock and Ludlow. Above the Stockdale Group is a 600 to 1000 m-thick sequence (Table 13; Figure 45) of Wenlock and lower Ludlow strata, dominated by dark grey, finely laminated, graptolitic, muddy siltstone. The base of this unit is marked by a transition from bioturbated siltstone to laminated siltstone; the probably diachronous top is marked by the incoming, during the lower *nilssoni* Zone, of numerous sandstone turbidites which swamp the laminated siltstone. However, the laminated siltstone persists as a background facies throughout the Gorstian Stage. Four formations are recognised, of which the lowermost (Brathay Formation) and uppermost (Wray Castle Formation) consist almost entirely of laminated siltstone. The cumulative thickness of laminated siltstone remains more or less constant at about 600 to 650 m, and variations in thickness are due largely to interbedded turbidites. The middle and upper Ludlow is represented by about 2000 m of sandstone-dominated turbidite units (Coniston Group) overlain by at least 2.5 km of mainly thinly bedded siltstone (Bannisdale Formation). The composition of representative Windermere Supergroup sandstones is summarised in Table 14.

ASHGILL: DENT GROUP

This part of the succession (Table 13) has traditionally been termed the Coniston Limestone (Sedgwick, 1845; 1846; Coniston Limestone 'Series' of Aveline and Hughes, 1872; 'Group' of McNamara, 1979a; 'Formation' of Moseley, 1984). The name Dent Group is adopted in recognition of the virtually unbroken sequence represented in the Dent and Cautley inliers (Figure 1) at the eastern margin of the Windermere Supergroup outcrop. Biostratigraphy is based upon the sequence of shelly faunal zones established there by Ingham (1966). The history of nomenclature of the upper Ordovician is discussed in detail by McNamara (1979a) and Lawrence et al. (1986, fig. 6) and is summarised by Kneller et al. (1994, fig. 3).

The Ashgill sequence comprises evidence of the foundering of the Borrowdale Volcanic Group with post-volcanic subsidence accompanied by marine planation, that produced a number of variably sized fault blocks which imposed local effects on sedimentation. Thus, in the Ambleside district, the Dent Group is subdivided into three parts by local unconformities. The sedimentary sequence starts with spatially restricted fluviatile conglomerate and sandstone beds composed entirely of locally derived volcanic detritus. These pass upwards abruptly into a marine succession in which shoreface deposits are preserved. The evolution of the basin is reflected in the succeeding strata.

Basal beds (Stile End Formation: Longsleddale Member)

Across the district, the Dent Group has a distinctive basal facies in which the dominant lithology is a medium- to coarse-grained, grey-green volcaniclastic sandstone. This ranges up to a maximum thickness of 40 m, and fills local topographical irregularities of tens of centimetres on the sub-Windermere Supergroup unconformity. The beds are

Table 13 Summary stratigraphical chart for the Windermere Supergroup (after Kneller et al., 1994.

Note that the Underbarrow, Kirkby Moor and Yarlside Volcanic (Y) formations are not present in the Ambleside district.

App/Hph Applewaite/High Pike Haw members.

	GROUP	FORMATION	MEMBER	GRAPTOLITE ZONE	STAGE	SERIES	SYSTEM
WINDERMERE SUPERGROUP		Kirkby Moor	Helm			PŘIDOLI	SILURIAN
WINDERMERE SUPERGROUP		Kirkby Moor		*leintwardinensis*	Ludfordian	LUDLOW	SILURIAN
WINDERMERE SUPERGROUP		Underbarrow		*leintwardinensis*	Ludfordian	LUDLOW	SILURIAN
WINDERMERE SUPERGROUP		Bannisdale		*incipiens*	Gorstian	LUDLOW	SILURIAN
WINDERMERE SUPERGROUP	Coniston	Yewbank		*scanicus*	Gorstian	LUDLOW	SILURIAN
WINDERMERE SUPERGROUP	Coniston	Moorhowe		*scanicus*	Gorstian	LUDLOW	SILURIAN
WINDERMERE SUPERGROUP	Coniston	Poolscar		*scanicus*	Gorstian	LUDLOW	SILURIAN
WINDERMERE SUPERGROUP	Coniston	Latrigg		*scanicus*	Gorstian	LUDLOW	SILURIAN
WINDERMERE SUPERGROUP	Coniston	Gawthwaite		*nilssoni*	Gorstian	LUDLOW	SILURIAN
WINDERMERE SUPERGROUP		Wray Castle		*nilssoni*	Gorstian	LUDLOW	SILURIAN
WINDERMERE SUPERGROUP		Coldwell	High Cross	*ludensis*	Homerian	WENLOCK	SILURIAN
WINDERMERE SUPERGROUP		Coldwell	Randy Pike	*nassa*	Homerian	WENLOCK	SILURIAN
WINDERMERE SUPERGROUP		Birk Riggs / Brathay		*lundgreni*	Homerian	WENLOCK	SILURIAN
WINDERMERE SUPERGROUP		Brathay		*ellesae*	Sheinwoodian	WENLOCK	SILURIAN
WINDERMERE SUPERGROUP		Brathay		*linnarssoni*	Sheinwoodian	WENLOCK	SILURIAN
WINDERMERE SUPERGROUP		Brathay		*rigidus*	Sheinwoodian	WENLOCK	SILURIAN
WINDERMERE SUPERGROUP		Brathay		*antennularius*	Sheinwoodian	WENLOCK	SILURIAN
WINDERMERE SUPERGROUP		Brathay		*riccartonensis*	Sheinwoodian	WENLOCK	SILURIAN
WINDERMERE SUPERGROUP		Brathay		*murchisoni*	Sheinwoodian	WENLOCK	SILURIAN
WINDERMERE SUPERGROUP		Brathay	Dixon Ground	*centrifugus*	Sheinwoodian	WENLOCK	SILURIAN
WINDERMERE SUPERGROUP	Stockdale	Browgill	Far House	*crenulata*	Telychian	LLANDOVERY	SILURIAN
WINDERMERE SUPERGROUP	Stockdale	Browgill		*griestoniensis*	Telychian	LLANDOVERY	SILURIAN
WINDERMERE SUPERGROUP	Stockdale	Browgill		*crispus*	Telychian	LLANDOVERY	SILURIAN
WINDERMERE SUPERGROUP	Stockdale	Browgill		*turriculatus*	Telychian	LLANDOVERY	SILURIAN
WINDERMERE SUPERGROUP	Stockdale	Skelgill		*sedgwickii*	Aeronian	LLANDOVERY	SILURIAN
WINDERMERE SUPERGROUP	Stockdale	Skelgill		*convolutus*	Aeronian	LLANDOVERY	SILURIAN
WINDERMERE SUPERGROUP	Stockdale	Skelgill		*argenteus*	Aeronian	LLANDOVERY	SILURIAN
WINDERMERE SUPERGROUP	Stockdale	Skelgill		*magnus*	Aeronian	LLANDOVERY	SILURIAN
WINDERMERE SUPERGROUP	Stockdale	Skelgill		*triangulatus*	Aeronian	LLANDOVERY	SILURIAN
WINDERMERE SUPERGROUP	Stockdale	Skelgill		*cyphus*	Rhuddanian	LLANDOVERY	SILURIAN
WINDERMERE SUPERGROUP	Stockdale	Skelgill		*acinaces*	Rhuddanian	LLANDOVERY	SILURIAN
WINDERMERE SUPERGROUP	Stockdale	Skelgill	Spengill	*atavus*	Rhuddanian	LLANDOVERY	SILURIAN
WINDERMERE SUPERGROUP	Stockdale	Skelgill	Spengill	*acuminatus*	Rhuddanian	LLANDOVERY	SILURIAN
WINDERMERE SUPERGROUP	Stockdale	Skelgill	Spengill	*persculptus*	8 Hirnantian	ASHGILL	ORDOVICIAN
WINDERMERE SUPERGROUP	Dent	Ashgill		*extraordinarius*	8 Hirnantian	ASHGILL	ORDOVICIAN
WINDERMERE SUPERGROUP	Dent	Ashgill	Troutbeck	*(pacificus sub-zone)*	7 Rawtheyan	ASHGILL	ORDOVICIAN
WINDERMERE SUPERGROUP	Dent	Appletreeworth Formation		*(pacificus sub-zone)*	6 Rawtheyan	ASHGILL	ORDOVICIAN
WINDERMERE SUPERGROUP	Dent	? ? ? Broughton Moor Formation		*anceps*	5 Rawtheyan	ASHGILL	ORDOVICIAN
WINDERMERE SUPERGROUP	Dent	Lumholm Member ? ?		*anceps*	4 Cautleyan	ASHGILL	ORDOVICIAN
WINDERMERE SUPERGROUP	Dent	Kirkley Bank	Torver	*anceps*	3 Cautleyan	ASHGILL	ORDOVICIAN
WINDERMERE SUPERGROUP	Dent	Kirkley Bank	App / Hph	*anceps*	3 Cautleyan	ASHGILL	ORDOVICIAN
WINDERMERE SUPERGROUP	Dent	Kirkley Bank	Kentmere	*(complexus sub-zone)*	2 Cautleyan	ASHGILL	ORDOVICIAN
WINDERMERE SUPERGROUP	Dent	Kirkley Bank	basal clastics	*(complexus sub-zone)*	2 Cautleyan	ASHGILL	ORDOVICIAN
WINDERMERE SUPERGROUP	Dent	Y		*(complexus sub-zone)*	2 Cautleyan	ASHGILL	ORDOVICIAN
WINDERMERE SUPERGROUP	Dent	Stile End	Longsleddale	*(complexus sub-zone)*	1 Cautleyan	ASHGILL	ORDOVICIAN

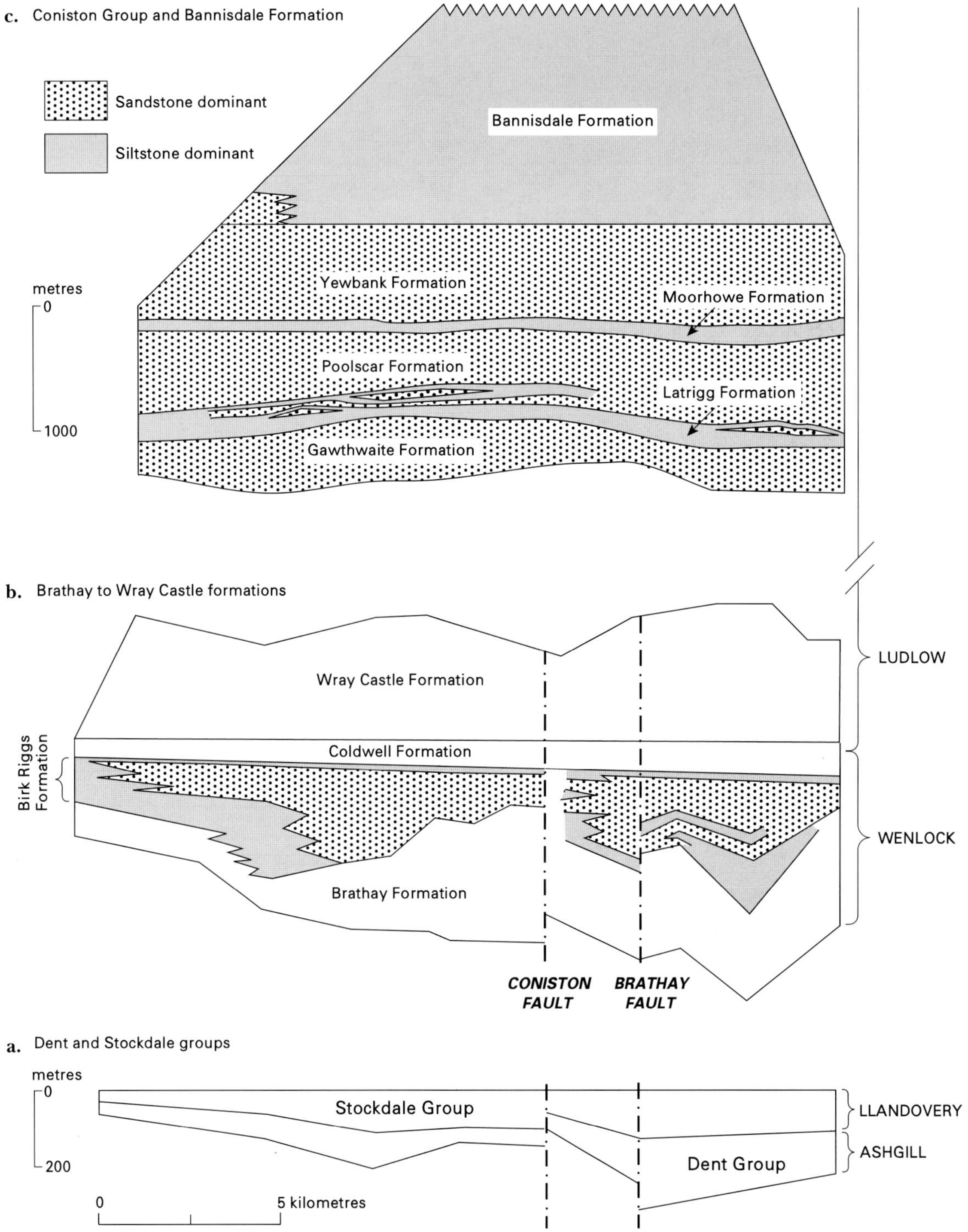

Figure 45 Schematic representation of thickness and facies changes within the Windermere Supergroup.

Note different vertical scale for Figure 45c.

a. Ashgill and Llandovery
b. Wenlock and Lower Ludlow
c. Upper Ludlow

highly variable, discontinuous and probably diachronous, but are here associated together and correlated with the Longsleddale Member of the Stile End Formation. The presence of the trilobites *Decoroproetus piriceps, Erratencrinurus cornutus* and *Toxochasmops marri* suggest deposition within Zone 2 of the Cautleyan Stage (McNamara, 1979a, b). The Longsleddale Member is commonly absent so that the overlying Kirkley Bank Formation rests directly on the eroded surface of the Borrowdale Volcanic Group.

Over much of the outcrop of the Longsleddale Member, bedding is more or less concordant with that in the underlying volcanic rocks, which are interbedded with lithologies similar to the sandstone of the basal facies. Where there is no marked lithological contrast or angular discordance, the unconformity can in many instances be placed only by virtue of the presence of fossils (usually fragmented) in the Longsleddale Member sandstone. Fossils recorded include brachiopods, molluscs and trilobites, well

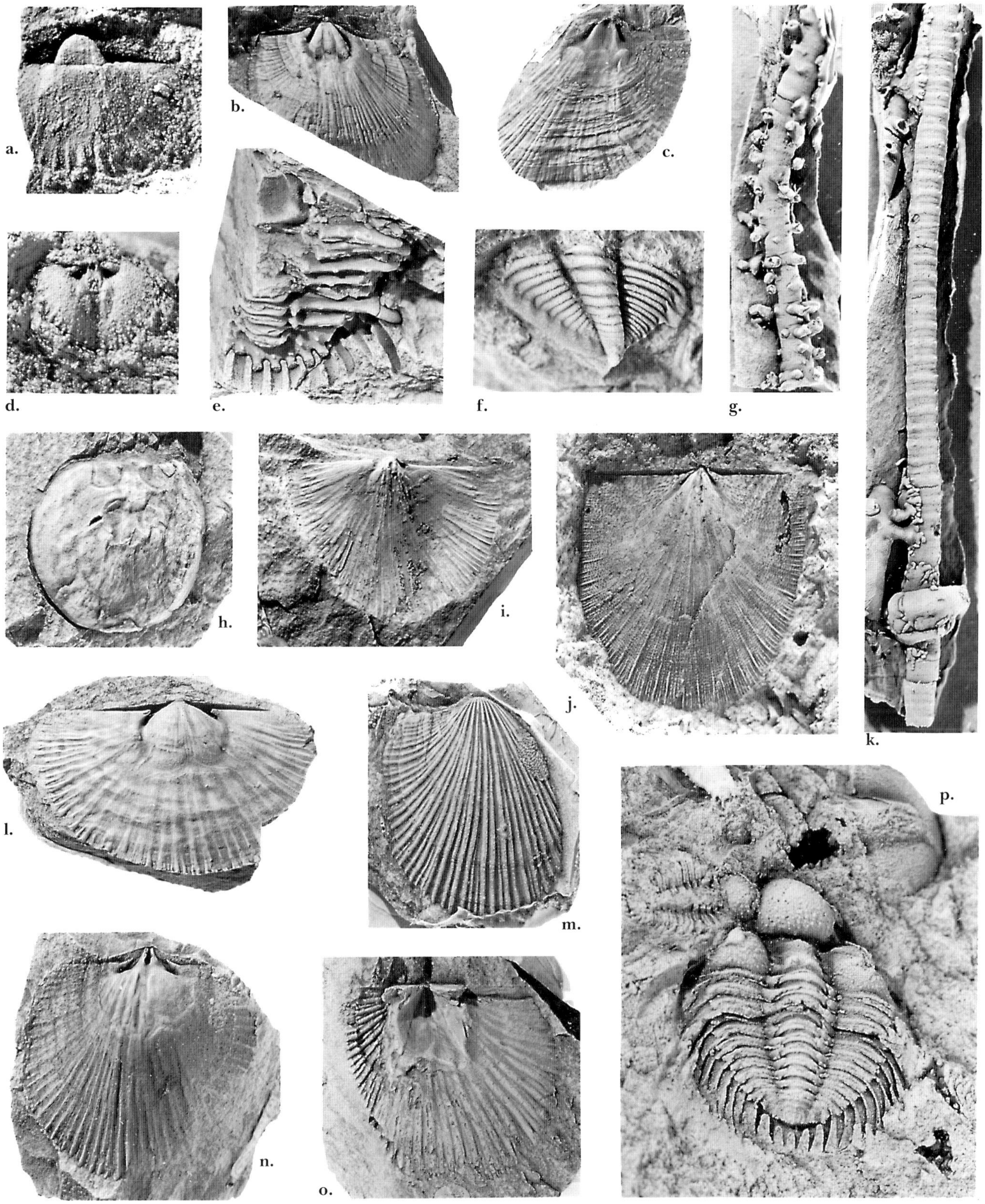
a.
b.
c.
d.
e.
f.
g.
h.
i.
j.
k.
l.
m.
n.
o.
p.

Plate 22 Examples of shelly fossils from the Ashgill Dent Group. All the specimens figured are in the collections of the British Geological Survey.

a, d. *Dedzetina* cf. *microstoma* Havlíček, internal moulds of pedicle and brachial valves, DJ363 and DJ376, both × 8; Broughton Moor Formation (Rawtheyan), near Appletree Worth [SD 2430 9243].
b, c. *Hirnantia sagittifera* (McCoy), internal moulds of pedicle and brachial valves, DJ626 and DJ624, both × 2.5; Ashgill Shale Formation (Hirnantian), 700 m north-east of Appletree Worth [SD 2488 9309].
e. *Diacanthaspis decacantha* (Angelin), pygidium with part of thorax, RU4204, × 4; Broughton Moor Formation (Rawtheyan), Ash Gill Beck [SD 268 954].
f. *Mucronaspis olini* (Temple), small pygidium Zs430,× 6; Ashgill Shale Formation, Troutbeck Member (Rawtheyan), 700 m north-east of Appletree Worth [SD 2488 9309].
g, k. Crinoid pluricolumnal fragments, latex casts. **g** shows the bases of numerous cirri; **k** is overgrown by the cirri of other columnals and also shows a gall-like swelling. Both on Zs943, × 2; Ashgill Shale Formation (Hirnantian), 700 m north-east of Appletree Worth [SD 2488 9309].
h. *Petrocrania* sp., internal mould of large pedicle valve, Zw8344 (R. W. Scott coll.), × 1; Kirkley Bank Formation, High Pike Haw Member (Cautleyan Stage), 1.1 km north-east of Appletree Worth [SD 2508 9341].
i. *Fardenia* (?) cf. *transversaria* Bancroft, internal mould of pedicle valve, Zw8345, × 1.5; Kirkley Bank Formation, High Pike Haw Member (Cautleyan Stage), 1.1 km north-east of Appletree Worth [SD 2508 9341].
j. *Rafinesquina* sp., internal mould of pedicle valve, RU4075, × 2; basal clastic beds of the Kirkley Bank Formation, High Pike Haw Member (Cautleyan Stage), 400 m north-east of Appletree Worth [SD 2465 9285].
l–o. *Plaesiomys inflata* (Salter), all thought to be from the Kirkley Bank Formation, High Pike Haw Member (Cautleyan Stage), all × 1.5. **l**, internal mould of transverse pedicle valve, Gsd461 (D. Sharpe coll.), from 'Coniston'; **m, n**, brachial valves, latex cast of external mould Gsd271 and internal mould Gsd272 (both D. Sharpe coll.), 'Coniston Water Head', probably near Fell End [c. SD 3222 9869]; **o**, lectotype (Cocks, 1978, p.51), external mould of pedicle valve, retaining shell material around umbo, GSM26039, from 'Coniston'. The coarsely costellate ribbing and convex brachial valve are very like those of *P. porcata* (McCoy), which may be therefore a senior synonym of *P. inflata*.
p. *Staurocephalus* cf. *clavifrons* Angelin, internal mould of exoskeleton, associated with a small individual and (top right) a pygidium of *Panderia* ?, Zw9286 (R. W. Scott coll.), × 6; Ashgill Shale Formation, Troutbeck Member (Rawtheyan Stage), Boo Tarn, 2.2 km west-south-west of Coniston [SD 2811 9673].

seen 400 m north-east of Appletree Worth [SD 2465 9285] where decalcified bedding planes contain numerous solitary corals, atrypoids, *Dalmanella*, *Rafinesquina*, bivalves, gastropods, *Decoroproetus piriceps*, harpetids, *Toxochasmops* and other fossils. Few sedimentary structures are seen with the exception of crude bedding and sporadic low-angle cross-stratification; only very locally is there well-developed parallel stratification, low-angle cross-stratification, scouring, and multi-directional cross-lamination, for example at [SD 2761 9628]. In the more conglomeratic horizons the clast lithologies are comparable with those of the underlying volcanic rocks, but also include rare jasper. These coarser beds are concentrated into the basal few centimetres where locally derived granule or pebble grade material is common. On Little Arrow Moor [SD 2832 9712] an exceptionally coarse deposit occurs with sandstone infilling the interstices between large (10 m scale) randomly orientated blocks of unfossiliferous volcaniclastic sandstone.

South-west of Coniston the basal facies is more variable, and the preserved strata include locally developed clast-supported conglomerate, an association of matrix-supported conglomerate and red, cross-stratified sandstone [for example SD 2945 9761, SD 2625 9485], and chert [SD 2832 9712], containing halysitid corals and lingulids. At one locality [SD 2780 9661], to the south-west of Coniston, is a well-sorted coarse-grained sandstone containing concave-up bedsets up to 0.5 m thick. Foresets indicate a flow direction towards the north-west. Clasts within the conglomerates are mainly of lithologies comparable with those within the underlying Borrowdale Volcanic Group. Also present, particularly in the sandstone units, are quartz clasts of resorbed phenocrystic character, perthitised and sericitised plagioclase and orthoclase feldspars, relict ferromagnesian minerals (probably pyroxenes) and rare rutile and apatite in a silicious and, in places, chloritic matrix (McCaffrey, 1991).

The Longsleddale Member assemblage includes strata accumulated in a variety of marginal marine, largely shoreface environments; the red lithofacies includes subaerial debris-flow, lacustrine and fluviatile deposits.

Kirkley Bank Formation

This formation (defined in Scott and Kneller, 1990) includes three constituent members: Kentmere, Applethwaite (or laterally equivalent High Pike Haw Member in the south-west), and Torver members, equivalent to formations similarly named and defined by McNamara (1979a). The transition between the High Pike Haw Member and the Applethwaite Member is considered by Scott (1992) to take place immediately to the east of, or across, the Brathay Fault. This may suggest control by movement along the fault line. The Kentmere and Applethwaite members equate with the *Calymene* Beds of Marr (1916) (so named from the abundance of the trilobite *Calymene* s.l. *subdiademata*). The formation consists of calcareous siltstone with varying proportions of fine sandstone and nodular limestone of diagenetic origin. A basal clastic facies is locally developed.

Kentmere Member

Overlying the Longsleddale Member clastic facies or locally lying directly on the unconformity [for example SD 2815 9658], the basal few metres of the Kirkley Bank Formation comprise the Kentmere Member. Semi-continuous decimetre-scale limestone horizons are separated by thin (less than 2 cm) bifurcating siltstone layers, but the limestones, though crudely layered in a style which mimics bedding, appear to be of diagenetic origin. Diagenesis and extensive recrystallisation have virtually obliterated all

Table 14 Representative modal and geochemical analyses of sandstone from the Windermere Supergroup.

	WDM34	WDM40	WDM53	WDM56	WDM65	WDM82	WDM84	WDM86
	Birk Riggs Formation		Gawthwaite Formation	Poolscar Formation	Yewbank Formation	Bannisdale Formation		
Grid ref. SD	65 940	285 961	268 936	415 011	293 936	284 918	358 956	396 912
vol. %								
MQP	32	41	48	45	45	33	37	31
MQV	8	1	1	1	1	2	0	1
PQ	3	4	2	5	4	1	1	4
PLG	1	1	5	5	6	8	10	11
KFL	6	4	10	10	15	21	27	18
LV	1	3	1	5	3	5	8	4
LMV	1	3	2	1	0	1	1	0
LS	0	2	0	2	0	0	0	0
LMS	0	4	1	0	3	1	0	1
Mica	4	1	2	2	3	7	2	11
PYRT	1	2	0	0	0	0	0	0
CCL	3	1	1	2	0	0	0	0
CCM	6	0	3	1	3	1	2	0
Clay	17	15	1	4	0	2	2	1
CHLR	17	18	20	16	16	13	8	14
wt. %								
SiO_2	66.81	70.99	66.66	70.01	68.54	70.31	68.66	69.01
TiO_2	0.63	0.65	0.92	0.69	0.64	0.66	0.61	0.67
Al_2O_3	8.85	9.91	11.64	11.33	11.57	11.64	10.37	12.35
Fe_2O_3	0.64	0.88	4.92	4.23	4.14	4.17	3.84	4.87
FeO	2.68	2.21						
MnO	0.12	0.07	0.06	0.07	0.06	0.07	0.10	0.07
MgO	3.93	2.87	3.23	2.32	2.44	2.31	2.27	2.67
CaO	4.50	2.86	3.02	2.72	3.38	2.02	4.07	1.56
Na_2O	1.52	2.02	1.86	2.93	2.61	2.24	2.02	2.36
K_2O	1.89	1.92	2.28	2.16	2.24	2.48	2.15	2.44
P_2O_5	0.14	0.15	0.20	0.15	0.13	0.13	0.13	0.13
LOI	8.36	4.50	4.49	3.34	4.07	3.31	4.74	3.33
Total	100.07	99.63	99.26	99.93	99.82	99.34	99.19	99.47
ppm								
Sc	12	9	17	9	14	6	10	14
V	58	58	97	78	74	80	63	83
Cr	368	234	420	129	110	113	105	114
Co	17	21	26	20	20	21	18	25
Ni	70	77	77	40	51	49	45	61
Cu	12	12	15	8	13	7	12	15
Zn	34	59	62	51	56	44	49	67
Rb	63	67	84	63	72	84	73	83
Sr	68	104	121	212	192	141	147	106
Y	29	27	38	26	25	29	27	23
Zr	356	287	607	284	181	228	210	175
Nb	13	12	16	11	12	13	12	14
Ba	229	311	404	605	489	529	449	501
Pb	4	10	12	12	16	5	12	7
Th	11	11	16	12	10	12	13	13
U	3	2	6	4	4	3	3	2

Component abbreviations: MQP monocrystalline plutonic quartz; MQV monocrystalline volcanic quartz; PQ polycrystalline quartz; PLG plagioclase feldspar; KFL potassium feldspar; LV lithic volcanic; LMV lithic metavolcanic; LS lithic sedimentary; LMS lithic metasedimentary; PYRT pyrite; CCL carbonate clast; CCM carbonate cement; CHLR chlorite.

All data from McCaffrey (1991); for additional geochemical data see appendices in Kneller (1990a, b) and Scott and Kneller (1990). Analyses made by X-ray fluorescence spectrometry at the University of Leeds. Modal compositions in per cent; major elements in wt%; trace elements in ppm. Where FeO is not shown total Fe is reported as Fe_2O_3.

original sedimentary structure. A few fossils, including crinoid columnals, disarticulated brachiopods and the ostracod *Uhakiella strangulata*, have been recorded (Scott, 1992). At High Pike Haw, McNamara (1979a, p.52) recorded the presence of solitary (streptelasmatid) corals and Orchard (1980) recorded ten species of conodonts. The member can be traced across country by the presence of pits and associated lime kilns.

APPLETHWAITE MEMBER

The Applethwaite Member and laterally equivalent High Pike Haw Member (in the south-west) constitute the bulk of the Kirkley Bank Formation, with thicknesses ranging from 8 to 35 m. The dominant lithologies are calcareous siltstone and fine-grained sandstone which are brown-weathered, blue-grey and thin to medium bedded. Varying proportions of nodular limestone are present, the nodules generally no more than 40 cm long and 10 to 15 cm thick. In places, more continuous layers of nodules are traceable for 10 m or more, parallel to bedding. The interlayered sandstone and siltstone beds are more continuous, enveloping the limestone which is partially secondary in origin. It overprints repeated decimetre-scale vertical sequences consisting of a graded bed which, where fully developed, is made up of basal shell lags, followed by a parallel to swaley or hummocky cross-stratified interval, and a pervasively bioturbated siltstone top. Bed bases are conformable to markedly erosional, and individual beds, although laterally persistent over tens to hundreds of metres, cannot be correlated between more widely spaced sections. Limestone nodules are developed principally at the contacts between successive units, nucleated on the basal lenses of winnowed bioclastic material, but extending into the overlying sandstone, and into the underlying bioturbated siltstone (where preserved) or sandstone (Scott, 1992). Burrow systems (*Palaeophycus* and *Planolites*) preserved within limestone nodules are less compressed than those in the surrounding siltstone, demonstrating that nodules formed before much of the compaction.

The sandstone and siltstone contain a rich fauna of colonial and solitary corals, bryozoans, brachiopods (both articulate and inarticulate), gastropods, more than 30 kinds of trilobites (of which *Calymene* s.l. *subdiademata* is the commonest), ostracods and crinoids. Shell lags consist principally of fragmental brachiopods, crinoids and corals. The faunas are indicative of Zone 2 of the Cautleyan Stage, with the top of the member ranging up into Zone 3 to the east of the district (McNamara, 1979a). Scott (1992) interpreted most of these faunas as *Nicolella*-type associations that inhabited muddy shelf sites at depths affected only by severe storms.

HIGH PIKE HAW MEMBER

The High Pike Haw Member differs from its lateral equivalent, the Applethwaite Member, in that the siliciclastic fraction is of coarser grain size, and the proportion of limestone is lower. McNamara (1979a) regarded the junction between the High Pike Haw and the Applethwaite members to be a lateral lithological gradation, well exposed over about 1.5 km between High Pike Haw [SD 264 949] and Torver Beck [SD 276 963]. However, on sedimentological grounds, Scott (1992) considered this transition to take place across, or to the east of, the Brathay Fault. Both members show a decreasing proportion of limestone upwards, and grade into the overlying Torver Member. The limestones have a packstone to grainstone texture, and in thin section the abundance of fine sand-grade skeletal carbonate fragments exceeds that of detrital quartz. The High Pike Haw Member yields rich coral–brachiopod faunas which Scott (1992) assigned to two faunal associations: '*Streptelasma*-trilobate brachiopod Associations' in the lower beds are suggestive of shallower water and higher energy conditions than the 'Ramose bryozoan Associations' which occur in the overlying beds. The trilobite fauna is much less diverse than that of the Applethwaite Member and is dominated by *Toxochasmops*, *Decoroproetus* and *Prionocheilus* (McNamara, 1979b).

TORVER MEMBER

The Torver Member equates with the *Phillipsinella* Beds of Marr (1916), so named after the abundance of the trilobite *Phillipsinella parabola aquilonia* (Ingham, 1970). The member consists of 3 to 8 m of grey-brown weathered, bluish, thickly bedded and slightly calcareous siltstone, locally with more calcareous zones that produce a patchy, honeycomb weathering effect. No sedimentary structures have been recognised in this unit, because of the rather poor exposure and general homogeneity caused by pervasive bioturbation. The fauna, listed by McNamara (1979a, p.56) and Scott (1992), is generally fragmentary, and occurs in coquinas throughout the member. McNamara (1979a, b) recorded the presence of *Tretaspis hadelandica deliqua* at several localities and assigned these beds to the Cautleyan, Zone 3. Scott (1992) interpreted the fauna as a *Nicolella*-type association inhabiting a fine-grained substrate in deeper water than that which covered the High Pike Haw depositional environment.

McNamara (1979 a, b) considered the Torver Member to be progressively overstepped to the north-east by the overlying Broughton Moor Formation. Between the two there appears to be a stratigraphical hiatus from mid-Zone 3 (Cautleyan) to about Zone 5 or 6 (Rawtheyan) (Table 13), and east of Troutbeck the Torver Member may have been partially removed by erosion prior to deposition of the Broughton Moor Formation. More recently, Scott (1992) has come to the contrary conclusion, that the Torver Member is present across the whole southern Lake District. He found no evidence for widespread erosion prior to deposition of the Broughton Moor Formation.

Broughton Moor Formation

The Broughton Moor Formation (Scott and Kneller, 1990) is equivalent to the White Limestone of Marr (1916). It consists of pale grey, bioturbated, interbedded nodular and micritic limestone, which weathers to a white surface; the micritic limestone contains skeletal fragments of trilobites and brachiopods. Chamosite ooliths occur very locally in the basal part (Scott, 1992). The formation maintains a thickness of 4 to 6 m across its outcrop

[SD 2297 9125 to SD 2761 9628], with sporadic outliers farther east in the Kendal district. The upper part of the formation is commonly silicified where overlain by the Appletreeworth Formation. Bioturbation is pervasive, and all other sedimentary structures have apparently been destroyed. Identifiable fossils are generally difficult to collect, but in addition to the restricted trilobite assemblage recorded by McNamara (1979a, p.58), *Diacanthaspis decacantha* has also been found. However, at a locality [SD 2430 9243] by Appletree Worth Beck, 650 m north-east of Hawk Bridge, the limestone is decalcified and a diverse fauna of over 30 taxa was obtained. The trilobites include a few *Tretaspis* which are comparable with *Tretaspis hadelandica* cf. *brachystichus* and suggest the Rawtheyan stage, Zone 5 or 6.

Lumholm Member

The Lumholm Member crops out sporadically in the extreme south-west of the district [south-west of SD 234 915] as a discontinuous matrix-supported conglomerate up to about 5 m thick. It overlies nodular calcareous beds which, on lithological grounds, were originally assigned to the Browgill Formation. Hence this new member was placed within the *griestoniensis* Biozone (Scott and Kneller, 1990) though locally the member is separated from the Borrowdale Volcanic Group by only 2 m of unexposed ground. More recently, conodonts have been recovered from both the matrix of the conglomerate and the calcareous bed immediately below (Armstrong et al., 1996) which all indicate a Rawtheyan age. On this basis the member is now included as part of the Broughton Moor Formation but could alternatively lie in the basal part of the Ashgill Formation. It contains a variety of clast types, including dark mudstone and calcareous rock resembling lithologies in the pre-Hirnantian part of the Dent Group, all contained within a calcareous mudstone matrix. An origin by debris-flow seems most likely, but the precise stratigraphical relationship and significance of the member remain uncertain.

Appletreeworth Formation

The Appletreeworth Formation (Scott and Kneller, 1990) is equivalent to the Rhyolitic Ash of Marr (1916, see also McNamara, 1979a). Its outcrop is discontinuous [between SD 236 918 and NY 348 019], and a maximum thickness of about 6 m occurs around High Pike Haw [SD 264 949]. East of Windermere, the Appletreeworth Formation has been entirely removed by intra-Ordovician erosion. It consists of highly altered felsic volcanic material. The original lithology was of fine to medium sand-grade vitric ash, but it is now so pervasively altered that the original textures are largely obscured. The ash was deposited by a series of gravity flows probably during the Rawtheyan stage, Zone 6, and now occurs in re-sedimented beds each a few tens of centimetres thick. Few sedimentary structures are visible in the field, other than crude bedding (in places disturbed by loading) and cross-lamination. However, in thin sections and on polished slabs, layers of apparently reworked coarse ash can be identified which are commonly associated with small flame structures. Thin sections from Broughton Moor (E61341, E61344, E61345) reveal a minutely banded, formerly glassy rhyolitic rock containing about 5 per cent angular, anhedral grains of quartz and alkali-feldspar. Slender seams of very fine-grained white mica form a discontinuous cleavage which transgresses the banding. Minor constituents include chlorite, after primary biotite, and small patches of secondary calcite. However, consistent analyses from three samples (Table 15) indicate that alteration has been limited.

Ashgill Formation

The Ashgill Formation is equivalent to the Red Gill Formation of Scott and Kneller (1990) and the Ashgill Shales as defined by Salter (1873; the Ashgill Group of Marr, 1892). Its total thickness varies from 4 m in the north-east to 25 m in the south-west. It includes the Troutbeck Member (*Mucronatus* Beds of Marr, 1916) at its base.

Troutbeck Member

A slight unconformity at the base of the member increases in magnitude eastwards so that the Troutbeck Member successively rests on strata of the Appletreeworth Formation (in the west of the district), the Broughton Moor Formation and the Kirkley Bank Formation (in the east of the district). It is defined similarly to the Troutbeck Formation of McNamara (1979a) as a grey variably calcareous siltstone, locally with a basal nodular limestone facies. Thickness varies from 2 m at the eastern limit of the district to a maximum of 7 m [SD 2945 9761]. In thin section it consists of skeletal calcite fragments less than 1 mm across, contained in a terrigenous silty mudstone with a wackestone to packstone texture. Sporadic laminae of mica flakes are present locally. The fauna consists of bryozoans, brachiopods, tentaculitids, trilobites, crinoids and cystoids. Fragmental skeletal material is distributed throughout; the majority of the fauna is found in the upper part of the member, but there is an upwards decrease in the overall carbonate content. The member is locally richly fossiliferous [for example NY 3484 0192], with monospecific bedding-plane assemblages of the trilobite *Mucronaspis* (most commonly *M. olini*) and the brachiopod *Hirnantia*. McNamara (1979a, b) recorded several other trilobites and assigned a late Rawtheyan age (upper part of Zone 7) to this unit.

Ashgill Formation mudstone

Above the Troutbeck Member in apparent conformity is a mudstone sequence equivalent to the Ashgill Shale Formation of McNamara (1979a). Its thickness decreases from about 20 m in the south-west to 2 m at the north-eastern limit of the district. The blue-grey silty mudstone differs from the Troutbeck Member in being slightly finer grained, less calcareous and less fossiliferous. It is generally pervasively bioturbated and virtually structureless, with only a nebulous and highly burrowed stratification visible in places. In thin section the silty mudstone appears as an homogeneous deposit with scattered, matrix-supported quartz grains of coarse silt grade. Sporadic calcareous nodules occur throughout the member, generally in the lower part, but forming a conspicuous band near the top.

Table 15 Geochemical analyses of tuff and meta-bentonite from the Windermere Supergroup.

	Appletreeworth Formation			Brathay Formation		Coldwell Formation
	Rhyolitic tuff			Metabentonitic tuff		
Registered no.	E61341	E61344	E61345	E61324	E61343	E61346
Grid ref. SD	2602 9448	2519 9326	2519 9326	2889 9697	2560 9382	2541 9304
Age	Ashgill	Ashgill	Ashgill	Wenlock	Wenlock	Late Wenlock
wt. %						
SiO_2	75.30	76.39	77.06	52.03	50.03	50.61
TiO_2	0.16	0.15	0.14	2.45	0.72	0.75
Al_2O_3	12.73	11.71	11.73	21.94	22.19	20.21
Fe_2O_3*	1.56	1.69	1.61	6.07	7.34	5.42
MnO	0.07	0.02	0.03	0.02	0.01	0.03
MgO	2.50	3.05	2.72	1.89	2.34	2.78
CaO	0.37	0.26	0.32	0.13	0.90	2.10
Na_2O	0.38	0.57	0.35	3.66	2.64	3.25
K_2O	3.50	2.87	3.03	5.70	6.45	5.94
P_2O_5	0.02	0.02	0.02	0.28	0.11	0.25
LOI	3.30	2.80	2.70	2.10	6.50	5.20
Total	99.89	99.53	99.71	96.27	99.23	96.54
ppm						
Sc	4.7	4.6	4.2	36.1	15	13.2
V	10	10	10	196	89	78
Cr	10	10	10	10	28	35
Co	1	1.3	0.9	20.1	17.3	12.7
Ni	10	10	10	22	37	39
Cu	5	5	5	33	64	70
Zn	31	30	18	131	87	99
Rb	139	116	124	218	237	209
Sr	40	44	34	121	149	143
Y	39	42	34	87	85	28
Zr	127	121	113	554	216	215
Nb	14	15	13	23	6	8
Cs	5.7	5.5	6.1	—	—	—
Ba	506	357	362	1415.	922	1034
La	34.1	35.7	31.5	0.1	12.3	21.8
Ce	87.1	91.8	79.6	94.3	35.2	52.7
Nd	31.5	31	28.7	45.8	17.7	22.2
Sm	6.5	5.7	6.4	0.1	3.1	4.5
Eu	0.23	0.26	0.21	2.28	0.82	0.97
Tb	1.05	1.02	0.84	1.81	0.65	0.60
Yb	3.16	4.16	3.06	6.00	3.85	0.10
Lu	0.01	0.63	0.40	1.04	0.10	0.27
Hf	4.4	4.5	3.6	14.4	6.4	6.2
Ta	1.77	1.59	1.68	1.7	0.54	0.89
Pb	10	10	10	10	10	10
Th	35.5	33.2	35.0	10.3	11.9	15.4
U	8.6	7.1	6.5	4.3	4.0	4.5

Major elements and trace elements Ba to Zn were analysed by X-ray fluorescence (Department of Geology, University of Sheffield); Th, U, Co, Sc, Ta, Hf, Cs and REE by Instrumental Neutron Activation Analysis (Technische Universitat München).
* total iron.

At the type locality in Ashgill Quarry [SD 2690 9550], a 1 cm-thick bentonite horizon occurs in loose blocks of distinctive Ashgill Formation mudstone, but has not been found in situ. The mudstone is generally less well cleaved than the underlying Troutbeck Member, and locally it has been quarried for roofing slates [for example SD 2468 9279]. The contained fauna consists mainly of a low-diversity *Hirnantia* brachiopod association with rare examples of the trilobite *Mucronaspis mucronata*; it is assigned to the Hirnantian Stage. A typical assemblage, including *Bracteoleptaena polonica, Eostropheodonta hirnantensis, Hirnantia sagittifera, Plectothyrella crassicostis* and *Mucronaspis mucronata*, together with crinoid pluricolumnals, has been collected from an exposure by Appletree Worth Beck [SD 2488 9309] and at other localities, for example near Hol Beck [NY 3974 0334, NY 3989 0344].

LLANDOVERY: STOCKDALE GROUP

The Stockdale Group (Table 13; Figures 45, 46) includes the 40 to 110 m-thick mudstone sequence that conformably

overlies the Dent Group, and is conformably overlain by the Brathay Formation. It broadly corresponds to the Llandovery Series, ranging in age from late Hirnantian (*persculptus* Biozone) to uppermost Telychian (*crenulata* Biozone) or possibly basal Wenlock (*centrifugus* Biozone). It includes two formations; the Skelgill Formation, of Rhuddanian to Aeronian age, is dominated by black graptolitic mudstones; and the Browgill Formation, of Telychian age, which consists predominantly of pale green mudstones with subordinate black graptolitic shale horizons. The graptolite faunas and graptolite biostratigraphy have been examined in detail by Hutt (1974, 1975) and zonal determinations cited herein are taken from her work (Figure 46).

Skelgill Formation

The Skelgill Formation, originally defined as the Skelgill Beds by Marr and Nicholson (1888), consists of approximately 20 m of black and dark grey graptolitic mudstone and silty mudstone with subordinate non-graptolitic grey mudstone bands; a thin calcareous unit, the Spengill Member, occurs at the base. Definitive estimates of the thickness are not possible since the formation is everywhere affected by an almost bedding-parallel tectonic detachment, commonly within strata of the *triangulatus* Biozone. However, all graptolite biozones from *persculptus* to *sedgwickii* are represented (Figure 46), the faunas being listed by Hutt (1975, 119–122).

The lower half of the formation (up to the *magnus* Biozone) is dominated by laminated and massive varieties of black and dark grey, carbonaceous, slightly calcareous and locally pyritic mudstone, in places bioturbated by rare burrows 1 to 2 mm in diameter, but lacking a shelly fauna. The mudstone is commonly rich in graptolites with rare orthocones and phyllocarids. The individual graptolites are preserved in relief as pyritic casts or are flattened as chloritic films (Hutt, 1974). Silt laminae are present throughout the laminated mudstone sequence and thicker horizons of bioturbated pale grey unfossiliferous mudstone alternate with black mudstone in the upper part of the *cyphus* Biozone. A local increase in the thickness of the formation to about 40 m in the Coniston area is due to the increased thickness of the mudstone relating to the *acuminatus* and *atavus* biozones, within which a sparse *Hirnantia* brachiopod fauna is preserved (Cocks, 1988).

The upper part of the formation (upper *magnus* to *sedgwickii* biozones) is of black graptolitic mudstone, less calcareous and more pyritic than in the lower part. The black mudstone is intercalated with a number of blue-grey mudstone horizons, individually up to about 4 m thick and generally slightly calcareous, with sporadic layers of siderite nodules. The blue-grey mudstones locally yield dwarf shelly faunas, including the trilobites *Cyphaspis brachypygus*, *Dalmanites* cf. *tenuimucronatus*, *Eophacops glaber*, *Leonaspis erinaceus*, *Proromma acanthodes*, *Raphiophorus aloniensis*, *Scotoharpes judex* and *Youngia moroides*, in addition to favositid corals and small plectambonitacean brachiopods. The margins of these horizons lack shelly fossils, and grade into the adjacent graptolitic mudstone. A distinctive 5 to 6 mm of pale green mudstone, the 'Green Streak' (also reported from the Welsh Basin by Rickards, 1978), occurs within the *argenteus* Biozone. Bentonites occur at intervals throughout the formation, but are most common in the upper part.

Spengill Member

A locally developed basal assemblage within the Skelgill Formation, the Spengill Member (Scott and Kneller, 1990), equivalent to the *Atrypa flexuosa* band of Marr and Nicholson (1888), consists of between 7.5 to 250 cm of pale to dark grey fossiliferous bioturbated mudstone, commonly calcareous, with thin bands of impure pyritic limestone or limestone nodules. It yields *Plectatrypa flexuosa* and other brachiopods together with trilobite fragments (Marr, 1878; Marr and Nicholson, 1888). Mudstone immediately overlying the Spengill Member is within the *atavus* Biozone in the north-east and south-west of the district, but ranges down to the *persculptus* Biozone in the Coniston area.

Browgill Formation

The Browgill Formation, defined as the Browgill Beds by Marr and Nicholson (1888) consists of 50 to 90 m of pale green mudstone and silty mudstone, with subordinate dark grey or black interbeds of graptolitic mudstone (Figure 46). The thickness of the formation decreases towards the south-west. The green mudstone generally has a wispy bioturbated fabric. Up to four of the dark mudstone horizons occurring near the base of the formation yield a *turriculatus* Biozone graptolite fauna (locally including the *maximus* Subzone), whereas *crispus* Biozone faunas occur in a 7 m-thick sequence of alternating green and black mudstone higher in the formation. A concentration of siderite nodules occurs approximately two thirds of the way up the formation, and a graptolitic band above this, exposed in Torver Beck [SD 2768 9607] and described by Hutt (1974, p.5), yields possible *griestoniensis* and/or *crenulata* Biozone faunas. Very scarce phacopid trilobites (Rickards, 1964b) and indeterminate brachiopods (Marr and Nicholson, 1888, p.677) have been recorded from the upper part of the formation, but the green mudstone is otherwise barren. The red beds, which are well developed in the upper part of the formation farther east, are represented within the district merely by a few red laminae [SD 2765 9612].

Far House Member

The Far House Member (Soper and Kneller, 1990; equivalent to the Grey Beds of Rickards, 1964a) consists of hard, dark grey, massive siltstone and forms the uppermost 2 to 3 m of the Browgill Formation, (thickening to over 10 m immediately east of the Coniston Fault). It is unfossiliferous in the Ambleside district, but has yielded trilobite fragments elsewhere (Wilson, 1954).

SILURIAN: WENLOCK TO LOWER LUDLOW FORMATIONS

The Wenlock succession ranges in thickness from 220 to 580 m (Figure 45) and spans three formations (Table 13).

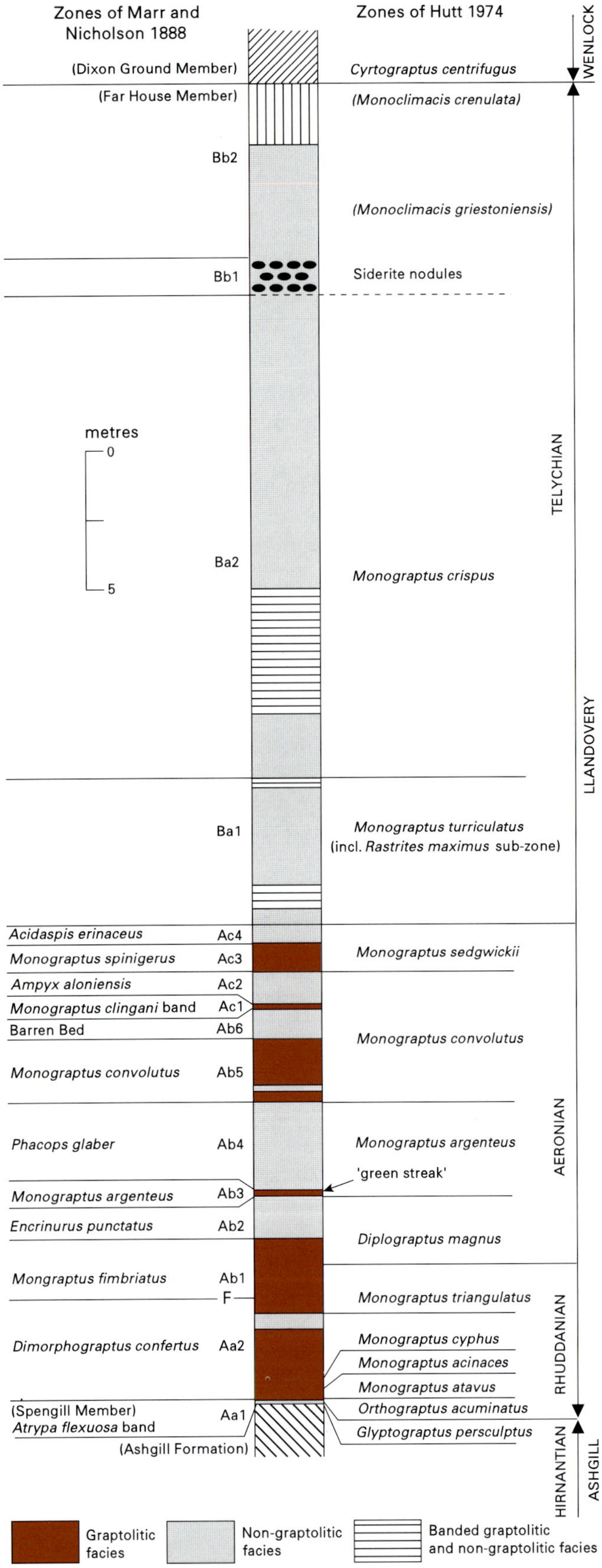

Figure 46 Llandovery (Stockdale Group) stratigraphy in the district.

The lowest, the Brathay Formation, is dominated by hemipelagic laminated siltstone which is graptolitic and carbonaceous (locally free carbon may exceed 1.5 per cent according to Rickards, 1964a). In the overlying Birk Riggs Formation, hemipelagite is generally subordinate to turbidite units of mud to very coarse sand. In the succeeding Coldwell Formation, the laminated hemipelagite is interbedded with two intervals of bioturbated, calcareous siltstone and passes back into hemipelagic beds in the lower Ludlow.

Brathay Formation

The Brathay Formation (Brathay Flags of Marr, 1878; redefined by Kneller, 1990a) consists of 100 to 320 m (Figure 45) of graptolitic laminated siltstone. The laminations consist of alternations (about 3 per mm) of terrigenous muddy siltstone and carbonaceous mudstone. Graptolites collected farther east in the Howgill Fells (Rickards, 1964a; 1967) and in the Kentmere area (Rickards, 1969; Lawrence et al., 1986) established that the formation ranges from the *centrifugus* to the lower *lundgreni* biozones, all of Wenlock age. The laminated siltstone, commonly slightly pyritic, is interbedded with subordinate thin horizons of homogeneous mudstone, or graded siltstone to mudstone, that are interpreted as the deposits of low-density turbidity currents. At the top of the *centrifugus* Biozone (about 15 m above the Dixon Ground Member, described below) is a distinctive 10 cm band of calcareous nodules. Graptolites (and less commonly, orthocones) in the lower part of the formation, up to *murchisoni* Biozone, are pyritised; above this level graptolites are carbonised and shelly fossils are only exceptionally recorded. However, some bedding surfaces reveal abundant crinoid debris. Siltstone/mudstone turbidites become more common towards the top of the formation, where there is generally a vertical and/or lateral transition into a mixed siltstone–mudstone facies at the base of the overlying Birk Riggs Formation. Calcareous nodules become increasingly common above the *linnarssoni* Biozone.

Bentonite horizons are present throughout the formation, in places as graded beds with mixed ash/mud tops suggesting resedimentation by turbidity currents. Several bands, mostly less than 1 cm thick, of pale grey, orange-weathered bentonitic tuff are exposed beside the Walna Scar road, about 1 km south-west of Coniston [SD 2889 9697]. A thicker band comprises an upper 3 cm of coherent pale yellow silty rock (E61324), and a lower 4 cm of pinkish brown, loose sandy material. Thin sedimentary lamination and cross-lamination are present, indicating a degree of reworking and possible mixing with the host sediments. Upward grading and weak, submillimetre-scale banding are present in the topmost few millimetres of the upper part. Throughout the bed the grains consist mainly of altered feldspar and mafic silicates. They are accompanied by a few opaque grains and granules, and rare grains of quartz. In the Broughton Moor area, at least one pale grey metabentonitic tuff band occurs in an old quarry [SD 2560 9382] interbedded with grey Brathay Formation siltstone. Alteration to white mica, calcite and

pyrite is pervasive, but the primary character remains clear; feldspar and lithic grains, accompanied by smaller numbers of quartz and apatite grains, reside in a fine-grained clay-rich groundmass. The thin section (E61343) shows both upward grading and an upward decrease in the grain content. Analyses of these rocks (Table 15) show relatively low SiO_2, but high Al_2O_3 and high alkali contents (both K and Na). These features are consistent with a high concentration of alkali feldspar. However, in one of the samples (E61324), high levels of Zr, Nb and Y, together with erratic concentrations of rare earth elements, suggest considerable alteration of the primary composition of the rock. In view of their high content of feldspar and other grains, these rocks are not metabentonites in the strict sense, but can be interpreted as thin, crystal-rich tuff bands which have a metabentonitic groundmass.

Dixon Ground Member

The Dixon Ground Member (Kneller, 1990a) includes the basal 10 to 30 m of the formation and consists of variable pale to dark grey muddy siltstone and mudstone. In the lower part of the member the lithology is thoroughly bioturbated and homogeneous; it becomes first streaky and then more clearly banded upwards, with the preservation of discrete burrow structures in the less bioturbated parts. In the upper part of the member, structureless or bioturbated zones alternate with laminated siltstone, the former becoming progressively less common upwards. A few calcareous nodules are present locally. The overlying laminated siltstone yields *centrifugus* Biozone graptolite faunas, and the member is presumed to represent the lower part of the *centrifugus* Biozone.

Birk Riggs Formation

The Birk Riggs Formation (Kneller, 1990a) consists of sandstone, siltstone and mudstone turbidites, with subordinate laminated siltstone; it overlies the Brathay Formation throughout the district. It is partly equivalent to Grit Band 1 of Aveline and Hughes (1872), the Lower Coldwell Beds of Marr (1878), but includes, in addition, the underlying mixed mudstone–siltstone sequence assigned by these authors to the 'Brathay Flags'. The formation varies in thickness from 55 to 360 m and can be subdivided into sandstone-rich facies and a mixed siltstone–mudstone facies (Figure 45). The sandstone-rich facies includes thin to very thick beds of fine- to medium- (rarely coarse-) grained quartz-rich greywacke. Commonly, the thinner beds are normally graded and interbedded with laminated siltstone. Thicker beds have reverse-graded bases and normally graded tops, but are otherwise massive; they are mainly amalgamated into composite sandstone beds up to 5 m thick. The sandstone is generally slightly calcareous, and thicker beds may contain layers of calcareous nodules. The siltstone–mudstone facies consists of thinly bedded, graded siltstone and mudstone intercalated with laminated siltstone. This facies is locally intercalated between units of sandstone (Figure 45), but in many parts of the area it forms a unit up to 200 m thick at the base of the formation, and grades laterally and vertically into the sandstone-rich facies. Elsewhere, the sandstone-rich facies rest directly on laminated siltstone of the Brathay Formation. The uppermost sandstones are everywhere thickly bedded, and are abruptly succeeded by a 21 to 27 m-thick unit of laminated hemipelagite which is included within this formation. Graptolites have been recovered [SD 2619 9371] from the siltstone–mudstone facies and from the hemipelagite unit at the top of the formation. Together with faunas described from outcrops farther east by Rickards (1969; 1970) these indicate a *lundgreni* Biozone age.

Texturally the Birk Riggs Formation sandstone is greywacke with up to 35 per cent matrix. The principal detrital components are summarised in terms of quartz, feldspar and lithic clast components (QFL, Figure 47; Table 14) and described by McCaffrey (1991) as follows:

Quartz This is the major detrital mineral (up to 45 per cent) with most grains subangular. Monocrystalline quartz is the most common with a plutonic variety exhibiting undulose extinction, fluid inclusion trails and heavy mineral inclusions, predominating. Rare polycrystalline quartz grains have subgrain texture and preferred axes of elongation.

Feldspar This mineral generally forms about 7 per cent of the rock with the feldspar grains typically showing turbid alteration through kaolinisation or sericitisation. Orthoclase is by far the most abundant variety of feldspar, and is commonly replaced by calcite. Simple twinning is rare and perthitic texture is due to exsolution of albite lamellae. Some microcline is present. Plagioclase usually exhibits lath or plate-like morphology with composition in the albite–andesine range.

Phyllosilicates Micas generally form around 2 per cent of the rock. White mica is the dominant phyllosilicate, occurring as both elongate laths and short plates, and most are detrital. Chlorite is present as recrystallised detrital laths and as small well-crystallised plates. Many of the elongate micas show evidence of post-depositional compaction. Biotite is very rare and when present is commonly chloritised.

Lithic fragments Of these, volcanic fragments are the most abundant. Varieties include spilite (with albite laths in either spherulitic or flow-orientated textures), sodic trachyte, rhyolite and rare porphyritic lithologies with mafic and plagioclase phenocrysts. The less common plutonic clasts include granite and granophyre. Among the sparse sedimentary fragments the commonest lithologies are mudstone flakes of probable autochthonous origin. Well-rounded chert grains are fine grained or cryptocrystalline and may contain rare radiolarian remains. They can be difficult to distinguish from some rhyolite fragments. Also present are some quartzite grains and primary organic carbonate clasts including bryozoan and crinoid fragments. Metamorphic fragments are rare, but some slate, phyllite and quartz–mica schist grains are present.

Coldwell Formation

The Coldwell Formation (Kneller, 1990a; equivalent to the Coldwell Beds of Aveline and Hughes, 1872; or Middle

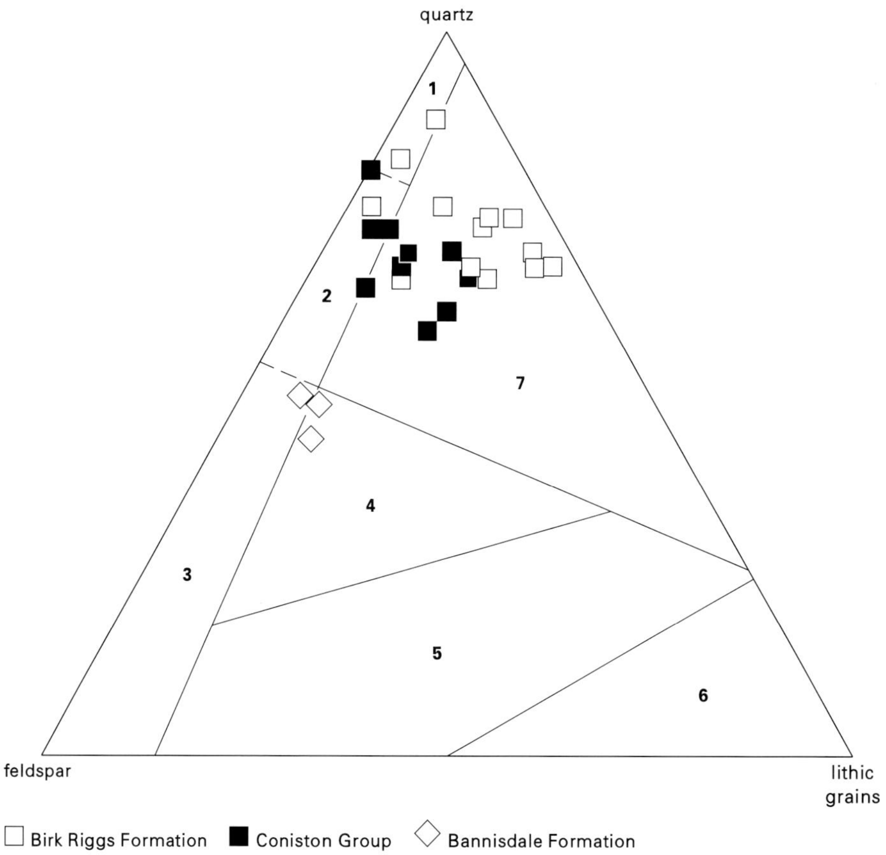

Figure 47 Detrital components of Windermere Supergroup sandstones, compared in terms of quartz, feldspar and lithic grains.

Numbers 1 to 7 indicate provenance fields (after Dickinson et al., 1983: 1 craton interior, 2 transitional continental, 3 basement uplift, 4 dissected arc, 5 transitional arc, 6 undissected arc, 7 recycled orogen.

Coldwell Beds of Marr, 1878) is from 55 to 83 m thick, with the thickest section in the north-east of the district. The middle part of the formation consists of 27 to 55 m of laminated siltstone, but the lower and upper parts of the formation (distinguished as members) consist of pale blue-grey, calcareous fine- to coarse-grained siltstone, with a mottled appearance due to extensive bioturbation. Where bioturbation is less intense individual *Chondrites* burrow systems are discernible. Thin graded beds of laminated or cross-laminated coarse-grained siltstone or fine-grained sandstone occur every few decimetres, passing up first into faintly laminated siltstone then into massive siltstone as the intensity of bioturbation increases. The more-massive siltstone contains a shelly benthic fauna that is also locally abundant on the bedding planes at the base of the coarser-graded beds. The pale grey siltstone and laminated siltstone facies are interbedded at the top of the lower and base of the upper calcareous units.

Sporadic bentonite horizons occur through the middle part of the formation. At High Cross Quarry [SD 3280 9850], bentonitic tuff bands, up to 3 cm thick, occur interbedded with grey siltstone beds. A thin section of one bentonite (E61325) indicates a submillimetre alternation of layers rich in white mica and alkali-feldspar, deformed during cleavage formation and invaded by calcite veinlets which themselves are sheared and internally deformed. Another example (E61326) preserves fine bedding lamination, and is the upper of a closely spaced pair, the lower of which has a conspicuously undulose base. The presence of quartz and muscovite grains indicates mixing with the host sediment and analyses record more than 65 per cent SiO_2. In the Broughton Moor area a bentonite horizon is exposed, interbedded with siltstone, in a disused quarry [SD 2541 9304]. A thin section (E61346) shows a sandy rock in which angular feldspar forms greater than 50 per cent of the grains, accompanied by quartz, chlorite, muscovite and accessory zircon. The grains are set in a fine-grained groundmass rich in white mica. The bed grades upwards as grain size and abundance decrease. Geochemically, the bentonite horizons in the Coldwell Formation are very similar to those described from the Brathay Formation (Table 15).

Randy Pike Member

The lower calcareous unit, the Randy Pike Member (Kneller, 1990a), maintains a thickness of about 14 m across the district. It contains a sparse fossil fauna which is most abundant in the area east of the Brathay fault. Crags near the base of the member [around NY 3672 0110], contain disarticulated specimens of the trilobite *Delops obtusicaudatus* (Rickards, 1964b), brachiopods including *Mezounia*, the bivalve *Cardiola interrupta*, a small high-spired gastropod (*Loxonema?*), orthocones and crinoid ossicles. A disused quarry at High Crag [NY 3550 0056] exposes strata containing particularly abundant orthocones (?*Kionoceras*), together with some disarticulated phacopid trilobites. Natural exposures [NY 3502 0048] contain *Delops obtusicaudatus*, gastropods (cf. *Loxonema*), and orthocones (*'Orthoceras'* sp.). This fauna contains elements that occur widely in late Wenlock mudstones in the Welsh Basin and elsewhere (Warren et al., 1984). The laminated siltstone overlying this member to the east of the district yields *nassa* Biozone graptolites at Scour Rigg [NY 449 036]; this fauna is described by White in Lawrence et al. (1986).

High Cross Member

The upper calcareous unit, the High Cross Member (Kneller, 1990a) is of similar thickness to the Randy Pike

Member, but differs in containing significant intercalations of laminated siltstone. It is less fossiliferous than the Randy Pike Member yielding only rare specimens of the trilobite *Delops nobilis marri?*. The laminated siltstone immediately underlying the High Cross Member contains *ludensis* Biozone graptolites, whereas graptolites from laminated siltstone within 3 m of the top of the member suggest the *nilssoni* Biozone (Lawrence et al., 1986; Rickards, 1969; 1970).

LUDLOW FORMATIONS

Wray Castle Formation

The Wray Castle Formation (Kneller, 1990b; equivalent to 'Upper Coldwell Beds' of Marr, 1878) consists of 280 to 320 m (Figure 45) of laminated siltstone, with subordinate thin graded beds of mudstone, siltstone and, rarely, fine-grained sandstone. In the eastern part of the district, the lower part of the formation is dominated by laminated siltstone, whereas the upper part includes a high proportion of graded beds, and contains scarce calcareous nodules; thin (less than 4 mm) bentonite horizons occur sporadically. In the south-west of the district the entire formation consists of laminated siltstone. The fauna is dominated by graptolites of the *nilssoni* Biozone and by orthocones but some crinoid debris is also present in places. The upper boundary of the formation is a transition, over about 10 m, into sandstone of the overlying Gawthwaite Formation of the Coniston Group.

Coniston Group

The Coniston Group (corresponding to the Coniston Grits of Sedgwick, 1845; 1846) comprises 1820 to 2150 m thick sandstone-dominated sequence overlying the laminated siltstone of the Wray Castle Formation (Figure 45). The base is marked by an abrupt increase, usually over about 10 to 25 m, in the mean grain size and bed thickness of sandstone, siltstone and mudstone intercalated with the laminated siltstone. The top is a vertical and lateral transition into the banded facies of the overlying Bannisdale Formation. The group spans the upper part of the *nilssoni* Biozone, and most or all of the *scanicus* Biozone (Rickards, 1967). The group is divided into five formations (Table 13; Figure 45), three being dominated by sandstone (the Gawthwaite, Poolscar and Yewbank formations) separated by two thinner, finer-grained units (the Latrigg and Moorhowe formations).

The laminated siltstone that dominates the underlying sequence persists as a background lithology intercalated with the turbidite sandstone throughout the Coniston Group, but becomes increasingly scarce towards the top. Typical Coniston Group sandstone is quartz-rich greywacke (Table 14; composition summarised in QFL plot, Figure 47); it is poorly sorted, matrix rich, dominantly of medium sand grade (locally ranging up to very coarse sand) and grades upwards into silt within individual turbidite beds. Subangular quartz is the most abundant clast type (generally over 25 per cent), with, in order of decreasing abundance, feldspar, sedimentary lithic grains (commonly including chert), acid volcanic grains, spilite fragments, detrital carbonate, white mica and chlorite. The varieties and relative proportions of the feldspar are the same as those of the Birk Riggs Formation, although it is slightly more abundant and forms, on average, about 10 per cent of the rock (Furness, 1965). Plagioclase composition is generally in the albite–andesine range (Norman, 1961; Furness, 1965; Redfearn, 1979) and rare grains of microcline occur in all sandstone units examined. Redfearn (1979) also noted compositional zoning in Coniston Group plagioclase grains. Commonly, the sandstone contains carbonate cement and, sparsely, includes bedding-parallel zones of carbonate nodules. The bedding characteristics and sedimentary structures of the sandstone are similar to those in sandstone of the Birk Riggs Formation. Thin and medium beds are usually graded, and some show Bouma-type sequences of sedimentary structures (Bouma, 1962). Thicker beds are generally structureless and ungraded through much of their thickness, with only the uppermost 10 to 30 cm being normally graded. Very thickly bedded sandstones are commonly amalgamated into composite beds several metres thick. The sandstone also displays a wide range of soft-sediment deformation structures, including dewatering dishes and pipes, and convolute bedding. The formation is only very sparsely fossiliferous. Graptolites recovered from the laminated siltstone interbeds beyond the Ambleside district suggest that the top of the *nilssoni* Biozone probably lies high in the Gawthwaite Formation (Rose and Dunham, 1977) and the top of the Coniston Group probably lies high in the *scanicus* Biozone (Rickards, 1967) (Table 13).

Gawthwaite Formation

The Gawthwaite Formation, 290 to 520 m thick, is dominated by thin- to medium-bedded sandstone, more rarely thickly bedded, with many intercalated thin beds either of graded siltstone and mudstone or of laminated siltstone. Mostly the sandstone is normally graded and individual beds are capped by mudstone, and separated by a few centimetres of laminated siltstone. Units of laminated siltstone, up to 2 m thick, occur within the formation, containing abundant graptolites and orthocones. In the eastern part of the district (east of Hawkshead) the sandstone of the lower part of the formation is more thinly bedded than in the upper part.

Latrigg Formation

The Latrigg Formation (Latrigg Member of Lawrence et al., 1986), is 120 to 260 m thick and consists mainly of laminated siltstone, commonly graptolitic. Numerous thin beds of graded siltstone and mudstone are also present and subordinate thin to thick beds of medium-grained sandstone occur sporadically. In well-exposed areas, the formation can be subdivided into a lower part, about 100 to 120 m thick, dominated by laminated siltstone, and an upper part containing a higher proportion of graded sandstone beds, of similar facies to the underlying Gawthwaite Formation, which form units up to 80 m thick. The upper part of the formation apparently interdigitates with the overlying Poolscar Formation.

POOLSCAR FORMATION

The Poolscar Formation is between 430 and 700 m thick. It consists primarily of thickly to very thickly bedded medium- to coarse-grained (rarely very coarse-grained) sandstone beds, commonly amalgamated into sequences several metres thick. Intercalated fine-grained material is restricted to the graded tops of the sandstone beds except in the basal part of the formation where rare, thinly bedded, fine-grained sandstone occurs. The fine-grained tops of graded sandstone beds near the base of the formation are locally bioturbated and an exceptionally coarse-grained unit in the central part of the formation locally contains a derived shelly fauna (Crewdson, 1915). The upper 50 to 60 m of the formation show a progressive decrease in grain size and bed thickness to merge with the overlying Moorhowe Formation.

MOORHOWE FORMATION

The Moorhowe Formation (Moorhowe Member of Lawrence et al., 1986) is 90 to 140 m thick and consists of thin to thickly bedded siltstone (generally well cleaved), commonly with thin interbeds of laminated siltstone. Some of the unlaminated siltstone beds grade upwards from fine sandstone at the base and have a thin mudstone cap, but the bulk of these beds is generally poorly graded and structureless apart from convolute lamination in the upper part. The upper boundary of the formation is a transition over 20 to 30 m into the overlying Yewbank Formation. The formation has yielded a single specimen of the Wenlock to Ludlow bivalve *Cardiola interrupta.*

YEWBANK FORMATION

The Yewbank Formation (Yewbank Sandstone Member of Moseley, 1984), 730 to 750 m thick, is dominated by sandstone. Thick- to very thickly bedded, massive sandstone is characteristic, but many thinner and graded sandstone beds and laterally persistent graptolitic siltstone horizons (*scanicus* Biozone) are interbedded with the massive sandstone. The graptolitic siltstone units range up to about 20 m thick. The upper 160 to 170 m of the formation consists almost exclusively of thin- to medium-bedded graded sandstone beds with siltstone or mudstone caps. The finer-grained lithologies show sporadic bioturbation. The upper boundary of the formation (and thus of the Coniston Group) is gradational into the overlying Bannisdale Formation (see below), with an abrupt decrease in the proportion of sandstone and a concomitant increase in both the number and thickness of banded, silty mudstone interbeds.

Bannisdale Formation

The Bannisdale Formation (Bannisdale Slates of Sedgwick, cited in Aveline and Hughes, 1872) consists of a thick sequence of generally thinly bedded silty mudstone and siltstone, with subordinate mudstone and fine-grained sandstone. In the northern part of the district, the base of the formation is marked by an abrupt transition from Coniston Formation sandstone into a mudstone-rich facies; in the south, and east of Coniston Water, the transition is more gradational. The formation has a characteristic banded appearance produced by repeated and abrupt variations in grain size. The true thickness of the formation is difficult to evaluate due to tectonic distortion, but assuming plane strain, observed strain ratios suggest an original stratigraphical thickness of 4.2 km, of which 2.6 km lie within the district. However, since part of the strain has been accommodated by volume loss within the cleavage, the assumption of plane strain yields only an approximate minimum estimate.

The silty mudstone of the formation is highly variable, and includes laminae and thin to medium beds of siltstone and mudstone. Some siltstone beds are normally graded, but much of the sequence displays unsystematic variations of grain size on a small scale. Thin beds of siltstone and fine-grained sandstone are generally parallel laminated or ripple cross-laminated, but throughout the formation, primary sedimentary structures have been extensively disrupted by small-scale liquefaction. Thus, the thinner siltstone laminae are commonly dismembered by loading, convolution and injection phenomena. The formation also contains very subordinate thin units, up to a few metres thick, of medium- to thickly bedded, fine- to medium-grained sandstone. These are made up of graded beds which locally develop into full Bouma-type sequences. The range of clast composition seen in the sandstone is similar to that recorded for the Birk Riggs Formation and Coniston Group, but feldspar is slightly more abundant relative to quartz (Table 14; Figure 47).

The Bannisdale Formation has yielded no fossils within the district. However, farther east within the Howgill Fells, a sparse graptolite fauna suggests a range from late *scanicus* Biozone to mid-*leintwardinensis* Biozone (Rickards, 1967).

A localised sandstone-rich sequence at the base of the Bannisdale Formation (compare with the Tottlebank Transition Formation of Norman, 1961) crops out to the south-east of Coniston Water. It contains alternations of the typical banded facies of the Bannisdale Formation, with units up to about 20 m thick consisting of medium to thickly bedded sandstone of similar facies to that in the underlying Coniston Group. This mixed lithological assemblage passes laterally into undifferentiated Bannisdale Formation as the sandstone pinches out northwards. Thickness ranges from about 250 m in the west to over 1000 m in the east.

DEPOSITIONAL ENVIRONMENT

Ordovician

The Dent Group represents shallow marine shelf sedimentation during the Ashgill epoch, with periods of emergence followed by periods of increasing water depth. Three such depositional cycles are recognised. Clastic material was locally derived, principally from the underlying Borrowdale Volcanic Group. The first depositional cycle is only poorly represented within the district, its deposits being now exposed mainly at the eastern end of the Lake District inlier in the Longsleddale area (Longsleddale Member and Stile End Member of Lawrence et al., 1986).

The second depositional cycle produced the sequence from the Kirkley Bank Formation to the top of the

Appletreeworth Formation; it records the inundation of the low-lying peneplain formed over the collapsed Borrowdale Volcanic Group massif (Scott, 1992). Within the basal clastic sequence beach and shore-face deposits are present. The succeeding carbonate-rich Kentmere Member indicates a period of restricted siliciclastic sediment supply. The High Pike Haw and Applethwaite members reflect contemporaneous deposition in a storm-dominated environment with the Applethwaite Member representing more distal deposition. The Torver Member records the effect of storm events on an environment with a low depositional rate. The period of maximum marine transgression is represented by the oolitic ironstone found at the base of the Broughton Moor Formation. However, the limestone of this formation possibly represents condensed, carbonate deposition during the Rawtheyan Stage. The Appletreeworth Formation records contemporaneous rhyolitic volcanism, during which the carbonate deposition was interrupted by subaqueous reworking of acid pyroclastic material, probably as a series of high-density turbidites derived from the south-east.

Following limited erosion, a rise in relative sea level initiated a third cycle of deepening. This is represented by the Ashgill Formation. The basal carbonate-rich facies seen in the Troutbeck Member (early Zone 7; late Rawtheyan) is analogous to the Kentmere Member of the second cycle. Abundance of fossils and intensity of bioturbation both reach a maximum in the upper part of the Troutbeck Member, but then decline abruptly upwards. The upper part of the Ashgill Formation contains a relatively deep-water variant of the *Hirnantia* brachiopod association of Brenchley and Cullen (1984). The appearance of this low-diversity, cold-water fauna is probably associated with the onset of a glacial episode (Brenchley, 1988) which reached its maximum during the Hirnantian (upper Zone 8 and late *extraordinarius* Biozone).

Silurian

The Llandovery Stockdale Group (Skelgill and Browgill formations) represents starved, relatively deep, marine shelf environments. The fine-grained sediment (predominantly mud and fine silt) was introduced by a combination of pelagic fallout, nepheloid plumes and very low-density turbidity currents. Changes in lithofacies and biofacies throughout the group reflect fluctuations in bottom-water oxicity caused, at least partly, by sea-level variations. The abrupt change from the underlying Ashgill Formation is associated with the widely recognised rise of sea level following the Hirnantian glaciation (Brenchley, 1988). The Spengill Member, at the base of the Skelgill Formation, is a condensed sequence; its *persculptus* to *acuminatus* Biozone brachiopods constitute a low-diversity deep-water assemblage.

The non-bioturbated, laminated black mudstones of the Stockdale Group, with free carbon up to 3 per cent or more, are deposits of the anaerobic zone (Thompson et al., 1985). Bioturbated pale, barren mudstones represent sedimentation in the dysaerobic zone, where moderate levels of dissolved oxygen allowed colonisation by soft-bodied infauna. Higher levels of oxicity permitted more extensive biological homogenisation of sediment (as in the green mudstone of the Browgill Formation) and development of a shelly benthos (central parts of the blue-grey mudstone units in the Skelgill Formation, and the underlying Ashgill Formation). However, the commonly dwarfed form of the shelly faunas, together with the trilobite ecology, suggest environments near the lower margin of the aerobic zone (Byers, 1977). This sequence of facies may represent a depth zonation as in modern oceans, or temporal variation in the intensity of the oxygen minimum zone (Scott and Kneller, 1990).

The Brathay Formation records a transition during the early Wenlock from the relatively aerobic conditions (and resulting extensively bioturbated strata) of the underlying Browgill Formation, through the dysaerobic environment which prevailed during deposition of the Dixon Ground Member, to anaerobic conditions indicated by laminated siltstone containing virtually no evidence of a benthic fauna. The laminated siltstone represents hemipelagic deposition in a deep, shelf environment, with minor sediment input from low-density turbidity currents represented by the homogeneous mudstone and graded siltstone beds. Minor scouring and orientation of graptolite rhabdosomes in the upper part of the Brathay Formation indicate limited current activity.

The Birk Riggs Formation records the growth of a series of small, coalescing turbidite fans during late *lundgreni* Biozone times. The thickly bedded sandstone, which is commonly lenticular over distances of a few hundred metres, was deposited from high-density turbidity currents in the more proximal parts of the fans; the siltstone–mudstone facies represents the deposits of lower density currents in the fan fringe or channel overbank areas (Pickering et al., 1989). Sole structures on sandstone beds indicate sediment dispersal principally towards the south-west; thinning and fining of sandstone beds towards the south-west, and lateral transitions from sand-dominant facies to siltstone–mudstone facies in this direction suggest that fan progradation was largely from north-east to south-west. Repetitions of this transition along strike imply the existence of several lateral supply points (upper fan channels), from which debouching flows were deflected predominantly to the south-west, perhaps by the Coriolis force (Pickering et al., 1989). Ripple cross-lamination almost invariably indicates currents flowing towards the south-east; the discrepancy between current directions deduced from sole structures and ripples may be a consequence of oblique reflection of the turbidity currents (Kneller et al., 1991).

Turbidite deposition ceased abruptly towards the end of the *lundgreni* Biozone as two episodes of eustatic regression re-introduced dysaerobic to aerobic conditions (Laufeld et al., 1975; McKerrrow, 1979; Scoffin, 1971). During these two episodes the Randy Pike and High Cross members were deposited (Kneller, 1990a). The faunas of these aerobic lithologies are nonetheless comparatively deep-water forms (Rickards, 1978). Bedding plane assemblages of the endemic shelly benthos within these members form lag deposits at the base of waning flow sequences, perhaps representing distal storm deposits. The ensuing basal Ludlow deepening induced a

return to hemipelagite deposition with accumulation of the Wray Castle Formation under anaerobic conditions.

The graded sandstone, siltstone and mudstone units of the Coniston Group are interpreted as the deposits of high and low-density turbidity currents, laid down during continuous Ludlow subsidence. Palaeocurrents determined from sole structures (Kneller, 1990b) indicate that sediment dispersal was predominantly from north-east to south-west in all the sandstone-dominated formations (with minor components towards the north-east and south-east), with no significant difference in dispersal patterns between formations. Individual sandstone beds are laterally impersistent, but packages of sandstone beds are laterally extensive, and display little evidence of systematic lateral (down-current) thickness or facies change. Each member probably represents a coalescence of several fan systems, similar to (but more extensive than) those of the Birk Riggs Formation.

The lateral extent of the finer grained formations implies extrinsic control of the depositional environment. Very low sand input during deposition of the anoxic Latrigg Formation (together with the generally thin, laminated siltstone intervals throughout the formation) may reflect periods of relatively high sea level. However, the fine-grained but relatively high-volume (thickly bedded) siltstone deposits in the Moorhowe Formation are probably no more 'distal' than the sandstone and simply indicate a change in the pathway and sorting mechanisms of sediment prior to final deposition.

The fine-grained Bannisdale Formation strata resulted from rapid sedimentation in a predominantly anoxic environment experiencing rapid subsidence, though levels of oxicity suggested by sporadic bioturbation show a slight increase over those in the underlying Wenlock to lower Ludlow laminated siltstone. The dominant facies probably represents sedimentation from relatively dilute, semi-continuous, sediment-gravity underflows, punctuated by incursion of sand-laden, 'normal' waning-flow turbidity currents. Palaeocurrent patterns of the sandstone turbidites are generally similar to those in the Coniston Group indicating sediment dispersal principally towards the south-west for west-south-west. Ripple cross-lamination in the banded silty mudstone facies almost invariably indicates currents from the north-west. A change from sand-rich facies to mud-rich facies during deposition of the lower part of the formation was diachronous, migrating progressively towards the south-south-east with time.

Silurian sandstone provenance has been investigated by McCaffrey (1991) and McCaffrey and Kneller (1996). The allochthonous grains reworked into the Windermere Supergroup sediments were thought to indicate derivation from a geologically heterogeneous source area. The combination of plutonic quartz, white mica and biotite suggested erosion of peraluminous granites though, because the quartz is not all angular, polycyclic contributions are possible. The relatively unaltered feldspars commonly exhibit lath and plate morphologies, suggesting a probable first-cycle origin. A possible arc source is indicated by the presence of volcanic lithic fragments, though the spilitic fragments, in conjunction with the radiolarian cherts, probably represent the erosional detritus of oceanic crust. The erosion of sedimentary sequences is required by the presence of sedimentary lithic fragments, whilst the occurrence of slate, phyllite and schist fragments indicates erosion of a metamorphic terrain. The lithic fragments are commonly more rounded than the quartz or feldspar grains.

A synthesis of this information, together with an appreciation of the large volume of material entering the depositional basin, suggests a rapidly uplifted source area producing dominantly immature first or second cycle sediments. Exposed granitic plutons would have probably supplied the bulk of the sediment, supplemented by first-cycle volcanic arc and ocean floor sources and by erosion of sedimentary sequences; minor metamorphic input suggests exposed basement. Any of the various allochthonous components may be found juxtaposed in the same sandstone bed requiring thorough mixing, either during initial transportation or in the temporary sediment traps which may have sourced the turbidites. Alternatively, most of this material could have been recycled through an immature sedimentary sequence, possibly an accretionary prism. In terms of quartz, feldspar and lithic clast components (Figure 47) the sandstone petrography indicates a likely recycled orogen source consistent with derivation from a destructive plate margin.

Nd model ages reported from the Windermere Supergroup by McCaffrey (1994) increase from the upper Ashgill to the upper Ludlow. In the Ordovician Dent Group, there is evidence for mixing between sediment derived from the underlying Borrowdale Volcanic Group and hemipelagic detritus, possibly sourced from the Laurentian continental margin. Model ages through the Silurian reflect the growing influence of an ancient basement source, possibly the Scandian Orogen. A reversal of the trend, with reduction of model ages, in the Přídolí was thought to indicate the re-establishment of a local provenance but these younger rocks are not present in the district.

The marked increase in subsidence and the corresponding increase in sedimentation rate during the Ludlow epoch has been modelled by Kneller (1991) and King (1992) in terms of a southward-migrating foreland basin. This is envisaged as a unified feature extending south from the southern part of the Southern Uplands and spanning the sutured Iapetus Ocean (Kneller et al., 1993b). The Southern Uplands thrust front constituted the north-western margin (Barnes et al., 1989) and the southward prograding thrusts have been described from the Skiddaw Group outcrop in the north of the Lake District (Hughes et al., 1993). The principal depocentre therefore migrated south with time and facies changes are correspondingly diachronous. These relationships and the development of the foreland basin are discussed more fully in Chapter 9.

REGIONAL CORRELATION

Ordovician

Regional correlations of the Ashgill Dent Group are summarised in Table 16. Facies similar to the nodular

calcareous siltstone of the Kirkley Bank Formation are widespread in sequences of Longvillian to Rawtheyan age elsewhere in the Windermere Supergroup outcrop. For example, the High Haume Mudstone of the Furness inlier (Rose and Dunham, 1977) may be a direct correlative of the Kirkley Bank Formation. Correlatives of the Broughton Moor Formation and/or Appletreeworth Formation are present in all the other Windermere Supergroup inliers, and in all but Cautley and Dent (which perhaps represent deposition in deeper water) they succeed a minor unconformity probably equivalent to that beneath the Broughton Moor Formation. The successions in Cautley, Dent and Furness include exact equivalents of the Ashgill Formation and the Troutbeck Member, and also follow a minor unconformity. Thin coarse clastic units occur near the top of Zone 8 in all of the other inliers except Cross Fell (where a synsedimentary breccia suggests emergence: Wright, 1985; Brenchley, 1988), but are not well developed in the district.

Silurian

Silurian correlations of the Windermere Supergroup are summarised in Table 17. The Skelgill and Browgill formations or correlatives are recognised in all Windermere Supergroup inliers, where they are of generally similar

Table 16 Regional stratigraphical correlation of the Dent Group (after Kneller et al., 1994).

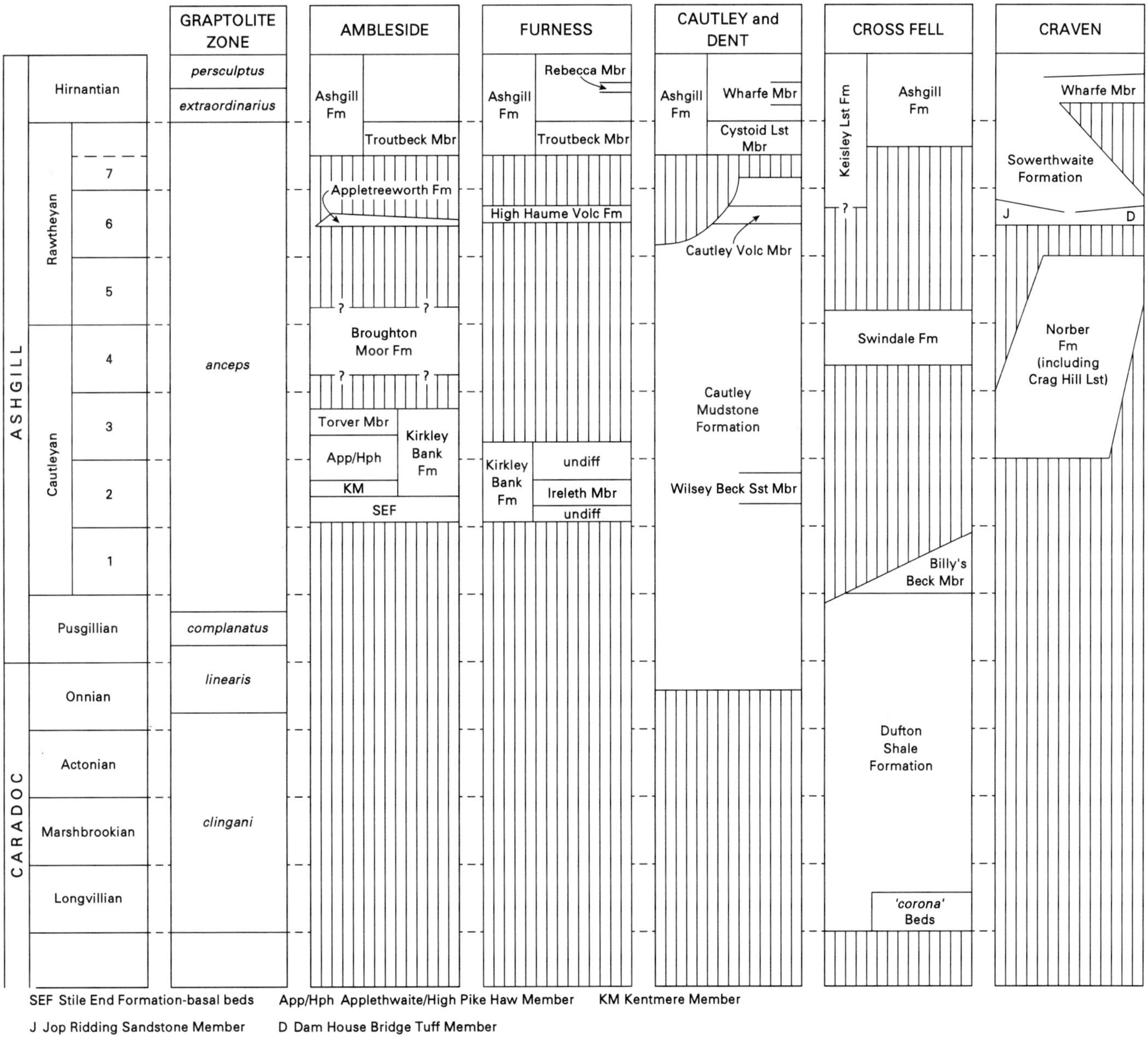

Fm Formation; Mbr Member, Lst Limestone.

Table 17 Regional stratigraphical correlation of the Silurian strata of the Windermere Supergroup (after Kneller et al., 1994).

Series	**Ambleside**			**Furness** (Norman, 1961; King, 1992)		**Craven** (Arthurton et al., 1988)		**Howgill Fells** (Rickards, 1967)
Ludlow	Bannisdale Formation			Bannisdale Slates		not present		Bannisdale Slates
	Coniston Group	Yewbank Fm			Coniston Grits			Upper Coniston Grits
		Moorhowe Fm		'slate'				
		Poolscar Fm				Neals Ing Formation		
		Latrigg Fm		Rosside Flags		? Studfold Sandstone	Horton Formation	? Sheerbate Flags
		Gawthwaite Fm						Lower Coniston Grits
	Wray Castle Formation			Horace Flags				
Wenlock	Coldwell Formation		High Cross Mbr			Arcow Formation		Middle Coldwell Beds
			Randy Pike Mbr					
	Birk Riggs Formation			Harlock Grits		Austwick Formation		Brathay Flags
	Brathay Formation		Dixon Ground Mbr	Brathay Flags				
Llandovery	Stockdale Group	Browgill Fm		Browgill Beds		Crummack Formation	Capple Bank Mbr	Browgill Beds
		Skelgill Fm		Skelgill Beds			Hunterstye Mbr	Skelgill Beds

Fm Formation; Mbr Member

facies to those described here from this district, and include a similar basal calcareous unit (Spengill Member). A slight variation is seen in the Craven inliers, where the Hunterstye Member (Rhuddanian and Aeronian stages) is dominated by black mudstone with graptolitic bands, whereas the Capple Bank Member (possible Aeronian to Telychian) includes calcareous beds with a shelly fauna comparable in composition with those from the Skelgill Formation.

An onset of laminated siltstone deposition occurs at the base of the Wenlock in all the Windermere Supergroup inliers. 'Brathay Flags' are described from Furness, the Howgill Fells and Cross Fell and are equivalent to the Austwick Formation of the Craven inliers. The Harlock Grits unit (*lundgreni* Biozone) of the Furness inlier (Rose and Dunham, 1977) is a close analogue of the Birk Riggs Formation, although there is no continuity of outcrop. Similar turbidite sandstone deposition began earlier (perhaps *rigidus* Biozone) in the Austwick Formation of the Craven inliers, but also ceased in the late *lundgreni* Biozone contemporaneously with the cessation of Birk Riggs Formation sedimentation. The calcareous facies of the Coldwell Formation spans the Wenlock–Ludlow boundary and is thus at least a partial correlative of the 'basal Ludlow' limestone of the Howgill Fells. The latter is an imprecise biostratigraphical designation (Rickards, 1969; 1970) and correlation may be more complete because the Howgill Fells sequence may also include beds of Wenlock age. The calcareous facies is also seen in the Arcow Formation of the Craven inliers and has been recorded from near the base of the laminated siltstone of the Horrace Flags in the Furness inlier. Lower Ludlow laminated siltstone, correlative of the Wray Castle Formation, forms the Horton Formation in the Craven inliers, and the greater part of the Horrace Flags in Furness.

Sandstone-rich units comparable to the Coniston Group occur in all Windermere Supergroup inliers containing strata of Ludlow age. Generally, a separation is possible into lower and upper units, divided by an interval of laminated siltstone which can probably be correlated with the Latrigg Formation. The lower unit is not present everywhere within the group, and lenses out both in the eastern part of the Lake District inlier and in the Craven inliers (Arthurton et al., 1988). Where the lower unit is present, the onset of sandstone sedimentation was apparently diachronous, following only 5 m of laminated siltstone in the Howgill Fells (Rickards, 1967), but overlying approximately 300 m of laminated siltstone which forms the Wray Castle Formation farther west in the Lake District inlier. The equivalent of the Moorhowe Formation is recognised in the Furness peninsula (Norman, 1961) and possibly within the Howgill Fells (King, 1992), but this stratigraphical level is not preserved in the Craven or Cross Fell inliers.

EIGHT

Intrusive igneous rocks

Two major intrusions, Eskdale and Ennerdale, are present in the west of the district, forming the surface expression of an Ordovician–Devonian batholith known to underlie the volcanic rocks of much of the central Lake District (Bott, 1978; Firman and Lee, 1986; see also Chapter 3). The Eskdale pluton has two major components, a well-exposed northern granite and a poorly exposed southern granodiorite. The contact between them is not exposed and intrusive relationships are unclear, but they represent discrete, albeit probably comagmatic, intrusions. The Ennerdale intrusion is predominantly a fine-grained, granophyric granite exposed in three separate areas to the west and north-west of Wast Water. Geophysical modelling suggests it is a tabular body no more than 2 km thick. Accounts of the field relations, petrology and geochemistry of the plutonic rocks are contained in Dwerryhouse (1909), Rastall (1906), Simpson (1934), Trotter et al. (1937), Firman (1978a) and O'Brien et al. (1985), together with unpublished PhD theses by Govinda-Rajulu (1959), Clark (1963) and Ansari (1983).

Numerous dykes or sheets intrude the Borrowdale Volcanic Group and Windermere Supergroup. They range in composition from basalt and dolerite to rhyolite (felsite) and microgranite, together with rarer lamprophyric varieties. These intrusions are sporadically distributed across the area but with some rock types considerably more abundant around the margins of the exposed batholith in the west. They range in age from Ordovician to Devonian with some possibly younger. Those believed to have a close genetic relationship with the Borrowdale Volcanic Group are discussed in Chapter 6.

ESKDALE PLUTON

The Eskdale pluton crops out over an area of 60 km^2. To the west, just outside the district, it is largely terminated at outcrop by a major north-trending fault system (Lake District Boundary Fault, including the Haverigg Fault; British Geological Survey, 1991). Gravity data (Lee, 1986a) show the western margin to dip steeply westward beneath Upper Palaeozoic and Mesozoic cover rocks. Evans et al. (1994) have shown from detailed seismic reflection studies that this margin is defined by the termination of a series of subhorizontal sheets, and it appears unlikely that the granite batholith extends beneath the Irish Sea. Elsewhere, the pluton is in contact with the Borrowdale Volcanic Group, except near Devoke Water [SD 152 972], where hornfelsed rocks of the Skiddaw Group intervene, and at the southern end of Wast Water where the granite apparently abuts the Ennerdale intrusion. The pluton is the westernmost exposed portion of the batholith, the top surface of which is at a depth of less than 2 km over much of the area underlain by the Borrowdale Volcanic Group (Lee, 1986a; 1989). A small isolated body of granite crops out near Wasdale Head [NY 195 090], which the evidence suggests is a cupola on the roof of the batholith (Lee, 1989). Gravity and magnetic modelling by Lee (1989) indicate the Eskdale granite to be a body of considerable thickness and extent, whereas the granodiorite is a relatively thin (less than 3 km), laccolithic body over most or all of its exposed area and is underlain by less-dense granitic material, possibly Eskdale granite, or a separate pluton.

Eskdale granite

The Eskdale granite is composed of three main facies which show complex intrusive relationships with each other. Most abundant is a white to pink, medium-grained, non-porphyritic muscovite granite, the normal granite; the next in abundance is microgranite which may be aphyric or megacrystic; and lastly, two small areas of coarse- to very coarse-grained granite. All three facies show features of two-phase granites in the sense of Cobbing et al. (1986). The normal and coarse-grained granites contain patches and veinlets of microgranite, whereas the megacrysts in the microgranite are xenocrysts or xenoliths of coarse granite. Microgranite is particularly common in the north-eastern part of the outcrop where the low-dipping contacts and inliers of hornfelsed volcanic rocks [for example NY 168 012, NY 185 016] suggest that there the roof zone is exposed. From Devoke Water north-east towards Burnmoor Tarn, then south-west to the Santon Bridge area, the granite contact is subparallel to bedding in the adjacent and overlying volcanic rocks and does not rise much above the base of the volcanic succession. Between Stony Tarn [NY 200 025] and Hardrigg Gill [NY 195 055], however, the contact climbs steadily up the Borrowdale Volcanic Group succession, and the Wasdale Head outcrop is within a few hundred metres of the base of the upper part of the Borrowdale Volcanic Group. Greisens are locally common adjacent to the contact (Young et al., 1988). Xenoliths of country rock are extremely rare and there is a largely undisrupted thin zone of microgranite at the pluton margin attributable to chilling. These features suggest that stoping was not a significant mode of intrusion for this granite, though locally the adjacent Borrowdale Volcanic Group rocks show brittle fracture with development of an intrusive vein-network breccia from which granite veinlets penetrate the country rock.

Several major faults offset the intrusive contact. In the Burnmoor Tarn area the Whillan Beck Fault is known from the Borrowdale Volcanic Group stratigraphy to have

a westward downthrow of 500 m and displaces the granite contact by some 2 km. The Greendale Fault, farther west at the southern end of Wast Water, has an easterly downthrow and also significantly offsets the granite contact. The combined effect across these faults indicates that the pluton roof has a gentle dip to the north in this area. A significant fault cuts the granite at Greathall Gill [NY 143 033], forming a deeply incised gully within which the granite is extremely friable because of kaolinitic alteration, and also locally exhibits silicification. Granite immediately outside the gully is not affected indicating that this may have been an early fault along which hydrothermal fluids were active. On Muncaster Fell and Garner Bank a north-north-west fault system contains vein quartz, breccia or haematite infill; some mineralised faults around Boot have been worked for haematite. A narrow mylonite zone has been recorded on the slopes west of Brantrake Crags [SD 143 983]. It is 1 to 3 m wide, trends north-north-east, but though contrasting lithologies are juxtaposed, it does not appear to mark any major movement. A spaced cleavage is developed locally in the granite (Allen, 1987). It is best developed around Smithy Mire [SD 165 995] where it trends 050 to 075°.

Normal granite

This lithology, which forms approximately 70 per cent of the granite outcrop, is predominantly a white to pink, medium-grained granite (average grain size 1 to 5 mm), locally with haematite spots. It is generally equigranular with sporadic perthite phenocrysts ranging up to about 1 cm. Small pegmatitic segregations and veins occur locally. The granite consists predominantly (over 95 per cent) of perthitic feldspar, sodic plagioclase and quartz, with minor biotite and muscovite. Textures vary from anhedral to more rarely subhedral granular, with complex, consertal crystal boundaries in places. Although generally equigranular, individual crystals may vary from less than 1 mm up to 5 mm within a thin section (Plate 23a).

Perthite usually appears dusty in plane-polarised light and commonly has simple Carlsbad twinning. Perthitic texture varies from very fine or cryptic to coarser, typically braided exsolution textures. Some very coarse perthite crystals exhibit aligned, squat albitic crystals in non-twinned orthoclase. Mesoperthitic feldspars also occur. Sodic plagioclase is usually smaller in grain size and may be poikilitically enclosed by perthite. It too appears discoloured and dusty in plane-polarised light and is commonly sericitised with fine white mica flakes distributed through the crystal in a reticular pattern. Quartz is characteristically strained and forms subhedral or rounded crystals in places enclosed by perthitic feldspar. Some rocks on Muncaster Fell contain abundant, conspicuous quartz pools, 1 cm or so in diameter mostly surrounded by fine-grained granite. Granophyric texture has been observed in a few localities (for example E68015).

Biotite is rarely fresh and generally occurs as altered, subhedral to lath-like pseudomorphs of chlorite or chlorite–muscovite intergrowths. Fine, granular elongate aggregates of probable rutile or anatase are usually interlaminated with the chlorite. Less commonly, biotite is replaced by interlayered muscovite and haematitic iron oxide. Muscovite occurs conspicuously as coarse single crystals or radial and sheaf-like clusters developed interstitially to quartz and feldspar. Rarely, single muscovite plates are enclosed in quartz or feldspar. Ansari (1983) suggested that all muscovite within the granite is secondary in origin, and clearly much muscovite/sericite has formed by replacement of earlier phases. Some coarser, clear, interstitial muscovite plates may represent primary late-stage magmatic or pneumatolytic crystallisation, though locally these recrystallised to finer-grained sericitic aggregates. Sericitic mica also replaces topaz.

The granites are poor in accessory minerals but tiny zircons, and their pleochroic haloes, are common in altered biotite. Tiny apatite crystals are also commonly associated with biotite. Topaz is sparse but its occurrence as relict grains within sericite masses suggests it may have been more abundant prior to its alteration. Traces of sphene are observed as alteration products within biotite. Small clusters of yellow-green tourmaline are sparsely present. Garnet has been recorded in granite close to the pluton margin (Ansari, 1983) and was identified in one sample (E70665). Therein it occurs as a 1 mm subhedral crystal, altered along fractures and on its rim to white mica, but enclosed by a single rounded quartz grain only slightly larger than the garnet itself.

Pegmatitic veins and segregations are mineralogically and petrographically similar to normal granite, though crystal sizes, particularly perthite, range up to more than

Plate 23 Photomicrographs of rocks from the Eskdale and Ennerdale intrusions.

a. Equigranular 'normal' medium-grained Eskdale granite; muscovite containing trails of Fe-oxide (lower right) probably after primary biotite (E70625). Crossed-polarised light; width of field 12 mm (PMS728).

b. Equigranular, fine-grained Eskdale granite (microgranite) with common platy muscovite as a late-stage crystallising phase subpoikilitically enclosing quartz and feldspar (E57614). Crossed-polarised light; width of field 12 mm (PMS729).

c. Xenocrystic, two-phase granite from the Eskdale granite. Medium-grained, subhedral to anhedral crystals or crystal aggregates of perthite, quartz and plagioclase derived from earlier phases of intrusion invaded by a microgranite matrix. Margins of the xenocrysts are locally intergrown with the finer matrix or can show overgrowths (top left) (E70593). Crossed-polarised light; width of field 12 mm (PMS730).

d. Eskdale granodiorite, containing common ragged platy biotite and usually cloudy (sericitised and saussuritised) plagioclase with later crystallising perthite and quartz (E70768). Crossed-polarised light; width of field 12 mm (PMS731).

e. Fine- to medium-grained granite from the Ennerdale intrusion showing characteristic granophyric intergrowth texture (E71121). Crossed-polarised light; width of field 7.4 mm (PMS732).

f. Blengdale diorite facies of the Ennerdale intrusion displaying characteristic texture of interlocking plagioclase laths with acicular chlorite pseudomorphs, probably after hornblende. Turbid epidotic partial alteration of plagioclase is common (E68113). Crossed-polarised light; width of field 12 mm (PMS733).

a.

b.

c.

d.

e.

f.

2 cm. Perthite and quartz tend to be intergrown, with the latter forming in some cases irregular, lobate lamellae, analogous to a coarse granophyric texture (for example at Green How [SD 1711 9934]; E70621).

The effect of strain in the granite is evident from the undulose extinction of quartz. In some samples (for example E70559, E70650) tectonic strain is indicated by buckling of mica and twin planes in feldspar, infill of cleavage fractures by quartz veinlets and minor cataclasis along crystal boundaries which are subsequently annealed or infilled by fine mica and oxides. In more extreme cases a strong planar fabric is present with relict foliae or lenses of normal granite in an inequigranular cataclastic matrix smaller in grain size. This may contain a weak foliation formed by mica stringers which wrap around rigid feldspar and strained quartz crystals (E70630). These features are related to areas of strong faulting within the granite, for example near Stony Tarn [NY 200 020]. One extreme example (E70686) from Bull How [NY 2071 0222] consists only of relict rounded, elongate quartz crystals in a cataclastically reduced matrix consisting of fine quartz pervaded by aligned anastomosing micaceous microveinlets.

COARSE-GRAINED GRANITE

Two areas of distinctly coarser-grained granite have been mapped around Eskdale Green and Knott End. These were probably contiguous but are now displaced by the Eskdale Fault. Petrographically, it is similar to normal granite though biotite and plagioclase may be more abundant. Perthite and quartz are characteristically greater than 5 mm in size. Many of the sections examined (E70788, E70789, E70802, E70807) show evidence of strain (cleavage) and fracturing with minor patches, veinlets and interstitial areas of fine-grained (0.03 mm and greater) recrystallised granite.

FINE-GRAINED GRANITE (MICROGRANITE)

Granites (with a mean grain size up to 1 mm) occur extensively throughout the outcrop of the Eskdale granite. They are particularly developed as a marginal facies adjacent to the granite contact, and within the known roof zone of the granite between Blea [NY 165 010] and Eel [NY 189 019] tarns. Large areas of microgranite also occur north of Devoke Water around Brantrake Moss and Hare Gill [SD 163 984] and on Muncaster Fell [SD 115 985]. Much of the microgranite is xenocrystic, with variable amounts of medium- and coarse-grained quartz and feldspar crystals and crystal aggregates occurring suspended within a microgranitic groundmass. Areas of aphyric microgranite are also common and grade into xenocrystic variants.

Mineralogically the fine-grained granites are identical to normal granite, consisting of quartz, perthitic feldspar and sodic plagioclase with lesser amounts of biotite and muscovite. Biotite is commonly altered to chlorite and/or muscovite with interlaminated iron oxide, but is partly fresh in some areas (E57601, E57609, E70658, E70661). Muscovite occurs as clear, interstitial single crystals as well as replacing feldspar and biotite (Plate 23b). Zircon and apatite are ubiquitous accessory minerals. Topaz and garnet occur rarely and tourmaline is developed as coarse interstitial aggregates close to the contact near Whillan Beck [NY 180 023].

Texturally the fine-grained granites are variable, ranging from essentially aphyric and equigranular to highly inequigranular and xenocrystic. The latter contain varying amounts (up to 50 per cent) of medium- to coarse-grained single crystals and clusters of quartz and feldspar set in a fine-grained, granular granitic matrix (Plate 23c). These megacrysts or megacrystic clusters, particularly those of quartz, in places have a subhedral form, but are more commonly anhedral with irregular grain boundaries intergrown, recrystallised and partly resorbed into the matrix of the rock. Marginal overgrowths on perthite megacrysts are observed in some examples. Megacrystic clusters are fragments of crystalline coarse granite rather than coalesced individual crystals and it is clear that these, as well as discrete megacrysts, represent xenoliths and xenocrysts derived from coarsely crystalline granite. In many samples (for example E57601, E57618) there is a distinct bimodal grain size between megacrysts and groundmass; the former up to 1 cm or more, the latter less than 1 mm and commonly less than 0.1 mm (Plate 23c). In other rocks (for example E57614, E70642) grain size forms more of a continuum with areas of coarser granite grading into finer matrix due to variable degrees of recrystallisation. Such two-phase granitic crystallisation textures are described by Cobbing et al. (1986) from south-east Asia and are interpreted as forming as a result of disruption of coarse granite by the infiltration of residual or later granitic fluids. On Green How [SD 171 993] normal granite shows incipient disruption by secondary granitic fluids. Samples from this locality consist of 80 to 90 per cent by volume medium-grained granite with a microgranitic component developed interstitially and along crystal boundaries, giving the appearance that the normal granite is undergoing disaggregation by the infiltration of microgranitic fluid. By contrast, some fine-grained granites, typically with grain size in the range 0.5 to 1 mm, show one-phase primary crystallisation textures (for example E70564, E70581).

Within the fine-grained granites graphic intergrowth textures are relatively common, patchily developed in the fine-grained matrices of the granites or marginal to xenocrystic perthite. Locally (for example E57602 at NY 1702 0156), virtually the whole rock exhibits beautifully developed granophyric texture with only sporadic pools of clear quartz.

Intrusive relationships within the Eskdale granite

The oldest granite phase demonstrable in the field is the coarse granite. In part, coarse granite appears gradational with normal (medium-grained) granite, but locally intrusive contacts show the coarser facies to be earlier. Both coarse and normal granite are extensively veined and intruded by microgranite, but there is also evidence that one or more phases of micrograniti intrusion predated the bulk of the normal granite. Intrusive contacts in which chilled normal granite cuts coarse granite are exposed by Linbeck Gill [SD 1438 9802] and on Brantrake

Crags [SD 1449 9818]. At the former locality, a 10 cm chilled zone of microgranite with sparse phenocrysts grades over 1 m into normal granite.

Microgranite that predates normal granite occurs on Muncaster Fell where it is sparsely xenocrystic. It occurs as xenoliths in normal granite ranging from a few centimetres to 400 m diameter enclaves, the normal granite chilling against the larger xenoliths [SD 1254 9899, SD 1219 9903] and locally penetrating the microgranite as veins. Abundantly xenocrystic microgranite occurs at Hooker Moss [SD 114 985] as 1 m xenoliths in sparsely xenolithic microgranite, whereas on Brantrake Crags abundantly xenocrystic varieties of microgranite are in contact with coarse and normal granite as well as microgranite. Xenocrystic microgranite may show complex interlayering with coarse granite [SD 1475 9823] with no clear evidence of age relationships, but it is present as xenoliths in normal granite [SD 1492 9843] and in many places is cut by veins of late sparsely xenocrystic microgranite.

Microgranite that is younger than the normal granite is widespread and distinguished by pegmatitic patches and 2 to 3 cm quartz-filled vugs. Xenocrysts and xenoliths of coarser granite increase or decrease in abundance toward the contact with normal granite. Contacts are mostly highly irregular with diffuse to angular xenoliths of normal granite up to several metres across, but usually less than 30 cm, suspended in microgranite. On Muncaster Fell [SD 110 980] microgranite veining of normal granite produces a crude intrusion breccia; microgranite veins are typically 10 to 20 cm wide with contacts gradational over 1 cm. On Irton Fell [NY 141 026] the transition from normal granite to microgranite is marked by the development of diffuse patches of microgranite containing disaggregated fragments or crystals of the former which merge into more coherent, sparsely phyric microgranite. In a microgranite body in Hooker Moss, multiple re-intrusion of microgranite is evident with veins and lobes of a sparsely xenocrystic variety cutting the xenocryst-rich margins.

In summary, the field relationships and the petrographic evidence indicate that intrusion of the Eskdale granite was multiphase. Medium- to coarse-grained, slowly cooling granite was repeatedly disrupted, disaggregated and remobilised by the intrusion and infiltration of later granitic fluids, giving rise to distinctly two-phase xenocrystic and xenolithic microgranites. These phases of intrusion and crystallisation were close in time, with much of the coarser granite probably in the form of a crystal mush which aided its infiltration by later granitic fluids. Despite the textural variety displayed by the granitic lithologies they are essentially mineralogically identical and clearly co-magmatic.

Seismic reflection profiles across the western margin of the Eskdale pluton in the area around Santon Bridge and Nether Wasdale (Evans et al., 1993; 1994) indicate that the granite was emplaced as a series of subhorizontal sheets. These sheets, each several hundred metres thick, appear to be interleaved with screens of country rock, most plausibly Skiddaw Group, and give this margin of the batholith a 'cedar tree' laccolith form (see Chapter 3).

Greisens associated with the Eskdale granite

Localised development of greisen is a feature of the Eskdale granite (Ansari, 1983; Young, 1985a; Young et al., 1988). Close to exposed contacts of the granite, quartz-mica greisens, with or without topaz, characteristically occur as near-vertical layers up to 2 m wide adjacent or parallel to joints and quartz veins. Greisens occur within both normal (medium–coarse-grained) and fine-grained granite facies. Details of greisen localities, field relations and mineralogy are given in table 1 of Young et al. (1988). Good examples of greisen occur at Brown How, near Boot [NY 1730 0201], on Dawsonground Crags [NY 2077 0239], Bull How [NY 2078 0203], Taw House [NY 2060 0159], and Birker Beck [SD 1761 9918]. In general, the greisens are white to buff coloured, of similar grain size to the host granite, and are clearly formed by metasomatic recrystallisation of the granite. Quartz and white mica are ubiquitous phases in the greisens with topaz an abundant constituent (up to 30 per cent) in some. Fluorite is a common accessory phase, and haematite is conspicuous at some localities. Biotite and feldspar have been completely replaced in these rocks. Less commonly, greisenised granite is in the main body of the pluton associated with, and adjacent to, quartz veins and prominent joints. These rocks do not contain topaz but consist of quartz and white mica with minor biotite and chlorite.

QUARTZ–ANDALUSITE ROCK

A distinctive andalusite-bearing rock is associated with topaz greisens on Water Crag [SD 1529 9733], near Devoke Water, where it forms a rib up to 1 m wide and 4 m long in granite close to its contact with hornfelsed Skiddaw Group (Young et al., 1988). Contact of the quartz–andalusite rock with the enclosing granite, however, is not exposed. This rock consists of a medium-grained quartz mosaic with clusters of pink to colourless andalusite up to 5 mm or so across. Individual prismatic crystals up to 2.5 mm occur. Minor sericitic alteration is apparent.

Eskdale granodiorite

An extensive blanket of till and other superficial deposits covers the granodiorite which is thus very poorly exposed. It is characteristically medium grained, locally rich in biotite and in the south displays strong argillic alteration. Mafic xenoliths are common and locally the granodiorite is conspicuously garnetiferous, especially in and around Waberthwaite quarries [SD 112 943]. A microgranodioritic facies is developed along the margins of this body which trends roughly southwards from the Devoke Water area. It is a more discordant intrusion than the granite reaching the lowest formations of the upper part of the Borrowdale Volcanic Group which crop out within the core of the Ulpha syncline.

In the extreme south-east of the granodiorite outcrop, south of Buckbarrow Beck [SD 138 910] and predominantly within the adjacent Ulverston district, Ansari (1983) delimited an area of 'adamellite'. He recorded no intrusive contacts with the granodiorite, nor significant visual differences in the field; the distinguishing criteria were

minor petrographical differences such as a slightly higher orthoclase to plagioclase ratio. There is considerable geochemical overlap between the 'adamellite' and the granodiorite (Ansari, 1983), suggesting minor variation within a single intrusive body.

Petrographically the granodiorite is distinct from the granites in the greater proportion of plagioclase and biotite, the presence of hornblende and lack of coarse, platy muscovite. Most of the granodiorite is medium grained with subhedral or euhedral plagioclase phenocrysts up to several millimetres across set in a granular matrix of plagioclase, perthite and quartz characteristically 1 to 2 mm in grain size. Plagioclase phenocrysts are commonly zoned and mainly highly altered to sericite/saussurite, the intensity of which commonly mimics primary zoning. Groundmass plagioclase tends to be less altered. Micro- to cryptoperthitic feldspar is slightly dusty in appearance but relatively fresh. Larger (up to 1 cm) anhedral perthite crystals occur sporadically, poikilitically enclosing biotite and plagioclase laths. Patchy development of granophyric intergrowth texture is not uncommon in the granodiorite. One sample (E70752), from close to the contact on Corney Fell [SD 1370 9207], is almost totally composed of graphically intergrown quartz and feldspar. Biotite generally occurs as small, squat tabular or lath-like to rather more irregular, anhedral crystals and is variably altered to chlorite. Locally, around Waberthwaite quarry, biotite is abundant and usually fresh (Plate 23d), and a biotite-rich facies of the granodiorite is shown on the map (British Geological Survey, 1996). Apatite, zircon and opaque oxide grains are associated with biotite. Alteration of oxide and biotite results in the formation of turbid, granular sphene or anatase/rutile. Minor green-brown hornblende occurs in some granodiorite specimens as small, subhedral to euhedral prismatic crystals or sporadic interstitial aggregates, though it is partly altered to chlorite or recrystallised to secondary pale green actinolitic amphibole. Epidote is common, occurring as fine saussuritic alteration of plagioclase, and as coarser crystalline aggregates forming crude veinlets and irregular patches. Allanite has been noted as a prominent accessory phase in some samples (for example E70529, E70530). Tourmaline is sporadically present within the granodiorite as ragged crystals (E70768).

Almandine garnet (Ansari, 1983) is locally prominent, particularly in rock exposed in Waberthwaite quarry [SD 1124 9430]. There it occurs as irregular, equant crystals, 1 to 2 mm in diameter, and intergrown with quartz and feldspar. Ansari (1983) described the garnets in some detail and noted that they also occur in some xenoliths. He suggested that garnet within the granodiorite is xenocrystic.

Aplitic granitic veins, up to 15 cm wide, are common within the body of the granodiorite. These are fine-grained equigranular and granitic (perthite exceeds plagioclase in abundance) with minor, usually chloritised biotite flakes. A small area of granite crops out in Samgarth Beck [NY 126 943], but its field relationship to the granodiorite is obscure. This granite is medium grained with dusty perthitic feldspar in excess of smaller sericitic plagioclase. Ragged biotite is partly chloritised.

The granodiorite is locally strongly deformed, for example close to a scheelite-bearing mineralised vein at Buckbarrow Beck [SD 137 910], where relict quartz is set in a chlorite and sericite matrix (E70784). Another zone of tectonised granodiorite trends north-west from near Whitestones [SD 111 933]. This consists of intensely strained and fractured quartz and feldspar crystals set in a matrix cataclastically reduced in grain size; turbid epidote, chlorite and mica form dark veinlets which anastomose around larger crystals and define a planar shear fabric (E70767, E70771).

ENNERDALE INTRUSION

The Ennerdale granitic intrusion is exposed over about 60 km^2 from Ennerdale and Buttermere lakes in the north to Mecklin Wood [NY 117 024] in the south. The southern extremity lies within the Ambleside district where it forms three outcrops in the north-west, jointly occupying about 16 km^2. Within the district, it is in contact with the Borrowdale Volcanic Group except at the southern end of Wast Water where there is an unexposed (possibly faulted) contact with the Eskdale granite. To the north and north-west of the district, however, the northern and western margins have intrusive contacts with the Skiddaw Group. Like the Eskdale granite, the Ennerdale intrusion does not apparently intrude high into the Borrowdale Volcanic Group stratigraphy; locally it is highly discordant with these strata. Gravity modelling and seismic reflection studies (Lee, 1989; Firman and Lee, 1986; Evans et al., 1993; 1994) suggest that it is a tabular body, less than 2 km thick, underlain by denser, less acidic plutonic rocks, ascribed by Lee (1989) to the concealed Buttermere granite (see Chapter 3).

Over much of the area, the contact of the Ennerdale intrusion with the country rock is poorly exposed though it can be located fairly precisely. However, the contact is well displayed on the side of Buckbarrow [NY 135 058] where it is subhorizontal across the crags with andesite exposed above the granite. On a local scale the contact is irregular with vein-like apophyses intruding the volcanic rocks. There is no obvious chilled margin.

Just east of Easthwaite Farm [NY 141 034] the Ennerdale and Eskdale granites are juxtaposed. The contact is covered by scree and, farther north-east towards Wast Water, by alluvial fan deposits; thus it is impossible to ascertain the intrusive relationships between the two granites. Micrograntic varieties of Eskdale granite occur close to the Ennerdale intrusion, but this is not conclusive evidence that the Ennerdale intrusion was earlier (see also Dwerryhouse, 1909). It is also possible that this junction is faulted as the thin slice of volcanic rocks between the two intrusions on Kilnhow Crag [NY 136 032] is mildly cleaved or sheared on a trend of about 050°. The contact of the Ennerdale granite with the volcanic rocks at this locality appears to be dipping gently (25 to 30°) to the south-east and is relatively planar and abrupt, though there is granite veining in the volcanic rocks. Locally, slickensides indicate that the contact may have been sheared or faulted.

The Ennerdale intrusion is predominantly a pink, fine-grained, slightly porphyritic granite, commonly with granophyric texture, giving it its colloquial name, the Ennerdale Granophyre. Two areas of a dioritic facies are present close or adjacent to the margin of the intrusion. These include the so-called 'needle rock' described by Rastall (1906). Descriptions of the intrusion have also been recorded in Trotter et al. (1937) and in two unpublished theses (Govinda Rajulu, 1959; Clark, 1963).

Granite

Most of the Ennerdale intrusion comprises a leucocratic, fine-grained or, locally, medium-grained granophyric granite. It is predominantly pink in colour and has a granular (saccharoidal) texture. Jointing is regular with two vertical and one subhorizontal orthogonal sets. Locally, for example on Caw Fell [NY 108 120], it is sparsely porphyritic with 2 to 4 mm white feldspar phenocrysts set in a dark pink, fine-grained matrix; generally it is fine to medium grained and equigranular. Small dark mafic crystals are sparsely distributed through the granite, but close to the dioritic portions of the intrusion these become patchily more abundant and give the granite a melanocratic appearance with scattered diffuse darker xenoliths. Govinda Rajulu (1959), Clark (1963) and Trotter et al. (1937) have also described textural variants amongst the granitic rocks, but in the field, due to poor exposure and gradational contacts, these cannot be accurately delimited.

In thin section the granite is equigranular to mildly inequigranular, consisting of subhedral, zoned plagioclase laths, up to 3 mm in length, set in a matrix of anhedral granular quartz and perthite or orthoclase showing a variable granophyric intergrowth texture (Plate 23e). Plagioclase crystals are commonly irregular and epitaxially overgrown by alkali feldspar. The granophyric matrix in places exhibits a radial crystallisation habit. Primary mafic phases are mostly totally altered to chlorite, which occurs as small irregularly shaped patches with inclusions of turbid granular sphene. Biotite was probably the main mafic component. Small acicular chlorite pseudomorphs possibly replace amphibole. Disseminated small, rounded to subhedral, opaque oxides also occur, and apatite and zircon are typical accessory minerals. Epidote is a common alteration product, forming interstitial blocky clusters, or developed patchily in feldspar. Feldspar is variably altered to sericite as well as epidote. Sphene is present as an alteration of biotite or amphibole in association with chlorite and also as a replacement of oxide (ilmenite).

Diorite, dolerite and hybridised rocks

Within the intrusion, two areas of conspicuously more basic granitoid rocks have been delimited; one is centred on the upper Bleng valley [NY 125 080] and one at the extreme south of the outcrop around Mecklin Wood [NY 118 026]. Rastall (1906) and others (Clark, 1963; Govinda Rajulu, 1959) have ascribed a hybrid origin to these rocks.

Bleng diorite

This mass of rock occupies a roughly elliptical zone approximately 3 km long and up to about 1 km wide. It is best exposed on Birk Crag [NY 125 081] and Raven Crag [NY 128 085] on the south side of the Bleng valley. The rock is typically fine grained to just medium grained, equigranular and greenish grey with dark mafic minerals and pink feldspars in approximately equal proportions. The dark minerals are conspicuously acicular (up to several millimetres) and this feature gave rise to the term 'needle-rock', coined by Rastall (1906) for this and other dioritic facies within the Ennerdale intrusion. In outcrop, the diorite is rather heterogeneous and commonly contains irregular veins and patches of pink granite, which in turn contain xenoliths of diorite. Consequently, the margin of this dioritic body with the main granite is poorly defined and gradational in a gross sense. Field relationships clearly indicate crystallisation of diorite prior to granite intrusion.

In thin section (E68112, E68113) the diorite is non-porphyritic and equigranular, predominantly composed of randomly orientated, interlocking plagioclase laths and secondary chlorite (Plate 23f). Elongate plagioclase laths, up to 4 mm, are locally prominent. Plagioclase is commonly dusty in appearance as a result of variable sericitic and saussuritic alteration. Minor anhedral quartz and K-feldspar occur interstitially, partly as an intergrown mesostasis. Chlorite occurs interstitially to plagioclase, commonly as equant patches, but also as elongate irregular pseudomorphs associated with trails of granular sphene. These account for the macroscopic needle-like fabric in the diorite, and probably result from the alteration of primary acicular amphibole or possibly pyroxene. Disseminated opaque oxides occur as elongate skeletal or subhedral equant crystals, partly altered to sphene. Some clear sphene is possibly primary. Epidotic alteration is locally intense, with some plagioclase laths overgrown by coarse aggregates. Acicular apatite prisms (up to 2 mm long) are a feature of this rock.

Interstitial quartz and K-feldspar may be more abundant closer to the margin of the mass, with the rock showing more coarsely crystalline granitic domains. This suggests that intrusion or infiltration of more evolved granitic fluids occurred in part before the dioritic magma was totally crystalline.

Mecklin dolerite and diorite

An area of poorly exposed, heterogeneous basic–intermediate granitoid rocks occurs at the southernmost margin of the intrusion from Mecklin Wood [NY 117 025] in the south a few hundred metres northwards toward Burnt House and High Coppice. A few isolated exposures in the wood consist of a medium- to coarse-grained dolerite, but most of the outcrop is of a fine- to medium-grained grey diorite or granodiorite which merges into a melanocratic pink granite. Locally, at Mecklin Wood [NY 1196 0260], angular blocks of grey diorite are suspended within a granitic matrix. Farther north, around High Coppice and in the few exposures near Burnt House, the dominant lithology is a grey, non-porphyritic fine-grained diorite

with irregular veins, stringers and patches of pink coarser granitic rock. Within these granitic patches and veins are diffuse xenolithic patches of grey diorite.

The dolerite exposed in Mecklin Wood (E70820, E70821, E70822) is a medium- to coarse-grained rock consisting of interlocking plagioclase laths with subophitic pale green amphibole (actinolite–hornblende) partly replacing clinopyroxene (Plate 24a). Relict clinopyroxene is visible as ragged cores within epitaxial pseudomorphs of hornblende. Some amphibole has recrystallised as feathery acicular sheaves. Plagioclase is generally sericitised, but is clear in patches and on rims. Common, 1 to 2 mm skeletal to embayed subhedral equant oxides are partly replaced by sphene. Minor chlorite is developed as interstitial pools, partly replacing amphibole or oxide. Quartz is essentially absent but traces of K-feldspar are present in the interstices of the rock. Epidotic alteration is common.

The dioritic rocks north of Mecklin Wood are finer grained than the dolerite but also consist of subhedral plagioclase laths with common interstitial green amphibole as ragged elongate crystals or acicular aggregates. However, interstitial areas of quartz and perthite are common, poikilitically enclosing plagioclase and amphibole. In addition, fresh ragged plates of biotite are conspicuous, particularly in exposures by the disused quarry just south of Burnt House farm [NY 1182 0282] (E70827, E70831). Chlorite occurs as alteration of biotite and amphibole. Disseminated anhedral to bladed oxides (ilmenite) are partly or completely altered to sphene. Accessory phases comprise small prismatic apatite crystals and less common zircons. A feature of these rocks is their somewhat heterogeneous texture with gradational domains richer in slightly coarser-grained quartz, biotite and perthite (Plate 24b). In places these envelop amphibole and plagioclase laths which show evidence of dissolution. Irregular, corroded plagioclase is, in places, epitaxially overgrown by orthoclase. At High Coppice [NY 115 031] the proportion of amphibole is much less and the rock is more granodioritic.

Within the area around Mecklin Wood and High Coppice, there appears to be a northward zonation from more basic rocks at the southern margin to more granitic rocks as the main body of the Ennerdale granite is approached. On the macroscopic and microscopic levels there is evidence of hybridisation with earlier, partially crystallised, mafic–intermediate magma infiltrated by slightly later granitic fluids.

AGE OF THE PLUTONIC ROCKS

Recent U–Pb isotope analyses of zircon from the Ennerdale granitic intrusion and the granitic component of the Eskdale pluton have given ages of 452 ± 4 Ma and 450 ± 3 Ma respectively (Hughes et al., 1996), indicating a Caradoc age of emplacement for the plutons. A supra-subduction zone setting, related to the final phase of the Borrowdale Volcanic Group magmatic episode, seems likely. These high precision results should effectively conclude a long-running debate concerning the timing of intrusion of the plutons, though considerable uncertainty remains on the nature of the process that caused resetting of the other isotopic systems.

Plate 24 Photomicrographs of intrusive rocks (*opposite*).

a. Doleritic facies of the Ennerdale intrusion, Mecklin Wood. Plagioclase laths subophitically enclosed by (clino)pyroxene which is epitaxially replaced by calcic amphibole (E70822). Crossed-polarised light; width of field 12 mm (PMS734).

b. Hybridised facies of the Ennerdale intrusion. Patchy granitic domains, locally showing granophyric texture, are enclosed within, and intergrown with microdiorite (E70819). Crossed-polarised light; width of field 8.2 mm (PMS735).

c. Basalt (dolerite) dyke of tholeiitic affinity showing subophitic texture. Clinopyroxene is replaced by actinolite due to contact metamorphism by the Eskdale granite (Dyke no. 12 from Macdonald et al., 1988. University of Lancaster thin section). Crossed-polarised light; width of field 12 mm (PMS736).

d. Sparsely porphyritic rhyolite dyke displaying characteristic fine-grained felsitic texture with spherulites (E67952). Crossed-polarised light; width of field 8.7 mm (PMS737).

e. Porphyritic microdiorite dyke of the 'Bastard Granite' suite with abundant phenocrysts of plagioclase and microphenocrysts of biotite grading into a finer intergranular feldspathic groundmass (seriate textured) (E68107). Crossed-polarised light; width of field 12 mm (PMS738).

f. Quartz–feldspar granite porphyry dyke, Wasdale. Abundant euhedral and subhedral phenocrysts of quartz, sericitised plagioclase and perthite occur within a microcrystalline granitic groundmass (E67944). Crossed-polarised light; width of field 12 mm (PMS739).

Eskdale pluton

It was once considered that the Eskdale granite was linked petrogenetically with the post-cleavage, Early Devonian Skiddaw and Shap granites (for example Brown et al., 1964; Firman, 1978a). However, recognition that the Eskdale granite is locally cleaved (Allen, 1987) and a whole-rock Rb–Sr isochron date of 429 ± 4 Ma (Rundle, 1979) indicated that it was an earlier, pre-deformation intrusion. A less reliable Rb–Sr isochron age of 429 ± 22 Ma obtained for the granodiorite (Rundle, 1979) suggested that these two portions of the pluton were essentially coeval, albeit compositionally distinct. Rb–Sr dating of the granite at Wasdale Head was less satisfactory, but selected data yielded an age of 431 ± 6 Ma (Al Jawadi, 1987). K–Ar ages on biotite from the granodiorite and Wasdale Head granite agree with the Rb–Sr dates (Rundle, 1979; Al Jawadi, 1987), though on muscovite from granite and greisen variable Ar loss was indicated. These isotopic ages for the Eskdale pluton indicated intrusion during the Silurian.

Firman and Lee (1986) cast doubt on the Rb–Sr dates as representing the emplacement age of the Eskdale pluton, and argued that the bulk of the Lake District batholith, of which Eskdale represents an exposed portion, was in place during the Ordovician. In their model the batholith controlled late Ordovician (Ashgill) and Silurian sedimentation and possibly the Caradoc deformation represented by the Ulpha Syncline. They regarded the Rb–Sr dates as possibly recording a metasomatic event. Branney and Soper (1988) also argued, on volcanological grounds, that emplacement of a batholithic body or magma

a.

b.

c.

d.

e.

f.

chamber occurred during Borrowdale Volcanic Group volcanism (pre-Longvillian), though they agreed that final intrusion occurred later than the extensive volcanism and comprised more evolved magma. The recent U–Pb zircon Caradoc age of 450 ± 3 Ma clearly supports these contentions.

Ennerdale intrusion

A ten-point Rb–Sr isochron obtained by Rundle (1979) yielded a date of 420 ± 4 Ma indicating emplacement during the Silurian. Rundle (1992) argued that, since the samples were collected from widely separated localities, and represented various facies, grain sizes and degrees of alteration, the high-quality isochron obtained must represent primary cooling and hence contemporaneous alteration. Subsequent resetting by later metasomatism, as argued by Firman and Lee (1986), would be expected to produce scatter on the isochron plot. Though Rundle's (1979) samples do not appear to include any that are severely depleted in Rb (see below) the Rb content of these rocks is known to have been disturbed considerably by metasomatic alkali exchange, analogous to that which affects felsitic rhyolite dykes. Rb–Sr dates of around 420 Ma have also been obtained for the Harestones Felsite from Carrock Fell and for the Stockdale Rhyolite (Rundle, 1979). However, recent attempts at dating the Caradoc Borrowdale Volcanic Group rocks by Rb–Sr (Rundle, 1992) have given dates also around 420 Ma which are clearly reset. The recent U–Pb age of 452 ± 4 Ma for the Ennerdale granite places the intrusion of this body during the Caradoc. The possibility that intrusion of the Ennerdale granite is responsible for 420 Ma resetting of Rb–Sr systematics in the Borrowdale Volcanic Group (compare with Rundle, 1992) therefore cannot now be upheld. Indeed, the Ennerdale intrusion and probably other felsitic rocks with 420 Ma Rb–Sr dates must have been subjected to a regional resetting event.

Regional resetting event

The evidence cited above confirms that the Rb–Sr ages for the Eskdale and Ennerdale plutons are reset, and hence are unrelated to intrusion and cooling of the igneous bodies. Evans et al. (1995) show that Rb–Sr whole-rock resetting in the comparable late Ordovician volcanic province of North Wales probably resulted from disturbance of the isotope systems by water–rock interactions triggered by pressure release at the onset of uplift. In terms of the foreland basin model for the Silurian development of the Lake District, Kneller et al. (1993b) describe intra-Ludlow tectonic deformation in the Windermere Supergroup as a 'rapid release of elastic strain during slip events on the basal detachment' prior to basin inversion in late Ludlow times. The ages of the lower and upper Ludlow boundaries are disputed (424 to 411 Ma according to Harland et al., 1990; 424 to 418.5 Ma according to Fordham, 1992), but permit linking the resetting of at least some Lake District Rb–Sr isotopes to the intra-Ludlow tectonism described by Kneller et al. (1993b). The same conclusion follows from the proposals of Hughes et al. (1993) that diachronous, thrust-related deformation migrated southwards from the Skiddaw Group during the late Silurian.

The 429 Ma (Llandovery–Wenlock) reset Rb-Sr age for the Eskdale intrusion is less obviously related to late Silurian deformation (though experimental error and degree of uncertainty in the ages could still allow a correlation). This was a time of pronounced subsidence across the Lake District and basin deepening, as recorded by the transition to deep-water sediments in the Windermere Supergroup (see Chapter 7).

GEOCHEMISTRY AND PETROGENESIS OF THE PLUTONIC ROCKS

Eskdale pluton

Ansari (1983) and O'Brien et al. (1985) noted the differences in geochemistry between the granodiorite in the south and the granites in the north. New analytical data for the mapped lithologies indicate that the granites contain 72 to 77 per cent SiO_2 compared with 63 to 67 per cent SiO_2 in the granodiorite (Table 18). Amongst the other major oxides the former have lower MgO, TiO_2, CaO and Fe_2O_3(total) contents emphasising their more evolved nature. Similarly, trace element concentrations (Figure 48b–d) reinforce the compositional hiatus between the two major units comprising the intrusion. Ansari (1983) concluded on this basis that there was no genetic relationship between the granodiorite and the 'muscovite–perthite granites' and that they were effectively separate intrusions. O'Brien et al. (1985), however, suggested that granodiorite and granite were petrogenetically linked by crystal–liquid fractionation of a common parental magma intermediate in composition to the granite and granodiorite. This is supported by similar initial $^{87}Sr/^{86}Sr$ ratios of 0.7076 ± 5 and 0.7073 ± 7 for the granite and granodiorite, respectively (Rundle, 1979).

Within the granites, immobile elements display covariation trends which project back toward the granodiorite field (Figure 48a–b). Potentially more mobile elements, such as Rb, Ba and Sr (Figure 48c, d) yield similar patterns with increased scatter that can be ascribed to the effect of late magmatic and hydrothermal fluids. There is little evidence for the dramatic variations in alkali contents recorded in the Ennerdale intrusion (see below). These plots indicate that both granites and granodiorite could lie on a co-magmatic fractionation trend, substantiating the view of O'Brien et al. (1985). The observed geochemical trends are consistent with hornblende–biotite–plagioclase fractionation together with the crystallisation and segregation of zircon and apatite.

Compositional differences between the three principal granite lithologies are evident from the bi-element plots shown in Figure 48. Early, coarse-grained, predominantly primary-textured granites show consistently lower levels of differentiation (i.e. higher Ti, Mg, Fe, V) compared with normal granite. By contrast, the microgranites, which record a spectrum of two-phase crystallisation textures, span or exceed the compositional range of the coarse and normal granites. Local variations within the micro-

Table 18 Representative geochemical analyses of Eskdale granite and granodiorite.

Sample no.	KDC 1506	KDC 1555	KDC 1560	KDC 1516	KDC 1535	KDC 1545	KDC 1502	KDC 1519	KDC 1549	KDC 1508	KDC 1513	KDC 1571	KDC 1583	KDC 1570
Grid ref.	1498 9950†	1527 0074	1489 0037	1625 9902†	1680 0097	1689 0075	1974 0316	1840 0165	1687 0065	1365 9090†	1343 9087†	1390 9477†	1122 9432†	1264 9425†
	Eskdale granite									Eskdale granodiorite				
Lithology	Coarse-grained granite			Normal granite			Fine-grained granite			Granodiorite			Biotite grano-diorite	Samgarth Beck granite
wt %														
SiO_2	75.61	74.07	74.75	75.99	76.82	76.71	75.96	75.45	76.17	66.2	65.26	65.99	64.70	76.12
TiO_2	0.17	0.17	0.17	0.10	0.08	0.11	0.04	0.16	0.04	0.62	0.64	0.60	0.45	0.10
Al_2O_3	12.98	13.19	13.36	12.87	12.73	13.02	13.64	13.11	13.27	15.69	15.85	15.62	16.11	12.67
Fe_2O_3*	1.79	1.76	1.77	1.50	1.32	0.94	0.54	1.63	1.15	4.77	5.00	4.68	4.38	1.34
MnO	0.05	0.04	0.03	0.03	0.03	0.02	0.01	0.03	0.03	0.10	0.15	0.13	0.12	0.03
MgO	0.23	0.25	0.28	0.09	0.09	0.12	0.03	0.26	0.11	1.19	1.30	1.19	0.88	1.12
CaO	0.34	0.29	0.35	0.25	0.21	0.26	0.38	0.19	0.32	2.64	2.73	2.37	2.82	0.19
Na_2O	3.36	4.07	3.58	3.58	3.19	3.32	3.97	3.11	3.37	3.43	3.44	3.54	3.42	3.52
K_2O	4.70	5.37	4.97	4.96	4.91	5.09	4.69	5.54	5.12	4.21	4.47	4.82	4.65	5.71
P_2O_5	0.17	0.18	0.17	0.19	0.2	0.2	0.26	0.06	0.23	0.17	0.18	0.17	0.19	0.01
LOI	0.80	1.05	0.97	0.64	0.88	0.63	0.64	0.88	0.71	1.26	0.89	1.23	1.22	0.62
Total	100.20	100.44	100.40	100.20	100.46	100.42	100.16	100.42	100.52	100.28	99.91	100.34	98.94	100.43
ppm														
V	9	9	15	5	5	6	1	8	1	57	58	49	39	5
Cr	17	16	12	9	12	12	7	4	8	14	20	12	20	9
Ni	7	5	5	4	2	5	5	5	5	11	9	9	10	4
Rb	265	275	263	367	357	373	374	242	395	132	167	143	158	212
Sr	38	56	42	10	14	20	29	43	11	214	267	269	205	32
Y	22	26	24	26	26	22	19	38	24	44	45	44	45	46
Zr	102	102	102	78	75	77	51	136	58	215	231	235	186	135
Nb	15	12	13	15	16	15	16	14	12	18	20	18	16	20
Ba	161	206	189	53	56	82	84	341	18	836	868	973	676	179
La	15	10	8	8	7	5	2	26	2	43	44	40	34	60
Ce	40	42	43	25	20	29	2	77	12	107	98	96	78	116
Th	9	8	8	7	6	6	5	28	3	8	15	12	11	19
U	3	3	2	3	7	2	4	2	2	3	2	< 2	2	3
Mol. ratios														
Al/(Ca+Na+K)	1.155	1.011	1.122	1.098	1.163	1.137	1.109	1.144	1.137	1.046	1.025	1.017	1.021	1.029
Al/(Na+K)	1.222	1.054	1.185	1.143	1.205	1.186	1.175	1.179	1.196	1.538	1.509	1.414	1.511	1.058
K/(Na+K)	0.480	0.465	0.478	0.477	0.503	0.502	0.438	0.540	0.500	0.447	0.461	0.473	0.472	0.516

* Total Fe

All grid references are in the rectangle NY apart from those marked † which are in the rectangle SD.

All analyses at University of Nottingham.

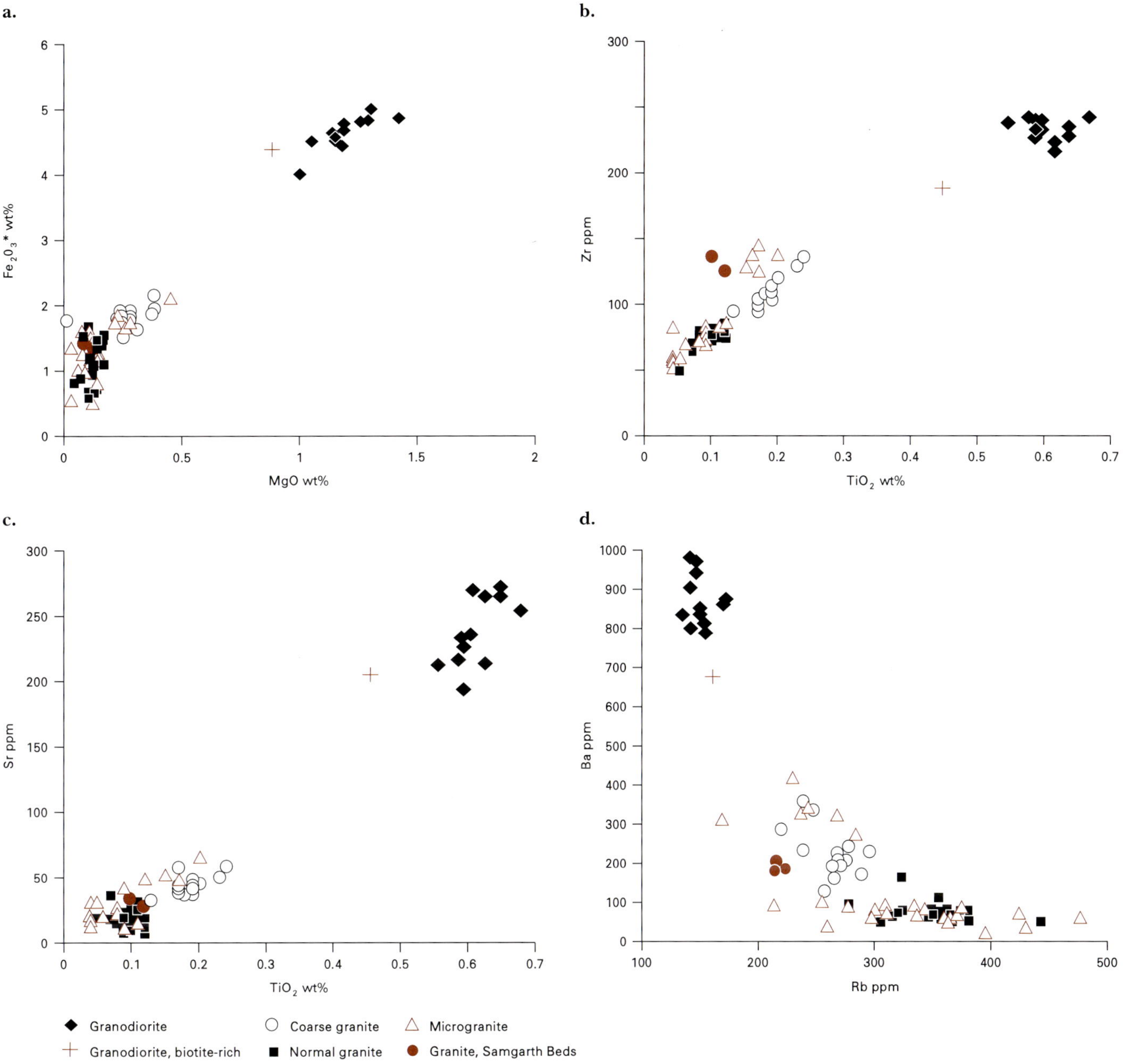

Figure 48 Co-variation diagrams for the Eskdale plutonic rocks.
a. Fe_2O_3–MgO **b.** V–TiO_2 **c.** Zr–TiO_2 **d.** Sr–TiO_2 **e.** Ba–Rb Fe_2O_3* total Fe as Fe_2O_3.

granites are common, for example samples from close to the roof pendant on Great Barrow are characterised by higher Zr, Th, Ce and Y (Table 18, no. 8) These patterns are entirely consistent with the mode of emplacement for the granite described above. Crystallisation of early magma, represented by the coarse granite, produced residual, more evolved melts that locally infiltrated, invaded and disrupted to varying degrees zones of partly crystallised granite, producing the more chemically variable mixed or hybridised compositions.

Granite at Samgarth Beck, within the outcrop of the granodiorite, is similar in composition to the Eskdale granites, except for generally higher Nb, Th, Y and Ce, and lower P_2O_5 (Table 18, no. 14). If it is related to the main body of the Eskdale granite this suggests that granite intrusion postdated emplacement of the granodiorite.

Ennerdale intrusion

The geochemistry of the Ennerdale intrusion was discussed briefly by Clark (1963), Govinda-Rajulu (1959), Ansari (1983) and also by O'Brien et al. (1985). Twenty seven new analyses covering the spectrum of lithologies within the Ennerdale intrusion are illustrated in Figure 49 as

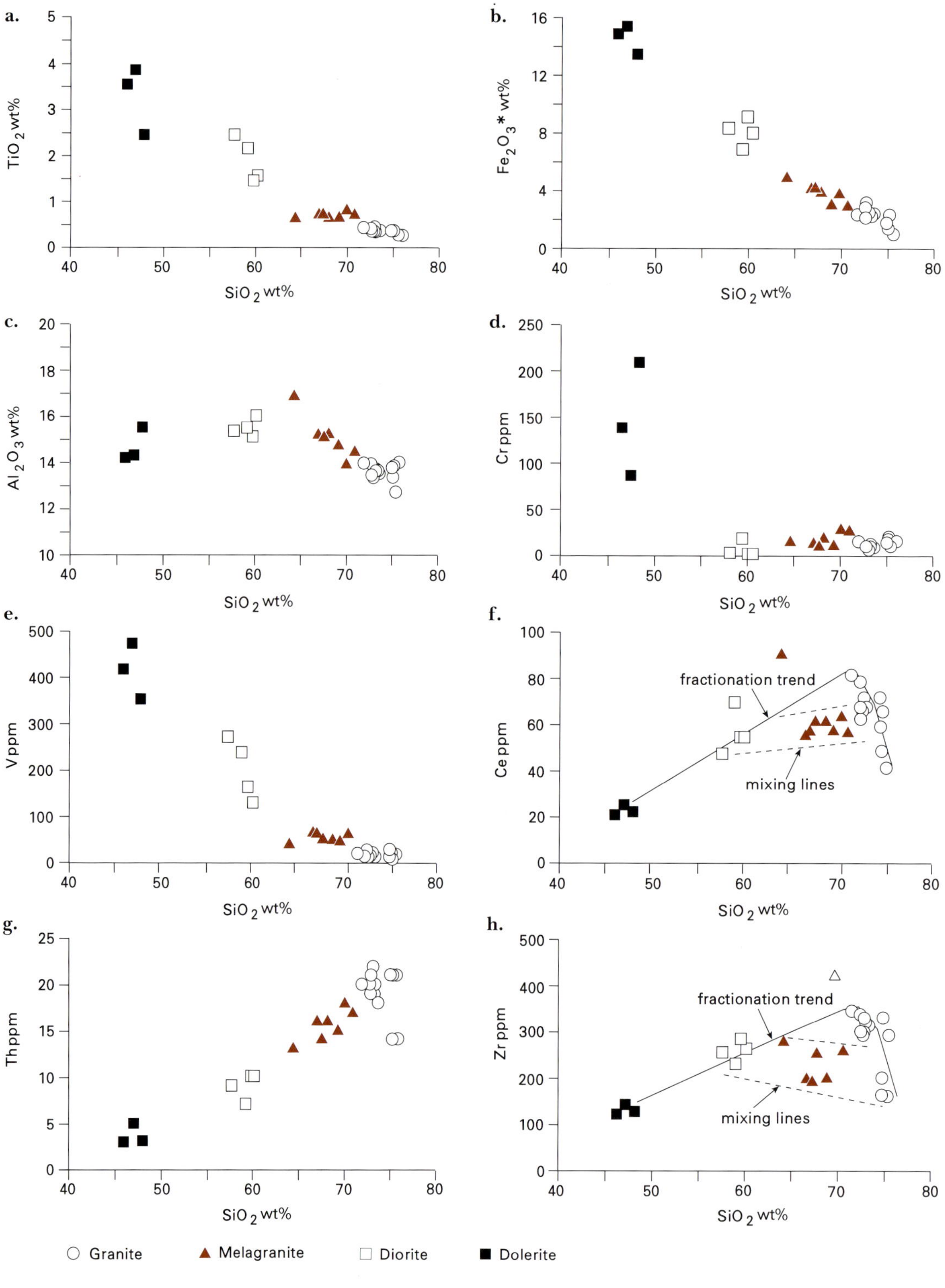

Figure 49
Harker variation diagrams for the Ennerdale plutonic rocks.

a. TiO_2
b. Fe_2O_3*
c. Al_2O_3
d. Cr
e. V
f. Ce
g. Th
h. Zr

The relation between magma mixing and crystal fractionation is illustrated in **f** and **h**. Major oxide wt.% and trace element ppm.

Fe_2O_3* total Fe as Fe_2O_3.

Harker variation diagrams. They show a range of SiO_2 content from 46 to 76 per cent, but with some apparent compositional gaps. A feature of a number of the analysed samples is an anomalously low K_2O content and correspondingly high Na_2O (Table 19). This has been noted before (Clark, 1963; Firman, 1978a) and can be attributed to metasomatic alkali exchange ($Na^+ \leftrightarrow K^+$), with Rb and Ba behaving similarly to K (Table 19). Less mobile, high field strength trace elements (Zr, Th, Y, Nb etc.) are largely unaffected. Overall the Ennerdale suite is geochemically relatively coherent, despite some compositional gaps and breaks dealt with in more detail below. In particular, Fe_2O_3, MgO and Th show highly correlated co-variation with SiO_2, strongly suggesting that the various mapped lithologies represent a co-genetic suite. The analysed samples fall readily in four geochemically distinct

Table 19 Representative of geochemical analyses of rocks from the Ennerdale intrusion.

Sample no.	NX 1588	NX 1589	NX 1631	NX 1630	NX 1627	NX 1629	NX 1649	NX 1670	NX 1653	NX 1623	NX 1669	NX 1660
Grid ref. NY	150 052	150 052	1130 0850	1170 0875	1287 0894	1202 0877	1208 0427	1214 0313	1220 0439	1240 0876	1159 0280	1184 0245
Lithology	Granite	Granite	Granite	Granite	Granite	Granite	Melagranite	Melagranite	Melagranite	Diorite	Diorite	Dolerite
wt %												
SiO_2	75.69	75.16	73.17	73.14	72.90	72.74	70.88	68.11	67.48	59.85	57.84	47.13
TiO_2	0.23	0.31	0.36	0.32	0.34	0.37	0.67	0.61	0.68	1.43	2.42	3.84
Al_2O_3	14.17	14.07	13.93	13.86	13.53	14.14	14.68	15.46	15.32	15.35	15.60	14.57
Fe_2O_3*	0.67	1.10	2.17	2.20	2.89	1.81	2.63	3.63	3.87	8.88	8.06	15.13
MnO	0.01	0.02	0.03	0.03	0.05	0.03	0.02	0.09	0.03	0.21	0.23	0.30
MgO	0.44	0.61	0.50	0.37	0.85	0.34	1.36	1.28	1.45	2.52	3.67	5.74
CaO	0.37	0.44	0.32	0.51	0.48	1.16	1.89	2.14	1.74	2.42	4.89	8.19
Na_2O	7.63	7.43	3.87	4.05	2.52	4.33	6.63	3.82	3.17	2.68	5.50	2.73
K_2O	0.35	0.34	5.55	5.17	5.81	4.65	0.35	4.21	4.71	4.53	0.62	1.53
P_2O_5	0.09	0.10	0.07	0.04	0.06	0.04	0.16	0.18	0.17	0.23	0.33	0.28
LOI	0.59	0.53	0.79	0.72	1.10	0.67	1.13	1.09	1.67	2.29	1.57	1.51
Total	100.24	100.11	100.76	100.41	100.53	100.28	100.40	100.62	100.29	100.39	100.73	100.95
ppm												
V	13	19	15	11	17	10	56	46	57	152	257	471
Cr	16	20	12	10	5	10	27	19	10	< 1	2	86
Ni	1	2	2	1	1	< 1	4	7	3	5	7	11
Rb	16	15	172	159	147	113	13	144	169	119	26	38
Sr	79	64	42	55	33	75	108	234	149	111	210	463
Y	18	17	26	29	25	32	25	30	24	29	48	27
Zr	160	164	311	335	304	341	262	256	194	286	257	145
Nb	11	11	14	15	14	15	14	15	12	13	16	12
Ba	28	17	432	464	582	504	29	417	432	512	95	239
La	13	16	18	24	17	26	23	35	27	21	21	16
Ce	41	48	65	65	62	67	63	61	57	54	47	25
Th	14	14	19	22	21	19	17	16	14	10	9	5
U	7	5	5	6	5	4	5	5	6	1	4	1
Mol. ratios												
Al/Ca+Na+K)	1.04	1.05	1.08	1.05	1.20	0.99	1.00	1.05	1.14	1.12	0.84	0.69
Al/(Ca+K)	1.10	1.12	1.13	1.13	1.30	1.16	1.30	1.43	1.49	1.65	1.61	2.37
K/(Na+K)	0.03	0.03	0.49	0.46	0.60	0.41	0.03	0.42	0.49	0.53	0.07	0.27

All analyses by the British Geological Survey
* Total Fe; LOI Loss on ignition.

groups that correspond to described lithological units (see Figure 49).

1. The *Mecklin dolerite* is characterised by less than 50 per cent SiO_2, and high TiO_2 and Fe_2O_3. The MORB-normalised spider diagram in Figure 50 (after Pearce, 1982) shows elevated levels of K, Rb, Ba and Th, a negative Nb anomaly and declining enrichments from La to Y. These are typical of a continental-margin subduction signature (Pearce, 1982) and generally compare with that of the Borrowdale Volcanic Group basaltic rocks described in Chapter 4. However, Ti forms a pronounced positive anomaly, which together with the high Fe and V in this unit may reflect a cumulus history with enhanced abundances of ilmenite/magnetite. A suite of possibly related tholeiitic dykes within the Eskdale–Wasdale area also exhibit these features and are known to both predate and postdate the Eskdale intrusion (Macdonald et al., 1988).

2. Four analyses of the *Mecklin and Bleng diorites* show that these rocks contain 58 to 60 per cent SiO_2, but also have elevated TiO_2 compared, for example, with Borrowdale Volcanic Group andesites of similar SiO_2 content. This suggests the diorite and dolerite may be genetically related, though as noted above, field relationships between the two at Mecklin Wood are obscure.

3. A number of analysed rocks have granodioritic compositions (65 to 71 per cent SiO_2). These are melagranitic lithologies from the margins of the Mecklin area, and also from the Low Wood–Whitesyke area [NY 122 042], which show microscopic evidence of magmatic mixing or hybridisation.

4. Granitic compositions (SiO_2 over 72 per cent) are characteristic of the bulk of the Ennerdale intrusion. They show a characteristic trend of steeply declining Th, Zr, Y and Ce levels (Figure 49) toward the most silicic compositions, reflecting the precipitation and fractionation of accessory phases such as zircon. For the major oxides Al_2O_3, Fe_2O_3(total) and MgO (Figure 49) this group of compositions falls on trends coherent with the melagranitic (granodioritic) variants. However, for Zr and Ce in particular, and less obviously for other trace elements and TiO_2 (Figure 49), there is a geochemical discontinuity, indicating that these two groups are not linked simply by fractional crystallisation (see below).

Earlier descriptions of the basic facies within the Ennerdale intrusion, including the 'needle rock' of Bowness Knott near Ennerdale, suggested that they are hybridised rocks, formed by the mixing of earlier (partly crystallised) basic (doleritic) magma with granitic magma (for example Rastall, 1906). Clark (1963), on the basis of uniform andesine plagioclase composition, concluded that more basic or hybridised varieties of the intrusion were originally diorite rather than dolerite. Within the Bleng dioritic body there is much gross 'mixing' of magma types with veining by granite, though much of the diorite is petrographically uniform on a smaller scale. In the Mecklin area, the margins of the mapped body are diffuse with a gradation between diorite and granite producing patchy melagranites. The variation diagrams (Figure 49) do not support formation of the Mecklin and Bleng diorites by mixing of granite and doleritic magmas. In particular, the Cr contents are far too low and the Al_2O_3 and Zr contents slightly too high in relation to linear mixing lines. However, the melagranites (group 3 above) have compositions broadly consistent with mixing (or hybridisation) between a diorite (58 to 60 per cent SiO_2) and a granite (74 to 76 per cent SiO_2) magma; most oxides and elements falling on linear (mixing) tie lines between these two end members. The effect is particularly well illustrated by Ce and Zr (Figure 49f, h) with inflection of the fractionation trend linking the diorite and granite compositions. The analysed diorites have slightly high Ti, Fe and V (Figure 49a, b, e) to be totally consistent with this model,

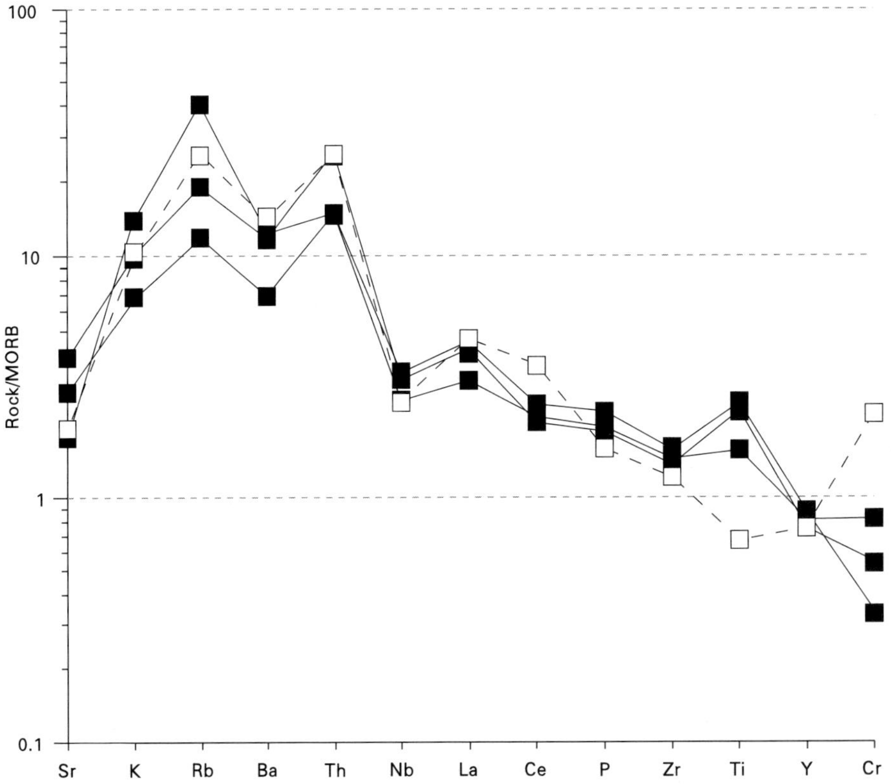

Figure 50 MORB-normalised trace element plot (following Pearce, 1983) for the Mecklin dolerite samples. Open squares indicate average Borrowdale Volcanic Group basalt for comparison.

but this could reflect accumulation of Fe–Ti oxide during the crystallisation of the diorites.

The relationship between the Mecklin dolerite and the rest of the pluton is not clear, though the geochemical evidence indicates a co-magmatic relationship, linked by crystal–liquid fractionation. The dolerites possibly represent a sidewall cumulate resulting from crystallisation of an early basic dioritic magma. Subsequent intrusion of more voluminous granitic magma, derived by differentiation, occurred relatively soon after and led to the development of melagranitic hybrid rocks due to reaction with hot and partially crystalline diorite.

Tectonic discrimination and source

On a Rb–(Y+Nb) discrimination diagram (Pearce et al., 1984) the Eskdale granite lies predominantly within the field for syn-collisional granites but overlaps slightly into the defined field for volcanic arc granites (Figure 51). A number of Ennerdale granite analyses show severe depletion in Rb (Figure 51), reflecting metasomatic alkali exchange described above, but most plot as a coherent group within the volcanic arc field, showing a clear difference between the Ennerdale and Eskdale intrusions. This is supported by the higher levels of Ba, Th, Ce, Zr and TiO_2 in the former. As O'Brien et al. (1985) noted, these differences, together with an enhanced negative Eu anomaly, possibly indicate an extended, more evolved fractionation history for the Eskdale pluton.

Molar $Al_2O_3/(CaO+Na_2O+K_2O)$ and $Al_2O_3/(Na_2O+K_2O)$ ratios for the Eskdale granite (Table 18) show it to be consistently peraluminous, and imply a post-orogenic or continental collisional granite signature using the criteria of Maniar and Piccoli (1989), whereas the Eskdale granodiorite is more unequivocally of continental arc affinity, being less peraluminous to mildly metaluminous. Compared with the Eskdale granite the Ennerdale pluton is less peraluminous to slightly metaluminous (Table 19) and has more of a continental arc signature.

Based on the well known S- versus I-type protolith nomenclature (Chappel and White, 1974) the Eskdale granites are transitional to just S-type; features include a restricted high SiO_2 content, peraluminous composition, occurrence of muscovite and garnet, and a K/(K+Na) ratio of about 0.5. In addition, as Harmon and Halliday (1980) noted, initial $^{87}Sr/^{86}Sr$ (0.7076) and $\delta^{18}O$ (about 11‰) are also indicative of an S-type affinity. From the combined Sr–O–Nd isotopic composition of the granites Halliday (1984), and Harmon and Halliday (1980) deduced that some interaction with, or melting of, Lower Palaeozoic sediment occurred during magma genesis. However, O'Brien et al. (1985) concluded that the Eskdale granite was not unequivocally S-type and that some of the isotopic features may be due to hydrothermal alteration. Cooper and Bradley (1990) have also concluded that the relatively high ammonium content of some Lake District granite samples is a product of secondary alteration and that evidence of a sedimentary protolith for the Eskdale pluton is equivocal. The lower initial $^{87}Sr/^{86}Sr$ ratio of 0.7057 ± 2 (Rundle, 1979) for the Ennerdale intrusion is in accord with its more I-type, volcanic arc characteristics, such as K/(NaK) <0.5, lower peraluminosity and its extended compositional range.

It is clear therefore that significant differences are present between the two major granitic bodies suggesting different tectonic environments. In summary, these are the rather more S-type Eskdale granite displaying features considered to be indicative of a syn- or post-collisional

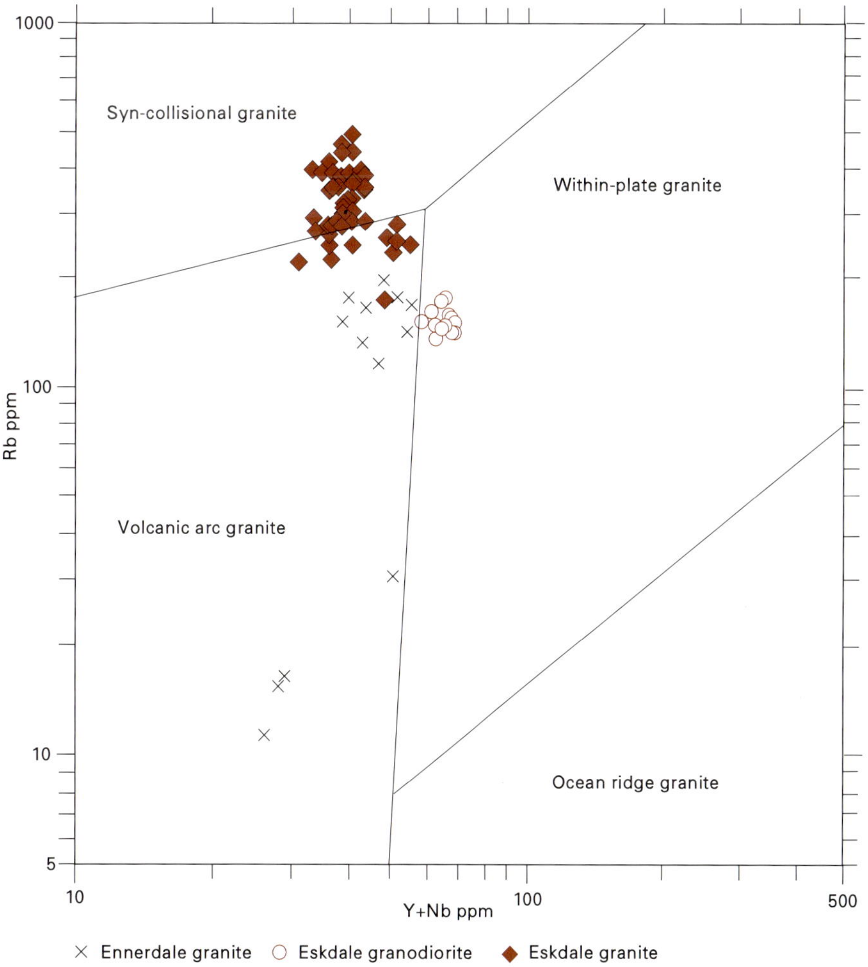

Figure 51 Rb versus Y+Nb tectonic classification diagram for the Eskdale and Ennerdale granitic rocks (after Pearce et al., 1984).

setting, and the Ennerdale granite consistent with a continental margin volcanic arc setting. Given that the two bodies were emplaced contemporaneously at 450 to 452 Ma, probably during, or close to the end of Borrowdale Volcanic Group volcanism, and substantially before the final closure of the Iapetus Ocean and the Acadian deformation, differences must be related to magma source contributions, such as greater crustal involvement in the Eskdale magmagenesis, and/or fractionation processes.

MINOR INTRUSIONS

On the basis of their areal distribution, petrography and geochemistry the minor intrusions can be divided into a number of suites, described below. They have received little systematic attention previously (for example Firman, 1978a) though descriptions are included in Dwerryhouse (1909), Firman (1957), Oliver (1961) and several unpublished PhD theses (for example Clark, 1963; Ansari, 1983; Al Jawadi, 1987; Davis, 1989). More recent accounts of particular suites have been published by Macdonald et al. (1988), and Beddoe-Stephens and Mason (1991). Those minor intrusions believed to have a close genetic association with the Borrowdale Volcanic Group are described in Chapter 6.

Basalt and dolerite dykes

Basic dykes occur throughout the outcrop of the Borrowdale Volcanic Group, but are particularly abundant around the margins of the Eskdale granite in the Eskdale to Wasdale area. They occur mainly as near-vertical intrusive sheets, commonly 0.5 to 6 m, but locally up to 15 m in width, with sharp, chilled margins. Some dykes are weakly flow foliated parallel to their margins. More rarely, intrusion-related autobrecciation is evident, for example in a 10 to 15-m wide dyke on Latterbarrow Crag [NY 128 029] where rounded angular blocks of medium-grained dolerite are suspended in a finer-grained doleritic matrix. They typically weather. dark green-grey to brown with bluish grey to dark green broken surfaces with a fine- to medium-grained doleritic texture. Generally they are aphyric, but pyroxene-phyric examples have been recorded. In the upper Esk valley area, east of Burnmoor Tarn, the dykes are strongly orientated west-north-west to east-south-east, swinging around to a more north-westerly trend either side of Wast Water. Though emplaced almost entirely within the Borrowdale Volcanic Group a few dykes belonging to this suite cut the Eskdale granite, for example near Boot [NY 179 007].

The basic dykes comprise clinopyroxene, plagioclase and Fe–Ti oxide with probable olivine in some examples. However, there is considerable development of secondary minerals resulting from low-grade and, locally, contact metamorphism. Most dykes are holocrystalline with subophitic to intergranular texture, though others are microporphyritic to seriate textured with microphenocrysts of plagiolcase and pyroxene grading into a granular groundmass. Hornblende–plagioclase hornfelsed textures occur close to the Eskdale pluton. Fresh, colourless to pale brown clinopyroxene is present in places and electron microprobe analysis shows it to be augite/salite, with significantly higher Ti and Al contents than Borrowdale Volcanic Group pyroxenes (Macdonald et al., 1988). More commonly, clinopyroxene is replaced by actinolite–hornblende and/or chlorite (Plate 24c). Amphibole replacement of clinopyroxene varies from marginal to complete epitaxial overgrowth to sheaf-like acicular or felted aggregates. Plagioclase is also variably altered to sericite or fine epidote and is locally albitised. Macdonald et al. (1988) report plagioclase compositions up to An_{68}. Patches of chlorite, usually associated with prisms of amphibole may be pseudomporphs after olivine. Opaque oxides include magnetite and ilmenite, and occur as common disseminated equant to ragged, skeletal and bladed crystals showing variable alteration to sphene. Epidote/clinozoisite is a late alteration phase and fine brown biotite is developed close to the contact with the Eskdale pluton.

The presence within this suite of minor intrusions of hornfelsed dykes as well as dykes cutting the Eskdale pluton clearly indicates an episode of basic magma intrusion spanning emplacement of the Eskdale intrusion. Macdonald et al. (1988) demonstrated that within this suite of intrusions two groups could be defined on geochemical criteria (see below); a high Fe–Ti tholeiitic subset and a calc-alkaline subset that was geochemically similar to the Borrowdale Volcanic Group. From the data available (Macdonald et al., 1988; Al Jawadi, 1987) the former subset appears to be more prevalent in the Eskdale area, whereas the latter is more common in the Wasdale area. Both appear to predate and postdate the Eskdale granite. Elsewhere, this distinction cannot be confirmed because the petrographic characteristics of each group are similar.

Rhyolite (felsite) dykes in the Borrowdale Volcanic Group

Pink to buff coloured, splintery, felsitic dykes are locally abundant particularly north-west of Wast Water and south of Devoke Water. In the former area, they form a conjugate north-east and north-west set, whereas near the latter they are generally aligned west-south-west. They occur predominantly within the Borrowdale Volcanic Group but also intrude the Eskdale pluton. On Buckbarrow [NY 135 058], felsite dykes cut the margin of the Ennerdale granite though Al Jawadi (1987) noted from the Wasdale area that there is no evidence of dykes of this suite passing into the granite. The dykes are generally near-vertical and 1 to 8 m in width, but in places they reach 15 m [for example NY 1220 0243]; they are commonly flow laminated parallel to their margins and tend to be intensely fractured. Internal folding of the flow banding is present locally, for example on Latterbarrow Crag in Wasdale [NY 1301 0294]. Where dykes are subparallel to the regional cleavage that fabric may continue into the dyke margins; good examples can be seen south of Devoke Water.

The dykes are aphyric to sparsely microporphyritic, with a fine-grained to cryptocrystalline felsitic texture (Plate 24d). Fan shaped, variolitic to well-developed spherulitic intergrowth textures are common, the latter up to 1.5 mm

in diameter. Primary mafic minerals are mostly absent, but disseminated chlorite, sericite and epidote flakes and grains form a minor component together with scattered altered oxide grains. In some dykes, chlorite and/or epidote occur as a decussate arrangement of tiny acicular structures, possibly as pseudomorphs after amphibole. Subhedral to irregular phenocrysts or glomerphyric clusters, up to a maximum of 2 mm, consist of alkali feldspar and plagioclase. Quartz microphenocrysts are rare.

Quartz–feldspar granite porphyry

A distinctive highly porphyritic dyke (described by Dwerryhouse, 1909) is mapped intermittently from Kirk Fell [NY 195 109] in the north, across the western flanks of Lingmell, over Illgill Head and the southern slopes of Whinrigg to Great Bank [NY 144 018] where it passes into the Eskdale granite, a distance of 11 km. In part, it is offset by faulting, but elsewhere intrusion is en echelon. The dyke varies in width from 5 to 20 m over most of its length, but is 30 to 35 m in width on Kirk Fell. It is massive, pink weathered and contains abundant quartz and tabular feldspar phenocrysts up to several centimetres in length set in a microcrystalline matrix. Major joints are parallel to its strike.

The granite porphyry is composed of up to 60 per cent phenocrysts which show a gradation in size from 0.1 to over 10 mm (E67953, E67954; Plate 24f). The largest phenocrysts comprise orthoclase displaying good euhedral form, simple twinning and fine to cryptoperthitic texture. They are ususally dusty in appearance due to incipient sericitic alteration. Some orthoclase phenocrysts show zones of graphic intergrowth with quartz. Quartz phenocrysts are generally smaller, rounded to euhedral, and some are embayed. Plagioclase occurs as small, squat, tabular, variably sericitised phenocrysts. Quartz and feldspar phenocrysts commonly form glomerophyric clusters. Biotite is present as rare microphenocrysts, but it is altered to chlorite or muscovite and haematitic oxide in the more reddened examples. Fresh biotite laths occur as inclusions within quartz, and both quartz and biotite laths may be poikilitically enclosed by orthoclase. The groundmass of this rock is composed of a very fine-grained (typically 30 to 50 μm) granular quartz–feldspar mosaic with much disseminated flaky white mica. Small zircons are a conspicuous accessory phase. In the more altered, reddened rocks, disseminated, vein-like or patchy haematitic oxide is present.

Microdiorite–microgranite porphyry minor intrusions

A geochemically variable suite of minor intrusions crops out in the Lingmell, Piers Gill and Scafell area. They occur as small bosses, stocks and dykes developed along major faults such as the north-west-trending Rest Gill Fault, and include the 'Bastard Granite' of Dwerryhouse (1909) together with the granite and other porphyries described and mapped by Oliver (1961) within the Central Fells. A similar suite of broadly south-west-trending microgranitic dykes occurs in the Duddon valley south from Troutal [SD 234 983] to Logan Beck [SD 173 914].

The minor intrusions of the Lingmell–Scafell 'Bastard Granite' suite are characteristically pink to reddish brown weathered and pale to dark green-grey or pink on fresh surfaces; they have a distinct mottled appearance because of the abundant feldspar and mafic phenocrysts, up to several millimetres (Plate 24e). Plagioclase and (less numerous) alkali feldspar are the most abundant phenocrysts; commonly they are strongly altered to sericite and turbid epidote. Quartz occurs as rounded to embayed phenocrysts in the more evolved variants (for example E68095). Biotite is an essential phenocrystic phase in these rocks and varies from fresh to chloritised with secondary sphene, epidote, carbonate and oxide. Small prismatic apatite and zircon crystals are included within the biotite. Pyroxene occurs in the more basic varieties, usually altered to fibrous amphibole and chlorite though relict kernels are sometimes preserved. Phenocrysts make up 25 to 70 per cent of the rock, and show a range of grain size (seriate textured) up to several millimetres. The matrix of these rocks comprises a fine-grained granular mosaic of quartz and feldspar. Garnet is conspicuous locally in rocks of this suite as a euhedral phenocryst; for example in the sheet-like intrusion at Cam Spout Crag, Scafell [NY 215 056] (Beddoe-Stephens and Mason, 1991; Al Jawadi, 1987). Some intrusions carry common rounded xenoliths.

Rare, 6 to 15 m-wide micrograpitic dykes intrude the Windermere Supergroup. Three discontinuously exposed dykes occur between Windermere and Coniston lakes, and are described by Soper and Kneller (1990). They strike roughly east-north-east, but at a small anticlockwise angle to the regional cleavage, and their margins carry a spaced cleavage continuous with the Acadian (Early Devonian) fabric in the country rock. Soper and Kneller (1990) concluded that these dykes were emplaced during episodic cleavage development within the main Acadian deformation and were related to the Shap intrusive suite. The dykes are pink weathered and have microporphyritic felsitic margins with phenocrysts of alkali feldspar and rarer rounded to embayed quartz and altered biotite, in a fine felsitic groundmass. The interiors of the dykes have similar mineralogy and microgranitic texture.

Lamprophyres

Lamprophyric dykes are restricted to the west of the district where they intrude the Eskdale granite and Borrowdale Volcanic Group. They form a northerly to north-north-westerly trending suite between Eel Tarn [NY 1905 0190, 1893 0136], Whin Rigg [SD 1715 9880], Brantrake Moss [SD 1530 9840] and south of Yoadcastle [SD 1562 9505]. They are generally less than 5 m wide and weather to a pink-brown to fawn crust. They are uncleaved.

The lamprophyres are classified as kersantites. In thin section they are sparsely phyric, fine grained and holocrystalline, composed of abundant, usually fresh biotite plates set in a granular matrix of turbid, sericitised feldspar (plagioclase and alkali feldspar) and minor interstitial quartz. Sparse phenocrysts of clinopyroxene occur; some are fresh in part but most are pseudomorphs of chlorite and/or carbonate. Small, squat chlorite pseudomorphs in the matrix may be after pyroxene. Resorbed,

rounded to embayed, quartz xenocrysts are relatively common and sporadic coarse granite xenoliths are also present. Carbonate is a conspicuous phase in these dykes, occurring in rounded to subangular 'ocelli' or vugs associated with quartz and alkali feldspar, commonly zonally arranged. Accessory apatite subhedra are generally present.

Age of the minor intrusions

The basic dykes from the Eskdale area were intruded both before and, less abundantly, after the emplacement of the Eskdale granite. Whole rock K–Ar ages (Rundle, 1992; Al Jawadi, 1987), however, range from 372 to 262 Ma, including one dyke from near Boot which intrudes the granite and gave an age of 365 ± 5 Ma, implying intrusion during Late Devonian. Oliver (1961) distinguished three phases of basic dyke intrusion in the Central Fells: syn-Borrowdale Volcanic Group dykes, composite intrusions associated with granite and granite porphyry, and finally, extremely altered dykes that represent the latest phase of intrusive activity in the area. It is possible that some of Oliver's 'late dykes' are lamprophyric. Dwerryhouse (1909) also considered the basic dykes to be the latest intrusive phase. Al Jawadi (1987) reported basic dykes cutting both rhyolite and microgranite porphyry dykes and considered a whole rock K–Ar age of 320 Ma as indicating a Carboniferous age of intrusion. The field evidence and age of the Eskdale granite at 450 Ma indicate that the whole-rock K–Ar ages quoted are incorrect. This is supported by whole-rock K–Ar ages on other major and minor intrusions reported by Al Jawadi (1987) that are clearly young as a result of Ar loss compared with Rb–Sr isochron or biotite K–Ar ages.

Rb–Sr isochron ages for the rhyolite (felsite) dykes (Rundle, 1992; Al Jawadi, 1987) are in the range 428 to 436 Ma, but in the light of evidence from the plutonic rocks, these ages are almost certainly reset. Their initial $^{87}Sr/^{86}Sr$ ratios, pre-cleavage aspect, distribution and geochemistry indicate a probable genetic link with the approximately 450 Ma Ennerdale and/or Eskdale plutons.

The quartz-feldspar granite porphyry dyke demonstrably postdates the Eskdale granite, and is texturally distinct from it, though Dwerryhouse (1909) believed that there was a genetic link between the two. Al Jawadi (1987) reported chilling of such a dyke against a rhyolite dyke on Middle Fell [NY 1528 0640]. A Rb–Sr isochron age of 393 ± 9 Ma was obtained for this dyke by Al Jawadi (1987), possibly suggesting a link with the Skiddaw–Shap intrusive episode.

Al Jawadi (1987) has also produced an Rb–Sr isochron age from samples of the microdiorite–microgranite suite from the Lingmell area ('Bastard Granite') of 392 ± 4 Ma. Dating of dykes from the Duddon valley is restricted to a single quartz porphyry dyke at Holehouse Gill [SD 1886 9254] by Rundle (1992) which gave an Rb–Sr age of 391 ± 19 Ma. This is consistent with the lack of thermal metamorphism in dykes from this suite close to the 450 Ma Eskdale pluton (for example at Whitrow Beck, SD 1419 9383). These ages are also close to the 395 to 400 Ma age of emplacement of the Skiddaw and Shap granites (Rundle, 1992) and the probably related microgranite dykes described by Soper and Kneller (1990) from the Windermere Supergroup.

Lamprophyre dykes postdate the emplacement of the Eskdale pluton and are also known to intrude Silurian strata of the Windermere Supergroup; they probably represent the youngest phase of igneous intrusion. From the available evidence, Macdonald et al. (1985) considered that lamprophyre dykes in northern England were Early Devonian in age.

Geochemistry of the minor intrusions

A Zr–SiO_2 plot (Figure 52) illustrates some of the available geochemical data for the minor intrusions within the district, excluding those sills and discordant sheets of Borrowdale Volcanic Group affinity. For comparison the fields of the Borrowdale Volcanic Group and major Lake District plutons are shown.

Basic dykes

Analyses of basic (dolerite) dykes have been reported by Macdonald et al. (1988) and Al Jawadi (1987) from the Eskdale and Wasdale areas, respectively (Table 20). They are clearly divisible into two groups based on FeO, TiO_2 and Zr contents (Macdonald et al., 1988) as shown by Figure 53. A TiO_2-enriched group, predominant in the Eskdale area, shows tholeiitic affinity and increasing Fe, V and Ti with differentiation. By contrast, a low-TiO_2 group shows level to declining FeO, TiO_2 and V with differentiation and has calc-alkaline affinity. As Macdonald et al. (1988) indicated, both groups converge towards a common parental composition, suggesting derivation from a common source. REE data (Macdonald et al., 1988) show a variation from flat MORB-like patterns to light REE-enriched patterns with small negative Eu anomalies. On the basis of Th/Y and Nb/Y ratios the dykes exhibit a variable subduction signature, indicated by elevated Th compared with the mantle array defined by within-plate magmas, and overlapping with that for the Borrowdale Volcanic Group. The Mecklin dolerite, which is apparently linked genetically with the Ennerdale intrusion, also shows this feature (Figure 50) and is geochemically akin to the high-Ti group of dykes (Figure 53) and reinforces the contention that some dykes at least are Ordovician in age.

Rhyolite

The rhyolite (felsite) dykes have a range of SiO_2 content from 72 to 78 per cent. Al Jawadi (1987) has shown that Na and K have undergone substantial secondary mobility (along with Rb) caused by alkali exchange, probably during vapour-phase processes during cooling. The Ennerdale granite (see above) has similarly been affected, as have some Borrowdale Volcanic Group ignimbrites (Millward et al., 1978). Rb–Sr ages and $^{87}Sr/^{86}Sr$ initial ratio (about 0.707) of these dykes (Rundle, 1992; Al Jawadi, 1987) are compatible with values for the Eskdale granite and a genetic link has been proposed. Geochemically, the dykes are distinct from the Eskdale pluton in having considerably higher Zr, Th, Y and Sr, and in this respect are more akin to the Ennerdale granite (Figure 52; Tables 19, 21).

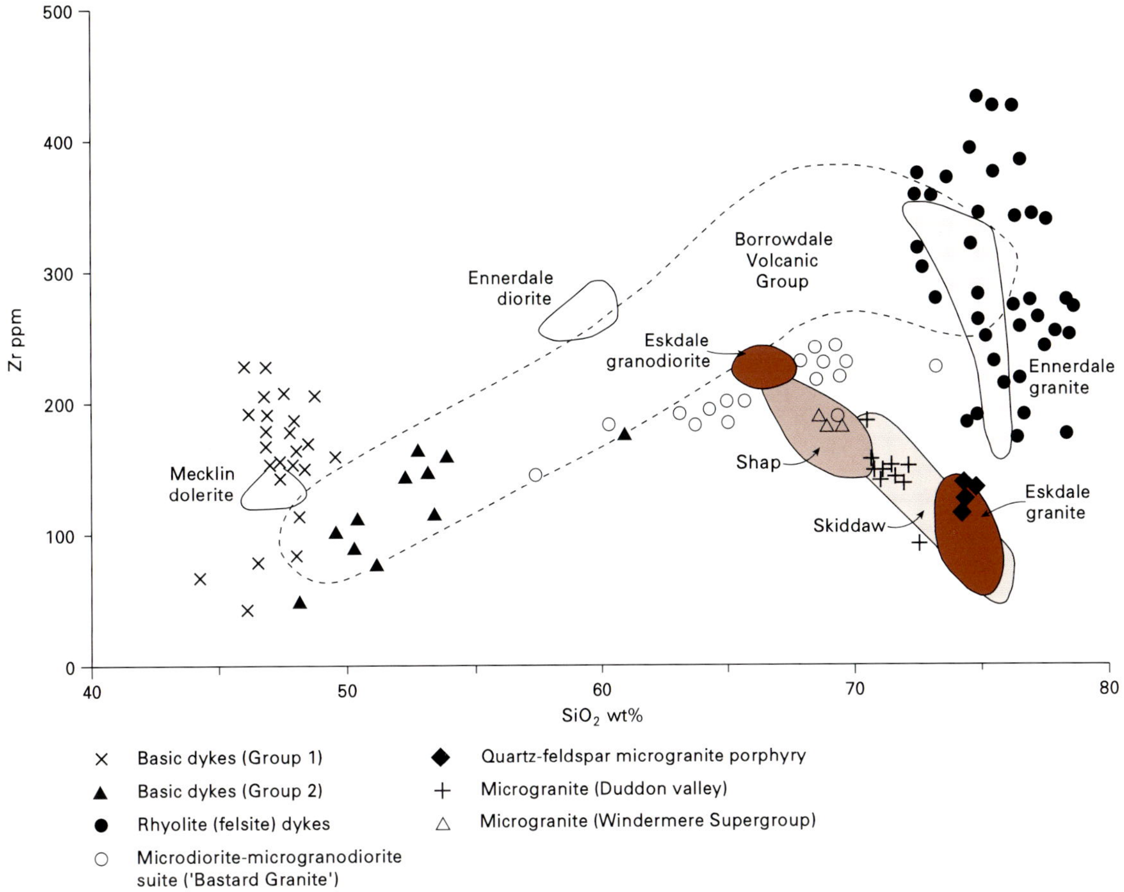

Figure 52 Zr–SiO_2 variation diagram summarising data for minor intrusions.

Fields of the Borrowdale Volcanic Group and various plutons shown for comparison. (Some minor intrusion data from Al Jawadi, 1987 and Macdonald et al., 1988; Shap and Skiddaw fields based on data of O'Brien et al., 1985; Webb and Brown, 1984).

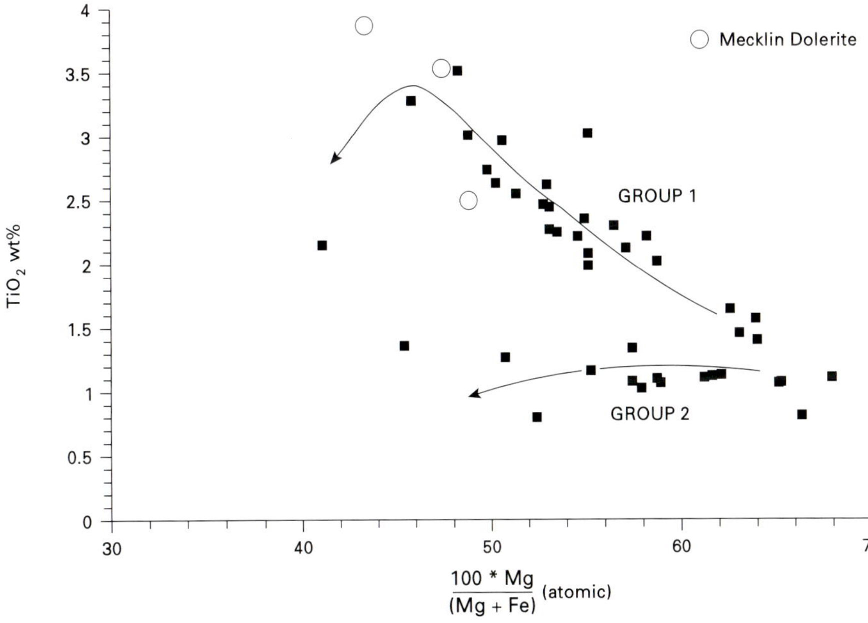

Figure 53 100*Mg/(Mg+Fe) versus TiO_2 showing the division of the basic dykes from the Eskdale–Wasdale area into a (Group 1) tholeiitic and a (Group 2) calc-alkali trend. Analyses of the Mecklin dolerite are also shown. (Minor intrusion data from Macdonald et al., 1988; Al Jawadi, 1987).

Table 20 Representative geochemical analyses of mafic dykes from the Eskdale area.

Grid ref. NY	1777 0080	2317 0206	2313 0202	2121 0082	2261 0332	2034 0255	1974 0359	2138 0462	2120 0494	1704 0433
wt %										
SiO_2	46.80	46.20	48.00	46.60	50.30	49.30	53.30	56.40	48.20	45.30
TiO_2	2.29	1.07	1.10	3.51	0.80	1.10	1.12	2.14	1.63	3.26
Al_2O_3	14.90	17.07	17.28	14.70	14.65	18.34	16.30	15.28	16.26	13.93
Fe_2O_3*	12.72	9.87	10.59	14.42	9.71	8.48	8.23	9.37	10.15	15.50
MnO	0.42	0.33	0.44	0.20	0.29	0.22	0.23	0.22	0.38	0.47
MgO	8.31	9.29	8.43	6.78	9.63	9.04	6.65	3.30	8.53	6.62
CaO	10.05	10.73	6.72	7.59	10.29	9.79	8.22	5.88	10.07	8.76
Na_2O	1.17	1.57	2.28	2.12	1.60	1.35	2.37	3.60	1.58	2.27
K_2O	0.55	0.36	1.53	0.70	0.58	1.69	1.48	0.87	0.73	0.63
P_2O_5	0.23	0.06	0.06	0.41	0.16	0.09	0.14	0.32	0.17	0.42
LOI	3.64	3.14	3.37	2.88	1.73	2.07	1.49	2.04	2.91	1.99
Total	101.08	99.69	99.80	99.91	99.74	101.47	99.53	99.42	100.61	99.15
ppm										
V	256	171	180	289	190	176	203	258	204	264
Cr	240	228	469	53	700	227	163	19	187	51
Ni	120	84	122	46	43	115	59	13	76	11
Rb	51	11	53	29	18	49	49	24	33	25
Sr	140	144	223	198	267	188	372	212	192	197
Y	38	24	22	44	21	22	26	47	27	47
Zr	165	44	47	226	89	67	118	285	116	226
Nb	9	3	2	14	8	3	8	17	7	15
Ba	64	202	305	77	180	72	140	189	212	162
La	—	2.75	2.44	—	—	—	13.09	—	—	17.19
Ce	—	7.74	7.17	—	—	—	28.36	—	—	43.75
Th	5	1	2	8	4	3	6	10	2	4

Analyses performed at the University of Lancaster (Macdonald et al., 1988)
* Total Fe; LOI Loss on ignitition.

Table 21 Representative geochemical analyses of rhyolite, microdiorite–microgranite suite and quartz–feldspar microgranite porphyry dykes.

Sample no.	0/26	NI	0/18	11	P2	P7	B8	AR16	AR17
Grid ref. SD								1963 9347	2154 9588
wt %									
SiO_2	75.10	74.50	75.20	74.20	69.50	64.60	60.00	71.80	71.20
TiO_2	0.49	0.34	0.33	0.19	0.71	1.00	0.84	0.35	0.45
Al_2O_3	13.50	14.20	13.30	15.00	15.30	16.40	16.20	16.31	16.41
Fe_2O_3*	1.69	1.95	1.85	1.13	4.18	5.88	6.95	1.82	2.08
MnO	0.04	0.01	0.02	0.04	0.08	0.12	0.11	0.06	0.07
MgO	0.54	0.55	0.50	0.36	1.46	2.93	4.04	0.95	1.07
CaO	1.53	0.99	0.16	0.42	0.42	2.63	4.80	0.58	1.38
Na_2O	3.49	5.91	2.63	3.98	4.67	2.81	2.60	2.74	3.16
K_2O	3.69	1.77	5.59	4.14	2.90	3.31	2.40	4.29	4.09
P_2O_5	0.05	0.03	0.03	0.15	0.14	0.17	0.20	0.13	0.17
LOI	0.31	0.19	0.62	0.50	1.30	1.01	2.33	—	—
Total	100.43	100.44	100.23	100.11	100.66	100.86	100.47	99.03	100.08
ppm									
V	—	—	—	—	—	—	—	52	64
Cr	5	8	5	10	18	67	149	15	19
Ni	< 5	7	< 5	5	7	22	49	9	12
Rb	104	39	165	319	112	195	90	189	193
Sr	94	104	80	80	121	242	278	342	317
Y	46	31	34	14	31	28	29	11	12
Zr	377	280	231	136	229	201	186	153	149
Nb	18	13	13	19	15	14	15	6	8
Ba	—	—	—	—	—	—	—	626	707
Ce	—	—	—	—	—	—	—	87	99
Th	26	32	35	14	20	22	16	8	11

* Total Fe
LOI loss on ignitition

Analyses
0/26, NI, 0/18 Rhyolite dykes from the Wasdale area (Al Jawadi, 1987).
11 Quartz–feldspar microgranite porphyry dyke, Wasdale (Al Jawadi, 1987).
P2, P7, B8 Microdiorite–microgranite porphyries ('Bastard Granite'), Lingmell area (Al Jawadi, 1987).
AR16, 17 Microgranite dykes Duddon valley (unpublished data, University of Lancaster).

MICRODIORITE–MICROGRANITE AND QUARTZ–FELDSPAR PORPHYRY

The members of this suite of intermediate to acidic minor intrusions (Table 21) yield similar radiometric ages at about 395 Ma. They are clearly distinct from the rhyolitic (felsite) dykes and have lower Zr compared with the field of Borrowdale Volcanic Group magmas (Figure 52). Above 68 per cent SiO_2 they define a trend of declining Zr with differentiation comparable to the consanguineous Skiddaw and Shap plutons. Soper and Kneller (1990) show on geochemical criteria that the microgranite dykes within the Windermere Supergroup may be related to the Shap granite and attendant dyke swarm (compare with Figure 54). The Duddon valley dykes are slightly more evolved than those in the Windermere Supergroup as indicated by higher SiO_2 and lower Zr, Th and TiO_2, but they generally form a coherent group overlapping for the most part the Shap–Skiddaw field, though Th and Nb levels are noticeably lower in the dykes. Analyses of the 'Bastard Granite' microdiorite–microgranite suite from the Lingmell area show that overall it is less evolved, but has significantly higher Zr, Nb and Y for a given degree of differentiation compared with the 'Duddon–Windermere' dykes (Figure 54). This suggests a different fractionation history and possibly parental magma type or source. The Wasdale quartz–feldspar porphyry dyke is the most silicic of this group of minor intrusions (Figure 52), comparable with the Eskdale granite. In detail, they are geochemically dissimilar, with higher TiO_2, Al_2O_3, Sr and MgO, and lower Y in the quartz–feldspar porphyry. Field relationships, Rb–Sr age and trace element variation (Figure 52) possibly indicate an affinity with the Skiddaw granite.

LAMPROPHYRE

No chemical analyses are available for these dykes in the district. However, data for other lamprophyres intruded into the Skiddaw Group and into the Borrowdale Volcanic Group of the eastern Lake District (Macdonald et al., 1985) show them to be ultramafic to mafic but with enriched Zr, Nb, Ce and P_2O_5 contents. Their mineralogical and geochemical characteristics place them in the 'calc-alkaline' class of lamprophyre as defined by Rock (1987), though their trace element geochemistry indicates an origin from an alkaline, enriched mantle source.

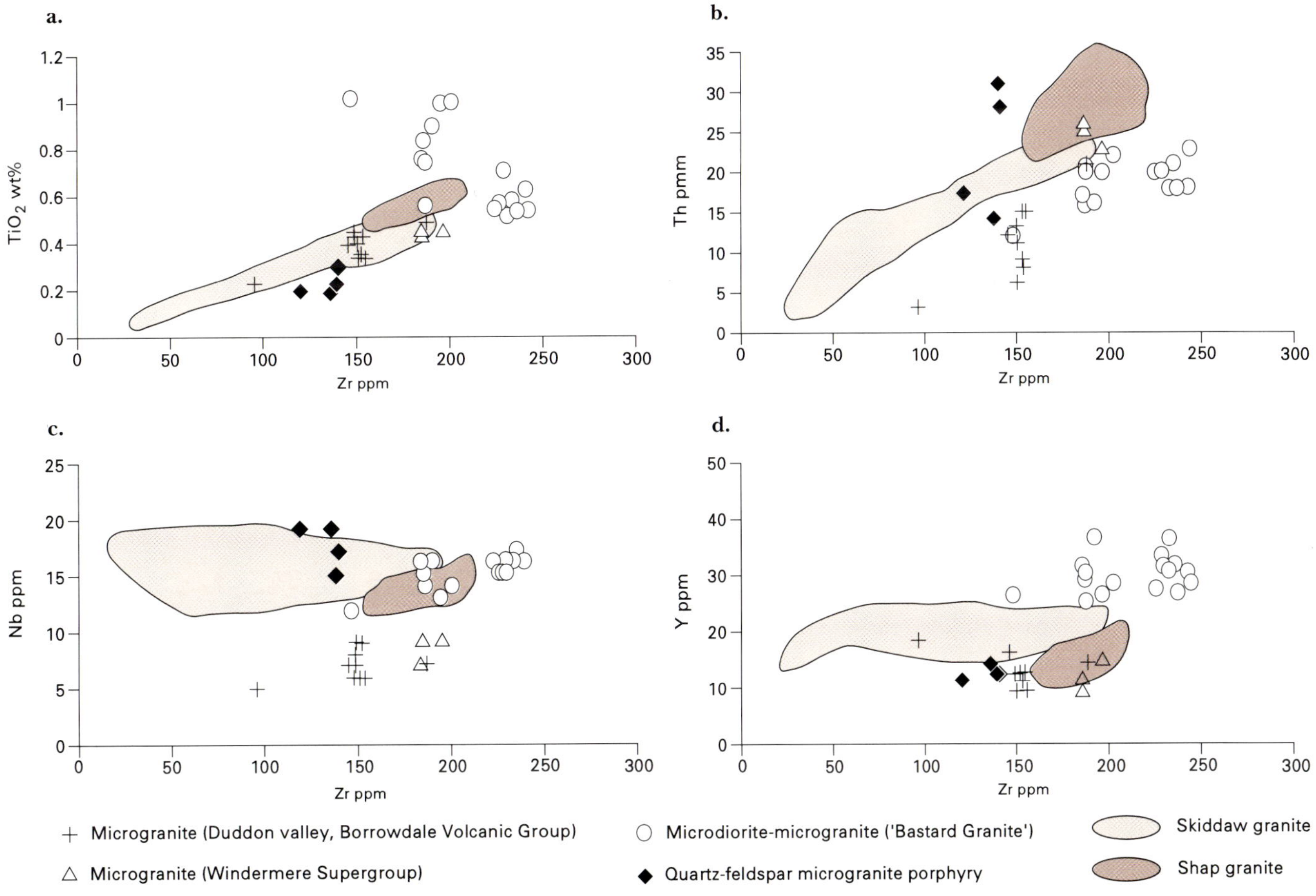

Figure 54 Variation diagrams for the (Devonian) microdiorite to microgranite porphyry suites within the Borrowdale Volcanic Group and Windermere Supergroup (Data sources as Figure 52).
a. TiO_2–Zr b. Th–Zr c. Nb–Zr d. Y–Zr.

NINE

Structure

REGIONAL DEFORMATION PATTERNS

The Lower Palaeozoic rocks of the Lake District and the smaller inliers in northern England (Figure 1) and the Isle of Man, record the Early Palaeozoic history of the northern margin of Eastern Avalonia. This microcontinent rifted from Gondwana and drifted north from high southerly latitudes during the Ordovician and early Silurian (about 60°S to 30°S; Torsvik and Trench, 1991). Structures now preserved in the region record events at the continental margin during that migration, a process which culminated in the destruction of the Iapetus Ocean and mid- to late-Silurian continental collision between Avalonia and Laurentia.

Within the Lake District, three major phenomena have contributed to the evolution of its distinctive geological structure (Cooper et al., 1993). The first involved uplift of the marine Skiddaw Group strata to form the subaerial basement for the volcanic assemblages of the Borrowdale and Eycott volcanic groups. A second phase generated mainly extensional structures during volcanism above a subduction zone. Thirdly, tectonic deformation and cleavage formation resulted from a combination of two processes: the southward propagation of a foreland thrust belt during the Ludlow as Avalonia collided with Laurentia, followed by the climactic Acadian orogeny in the Early Devonian. Windermere Supergroup sedimentation was controlled largely by foreland basin development, but the final inversion of that basin and the imposition of the regional cleavage, at least in the southern Lake District, seem more related to the Acadian event. The cause of the Acadian deformation remains uncertain.

The later structural history of the area is dominated by extensional tectonics. During the Early Carboniferous, a relatively stable, buoyant Lake District Block was established, adjacent to rapidly subsiding basins. Variscan inversion during the latest Westphalian and Early Permian caused renewed uplift of the Lake District massif followed by a further episode of extensional faulting and thermal subsidence from the Late Permian into the Early Jurassic. Regional, flexural uplift of the Lake District block of approximately 1750 m (Chadwick et al., 1994) occurred from about 65 Ma (Lewis et al., 1992).

Skiddaw Group uplift

The small inlier of Skiddaw Group within the Ambleside district is representative of a much wider outcrop in the northern Lake District (Figure 1). The complex, apparently polyphase structure of the Skiddaw Group there (Hughes et al., 1993, and references therein) led Simpson (1967) and Helm (1970) to propose a mid-Ordovician, pre-volcanic tectonic event. The regional cleavage, common to both the Skiddaw Group and the overlying volcanic sequence, seemed to preclude such an event, but the debate was revived by the appreciation of the magnitude of the unconformity beneath the Borrowdale and Eycott volcanic groups (Millward and Molyneux, 1992; Cooper and Hughes, 1993). Strata ranging in age from Late Cambrian to early Llanvirn are present subjacent to the volcanic rocks and these relationships necessitate substantial pre-volcanic uplift, disturbance and erosion of the Skiddaw Group. Regional, pre-volcanic uplift is also indicated by the change from the deep marine depositional environment of the Skiddaw Group to the subaerial environment in which the Borrowdale and Eycott volcanic groups were erupted.

No definitively pre-volcanic tectonic folds and fabrics have been identified in the Skiddaw Group (Hughes et al., 1993) and a pre-volcanic, compressional deformation event related to large-scale terrane collision or accretion seems unlikely. Large volumes of magma must have been generated as precursors to the volcanism of the Borrowdale and Eycott volcanic groups, and associated uplift would have been inevitable. Hughes et al. (1993) discussed a number of possible models for uplift, related in general terms to the production and rise of magmatic liquids above subducted oceanic crust (Gough, 1973; Fyfe and McBirney, 1975; Branney and Soper, 1988).

Deformation associated with volcanism

Folding within the competent volcanic rocks appears relatively simple compared with that of the weaker Skiddaw Group. The Borrowdale Volcanic Group is deformed by broad, open folds with approximately north-easterly and east-north-easterly trending axial planes (for example Scafell and Ulpha synclines, Figure 55) and by a major south-facing monocline adjacent to the Windermere Supergroup outcrop in the south (Figure 56) (Johnson et al., 1979; Kneller and Bell, 1993). The monocline is a compressional Acadian (Early Devonian) feature, but the other folds have ambiguous structural relationships and their ages are uncertain (Soper and Moseley, 1978; Johnson et al., 1979). Branney and Soper (1988) interpreted several small folds in the central Lake District as originating from a combination of fault-block rotation during volcanotectonic collapse caused by emplacement of the large-volume pyroclastic deposits in the Caradoc, and subsequent compression during the regional, Acadian cleavage-forming event early in the Devonian (Soper et al., 1987). A comparable Acadian overprint on a volcanotectonic basin is likely for the Scafell and Ulpha synclines. The latter structure occupies the south-west part of the outcrop in the Ambleside district (Figure 55) and is the only fold truncated by the pre-Ashgill unconformity at the base of the Windermere Supergroup; it is thus unequivocally of Ordovician age. Branney and Soper (1988) considered the

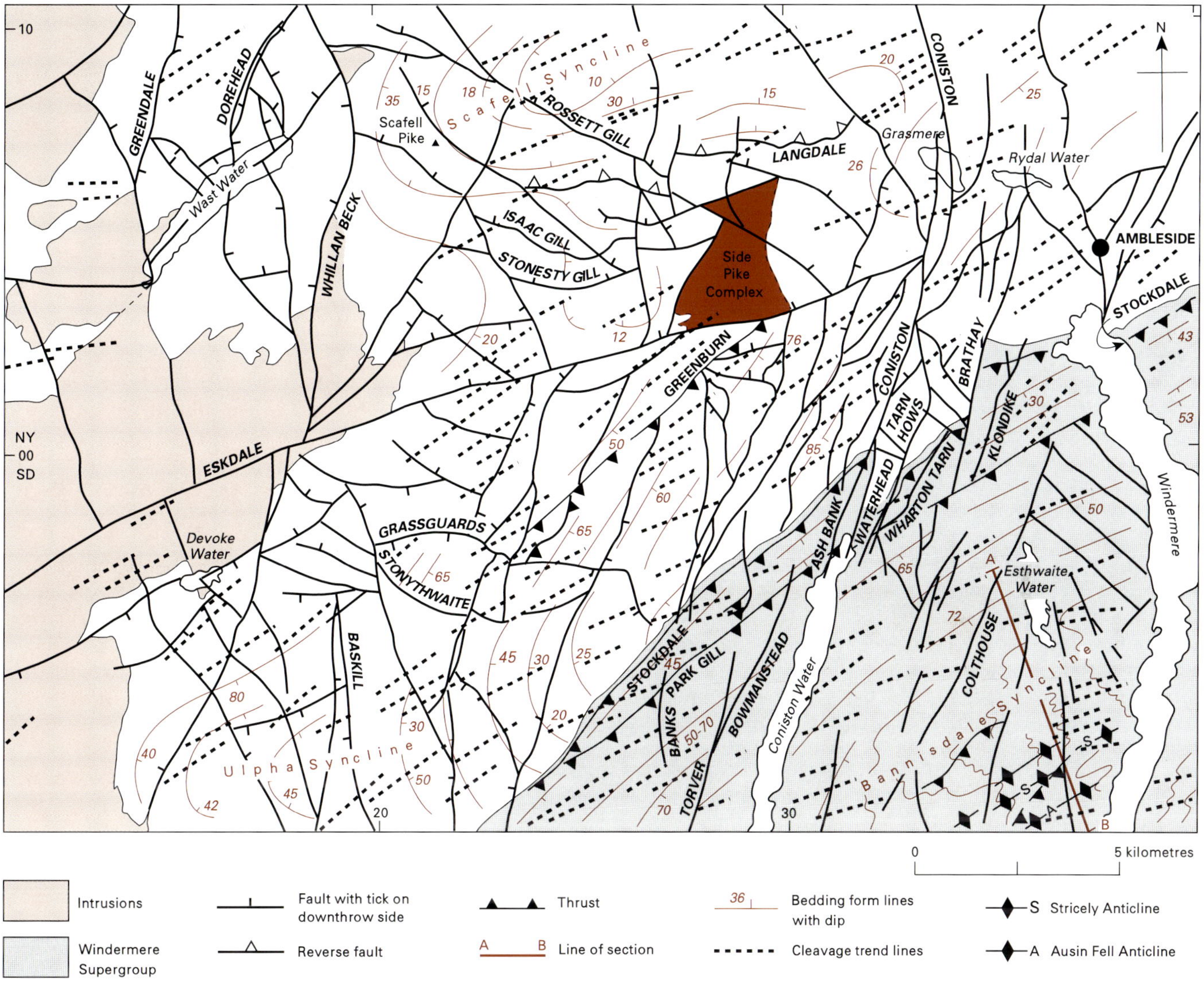

Figure 55 Synoptic structural map of the district.
A–B Line of section in Figure 56c.

pre-Ashgill geometry of this structure to be largely the result of volcanotectonic subsidence whereas previous interpretations (see Soper and Moseley, 1978 for review), invoked a pre-Ashgill episode of tectonic folding which is now considered unlikely.

The volcanic rocks have been intensely faulted. North-west-trending structures are dominant; but there are also significant north, north-easterly and easterly trending fault sets. Some of these faults exert considerable local control on facies within the Borrowdale Volcanic Group and are demonstrably of volcanotectonic origin (Branney and Soper, 1988; British Geological Survey, 1991; Branney and Kokelaar, 1994). Syndepositional deformation of volcanic strata is commonly concentrated in the vicinity of such faults.

A particularly complex structural zone occupies part of the north-west margin of the district, in the area between the main masses of the Eskdale and Ennerdale plutons. The zone is poorly understood at present but a sequence of fault movements appears to predate the Ennerdale intrusion and to be associated with the generation of intense cataclastic fabrics. These are locally overprinted by the hornfelsing caused by intrusion but have broadly the same regional attitude as the demonstrably Acadian cleavage elsewhere. A multiphase veining history (mainly quartz–epidote–chlorite followed by quartz–carbonate–haematite) further complicates interpretation. The zone's structural evolution remains a matter of conjecture, but the most likely explanation would seem to involve faulting and cataclasis during, and possibly controlling, intrusion of the major plutons at about 450 Ma.

Acadian deformation

During the final stages of closure of the Iapetus Ocean the leading edge of Avalonia collided with the margin of

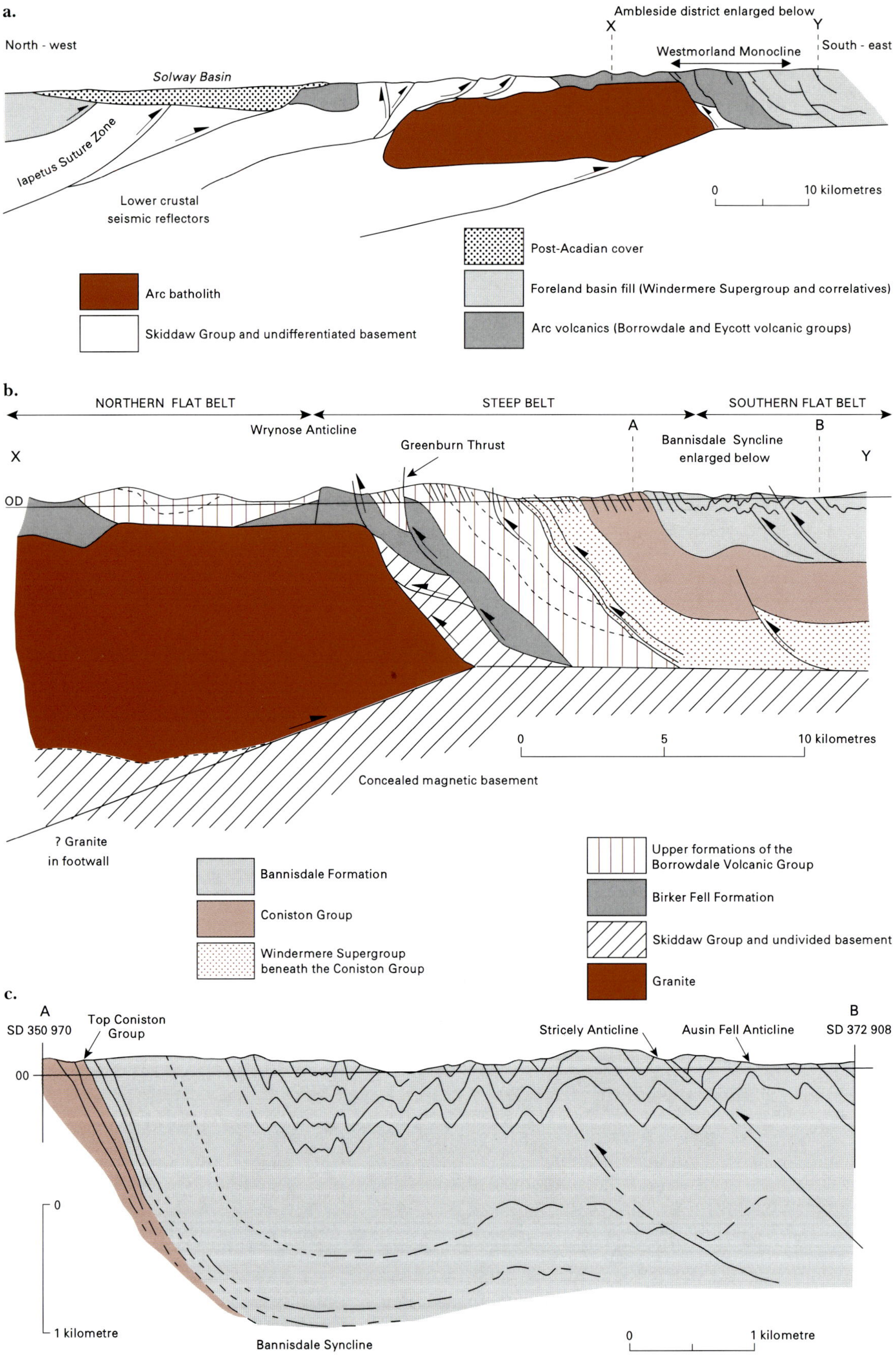

Figure 56 Schematic cross-section from north-west to south-east.

a. Iapetus Suture Zone and the Lake District (after Kneller et al., 1993b).

b. Southern Lake District (after Kneller and Bell, 1993).

c. Bannisdale Syncline (after Kneller and Soper, 1990).

Laurentia and a thin-skinned, south-east-propagating thrust sequence was one of the results (Figure 56). This linked the Southern Uplands accretionary complex with a foreland basin developed across the suture zone ahead of the thrust belt (Kneller, 1991; Kneller et al., 1993b; Hughes et al., 1993). These relationships involve diachronous deformation (Barnes et al., 1989) becoming sequentially younger southwards. However, the Early Devonian inversion of the Windermere Supergroup basin and the accompanying imposition of a regional slaty cleavage indicate a climactic orogenic event dated by Merriman et al. (1995) at about 400 Ma (i.e. Early Devonian, Emsian; see also discussion in Soper et al., 1987). The relationship between these two phenomena is not understood fully.

The Acadian orogeny is here taken to include both the late Silurian foreland thrust-belt development and the Emsian climactic event. The thrust-related shortening of the Skiddaw Group initiated the regional cleavage and associated folds; subsequent strain increments then produced a set of southward–directed thrusts and crenulation cleavages. There is only local evidence of southerly directed thrusts (with related folds and crenulation fabrics) in the Borrowdale Volcanic Group, and it is probable that the thrusts in the Skiddaw Group failed to propagate into, and through, the rigid mass of volcanic rocks. The resultant increasing strain in the Skiddaw Group might have been accommodated by the formation of domainal crenulation cleavages, and possibly by localised back-thrust movement along the contact between the Skiddaw and Borrowdale Volcanic groups. Thrust planes would have acted as the domainal boundaries during this deformation, controlling the distribution of crenulation fabrics. In the Borrowdale Volcanic Group a single regional cleavage has been identified forming an arc across the Lake District. Strike trends are north-north-east in the south-west, but become increasingly north-easterly and east-north-easterly trending in the central and eastern areas. Intensity and nature of the cleavage varies considerably, and is dependent on both lithology and strain state with some zones of intense cleavage development. These zones have been the focus of the modern slate industry (Chapter 2).

As the deformation front moved southwards a flexural basin was produced ahead of the thrust front by the orogenic load. In this foreland basin most of the Windermere Supergroup was deposited, with sediment accumulation rates indicating accelerating subsidence through the later half of the Silurian (Kneller, 1991). Some deformation of this sequence was related to continued southward thrust propagation, but the principal folding and cleavage generation probably occurred during basin inversion, the Emsian climax to the Acadian orogeny. Overall, the structure of the southern and central English Lake District is a south-east-facing monocline (the Westmorland Monocline of Kneller and Bell, 1993). It is defined by a 10 km-wide zone of highly cleaved, steeply south-east-dipping rocks which crosses the district (Figures 55, 56), and separates gently dipping, relatively poorly cleaved Borrowdale Volcanic Group strata to the north from extensively folded and cleaved, but regionally subhorizontal, Windermere Supergroup rocks to the south. The monocline accommodates at least 8 km of uplift, coincident with the steep, concealed southern margin of the Lake District batholith (Firman and Lee, 1986).

Some constraints on the timing of deformation are provided by the relationships of the structures to a dated igneous intrusion. Soper and Kneller (1990) suggested that cleavage formation in the southern Lake District coincided with the later stages of intrusion of the Shap granite (U–Pb zircon age of 390 ± 6 Ma; Pidgeon and Aftalion, 1978), thus confirming a broadly Emsian (Early Devonian) age. If deformation was diachronous, thrust and fold formation farther north in the Lake District may have spanned the late Silurian and earliest Devonian (Hughes et al., 1993), but evidence for this is permissive rather than definitive. Subsidence analysis of Ludlow–Emsian sequences in Wales and the Welsh Borderlands suggests that the Windermere basin continued to migrate southwards on to the Avalonian foreland through the Přídolí and Early Devonian (Kneller et al., 1993b). However, such an interpretation remains difficult to reconcile with the compelling evidence for a climactic Emsian deformation in the southern Lake District (Merriman et al., 1995).

SYN-VOLCANIC STRUCTURES

The term volcanotectonic fault has been defined by Branney and Kokelaar (1994) with reference to fractures with displacements resulting directly from the movement and eruption of magma. This includes gravitational collapse structures that have been triggered by magma movements. The faults whose movements are related to basin extension are specifically excluded from the usage, though they recognised that these faults in part could have a volcanotectonic component to their displacement history. The distribution of volcanotectonic and basin-extension faults within the district is shown in Figure 57.

The Scafell and Ulpha synclines were superimposed on original, extensional, volcanotectonic structures (Branney, 1988b; Branney and Soper, 1988; Davis, 1989). Facies evidence for the origin of the basin has been outlined in Chapter 4, and such an interpretation is further supported by palaeomagnetic work (Channell and McCabe, 1992), which indicates that some rotation of the limbs of the Scafell structure occurred prior to the Silurian and so predates the regional tectonic episode.

The Scafell area is dominated by a structural half-basin. The floor of the Scafell Caldera is intensely fragmented into 100 to 2000 m-sized blocks that subsided in an irregular, piecemeal fashion (Branney and Kokelaar, 1994; Petterson et al., 1992, fig. 8). Highly irregular caldera-floor topography is evident from dramatic thickness changes in successive ignimbrites of the same eruption episode. Most stratigraphical sections in the Central Fells area include every ignimbrite, but successive ignimbrites thicken in different, adjacent or overlapping, irregular-shaped fault-bounded depressions. The sense of displacement on some faults was reversed during the course of the collapse-related eruption episode, so that one flow unit is ponded where another thins (Davis, 1989). The complexity of the

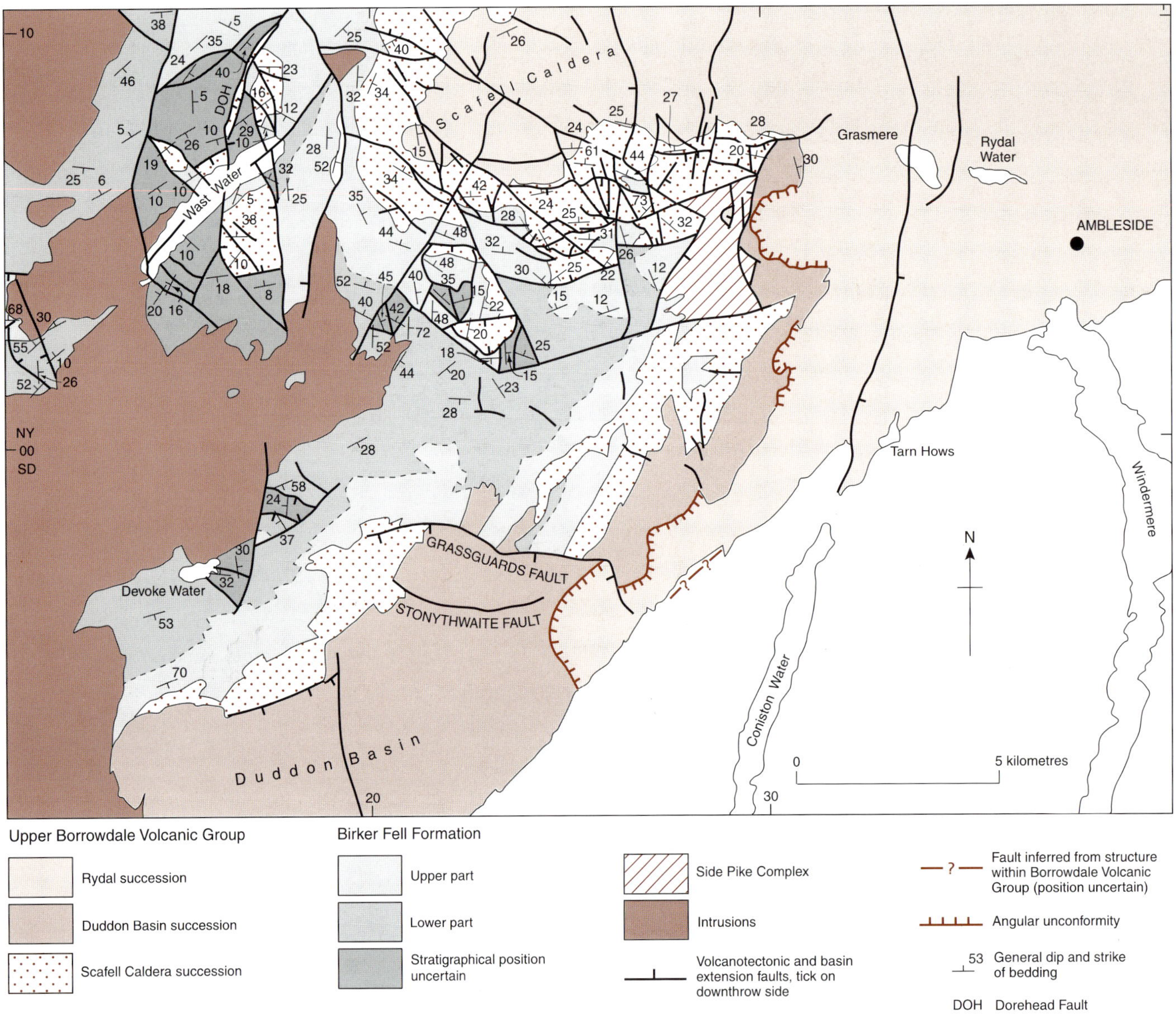

Figure 57 Distribution of volcanotectonic and basin-extension faults within the Borrowdale Volcanic Group.

collapse-related fault pattern and the absence of a single caldera-bounding ring-fault, make the caldera boundaries difficult to define, though to the west and south there is an arcuate system of volcanotectonic faults against which ignimbrite is ponded (Davis, 1989). The caldera floor did not undergo resurgent doming on a large scale, though eruption of the Scafell Dacite probably caused a localised topographical high.

The Duddon Basin, volcanotectonic precursor to the Ulpha Syncline, appears to have developed as an extensional half-graben. There are no closely spaced volcanotectonic faults, nor is the adjacent Birker Fell Formation broken up in the same way as it is in Eskdale. The sequence in the Duddon Basin records successive major ignimbrite eruptions along with intervening sediment accumulations; these are ponded towards the axis of the basin. The result, prior to Dent Group deposition, was a weak flexure with relative downward displacement to the north and east, a structure compatible with sagging during emplacement of the volcanic sequence (Branney and Soper, 1988) and controlled by the Grassguards Fault which marks the north-east boundary of the basin.

Volcanotectonic faulting

A key feature in the recognition of these faults within the ignimbrite successions is a change in the thickness or facies of stratigraphical units across the fault. For example, on Crinkle Crags the thickness of the Long Top Member is generally about 180 m, but increases abruptly to 300 m on the downthrow side of the Isaac Gill Fault. In Langdale, the Bad Step Tuff thickens from 40 m to more than 400 m

north-eastwards across two volcanotectonic faults [NY 2660 0603 to 2720 0610]. These faults now have a combined downthrow of 600 m to the south-west, opposite in sense to the downthrow associated with emplacement of the ignimbrite. In other cases of fault movement reversal, successive ignimbrites have been ponded on opposite sides of a fault.

The typical complexity of volcanotectonic fault movement in the Scafell Caldera is illustrated by the fault at Grave Gill, Langdale (Figure 58). There, the deformation fabrics and thickness variations in successive ignimbrites indicate a large displacement across this near-vertical fault. First, ponding of the Bad Step Tuff indicates a downthrow of at least 190 m to the south-east, prior to, or during, its emplacement. Re-activation of the fault is then indicated by a wedge, comprising several ponded massive units of the suprajacent ignimbrite, which extends down 40 m into the top of the Bad Step Tuff. There are two possible explanations for these relationships. Firstly, a downthrow of 30 m to the north-west was accompanied by the opening up of a surface fissure, more than 60 m deep, which partially collapsed and was partially filled with Bad Step Tuff blocks and unlithified ash, leaving a 40 m trench later filled by the next rheomorphic ignimbrite. Alternatively, a 20 m downthrow to the south-east formed a small scarp which rapidly degraded (because of the fault brecciation) leaving a shallow scarp-foot step and hollow, and a subsequent 40 m downthrow to the north-west happened during emplacement of the next rheomorphic ignimbrite.

In either interpretation, the contrast between brittle fracture of the Bad Step Tuff and the steep attenuation of the rheomorphic fabrics in the overlying ignimbrite is due to the timing of the re-activation, that is, after cooling of the Bad Step Tuff, but prior to cooling of the overlying ignimbrite. Though at Grave Gill the Bad Step Tuff was completely buried by the upper ignimbrite, matrix-poor mesobreccia lenses emplaced within and on top of the upper ignimbrite (Figure 58) are composed predominantly of Bad Step Tuff blocks and show that the Bad Step Tuff was exposed elsewhere. The breccia blocks are angular, reach 3 m in size, and reflect avalanching from the scarp of another fault nearby. A mineralised, brittle fault plane close to, but not precisely along, the plane of the volcanotectonic fault indicates further, post-caldera, re-activation with downthrow of 6 to 15 m to the north-west. Offsets of the mesobreccias indicate possible strike-

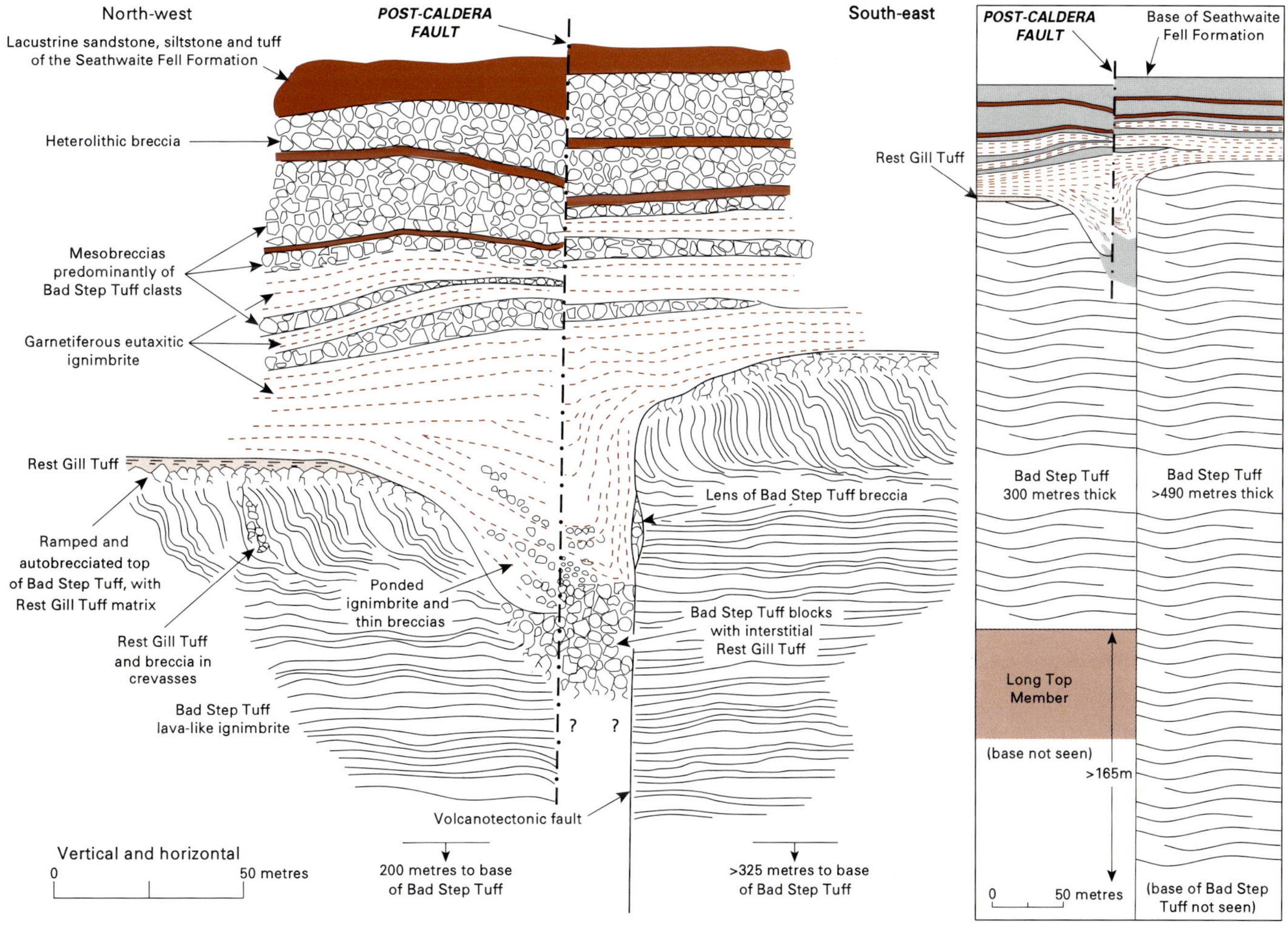

Figure 58 Detail of the volcanotectonic Grave Gill Fault in Langdale (after Branney and Kokelaar, 1994).

slip motion but this movement may have been relatively late and purely tectonic in origin.

CALDERA MARGIN FEATURES OF THE SCAFELL CALDERA

Locally, near-vertical and discordant bodies of sheared, high-grade tuff with steep eutaxitic fabrics occur enclosed, or hosted, by older stratigraphical units. They may represent surface fissures that opened up during subsidence and filled with tuff from above. However, other interpretations are possible and the discordant tuffs may represent tuff agglutinated in dyke-like pyroclastic conduits that fed the ignimbrite eruptions, or they may represent vertically attenuated tuff marking an early volcanotectonic fault scarp where the displacement direction was reversed during subsequent re-activation (Branney and Kokelaar, 1994, fig. 16). Large-scale surface fissuring in the hinge zone between adjacent volcanotectonic fault blocks with divergent dips is indicated at Crinkle Crags. There, a 400 m-wide zone of particularly complex deformation is seen where faults within pre-caldera rocks pass up into abundant, ramifying, closely spaced faults and rheomorphic shear zones within the ignimbrites. A vertical fault-bounded wedge of anomalously thick (300 m) and massive ignimbrite occurs within this complex zone [NY 2570 0470]. It represents an uncollapsed portion of a surface fissure that was filled by the overlying Long Top Member either during initial deposition of the ignimbrite, or by its subsequent slumping. A second example of this phenomenon is seen on Yewbarrow [NY 175 095] as a vertical, fault-bounded wedge of welded tuff between areas of pre-caldera rocks with divergent dips. It is well exposed over a length of 3 km adjacent to the Dorehead Fault. The wedge is 100 to 200 m wide, broadens upwards, and is filled with anomalously thick (over 350 m) ignimbrite of the Long Top Member that has steeply dipping (50 to 70°) and highly attenuated rheomorphic fabrics.

Both the tuff-filled fissures described above may lie along a putative arcuate system of faults that surrounds the inward-dipping strata of the caldera floor and fill (Figure 57). The faults have a cumulative downthrow of more than 800 m towards the centre of the caldera and Davis (1989) suggested that the arcuate fault system may represent part of the caldera's structural margin. However, down-faulted blocks of Airy's Bridge Formation lie to the west of the Dorehead Fault on Middle Fell and on the shores of Wast Water [NY 153 070 and NY 164 068]. The boundary to the zone of volcanotectonically disturbed Birker Fell Formation (Figure 57), may reflect part of the true margin of the Scafell Caldera.

FAULT-BLOCK ROTATIONS

Rotations of caldera-floor fault blocks are inferred from, firstly, lateral increases in thickness in successive intracaldera ignimbrites, indicative of abruptly changing palaeoslopes, and, secondly, from the present attitudes of blocks bounded by volcanotectonic faults. Illustrative of the first of these phenomena is a high-grade ignimbrite overlying the Bad Step Tuff which gradually thickens from 40 to 200 m, over a lateral distance of less than 500 m [NY 2785 0700 to NY 2735 0705]. Removal of the dip of the conformably overlying sedimentary rocks of the Seathwaite Fell Formation, which must have been deposited more or less horizontally, shows a rotation of the block of at least 20° immediately before and/or during the emplacement of the ignimbrite. Within the tilted area, the ignimbrite shows particularly intense rheomorphism. The change in block orientation across volcanotectonic faults is seen for example on Crinkle Crags [NY 2520 0425], where a southward dipping fault block is inclined about 45° relative to an adjacent north-dipping block to its north [NY 2990 0550]. It is also inclined 15° relative to a north-dipping block to its south [NY 260 035]. The two faults (Isaac Gill and Stonesty faults) that bound the southward dipping block both show soft-state deformation of phreatomagmatic tuff and steep rheomorphic fabrics in the silicic ignimbrites, suggesting that the block rotation probably occurred during volcanotectonic subsidence.

Block rotation is also present in the Coniston Fells. On the flanks of The Old Man of Coniston, north of the disused Cove Quarries [SD 271 974], the basal, laminated sandstone beds of the Seathwaite Fell Formation overstep the underlying Lag Bank Formation with an angular discordance of 5 to 10°. About 200 m to the east, the underlying Paddy End Member dips very steeply (Plate 17). Restoration of depositional dips in the Seathwaite Fell Formation leaves substantial residual dip to the southeast on the Paddy End Member. Angular displacement of the block containing the Paddy End Member clearly occurred prior to onset of sedimentation within the Seathwaite Fell Formation.

Many of the volcanotectonic faults identified within the Scafell Caldera system can be traced into the underlying Birker Fell Formation (Figure 57). The effects of piecemeal collapse within the caldera can be demonstrated also by the chaotic arrangement of bedding strike and dips between fault blocks in the Birker Fell Formation, particularly in the Wasdale and upper Eskdale areas (Petterson et al., 1992, fig. 8). Bedding dip and strike within individual fault-bounded blocks are consistent, but between blocks the bedding strike may vary by as much as 90° and the dip may change from shallow to very steep. Such rotations cannot be explained readily by normal extensional tectonic means. A number of small fault blocks on the south side of Hard Knott illustrate this [NY 235 018]. Dip azimuth differences across the blocks up to 90° and downthrow in excess of 1000 m may be indicated by the juxtaposition of the Throstle Garth Member from near the top of the formation, against andesite just above the Cockley Beck Tuff. The boundary to the chaotic zone of block faulting is located south of Eskdale where a coherent stratigraphy and structure is present (Figure 57).

In the west of the district, dramatic changes in dip occur across the Mecklin Fault [NY 120 025]. On its west side, strata strike normal to the fault but dip increases from south to north from moderately north-dipping to almost vertical within the Craghouse Member and in the tuffs underlying it. On the east side of the fault, the strike is similar but strata are gently to moderately dipping. This fault-block rotation clearly predated the Ennerdale intrusion and it is possible that rotation was related to subsidence as a consequence of the eruption of the 800 m-thick Craghouse Member.

POST-COLLAPSE STRUCTURES

Continued movements on volcanotectonic faults after emplacement of the main Scafell Caldera ignimbrites is indicated by abrupt thickness change across faults within the post-collapse sedimentary sequence. For example, on Side Pike in Langdale the sedimentary interval between two distinctive tuff units thins westwards from over 60 m to less than 15 m across two faults spaced 100 m apart (Branney and Kokelaar, 1994, fig. 15). In the vicinity of the faults, the exposed top of the welded ignimbrite, upon which the lacustrine strata rest, shows no evidence of an erosional topography, and a syndepositional downthrow to the east seems likely. The two faults form part of a set of several closely spaced northerly trending volcanotectonic faults characterised by soft-sediment deformation and hot-state deformation of welded tuffs to produce domain breccia and steep, sheared eutaxitic fabrics. Across the complete fault set, changes in thickness and facies of the sedimentary interval are so marked that correlation becomes increasingly difficult. The welded ignimbrite that underlies the lacustrine strata also thins dramatically from over 200 m in the east to only 33 m in the west, requiring a significant syndepositional downthrow to the east across five faults and a lateral distance of only 300 m. Later, reverse movements on these faults has produced the present-day, net downthrow across the set of faults of 500 m to the west.

Though large-scale volcanicity continued after aqueous transgression of the caldera and deposition of the Seathwaite Fell Formation, and the resulting intense soft-state deformation on all scales, several of the sedimentary units are widespread and relatively uniform in thickness and facies indicating that differential movement of the underlying fault blocks had decreased. Though less movement is apparent, many of the early faults continued to be active as magma conduits during later dyke emplacement.

Catastrophic slope failure

The climactic eruptions that produced many of the densely welded ignimbrites such as the Crinkle Member were accompanied by large-scale landslides. The large vertical displacements on volcanotectonic faults in the Central Fells area created unstable scarps, commonly of hot and plastic welded tuff. This resulted in landslides and the accumulation of mesobreccias. Such processes may explain the relationships seen on the east side of Hard Knott [NY 2333 0225] and on Yewbarrow [NY 1730 0915] where the Long Top Member is absent from part of the outcrop. Localised non-deposition of these ignimbrites is an unlikely explanation for the missing strata, because the Long Top Member includes material that would have blanketed the entire caldera and is preserved as condensed sequences on topographic highs (Davis, 1989). Removal of the welded tuffs by erosion is not consistent with the field evidence since there are no erosional remnants or erosion surfaces within the Long Top Member. In both cases, large-scale sliding or slumping is supported by evidence from the nearest exposures of the Long Top Member. These show internal slide surfaces and dips that are discordant to underlying strata, suggesting that the deposits are probably not in situ.

Further evidence for large-scale slope failure is provided by a discordant, valley-shaped feature on Scafell [NY 212 055], interpreted as a palaeotopographic depression left by the removal by slumping of the Long Top Member prior to emplacement of the Crinkle Member (Branney and Kokelaar, 1994, fig. 11). It is a steep-sided, U-shaped, elongate depression, 300 m wide, 150 m deep and more than 1000 m long, cut into the Long Top Member and filled with strongly rheomorphic Crinkle Member ignimbrite. Its axis trends north-north-west, the rheomorphic tuff filling it has fold axes and lineations parallel to this with steeply dipping foliations parallel to its walls. The wall rocks show abundant evidence of hot, soft-state shearing, with extensive domain breccias, slide surfaces and rheomorphic deformation structures. The location of the slumped mass of Long Top Member is uncertain, but it may now lie 1.5 km to the north-west [NY 202 070] where slump-thickened Long Top Member ignimbrites with numerous internal slide surfaces and local domain breccias crop out.

MESOBRECCIA

Abundant mesobreccia interstratified within intra-caldera ignimbrite was probably derived from local, ephemeral, fault scarps that either collapsed, became buried beneath tuff, or disappeared because of reversed-sense re-activation soon after they were formed. The mesobreccia contains blocks, up to about 3 m across, and commonly occur either as isolated lenses interstratified with welded ignimbrite or as stacks of lenses up to 70 m thick. Individual lenses are massive and vary in thickness from over 10 m down to layers one block thick that pass laterally into trains of isolated blocks; lenses rarely persist laterally for more than 400 m. Though heterolithic lenses have been recorded, most are composed primarily of blocks of a single lithology, each lens derived from a specific source. This, together with the lack of lateral continuity of the lenses and the high angularity of many of the blocks, indicates that they are local accumulations. Successive lenses in a stack may have different block compositions indicating an abruptly changing source.

In some cases, lenses contain rounded blocks indicative of abrasion or derivation from a hot ignimbrite source by thermal spalling within hot rock-fall avalanches. Many mesobreccia lenses are framework supported, with fine ash infiltrated into the upper interstices, and locally they are loaded into the top of subjacent rheomorphic ignimbrite, indicating rapid deposition on to hot tuff. Others comprise angular blocks supported in a welded ignimbrite matrix, with fiamme that show hot-state compaction around the blocks. Upper surfaces of the lenses may be irregular and hummocky; basal scours, inverse-graded basal layers, and normally graded lenses occur locally, though many lenses are ungraded and sorting is generally poor.

Mesobreccia was emplaced concurrently with volcanotectonic faulting because they are cut by faults that exhibit soft-state and rheomorphic deformation, they are ponded entirely within grabens bounded by such faults, and they

overstep some faults with corresponding thickness changes. There is a general increase in mesobreccia abundance towards the top of the Airy's Bridge Formtion, possibly reflecting increasing rates of piecemeal subsidence.

SIDE PIKE COMPLEX

The Side Pike Complex is an exceptionally coarse and chaotic breccia, covering an area of more than 5 km^2 in the south-east part of the Scafell Caldera (Figure 57). The complex is bounded to the north and west by faults, and grades southward and eastward into structurally more coherent stratigraphy cut by volcanotectonic faults. It comprises blocks, up to more than 500 m in size (megabreccia in the sense of Lipman, 1984), composed of andesite, rhyolite, welded tuff, phreatomagmatic tuff, and subaqueously deposited volcaniclastic sedimentary rocks including debris flow deposits (Branney and Kokelaar, 1994, fig. 12). Despite internal deformation many of the larger blocks have a coherent stratigraphy that can be matched with parts of the pre-caldera and caldera-fill successions elsewhere. However, though the internal stratigraphy of some other large blocks can be correlated from one block to another, they cannot be matched with any part of the regional stratigraphy. These 'exotic' blocks were derived either from formations or from facies variations not otherwise preserved in the district.

The coherent internal parts of the large blocks are increasingly deformed towards their margins, passing into a chaotic zone of disaggregated and locally comminuted material. In some places [for example NY 2825 0405 and NY 2870 0435], angular blocks of competent welded tuff appear to have been driven into adjacent, unlithified megablocks which accordingly exhibit particularly severe soft-state deformation and disaggregation structures adjacent to the point of impact.

Throughout, the complex deformation during deposition is indicated by fault and fold style. Deformation clearly predated lithification of the sedimentary and the non-welded pyroclastic units, and in the welded ignimbrites local domain breccias and rheomorphic deformation structures indicate that the deformation in some cases preceded or accompanied cooling. Most stratigraphical units, even within the larger, generally coherent blocks have undergone spectacular soft-state deformation [for example at NY 2856 0414] with abundant soft-state slide surfaces, clastic intrusions, faults, folds and chaotic convolutions. Locally, these produce vertical and even overturned strata [for example NY 2907 0533]. However, in most of the blocks the strata have a similar overall dip to the east that might be the result of systematic listric rotation from a large, west-facing caldera scarp.

Soft-state deformation

Soft-state deformation structures occur either associated with faults that are inferred to be volcanotectonic, or in laterally extensive areas of gravitational sliding and spreading. They predate lithification and are characterised by disaggregation, mobilisation and ductile strain. Their origin is clearly indicated where mud and silt-grade layers have behaved in a more competent fashion than sand-grade layers and also by an absence of open or mineralised discontinuities, slickensides, fault-breccias, and grain granulation.

STRUCTURES DEVELOPED IN HOT IGNIMBRITE

The localised shearing of eutaxitic fabrics associated with steeply inclined planes may have developed in a variety of ways: by differential welding-compaction of thick ignimbrite ponded against, and possibly burying, a cliff; as agglutinated pyroclastic flow, surge or fall deposit draped over, or accreted to, a steep cliff; or by movement of a fault after an ignimbrite was emplaced, but before its final cooling, so that the hot ignimbrite sheared in a ductile manner (Branney and Kokelaar, 1994, fig. 10).

Many of the ignimbrites in the Borrowdale Volcanic Group contain rheomorphic structures caused by deformation of the deposits while they were still hot. Though not generally rheomorphic, the Long Top Member is cut by steep planar zones, 1 to 10 m wide, in which the eutaxitic foliation has been deflected to become near-vertical. Locally, there are minor folds, and fiamme are attenuated and lineated. Vitroclastic fabrics within these zones are locally so deformed that the rock resembles flow-laminated silicic dykes [for example NY 2472 0464]. Such features indicate ductile shear and attenuation prior to cooling of the ignimbrites, and the zones are interpreted as recording volcanotectonic faulting. Vertical displacements are commonly up to, and locally exceed, 400 m, and some of the zones can be traced across continuously exposed ground downwards into brittle caldera-floor faults.

Hot-state brecciation accompanied the ductile ignimbrite deformation in most of the volcanotectonic fault zones. The resulting breccias are typically non-mineralised, resembling autobreccias rather than fault breccias; they grade laterally and highly irregularly into undisturbed flat-lying eutaxitic ignimbrite. A distinctive facies comprising fused breccia clasts is common and is called 'domain breccia' (Branney and Kokelaar, 1994). Each domain originated as a rotated block of eutaxitic tuff, irregular in shape and 5 to 50 cm across, the margins of which have been obscured by viscous compaction and local fusion. The domain breccia may grade through breccia with discernible margins bounding the rotated blocks, and jigsaw-fit breccias with non-rotated blocks, into eutaxitic tuff with kink folds and ductile (rheomorphic) fabrics (Branney and Kokelaar, 1994, fig. 9). Minor rheomorphic folds within the steep deformation zones vary from open to nearly isoclinal with kink bands and chevron folds in places. At some localities two or more eutaxitic ignimbrites, with thin interstratified non-welded, fall and/or surge layer(s), are deformed in a single fold. These examples indicate that the folding did not occur during emplacement or slumping of individual rheomorphic ignimbrite sheets, but that the ignimbrites were stacked and stable prior to volcanotectonic faulting and tilting.

Within the Airy's Bridge Formation the Oxendale Tuff, Crinkle Member, and particularly the Bad Step Tuff are widely rheomorphic. Many of the folds probably relate to tilting of the ground during volcanotectonic subsidence and consequent slumping of the still-hot ignimbrite.

However, rheomorphism is particularly intense adjacent to contemporaneous faults where eutaxitic fabrics are highly sheared and steepen to become subparallel to the fault-planes, for example on The Band [NY 2610 0620]. Several of the volcanotectonic faults, including those exposed at Dungeon Gill [NY 2610 0620] and Adam-a-Cove [NY 2472 0464], are not brittle fractures, but steep and narrow rheomorphic shear zones, 0.5 to 2 m wide, in which fabrics are so intense that the affected tuff cannot be correlated easily with any specific part of the ignimbrite succession. These highly attenuated tuffs may be finely brecciated in places.

Slump structures in bedded pyroclastic rocks

The phreatoplinian tuff phase of the Whorneyside Formation (Branney, 1991) contains abundant examples of compressional and extensional deformation structures (Figure 59) thought to have been induced by the volcanism, and caused by gravity sliding and spreading down slope. There are abundant normal faults and tension cracks filled with homogenised coarse tuff (Figure 59a, b). Some normal faults have a listric geometry, rooting out in bedding-parallel slides; others form sets of closely spaced, very steep to vertical faults that record dominantly centimetre- to decimeter-scale differential subsidence of adjacent blocks of ash. The style and geometry of the deformation was influenced by competence contrasts between wet ash layers during and subsequent to deposition. Rapid cohesion and hardening of fine-grained volcanic ash can be caused by a combination of electrostatic attraction between fresh or altered damp surfaces of ash particles, chemical alteration and adherence of vitric dust particles, and precipitation of mineral cements during evaporation of pore water (Segerstrom, 1950; Malin et al., 1983; Tomita et al., 1985). Cohesion clearly began prior to deposition in the case of the abundant accretionary lapilli that survived impact with the ground. By contrast, the coarse tuff layers were non-cohesive and so there are no intraclasts composed of coarse tuff within the syn-eruptive sedimentary rocks; sand intrusions and dewatering structures in sand-grade beds are common, reflecting the relative ease of disaggregation of newly deposited coarse ash.

Microfaults are abundant with spacings of 1 to 10 cm in the fine-grained layers. Where they can be traced into enclosing sand-grade layers they pass into diffuse shear and homogenised zones, commonly losing their identity. Fault planes in the coarse tuff are typified by closely spaced, ramifying and cross-cutting shear laminae, less than 5 mm thick, each with a fine-grained centre grading out to coarser-grained margins. This indicates kinematic segregation of cohesionless grains caused by granular shear (Bridgewater et al., 1985), rather than brittle fracture of lithified material.

Large competence contrasts between layers influenced the pattern of microfaults which tend to cut fine-grained layers at high angles with their spacing influenced by the thickness of the fine-grained layers. Thin layers of fine tuff display more numerous (and more closely spaced) microfaults for a given net amount of throw than do thicker layers of similar tuff. Some ductile deformation of fine tuff layers has produced pinching and swelling. Sites where voids might have formed, had the ash been more competent, have been filled entirely with disaggregated sand-grade ash (Figure 59c). This is taken as an indication of grain flow and/or localised water fluidisation (Allen, 1982), following shock or deformation-induced liquefaction of porous sand layers enclosed by impermeable fine-grained layers.

Some fine-grained layers fragmented by extension show block rotation with domino or 'bookshelf' structure (Wernicke and Birchfield, 1982), indicating low competence of the enclosing coarse-grained layers (Figure 59d). Larger-scale faults form complex zones up to 2 m wide; fault planes are very uneven, and are refracted around bedding or may be splayed and branched. Sheared and disaggregated sand-grade tuff in these zones appears to have flowed around small, lenticular fault blocks of fine-grained tuff up to 20 cm across, and any remaining layers of fine-grained tuff are attenuated by closely spaced extensional microfaulting. Steep zones of small-scale, extensional step faults are occupied by vertically laminated sand-grade tuff indicating incremental extension and infill.

Soft-state disruption of the bedded tuff increases in intensity adjacent to several major intra-caldera faults. A small horst on the west side of Crinkle Crags [NY 2432 0512], shows intense soft-state faulting and bed rotation towards the margins, whereas low dips and more widely spaced soft-state faults reflect minor gravitational spreading of the tuff at the centre of the horst. Adjacent to the bounding faults, brecciation of fine tuff layers and homogenisation of coarse tuff are evident. Small grabens, for example on Teighton How [NY 2770 0353], show virtually the same features as the horst, but with inverse sense (Branney and Kokelaar, 1994, fig. 5). Elsewhere, good examples of soft-state deformation are preserved adjacent to the steep, north-east-trending Isaac Gill Fault (Figure 60). It is one of a large number of intra-caldera faults that moved in response to eruption of the Whorneyside Formation. Some distance to the south of the fault, bedded tuff dips south-east away from the fault between 5° and 25°, but, within 100 m of the fault, bedding is slumped so that the dip is steep and to the north; that is, towards the fault. The most severely disrupted tuff units occur within 15 m of the fault plane from [NY 2571 0458 to 2631 0483] where relict bedding dips as steeply as 80°. Slump dislocations containing homogenised tuff (Figure 60, inset b), and numerous minor soft-state faults whose downthrow sense mimics that of the main fault, all dip steeply within the hanging wall of the main fault. Whorneyside Formation ignimbrite in the footwall is generally undisturbed, but locally [NY 2562 0462] fiamme have been rheomorphically deflected downwards parallel to the fault plane (Figure 60, inset a). Elsewhere the ignimbrite is brecciated and angular blocks up to 40 cm across occur in a matrix of homogenised phreatomagmatic tuff. This breccia passes laterally into slumped but coherently bedded tuff (Figure 60, inset c).

In a number of places thickening of the volcanic sequence has been caused by stacking of slump sheets (Figure 59e, f). At Stonesty Pike [NY 2487 0402] for

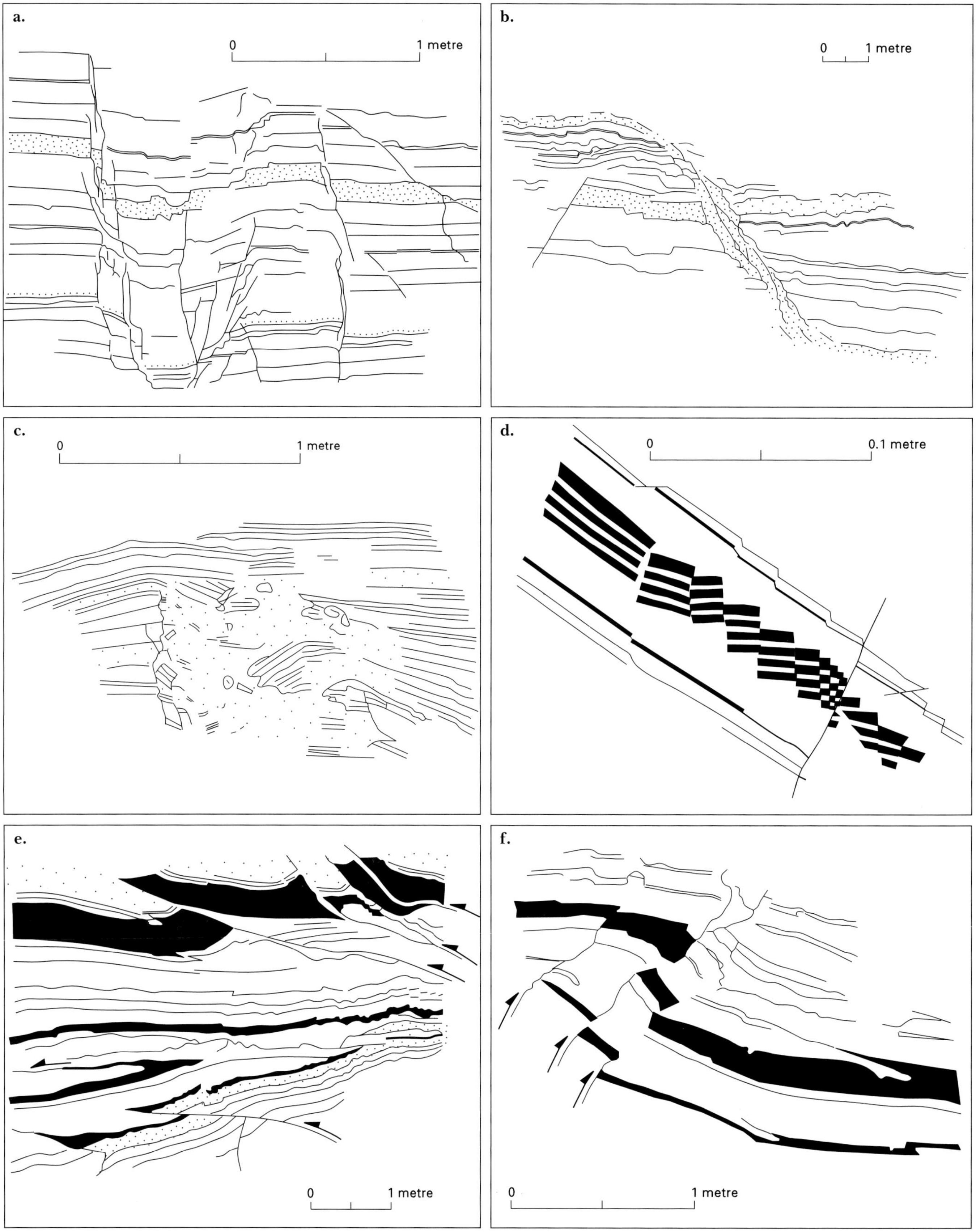

Figure 59 Field sketches of compressional and extensional deformation structures in the phreatomagmatic tuff of the Whorneyside Formation thought to reflect volcanotectonically induced gravity sliding and spreading prior to complete lithification (after Branney and Kokelaar, 1994) (a–f see text for description).

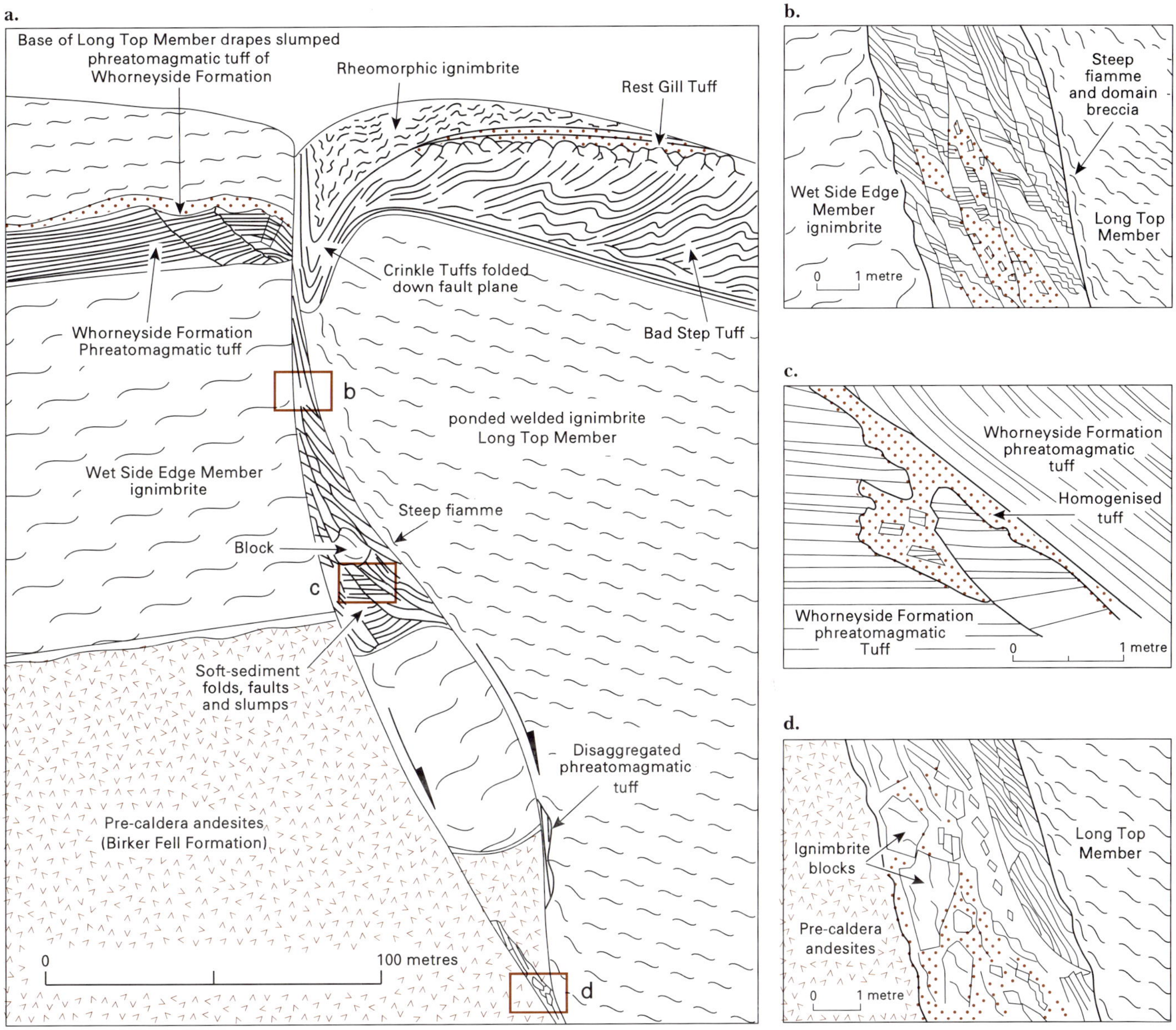

Figure 60 Details of structures within the volcanotectonic fault at Isaac Gill (after Branney and Kokelaar, 1994).

a. Fault viewed facing south-west.

b. Disrupted phreatomagmatic tuff of the Whorneyside Formation with soft-state faulting, shearing and partial homogenisation.

c. Homogenised tuff along one of the many slide surfaces in the phreatomagmatic tuff also occupies pull-apart structures and locally contains derived intraclasts.

d. Brecciated ignimbrite (Wet Side Edge Member) mixed with slumped and homogenised phreatomagmatic tuff.

example, a prominent accretionary lapilli-bearing layer is repeated four times by low-angle thrust faults, and one thrust in a stack of ten superimposed thrust sheets shows more than 17 m of shortening. The low-angle thrusts are accompanied by homogenisation and shearing out of sand-grade layers, and abundant steep microfaults in fine-grained tuff layers. This assemblage of compressional structures probably formed in the toe regions of down-slope gravity slides.

BASAL UNCONFORMITY OF THE WINDERMERE SUPERGROUP

The unconformity at the base of the Windermere Supergroup is most pronounced south-west of Torver High Common [SD 260 950], where Ashgill marine strata rest on a variety of older Borrowdale Volcanic Group formations within the Ulpha Syncline. The discordant relationships at the unconformity record a marine transgression

across the eroded volcanotectonic basin prior to Acadian folding. Farther north-east within the district, beyond Torver [SD 285 945], the strike directions of the Borrowdale Volcanic Group and the Windermere Supergroup are similar, but with the amount of dip at least 10° less in the Windermere Supergroup immediately above the junction (Plate 25). The unconformity has been interpreted previously as the result of a post-Borrowdale Volcanic Group orogenic event, predating the onset of Ashgill sedimentation above the unconformity (Aveline, 1872; Marr, 1916; Green, 1920; Mitchell, 1956a; Soper and Moseley, 1978; Moseley and Millward, 1982; Webb et al., 1987b). However, Branney and Soper (1988) considered that the pre-Ashgill Ordovician structure in the volcanic rocks was dominated by block faults with the folds resulting from volcanotectonic processes. They concurred with Ingham et al. (1978) that the unconformity represented a marine transgression across a subsiding, subaerial volcanic field. Firman and Lee (1986) suggested that erosion prior to that subsidence may have resulted from the emplacement of an Ordovician component of the Lake District batholith which uplifted and tilted the volcanic rocks eastwards. This seems likely, given the 450 Ma age (Hughes et al., 1996) and the laccolithic form to the Eskdale intrusion (Evans et al., 1993; 1994). Consequently, westernmost parts of the volcanic field were eroded to a deeper level before the late Ordovician transgression. Soper (1987) noted that the transgression appeared to be independent of any known eustatic changes in sea level and suggested that it could reflect subsidence brought about by cooling and contraction of the batholith and its volcanic superstructure.

ACADIAN DEFORMATION

The Acadian orogeny in the southern Lake District was an Early Devonian event (Emsian; Soper et al., 1987). True tectonic structures were superimposed on the volcanotectonic framework of the Borrowdale Volcanic Group, while inversion of the Windermere Supergroup basin generated folds and thrusts. A single regional cleavage is common to both lithostratigraphical units.

Regional cleavage pattern

The widespread Acadian cleavage affecting both the Borrowdale Volcanic Group and the Windermere Supergroup varies from spaced to slaty and penetrative, and is locally refracted through different lithologies; it is best developed in volcaniclastic sandstones and phyllosilicate-rich rocks (including altered andesite) (Plate 26). Most of the Central Fells area is underlain by a granitoid batholith above which strain was generally low with low axial ratios in sensitive strain indicators such as accretionary lapilli. Elsewhere, regions of relatively high cleavage strain occur, particularly above the margin of the buried batholith and in the belt of steeply dipping strata in the Coniston Fells. In such areas all lithologies possess a steeply inclined cleavage which usually trends about 060 to 070°. The cleavage strike is relatively consistent across the district (Figure 55), but the dip azimuth shows systematic spatial variations in association with the more significant thrust structures of the district.

On the north-west limb of the Scafell Syncline, largely to the north of the district, cleavage dips are mainly to the south-east. An abrupt transition to north-west-dipping cleavage occurs across a line which is parallel to the axis of the syncline but which lies to the north of the axial region (Figure 61). Farther south (Figure 55), beyond the Eskdale Fault, there is a reversion to mainly south-east-dipping cleavage between that structure and the north-west-directed Greenburn Thrust. Cleavage in the steep limbs of the regional monocline is generally vertical or dipping steeply to the north-west; zones of south-east-dipping cleavage developed adjacent to the north-west-directed Stockdale and Park Gill thrusts. Southwards, through the outcrop of the Windermere Supergroup cleavage dip continues to be steep (rarely less than 70°),

Plate 25 Junction between the Borrowdale Volcanic Group (Lincomb Tarns Formation — LTa — to left) with the Windermere Supergroup at Timley Knott [SD 281 967], near Boo Tarn, Coniston.

Ignimbrite (LTa) in the Borrowdale Volcanic Group is overlain by about 3 m of sandstone (sst) and well-bedded calcareous siltstone, sandstone and limestone of the Kirkley Bank Formation (KkB; Dent Group) (D 4046).

but the direction of dip, either to the north-west or to the south-east, is variable. A spaced cleavage is developed locally in the Eskdale granite (Allen, 1987); it is most apparent around Smithy Mire [SD 165 995] and north of Wha House Farm in Eskdale [NY 200 016], where it trends 050 to 075°.

Structures affecting the Borrowdale Volcanic Group

Extensional and volcanotectonic structures in the Borrowdale Volcanic Group have been overprinted or re-activated by later Caledonian tectonic deformation. The main volcanotectonic basins were tightened with folding of the contained strata and a cleavage locally superimposed. The geometry of the Acadian structures is superficially congruous with that of the original volcanotectonic basins, with two examples of regional importance: the Scafell and Ulpha synclines (Figure 55).

Scafell Syncline

The large-scale structure of the Central Fells is dominated by this synclinal structure of north-east–south-west Caledonoid trend. The north-west-trending Rossett Gill Fault defines two sub-areas (Figure 61). In the first, north-east of the fault, bedding dips are commonly 45° to 55° in the lower part of the sequence, exposed in the syncline limbs, decreasing towards the axial region of the syncline but maintaining a consistent east-north-east strike. The compound axial region and poorly defined south-eastern limb are broken up by a number of dominantly north and north-west-trending faults which were possibly volcanotectonic in origin. These parts of the syncline are marked by variable bedding strike and dips of 15 to 30°; there is no continuous discrete fold trace, and though there are minor culminations and depressions, the structure as a whole has no significant plunge. The south-eastern limb is cut by the reverse Langdale Fault. In the second subarea, south-west of the Rossett Gill Fault, the synclinal hinge is exposed. It has a consistent north-easterly plunge decreasing towards the Rossett Gill Fault from about 30° to 16°. The Langdale Fault extends into the hinge zone of the syncline.

Plate 26 Cleavage in volcaniclastic sandstone of the Caw Formation, east side of Dunnerdale, south-east of Stainton Ground Quarries [SD 2237 9136] (D 4145). Hammer is 37 cm long.

Ulpha Syncline

The Ulpha Syncline formed as Acadian sinistral transpression was superimposed on the volcanotectonic–extensional Duddon Basin. It is a poorly defined, eastward-plunging fold within a high strain-zone on the south side of the Lake District batholith. It is the only major structure within the Borrowdale Volcanic Group that is truncated by the basal beds of the Windermere Supergroup. Webb et al. (1987b) recognised that the trend of the syncline was anomalous in being more easterly than that of the later, Acadian folds. However, anticlinal folds in the Borrowdale Volcanic Group, like those affecting the overlying Windermere Supergroup strata, are periclinal structures. This geometry allows, for example, the axial plane trace of the Ulpha Syncline to converge with that of an adjacent, more northerly trending, anticline so that in the Duddon valley, near Ulpha [SD 190 930] the two folds die out leaving no more than a sharp change in the strike of the volcanic rocks (British Geological Survey, 1991). There is, therefore, no good evidence to indicate that the trend of the original basinal axes differed from that of the later folds.

Structures affecting the Windermere Supergroup

Within the Windermere Supergroup strata of the district two contrasting Acadian structural domains may be recognised. In the north-west is a broadly homoclinal region of south-east to south-south-east dips which forms the north-west limb of the Bannisdale Syncline and the steep limb of the monocline (Figures 55, 56); strata from the Ashgill unconformity up to the upper Ludlow, Bannisdale Formation are affected. The homoclinal region contrasts with an area of widespread minor- to intermediate-scale folding situated farther south-east and bounded to the north-west by the axial region of the Bannisdale Syncline. In the folded area, which within the district is virtually confined to the outcrop of the Bannisdale Formation, the enveloping surface is subhorizontal. The overall Acadian structure is thus a major south-east-facing

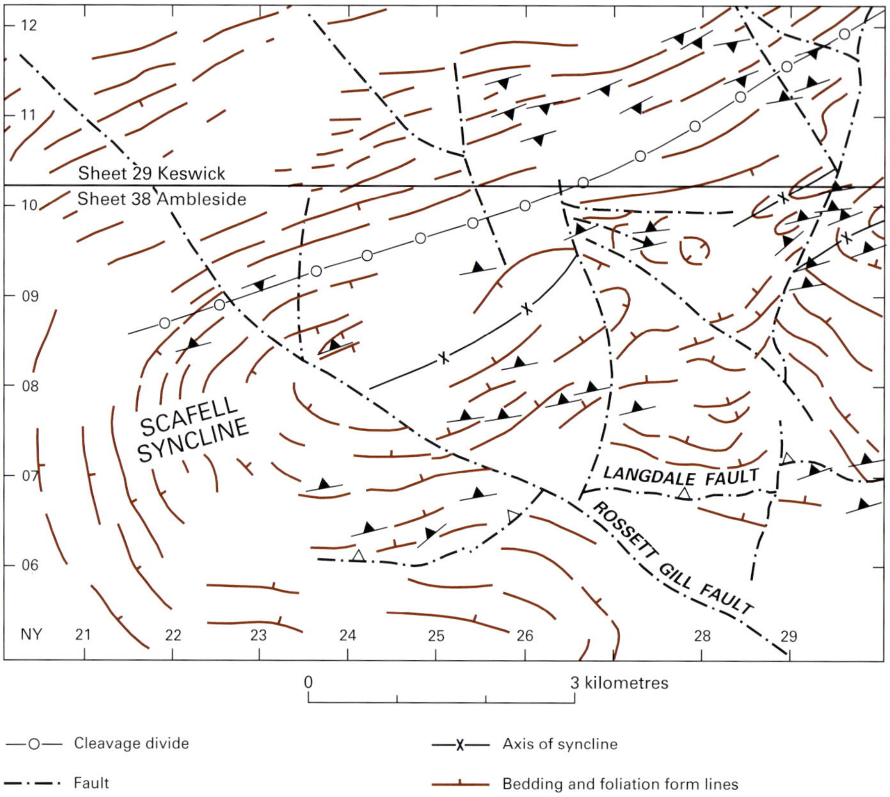

Figure 61 Structural summary for the Scafell Syncline.

monocline with an amplitude in excess of 8 km. Two major detachments, the Stockdale and Park Gill thrusts, occur within the homoclinal region, the structure of which can be described by reference to two subareas. To the north-west, from the Ashgill strata unconformably overlying the Borrowdale Volcanic Group to the Park Gill Thrust, the structural geometry is influenced by the Coniston and Brathay Faults which lose displacement up-sequence into the Coniston Group. To the south-east of the Park Gill Thrust a simpler structural pattern pertains. The main structural elements are identified in Figure 55.

Homocline north-west of the Park Gill Thrust

Most rocks up to coarse silt grade are affected by a near- vertical, slaty cleavage. Where the sandstone is especially matrix-rich, for example in the Kirkley Bank and Birk Riggs formations, or in areas of high strain associated with thrusts, cleavage is present as an anastomosing pressure-solution fabric. Elsewhere the coarser-grained rocks are only locally affected by a spaced fabric locally. The overall intensity of the cleavage decreases southwards, towards the Park Gill Thrust, such that the laminated siltstone within the Coldwell Formation rarely shows any obvious cleavage apart from a strong bedding-parallel fissility.

Bedding in this area is rarely folded and generally dips south-east or south-south-east at 40° to 45° close to the unconformity, steepening to 60° to 65° within the Ludlow sequence. The folds present are mostly controlled by more competent layers in the Birk Riggs and Coldwell formations. They are upright, with inter-limb angles of 80° to 120°, and half-wavelengths up to 75 m. Cleavage is axial planar or shows very slight clockwise transection, and is generally upright; in a few cases it dips moderately to the south-east on overturned north-west-dipping fold limbs, some of which are partially cut out by south-east-dipping minor thrusts.

West of the Coniston Fault bedding strikes approximately 230°, fold hinges and cleavage trend about 070°, clockwise of bedding, and both folds and bedding–cleavage intersection plunge gently to the north-east (Figure 62a). In the area between the Coniston and Brathay faults, bedding shows an approximately 15° northward-convex swing in strike from about 230° in the south-west to about 045° in the north-east. The cleavage trend shows an inflexion in the opposite sense in the same area, swinging from 250° (clockwise of bedding strike) in the south-west to about 240° (anticlockwise of bedding) in the north-east, producing a plunge depression in the area between the Coniston and Brathay faults. This is illustrated by the contrast between fold-axis plunge and bedding–cleavage intersection seen in the stereographic summary plot in Figure 62b. East of the Brathay Fault, bedding strike is somewhat variable, but cleavage and folds trend consistently anticlockwise of bedding, with folds and bedding–cleavage intersection plunging south-west (Figure 62c); the plunge depression between the Coniston and Brathay faults thus appears to be of regional significance.

Homocline south-east of the Park Gill Thrust

Cleavage is only weakly developed in most sandstone beds of the Coniston Group. It is largely confined to the fine-grained tops of graded sandstone beds, and to graded siltstone and mudstone. Development of the near-vertical cleavage in laminated siltstone south-east of the Park Gill Thrust is very localised; commonly only a bedding-parallel fissility is present. Within the Bannisdale Formation outcrop, cleavage intensity varies with lithology; it is near-vertical except within a few tens of metres of north-west-vergent thrusts, where it dips south-east.

Bedding dips are variable to the south-east, between 50° and 90°, averaging 60° to 65°, and striking 230° to 235° across the area. Very little folding is present and the rare minor folds, with half-wavelengths up to 100 m which affect sandstone of the Coniston Group, commonly pass laterally into minor north-west-vergent thrusts. Cleavage trends 055° to 065° and is virtually axial planar to the

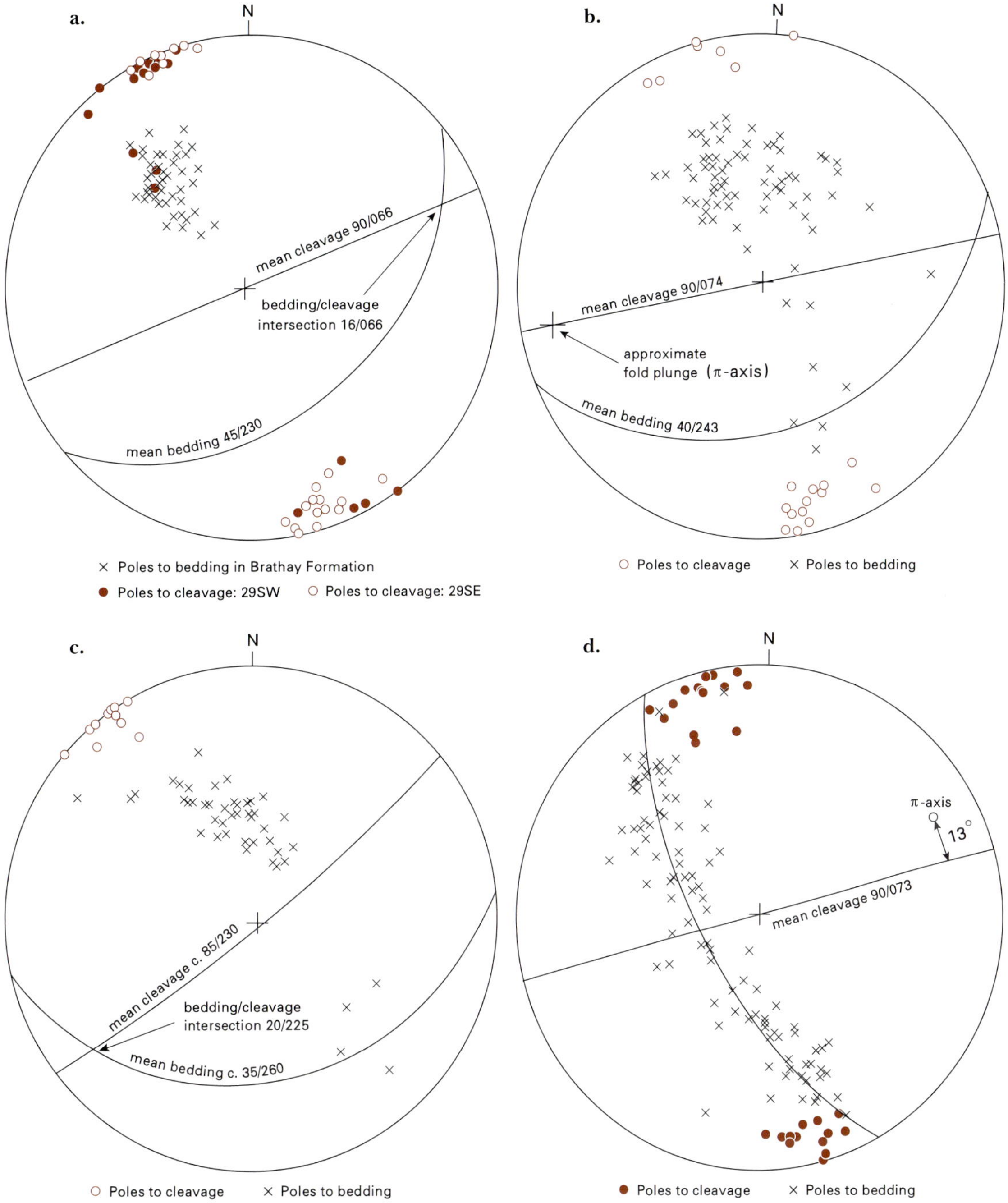

Figure 62 Stereograms illustrating structural variation within the Windermere Supergroup (after Kneller, 1990a, b; Kneller and Soper, 1990).

a. Wenlock strata west of the Coniston Fault.
b. Wenlock strata between the Coniston and Brathay faults.
c. Brathay Formation, east of the Brathay Fault.
d. Bannisdale Formation.

folds; both folds and bedding-cleavage intersections are subhorizontal or plunge very gently.

Folded region

The Bannisdale Syncline is the major, asymmetric synclinorial structure that bounds the monocline on its south-east side. Towards the axis of the Bannisdale Syncline, in the south-eastern part of the district, the Bannisdale Formation is increasingly affected by subhorizontal or gently north-east-plunging upright folds with a north-westerly vergence. These trend about 060° with a near-vertical cleavage approximately axial planar to the folds. Close to the axis of the Bannisdale Syncline, in the area around Sawrey [SD 374 951], the south-east-younging limbs of minor folds are attenuated and locally overturned. Farther south-west, the onset of minor folding is more abrupt and occurs close to the axial trace of the Bannisdale Syncline. This lies within the outcrop of the Bannisdale Formation which is affected by numerous minor folds. In the extreme south-west of the district, the syncline impinges on the upper part of the Yewbank Formation (Coniston Group) and minor thrusts replace most of the folds.

Fold hinges generally have a north-easterly plunge that decreases from about 20° in the south-west to almost zero in the north-east. Cleavage strike shows an abrupt and systematic clockwise swing across the axial zone of the

Bannisdale Syncline; north-west of the syncline it trends about 060° and is virtually axial planar to the folds, whereas to the south-east of the syncline it trends between 078° in the south-west and 085° in the north-east. In the south-eastern zone, moderate clockwise transection of the folds is common with the cleavage 10° to 20° oblique to axial planes.

On its south-east side, the Bannisdale Syncline is separated from the next synclinal region (Crook Synclinorium, Lawrence et al., 1986) by two compound and poorly defined anticlinal structures, the Stricely and Ausin Fell anticlines, which are in places separated by a north-west-vergent thrust (Figure 55). These folds have wavelengths of about a kilometre and are associated with minor congruous folds with half-wavelengths typically of less than a hundred metres. Individual folds are typically periclinal, persisting for only a few hundreds of metres. However, there is a regional north-easterly plunge which varies in magnitude across the area; fold plunges are as little as 10° in the south-eastern part of the district and locally as high as 34° along the trace of the Ausin Fell Anticline, where fold amplitude increases abruptly towards the south-west. The minor folds are transected by the cleavage (usually by between 10° and 20°) in a clockwise sense throughout the area (Figure 62d; Kneller and Soper, 1990).

FAULTING

The complex fault pattern of the Ambleside district has a polyphase evolution involving re-activation of some volcanotectonic faults, Acadian deformation and later, Late Palaeozoic extensional tectonism. Two major faults, the Coniston Fault Zone and the Brathay Fault, form part of a major and long-lived northerly trending fault system (Figure 55), but there are numerous smaller faults, many of which have components of strike-slip displacement, and some of which are linked to thrusts. Reverse faults and thrusts are present in Langdale and Greenburn. The principal fault structures are identified in Figure 55 and some are discussed below.

The presence of strongly discoloured fault breccias with closely spaced jointing and haematite, quartz or carbonate mineralisation is taken to indicate post-Borrowdale Volcanic Group activity. These features characterise, for example the Eskdale, Rest Gill, Rossett Gill, Browney Gill, Red Tarn, Redacre Gill, Stonesty Gill and Kettle Crag faults as well as the upper parts of the Isaac Gill Fault and some fractures within the Side Pike Complex. Their well-developed fault breccias, some up to 100 m across, commonly form prominent linear gullies across the glaciated landscape. In the Central Fells, faults with late displacements (commonly superimposed on volcanotectonic movements) trend predominantly north-west. Sinistral strike-slip movement occurred on the Broad Stand Fault, which displaces the volcanotectonic How Beck Fault to the south-east of Scafell by over 75 m laterally. Dextral strike-slip movement occurred on a fault to the south-west of the summit of Yewbarrow [NY 1744 0818] which displaces the volcanotectonic Yew Crag Fault by 50 m laterally.

The fault structure within the Windermere Supergroup of the district is dominated by north-east-trending thrusts and a north-north-east-trending swarm of normal and wrench faults. The thrusts, principally the Stockdale and Park Gill thrusts, are north-west-directed but the north-north-east-trending structures have a wide variety of observed effects. Both sets are interlinked with merging and thrust-transfer relationships developed locally.

South-east of the Park Gill Thrust, the eastern part of the district is dominated by four north-west-trending normal faults affecting the Coniston Group on Claife Heights, each throwing down between 200 and 300 m to the south-west. Three of these faults terminate against the Colthouse Fault, which forms a major north-north-east-trending lineament parallel to the Brathay–Coniston Fault Zone, but which produces little displacement over much of its length. Farther west, the Bowmanstead Fault, of similar trend, also produces little displacement; it marks the western limit of a zone dominated by north-north-east-trending faults, including a number of interconnected faults throwing down to the west at the head of the Grizedale valley. Some north-west-trending splays off these faults apparently terminate against the Park Gill Thrust. West of the Bowmanstead Fault, faults dominantly trend slightly east of north, mostly with very small displacements. The exception to this is the Torver Fault which offsets boundaries within the Coniston Group up to 300 m to the left. This displacement is linked to a thrust at the base of the Gawthwaite Formation.

In the south-east of the district, most faults within the folded region are nearly vertical, trend between north and north-east and have apparent sinistral displacement. Changes across them in fold pattern and cleavage transection angle indicate an active role during the Acadian ductile deformation. Two north-west-verging thrusts occur, one cutting the axial region of the Bannisdale Syncline, the other cutting the syncline between the Stricely and Ausin Fell anticlines. In places, the thrusts have small hanging-wall anticlines in which, on the north-west limbs, the cleavage has a shallow south-easterly dip. Both thrusts terminate at their north-eastern ends against sinistral strike-slip faults.

Eskdale Fault

The east-north-east-trending Eskdale Fault cuts the Borrowdale Volcanic Group and Eskdale pluton in the western part of the district. It terminates eastwards against the Coniston Fault, but extends to the west of the district. Some major fractures, including the Rossett Gill Fault and Greenburn Thrust, terminate against the Eskdale Fault. In the Wrynose Pass to Little Langdale area [NY 275 027 to 310 034], there is an abrupt change in dip and dip azimuth of bedding across the fault (Figure 55). North of the Eskdale Fault, dips are mostly shallow and approximately to the north, whereas south of the fault dips are steep and to the south-east. This structure was referred to as the Wrynose Anticline by Moseley (1993). There are also changes in fault orientation within the Birker Fell Formation across it, particularly in the Mosedale–Cockley Beck area [NY 244 017]. A very clear linear feature on

satellite images of the Lake District marks the course of the fault zone (British Geological Survey, 1992, fig. 4).

The fault is exposed in the streams on the west and east sides of the summit of Hard Knott Pass and on the west side of Wrynose Pass. In these locations a zone, more than 50 m wide, of intensely fractured rock heavily stained with haematite, contains anastomosing seams of crush breccia, fault breccia, and gouge (the 'Hardknott shatter belt' of Firman, 1957). Where the Eskdale Fault cuts the Eskdale pluton crush zones form anastomosing networks locally, and adjacent north-north-west-trending fractures commonly contain haematite-cemented breccias. Mylonitic fabrics have been generated adjacent to the fault west of Brantrake Crags [SD 143 983] but no significant movement is evident. Within the Birker Fell Formation at Hard Knott Pass [NY 226 014] and Wrynose Pass [NY 275 026] displacement on the fault is about 70 m down to the north. The main movement on the Eskdale Fault postdates that on the Whillan Beck Fault which it apparently offsets dextrally by about 500 m.

Whillan Beck and Greendale faults

The Whillan Beck and Greendale faults trend approximately north–south and in the Wast Water area create a graben structure cutting the northern margin of the Eskdale pluton. The Whillan Beck Fault has a westerly downthrow of 500 m, established from displacements in the Borrowdale Volcanic Group stratigraphy; it is exposed on Rakehead Crag [NY 195 068] as a 30 to 40 m wide gully in intensely fractured, haematised rock. The Greendale Fault has an easterly downthrow; because the fault largely cuts through the uniform Birker Fell Formation sequence the magnitude of downthrow is difficult to assess. The volcanotectonic Dorehead Fault of similar trend is located between these faults. Petterson et al. (1992) considered that westerly downthrow on the Greendale Fault controlled preservation of the Craghouse Member within the Birker Fell Formation. Davis (1989) contended that the Dorehead Fault marked the margin of the Scafell Caldera and south of Wast Water it is continuous with a fault which forms the south-west margin of a downthrown block of Airy's Bridge and Lingmell formation rocks. Though this fault was re-activated and is haematised, within the Airy's Bridge Formation on Illgill Head, an avalanche breccia up to 100 m thick, is banked against it indicating a volcanotectonic initiation. South of their intersection with the Eskdale Fault, the combined effect of these faults is to sinistrally offset the margin of the Eskdale pluton by some 2 km.

Greenburn and adjacent thrusts

A series of north-west-directed thrusts within steeply dipping Borrowdale Volcanic Group strata forms a 2 km wide zone that extends south-west from Little Langdale [NY 300 030] along the Greenburn valley and via Troutal Fell, north of Seathwaite Tarn, into the Duddon valley (Figure 55). Kneller and Bell (1993) have suggested that these structures lie above the south side of the granitic batholith and are part of the system of back-thrusts associated with the monocline (Figure 56). The main displacement occurs on the most laterally persistent fault, the Greenburn Thrust, which follows the course of Greenburn Beck, forms Broad Slack [NY 271 008] between Great Carrs and Swirl How and from Troutal Fell [SD 250 990] to the Grassguards Fault [SD 230 970]. It cuts out the common limb of an anticline–syncline fold pair and so locally has normal fault geometry. Other, less laterally persistent faults occur en echelon. A good example occurs on Birk Fell [NY 295 016], north-east of Wetherlam, where anastomosing, north-westerly directed thrusts repeat slices of mostly Paddy End Member rocks. Contacts between the slices, each only up to 200 m thick, are intensely sheared. All the thrust faults terminate to the north-east against the Eskdale Fault to which their displacement may have been transferred. Farther southwest, the thrust faults probably continue into the Ulpha Syncline as high strain zones in the steeply dipping and north-north-east-striking strata.

Where good stratigraphical markers occur adjacent to the thrust the cumulative displacement across the zone appears to be at least 4 km. However, shortening of up to 70 per cent occurs within the high strain zones (Bell, 1975) and bedding, cleavage and the fault planes within the thrust zone, are broadly co-planar. Consequently, it is not possible to discriminate between displacements that occurred on fault planes and the cumulative effect of shear. Hence the overall displacement may be considerably greater than the 4 km calculated. Assessment of thrust displacement is complicated further by facies variations within the Airy's Bridge Formation across the thrust zone, which suggest that some of the Acadian displacement may have occurred on pre-existing, syn-Borrowdale Volcanic Group extensional or volcanotectonic faults.

Langdale Fault

On the north side of Great Langdale the upper part of the Airy's Bridge Formation and lower part of the Seathwaite Fell Formation are duplicated by a reverse fault which dips approximately northwards at 40 to 50°. This structure can be traced from north of Chapel Stile to Mickleden, and an analogous structure can be demonstrated east of Mickleden as far as Yeastyrigg Gill, a distance of approximately 8 km. The hanging-wall rocks consist of Bad Step Member except at the eastern limit, apparently dipping concordantly with the fault plane. The fault is discordant to bedding in the footwall rocks, and climbs the stratigraphical sequence southwards; footwall rocks adjacent to the fault range from Bad Step Member to the Cockly Pike Member of the Seathwaite Fell Formation.

It is not possible to reconstruct the geometry from balanced sections since stratigraphical thickness variations are largely unconstrained. The Crinkle Member apparently thickens abruptly northwards in the hanging wall, and hanging-wall dips above the Crinkle Member are similar in both the hanging wall and the footwall; different facies of the Seathwaite Fell Formation are juxtaposed across the fault, with breccias much more prominent in the footwall sequence, which is also, in north Langdale,

attenuated. Horizontal displacement must be in excess of 900 m, based on the position of the footwall cut-off of the Bad Step Member. It seems likely that this structure arises from Acadian re-activation of a volcanotectonic fault.

Grassguards Fault

The Grassguards and Stonythwaite faults form an extensional, arcuate fracture system orientated west to east and marking the northern margin of the Duddon Basin between Wormshell How [SD 200 977] and Brown Pike [SD 263 965] (Figure 55). The fault system was apparently only active during accumulation of the Duddon Basin succession. Thickness contrasts across the faults suggest combined displacements across the system in excess of 2 km between deposition of the Airy's Bridge and the Lincomb Tarns formations. There is no evidence for significant re-activation following cessation of volcanism.

Baskill Fault

In the Duddon valley, the Baskill Fault trends north and throws down to the west [SD 190 966 to 198 912]; it originated as a synvolcanic extensional structure. Substantial thickening and facies changes occur within the Duddon Basin succession across the fault, in particular in the Holehouse Gill and Dunnerdale formations, but there is no evidence for subsequent activity during deposition of the younger parts of the Borrowdale Volcanic Group. Opposing dip directions across the fault north of Ulpha result from oblique slip compression during Acadian (Early Devonian) and/or later re-activation.

Park Gill and Stockdale thrusts

The Park Gill Thrust is present throughout the Windermere Supergroup of the district. It ramps gently down-sequence to the south-west, lying close to the base of the Coniston Group in the north-east, following the top of the Coldwell Formation south-west for some distance and finally cutting down to the Skelgill Formation. Though it is a contractional structure, it causes stratigraphical omission within the Coldwell and Wray Castle formations, and must therefore dip less steeply than bedding; the transfer of some displacement to the Brathay Fault (see below) produced an oblique slip direction plunging less steeply than the intersection of bedding on the fault plane, so producing a net sinistral offset.

The outcrop of the Skelgill Formation is everywhere affected by a bedding-parallel detachment, the Stockdale Thrust (Figure 55). It rarely produces any stratigraphical offset, and whereas a contractional geometry can be proved only in the south-west of the district, it is generally implied by analogy with the Park Gill Thrust. The Stockdale Thrust generally has *triangulatus* Biozone and younger rocks in the hanging wall, but commonly produces minor stratigraphical omission so that the youngest rocks in the footwall range from *triangulatus* Biozone rocks of the Llandovery Skelgill Formation to Ashgill strata of the Kirkley Bank Formation. Many minor faults affecting the Ashgill unconformity do not apparently propagate farther up-sequence than the Skelgill Formation, implying that displacements are linked to movement on the Stockdale Thrust. The Stockdale Thrust is itself offset by faults which are linked to the Park Gill Thrust, and thus appears to have been initiated earlier. An overstep thrust sequence is thus implied.

The Long Haws Fault (Figure 55) produces dextral offsets of both the Ashgill unconformity and the Stockdale Thrust by about 300 m. It is linked to a minor thrust within the Birk Riggs Formation and curves westward to merge with the Park Gill Thrust; the net effect is stratigraphical repetition and a transfer relationship is implied. However, the thrust dip is steeper than that of the bedding and the slip direction therefore plunges more steeply than the trace of bedding on the Long Haws Fault.

Coniston, Brathay and associated faults

The Coniston Fault Zone is one of the most substantial fracture systems within the central Lake District, cutting through the entire Borrowdale Volcanic Group from St John's in the Vale, west of Keswick via Thirlmere, Dunmail Raise and Grasmere to Coniston, a distance of some 40 km (Moseley, 1993 and references therein). North of Grasmere, the fault zone has a north-north-west trend curving to the south to a south-south-west trend. At the head of Coniston Water, the fault zone is bounded to the east by the Waterhead Fault and to the west by the Ash Bank Fault (Figure 55), the latter structure then continuing to the south through the Blawith Fells. These two main strands are connected by numerous splays suggestive of sinistral strike-slip movement. However, the system offsets the basal Windermere Supergroup unconformity dextrally by about 1.5 km, the main displacement being transferred northwards from the Waterhead Fault to the Ash Bank Fault along one of the north to south splays. This contradictory evidence illustrates the complexity of likely polyphase fault movement.

The line of principal displacement is always marked by a major topographical feature and, in contrast to other substantial Acadian faults in the area, by zones of brecciation, silicification and filling of open cavities by quartz. These high-level features suggest that the main fault strand was re-activated in post-Acadian time when a major extensional phase affected the region from the Late Devonian, through the Carboniferous and into the Early Permian. North-north-west-trending growth faults, characteristically throwing down to the west, produced half-graben structures such as the Vale of Eden to the east of the Lake District. Within the district, the observed displacement by the Coniston Fault of the unconformity at the base of the Windermere Supergroup is probably the result of northward-increasing, down-to-the-west Permian displacement, re-activating the earlier Acadian sinistral system.

Pre-Ashgill downthrow to the east on the Waterhead Fault is evident from the preservation of the Tarn Hows Formation on the east side. Possible further evidence for displacement on the Coniston Fault during the Caradoc includes substantial thickening of the Pavey Ark Member within the Seathwaite Fell Formation from west to east

across the fault, and control on the location of a thick andesite sill that occurs at the base of the Lincomb Tarns Formation only to the west of the fault. There is no evidence from the district to support Moseley's (1993) suggestion that the Coniston Fault was a volcanotectonic structure in the strict sense, but the continuity of the fault suggests that it may have been originally an extensional structure within the intra-arc rift zone.

The Brathay Fault offsets the Windermere Supergroup's basal unconformity, apparently sinistrally, by 1.7 km, but loses displacement up-sequence so that the base of the Coniston Group is not offset; the intervening formations are thicker to the east of the Brathay/Klondike fault system than to the west. This is partly a stratigraphical effect, because the Kirkley Bank Formation thickens east of the Brathay Fault, whereas the Broughton Moor Formation is absent immediately west of it, indicating that the fault was active (throwing down to the east) during the Ashgill. However, the major control of outcrop width is tectonic and two factors are involved: firstly, tectonic thickening is greater to the east of the fault, controlled principally by folding within the Coldwell Formation and thrusting within the Brathay Formation; and secondly, dips (to the south-south-east) are greater to the west of the fault. These two factors account for most of the apparent offset, and produce a northward-increasing dip-slip component of displacement on the Brathay Fault. The remaining displacement was transferred into the Park Gill Thrust, a fault almost parallel to bedding and dipping moderately to the south-east.

Between the Brathay Fault and the Coniston Fault are two north-north-east-trending faults, the Wharton Tarn and Tarn Hows faults, both with apparent sinistral offsets of some 120 m. The former terminates to the south against a strike fault parallel to, and immediately north of, the Park Gill Thrust; a similar oblique-slip/thrust transfer relationship to that described for the Brathay Fault and Park Gill Thrust is likely. The Tarn Hows Fault bifurcates to the south, offsetting the Park Gill Thrust, before losing displacement within the Coniston Group. The area between the Wharton Tarn and Tarn Hows faults is affected by many north-westerly to north-north-westerly trending faults with small, normal displacements.

East of the Brathay Fault a pattern of north-north-easterly and north-westerly trending faults have apparent sinistral and dextral offsets respectively and appear to form a conjugate set; many of the north-west-trending faults also have a component of dip-slip displacement. Prominent among the north-north-east-trending fault set is the Klondike Fault (Figure 55), a braided sinistral zone within which a microgranite dyke has marginal cleavage. This suggests that fault movement spanned the period of dyke emplacement and cleavage formation because elsewhere cleavage is cut by faulting. The dyke may be associated with other cleaved members of the Shap swarm and hence were intruded at about 390 Ma (Soper and Kneller, 1990).

Stratigraphical boundaries are offset to the right across the north-west-trending part of Lake Windermere which is probably underlain by a major north-west-trending fault zone. Some major faults within the Borrowdale Volcanic Group and minor north-westerly faults with small dextral offsets on both shores of the lake converge on the northern part of the lake.

TEN

Metamorphism

Previous workers, including for example Firman (1957) and Thomas (1986) recognised that rocks of the Borrowdale Volcanic Group had been subjected to both contact metamorphism related to emplacement of the major granitic intrusions, and the effects of very low-grade regional metamorphism. In the Ambleside district an extensive aureole of contact metamorphism surrounds the late Ordovician Eskdale and Ennerdale granitic intrusions (Figure 63). The aureole consists of a biotite and hornblende-rich inner zone, and an outer zone characterised generally by an actinolite–epidote-bearing mineral assemblage but containing some hornblende. In contrast, the Windermere Supergroup strata show only the effects of the regional metamorphism. This metamorphism, to anchizonal and epizonal grades, was associated with the development of an Acadian slaty cleavage, but the pattern

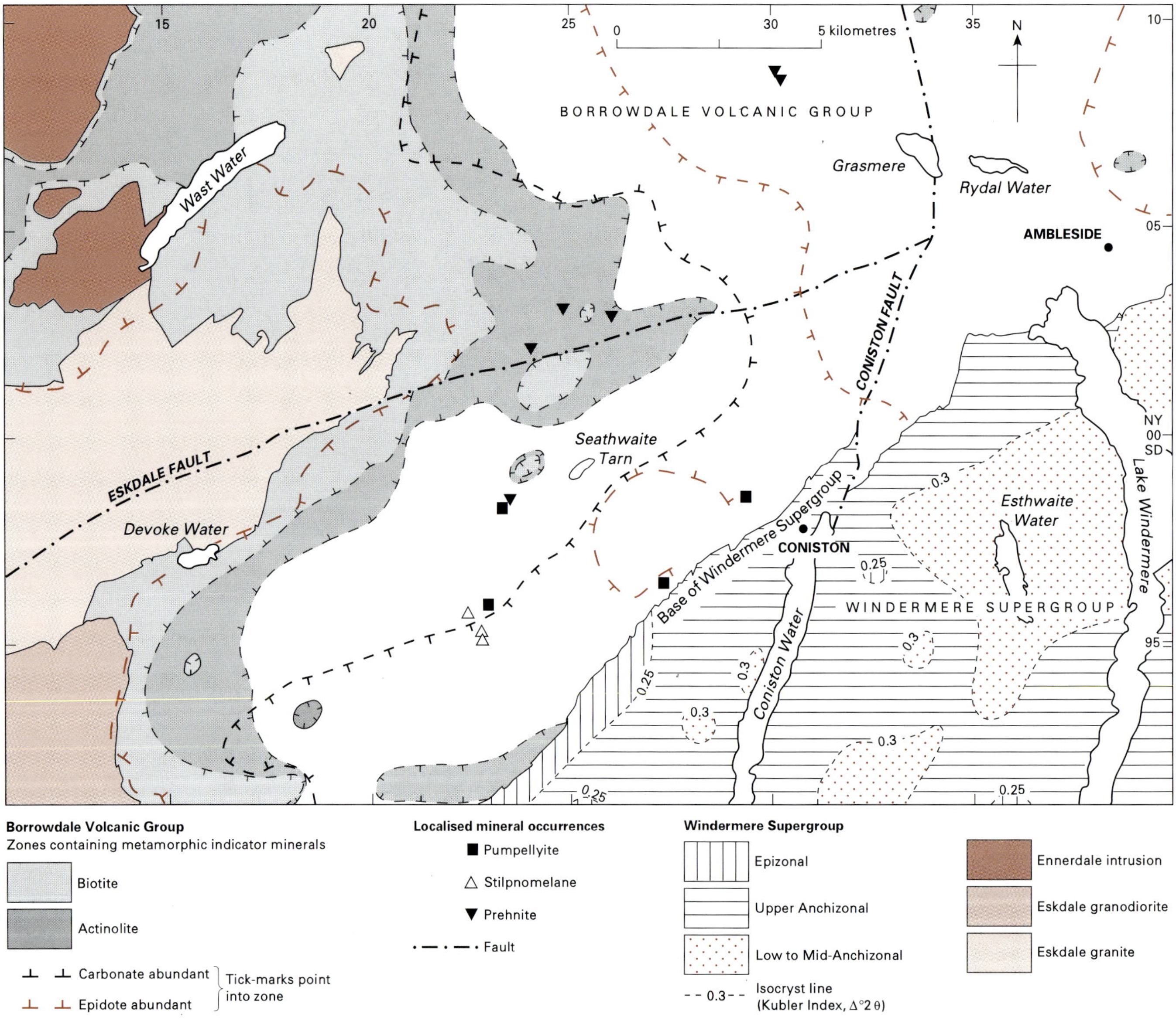

Figure 63 Map of the district showing the pattern of metamorphic mineral distribution and white mica crystallinity patterns. Widely spaced dashes indicate uncertainty caused mainly by small number of data points or uneven occurrence of indicator mineral.

may also require the presence of an appreciable cover of latest Silurian and Early Devonian sediments, which were eroded during late orogenic uplift.

BORROWDALE VOLCANIC GROUP

Contact metamorphic mineral assemblages have been described from an extensive composite aureole enclosing the Eskdale granite (Oliver, 1961; Firman, 1957), and the Ennerdale intrusion (Clark, 1963; 1964). In the former, green amphibole extends farther from the granite contact than biotite, and Firman (1957) considered there to be separate inner (biotite) and outer (hornblende) zones. Clark (1963) described a similar zonal pattern within the narrower aureole around the Ennerdale intrusion.

Thomas (1986) confirmed the extensive development of epidote and chlorite outside the contact aureoles, but recorded facies-indicator minerals such as prehnite and pumpellyite in a few samples. Allen et al. (1987) identified prehnite within fracture veinlets in a hornblende–actinolite-bearing andesite of the Birker Fell Formation on Little Stand in the Wrynose area [NY 249 032] (E70839, E70842), and suggested that the rock had undergone low pressure, sub-greenschist metamorphism in the prehnite–actinolite facies of Liou et al. (1985). Thomas et al. (1985) concluded that δD and $\delta^{18}O$ values for epidote and chlorite indicated isotopic exchange with sea water, and that alteration by the D-enriched syn-metamorphic fluid was terminated by loss of the fluid phase at 250 to 350°C.

The pattern of both regional and contact metamorphic mineral distribution in the Ambleside district (Figure 63) is based on examination of about 900 thin sections of Borrowdale Volcanic Group rocks, together with observations made by Clarke (1963; 1964) on metamorphism around the Ennerdale intrusion. Three mineral zones are recognised in which zone three is subdivided (Table 22):

Zone I based on the presence of metamorphic biotite plus hornblende/actinolite
Zone II hornblende and/or actinolite in the absence of biotite
Zone III the absence of both biotite and amphibole
IIIA epidote rich
IIIB carbonate and phyllosilicate rich

In Figure 63 the boundaries that indicate the extent of metamorphic biotite and amphibole probably approximate to isograds, though variations in host-rock composition and the degree of alteration necessarily render them imprecise. Other boundaries have been constructed to delineate the areas in which epidote and carbonate are commonplace. The distribution of epidote impinges extensively upon the biotite and amphibole zones (I and II above).

Igneous textures remain clear in most rocks, with common partial preservation of primary minerals including augite, plagioclase and magnetite–ilmenite phenocrysts. Metamorphic equilibrium was only approached and rarely, if at all, attained even on the scale of the thin section. Thus, for instance, the epidote-group phase formed within plagioclase may be clinozoisite whereas that in the enclosing groundmass is iron-rich epidote, and epidote grains are commonly zoned towards iron-deficient rims. Plagioclase has suffered variable degrees of alteration to one or more of albite, white mica, chlorite and epidote (or clinozoisite). Metamorphic amphibole displays wide textural and compositional variation from coarse-grained hornblende to acicular fine-grained actinolite both across a thin section and within individual pseudomorphs after pyroxene.

Zone I: biotite zone

The biotite zone forms a continuous aureole surrounding the outcrops of the Eskdale and Wasdale Head intrusions in particular. Biotite was also recorded in rocks from close to the two southern outcrops of the Ennerdale intrusion but, probably as a consequence of bulk rock composition, not in the few samples examined from close to the largest exposure of this intrusion. Clark (1963) recorded a biotite zone around the southern outcrops of the granophyre and also a narrow zone surrounding the boundary

Table 22 Metamorphic mineral assemblages in the Borrowdale Volcanic Group.

Zone		Mineral assemblages	
		Basaltic andesite–andesite	Dacite–rhyolite
I	'Biotite'	ab - hbl - bio - qz - chl ± act ± ep ± w.mica ± sph ± opaque	ab - qz - bio - chl ± w.mica ± opaque ± ep ±sph ± carb
II	'Actinolite'	ab - chl - ep - act - hbl - qz - sph ± w.mica ± opaque	ab - qz - ep - chl - w.mica - act - sph
IIIA	'Epidote'	ep - chl - ab - qz-sph ± w.mica ± opaque ± pump	ab - qz - chl - ep - w.mica - sph ± opaque
IIIB	'Carbonate-phyllosilicate facies'	carb - chl ± ab ± w.mica ± sph ± opaque	carb - ab - chl - qz - w.mica ± stilp ± sph

Abbreviations: ab albite; act actinolite; bio biotite; carb carbonate; chl chlorite; ep epidote; hbl hornblende; opaque 'opaque' minerals including secondary Ti oxides; pump pumpellyite; qz quartz; sph sphene (titanite); stilp stilpnomelane; w.mica white mica.

Table 23 Representative electron microprobe analyses of metamorphic minerals in the Borrowdale Volcanic Group.

	Biotite		Hornblende			Actinolite			Epidote		Prehnite
	1	2	3	4	5	6	7	8	9	10	11
Number	E69289	E69442	E69289	E69368	E67763	E69428	E70842	E70865	E69245	E70867	E692345
Grid ref. NY	2171 0145	2309 0474	2171 0145	2287 0276	2085 0298	2167 0409	2490 0320	2770 0280	2400 0214	2770 0280	2400 0214
SiO_2	36.77	38.19	48.42	44.45	46.81	53.18	53.83	51.55	37.49	37.59	44.29
TiO_2	0.77	1.73	0.26	0.23	0.26	0.03	nd	nd	nd	nd	nd
Al_2O_3	16.77	13.91	6.98	7.46	6.63	2.15	2.10	2.99	22.89	23.09	24.99
Fe_2O_3	—	—	—	—	—	—	—	—	14.45	13.39	—
FeO	20.52	18.70	25.56	22.99	21.45	13.60	18.01	15.66	—	—	0.64
MnO	0.44	0.58	0.58	0.76	0.60	0.56	0.73	0.55	0.66	nd	0.17
MgO	11.28	13.31	12.63	8.45	8.40	15.47	11.33	14.05	nd	nd	0.94
CaO	0.13	0.23	12.05	11.64	12.38	12.31	12.72	12.67	21.96	23.86	26.51
Na_2O	nd	1.25	nd	1.22	0.87	nd	0.21	0.49	nd	—	nd
K_2O	9.66	9.02	0.23	0.18	0.68	nd	nd	nd	nd	—	—
Total	96.33	96.92	96.71	97.38	98.08	97.30	98.93	97.56	97.45	97.93	97.54

Analyses by Cambridge Instruments Microscan V at the University of Edinburgh (analyst B Beddoe-Stephens) and BGS (analyst N J Fortey). Total iron is given as FeO except for epidote in which it is given as Fe_2O_3.

Abbreviations: nd not detected.

Samples 1, 2 and 6 are andesites from within the biotite-grade aureole in upper Eskdale. Other samples are from within the actinolite zone: 9 is andesite on the eastern side of Esk Pike; 4 and 7 are from a composite andesite unit at Little Stand [NY 249 032] (Allen et al., 1987); 5, 8 and 11 are from a composite andesite unit near the Three Shires Stone [NY 277 028] (Allen et al., 1987).

of the main outcrop of the intrusion. In a subsequent paper (Clark, 1964), the zone is shown only around part of the main outcrop.

In addition to the principal, continuous biotite zone, isolated outlying occurrences of fine-grained, metamorphic biotite are indicated on Figure 63. Three lie within the actinolite zone (E68880, E70957, and a group of samples near Cockley Beck), and may reflect underlying high-level granitic intrusions or cupolas of the main batholith. The other two (E69659, E68169) are more remote from known granites, and may indicate proximity to concealed minor intrusions. The biotite is in all cases fine grained and pleochroic through shades of tan-brown. No relict primary biotite has been recorded.

The metamorphic mineral assemblages vary in relation to lithology as indicated in Table 22, hornblende being more abundant in basic rocks, biotite in more acid varieties. In thin section, colour of the amphibole varies according to composition, from deep green hornblende with more than 12 per cent Al_2O_3, to very pale green actinolite with less than 3 per cent Al_2O_3 (Table 23). In the basic rocks, green to blue-green hornblende occurs in the groundmass as well as replacing pyroxene phenocrysts and infilling fractures. Mimetic replacement of pyroxene resulted in coarse-grained, commonly single crystal, hornblende pseudomorphs that preserve the original lattice orientation as indicated, for instance, by replication of the simple twinning on {100} of the original pyroxene. Accompanying plagioclase has altered to albite, fine-grained white mica and granules of clinozoisite. In basalt and andesite, biotite occurs in the groundmass and is intergrown with hornblende replacing pyroxene phenocrysts. In dacitic rocks, very fine-grained biotite has replaced primary biotite phenocrysts and other mafic phenocrysts; it occurs throughout the groundmass, and may also accompany white mica and clinozoisite in plagioclase replacement, whereas hornblende is a minor constituent or is absent.

Epidote, chlorite, actinolite, sphene and minor quantities of carbonate occur widely in the biotite zone. In some areas, the distribution of epidote and chlorite extends up to the boundary of the granite itself (Figure 63). Hornblende pseudomorphs after primary pyroxene display a complex pattern of alteration to pale green, actinolitic amphibole accompanied by epidote granules and pockets of chlorite. Minute, epitaxial prisms of actinolite occur within the chlorite and, in places, have overgrown the hornblende. Biotite-rich hornfels contains varying amounts of patchy, pale green chlorite accompanied by opaque granules, which appear to have replaced the mica. The textural relationships are in places equivocal. Where fine-grained actinolite prisms have grown within chlorite in the absence of coarse-grained amphibole, as described by Clark (1963), it is possible that the chlorite is a remnant of early, pre-granite, burial metamorphism.

Zone II: actinolite zone

The distinguishing features of Zone II are pale green actinolitic amphibole in most samples, but the absence of

biotite. Plagioclase phenocrysts display variable, commonly intense alteration to fine-grained white mica accompanied by epidote or clinozoisite. Augite phenocrysts are preserved in some porphyritic andesites, but in most samples both the augite and orthopyroxene have been replaced. Pseudomorphs after pyroxene range from coarse, mimetic green amphibole, generally paler than in the biotite zone even though hornblende is present locally (Table 22), to chlorite–epidote accompanied by minute actinolite prisms, turbid granules of sphene and opaque granules. Intermediate stages are common, in which chlorite, epidote and fine-grained actinolite progressively replace the coarser-grained, mimetic amphibole which becomes restricted to relict shreds.

Carbonate granules occur as a minor constituent, but are commonplace in only a limited part of the actinolite zone (Figure 63). Pumpellyite was seen in a single sample from close to the outer boundary of the actinolite zone. Fractures host veinlets of epidote and actinolite, chlorite, quartz, alkali feldspar and, in rare examples, prehnite. These are accompanied by intense epidotic alteration in the adjacent host rock. Identification of prehnite was confirmed by electron-probe analyses of two samples (E70839, E70842; see Table 23 and Allen et al., 1987).

Zone III A and B: epidote and carbonate–phyllosilicate subzones

Rocks lacking both biotite and green amphibole form most of the outcrop of the Borrowdale Volcanic Group in the district (Figure 63). A cleavage, which is penetrative in non-welded tuff and volcaniclastic sedimentary rocks, is present in most samples. It is possible to recognise within Zone III large areas in which either epidote or carbonate is common and the other is subordinate. The distinction is not sharp, however, as a large part of this zone contains rocks in which both minerals are widespread. The distribution of carbonate lies largely outside the contact aureole and corresponds to the outcrop of the volcaniclastic successions, except in the south-west where the volcaniclastic rocks have been thermally recrystallised adjacent to the Eskdale granodiorite. Conversely, the distribution of epidote is largely that of a belt extending out from the contact aureole, with extensive areas of epidote-poor rocks within the cleaved, carbonate-bearing upper part of the Borrowdale Volcanic Group.

Alteration styles vary considerably. In andesite within the epidote subzone, pyroxene phenocrysts are replaced by chlorite, epidote, turbid sphene and opaque granules; plagioclase by fine-grained white mica, albite, chlorite and epidote; and opaque grains are altered to haematite and brown, nearly opaque and poorly resolved Ti-oxide material ('leucoxene'). Formerly glassy, porphyritic, rhyodacite and tuff display patchy epidote replacement of feldspar crystals and groundmass, particularly in and around fractures, with chlorite and epidote also replacing mafic phenocrysts; white mica is rare in these rocks. The dominant tuff and sedimentary lithologies which occupy much of the carbonate-rich subzone are extensively altered to very fine-grained white mica and chlorite, variously accompanied by epidote and/or carbonate. The epidote there is conspicuously dark brown and turbid. Slender bands and aligned arrays of white mica and chlorite crystallites in the matrix, and orientated chlorite replacement of mafic clasts, define the penetrative cleavage fabric. The cleavage fabric is deflected around the more resistant clasts. Conspicuous chlorite replaces basaltic clasts and forms amygdales in pumice clasts. The cleavage may lie approximately parallel with primary clast alignment, but may also lie at a high angle to the primary fabric depending on the dip of the strata.

Prehnite and pumpellyite are rare in rocks of this zone in the district. Prehnite has been recorded only rarely in fractures and replacing plagioclase (E68209, E68210) (Figure 63). Pumpellyite occurs in andesite sills intruding the Seathwaite Fell Formation and in the Lincomb Tarns Formation in Torver Beck [SD 270 960] and on Long Crag [SD 290 980], close to the base of the Windermere Supergroup in the Coniston area. In addition, Thomas (1986) recorded prehnite from a veinlet cutting andesite in a sample from Dunnerdale; this was accompanied by brown, spongy material which gave an electron-microprobe analysis consistent with pumpellyite. In carbonate-bearing rocks, the formation of these minerals may be retarded by high $X(CO_2)$ in the metamorphic fluid, but the case remains unresolved for the extensive areas of epidotic rocks containing little or no carbonate mineral. Stilpnomelane was recorded in several andesite samples from Dunnerdale (E65446, E68998, E69009, E69010, E69013).

WINDERMERE SUPERGROUP

The first indication of anchizonal metamorphism in the Windermere Supergroup was provided by conodont colouration indices measured by Bergström (1980; and also see Oliver, 1988). Further confirmation was provided by white-mica crystallinity studies (Thomas, 1986) and graptolite reflectance (Oliver, 1988; King, 1992). In this memoir the results of a white-mica crystallinity survey are presented, using techniques described by Fortey (1989) and Roberts et al. (1991). Kubler Indices for the less than 2 μm fractions separated from 65 samples of mudrock have been contoured by hand in Figure 63, on which isocrysts have been constructed at 0.25 $\Delta°2\theta$ to represent the upper limit of the anchi-metamorphic zone, following Roberts et al. (1991), and at 0.30 $\Delta°2\theta$ to divide the anchizone into lower and higher grade parts. Results for 18 fine-grained sandstones and silty sandstones range from 0.24 to 0.42 $\Delta°2\theta$, and are excluded from Figure 63 to avoid possible lithological bias. Many of these differ by no more than 0.02 $\Delta°2\theta$ from the nearest mudrock values.

The pattern of white-mica crystallinity (Figure 63) is restricted by the limited outcrop of Windermere Supergroup rocks in the Ambleside district, though it is constrained by additional, unpublished data from adjoining areas to the east and south. The metamorphic grade increases from lower anchizonal (Kubler index within the range 0.31–0.42 $\Delta°2\theta$) in the east of the district to upper anchizonal (0.30–0.25 $\Delta°2\theta$) and, locally, epizonal (<0.25 $\Delta°2\theta$) towards the west and south. A zone of epizonal grade

extends along the base of the supergroup from Appletreeworth to Torver Common [SD 206 915 to SD 240 920]. Local occurrences of epizonal rocks in the south and south-eastern extremes of the district (Figure 63) extend into the complex metamorphic pattern of the Windermere Supergroup outcrop in the adjacent Ulverston district.

The mineral assemblages observed show only limited variation. X-ray diffraction of the less than 2 μm fractions of samples of lower anchizonal grade indicate $2M_1$ mica, chlorite and lesser quartz. Additionally, most contain albite and calcite, and about half show evidence of rutile and/or Na–K mica. Indications of the 1M mica polytype were recorded in only two cases. Upper anchizonal rocks contain a similar mineral assemblage, with only one sample containing 1M white mica. Minor X-ray diffraction peaks, suggesting accessory quantities of dolomite, pyrite and haematite, were recorded in a small number of samples. Mudrocks in the Windermere Supergroup contain rare chlorite–mica stacks derived by diagenetic alteration of detrital biotite and, possibly, other mafic silicate grains (Milodowski and Zalasiewicz, 1991), in marked contrast to the Skiddaw Group.

METAMORPHIC HISTORY

Figure 63 shows that the biotite and actinolite zones (Zones I and II in Table 22) surround the outcrops of the granitic intrusions. The aureole, as defined by the boundary of the actinolite zone, is 2 to 3 km wide in the south-west of the district, but north-east of the Eskdale granite and around its cupola at Wasdale Head it is more than 4 km wide, with hornblende-bearing rocks in the Wrynose Pass area lying some 7.5 km east of the nearest exposed granite. The pattern broadly accords with the model of the form of the granite batholith beneath the Borrowdale Volcanic Group given by Lee (1986a) in which the roof of the batholith lies at less than 2 km depth beneath a large area north-east of the Eskdale granite (Figure 8). Moreover, as Firman and Lee (1986) intimated, the broad structure of the volcanic sequence is arched around the form of the underlying granite, whose emplacement entailed a degree of updoming of the volcanic rocks. It may be significant that the Eskdale Fault is aligned along the conspicuous eastward extension of the thermal aureole in the Wrynose area (Figure 63), suggesting axial breaching of the updomed ridge above the underlying granite.

Some minor discrepancies between the metamorphic zonation and the gravity model for the batholith are apparent. Lee (1986a) indicated that granite lies at less than 1 km depth beneath an area of several square kilometres in Great Langdale [NY 270 070], but there is no expression of this in the, albeit, thinly distributed metamorphic data from this area. Conversely, isolated recordings of biotite and actinolite-bearing rocks, notably east of Grasmere (E68207) [NY 3445 0764] and east of High Sweden Bridge (E68169) [NY 3874 0670], do not correlate with high points on Lee's model of the batholith surface, and may represent local contact metamorphism around minor intrusions.

Andesites containing metamorphic actinolite occur at the southern margin of the Borrowdale Volcanic Group east of Great Stickle [SD 215 915] (Figure 63). The actinolite occurs as mimetic replacements of pyroxene phenocrysts, and has been chloritised and kinked. These metamorphic rocks are truncated by the unconformity at the base of the overlying Windermere Supergroup. Though the area is not considered to be underlain by granite (Lee, 1986a), the occurrence may be taken to support a pre-Acadian age for the metamorphism.

Outside the thermal aureole, the Borrowdale Volcanic Group rocks possess a secondary mineral assemblage indicative of sub-greenschist facies metamorphism (Zone III in Table 22), comparable in grade with the broadly anchizonal metamorphism recorded in the Windermere Supergroup. Within these rocks, rarity of grade-indicator minerals limits the scope for a precise delineation of the metamorphic pattern, but the sparse records of pumpellyite and prehnite support the contention that these rocks belong within the prehnite–pumpellyite facies (Thomas, 1986). The wide distribution of cleavage and the preferred orientation of many metamorphic minerals within the cleavage planes, particularly in the carbonate-rich subzone IIIB, indicate that much of the Zone III metamorphism took place during the Acadian orogeny. However, previous workers (Firman, 1957; Oliver, 1961; Clark, 1963) have considered that epidotisation intensifies towards the granites, and may be, at least in part, of granite-related, hydrothermal origin.

The period of greatest sedimentary burial was during the Ludlow and earliest Devonian, prior to the Acadian orogeny and approximately coincident with the resetting of the Rb–Sr ages of the Ordovician igneous rocks (Rundle, 1979), which may thus represent isotopic re- equilibration during burial diagenesis. The Borrowdale Volcanic Group and the Windermere Supergroup (Kneller, 1991) successions are each greater than 8 km thick. Though estimation of the aggregate late Silurian thickness must make allowance for subsidence and erosion of the volcanic edifice before its burial beneath the Windermere Supergroup, the deeper parts of the Borrowdale Volcanic Group were clearly buried to greater than 8 km depth prior to the Early Devonian orogenic phase. This in turn implies a temperature difference of at least 200°C through the succession during pre-tectonic burial metamorphism (assuming a 25°C/km gradient; after King, 1992). Thus, temperatures suitable for prehnite–pumpellyite facies burial metamorphism and isotope resetting may have existed in the Borrowdale Volcanic Group during the late Silurian. Also, the lower parts of the volcanic pile may have been buried during their aggradation. At present, it is not possible to differentiate the full metamorphic history.

Metamorphism in the Windermere Supergroup should be interpreted in relation to the development of the sedimentary basin, in which most of the strata accumulated during rapid late Silurian subsidence prior to development of the Lake District monocline and other Acadian folds and faults. The low to mid-anchizonal rocks in the east and south of the district (Figure 63) correspond in part with the core of the Bannisdale Syncline, though

they also extend north-west into the common limb with the monocline. It is arguable that the metamorphic gradients indicated by the white mica crystallinity are incompatible with a pattern of sedimentary burial metamorphism developed before the orogenic period. Thus, in the area west of Lake Windermere Kubler Indices vary from 0.34 to 0.27 $\Delta°2\theta$, suggesting a metamorphic temperature variation from about 225°C to 275°C across a sedimentary thickness of 6 km or more. If this pattern arose solely from sedimentary burial then it would require an improbable geothermal field gradient of less than 10°C/km. Considering the prevalence of tectonic cleavage in the rocks discussed above, it is possible to conclude that the metamorphic pattern developed during the Acadian orogeny, and that the differences in grade may have resulted from variations in strain rather than, or in addition to, variation due to the pre-metamorphic pattern of burial. The epizonal rocks within the monocline coincide approximately with the area of steep dips and repeated back-thrusts in the Appletreeworth area.

It is also arguable that the overall metamorphic grade attained in the Windermere Supergroup took place beneath a considerable overburden of uppermost Silurian to Lower Devonian sediments, which were then eroded during late-tectonic uplift. Estimation of the thickness of this overburden is beset with uncertainty regarding the field gradient and the roles of tectonic thickening and strain, but an effective overburden thickness of at least 8 km is required if the 25°C/km gradient is accepted, in order to attain anchizonal grades of metamorphism particularly in the highest parts of the succession. The implication is that Windermere Supergroup sedimentation continued after deposition of the youngest Přídoli strata that have been preserved (Merriman et al., 1995).

ELEVEN

Metalliferous mineralisation

Within the Lake District Lower Palaeozoic inlier, rocks of the Skiddaw, Eycott Volcanic and Borrowdale Volcanic groups, and the major granitic intrusions, host a large number of mineralised veins containing a wide variety of introduced minerals. Metalliferous veins are almost completely absent from the rocks of the Windermere Supergroup. Ores of copper, lead and zinc predominate, though small but locally significant concentrations of antimony, arsenic, bismuth, cobalt, manganese, molybdenum, nickel and tungsten also occur. Silver almost invariably accompanies lead, and in some deposits small amounts of gold have been reported. Baryte is locally an important component of some deposits. Iron ore veins are also locally common; most of these consist principally of haematite, though significant concentrations of magnetite are found in a handful of localities.

The Ambleside district includes much of the southern portion of the Lake District metalliferous orefield. It has been recognised for many years that a considerable proportion of the numerous mineral veins of the Lake District may be grouped into 'suites' based primarily on their principal metal content. Most abundant of the deposits are those characterised by carrying major lead–zinc, major copper, or major iron (haematite) mineralisation; other veins, distinguished by assemblages of tungsten, antimony, manganese and baryte mineralisation, are also present. In the district most of the known veins are either copper or haematite bearing, though examples of other types are present.

The origin of mineralisation within the Lake District has attracted investigation from an early date. Prominent among authors are Kendall (1884), Postlethwaite (1913), Eastwood (1921), Dewey and Eastwood (1925), Rastall (1942), Dunham (1952), Shepherd et al. (1976), Stanley (1979), and Stanley and Vaughan (1980). Ineson and Mitchell (1974) assembled evidence for the ages of individual deposits. Stanley and Vaughan (1982a, b) reviewed the area's mineralisation and suggested a genetic classification based on overall mineralogy, geological setting and age. Several distinct mineralising episodes were recognised, in some cases within the same vein. Suggested ages range from Early Devonian for the copper and tungsten-bearing veins, through Early Carboniferous for the main lead–zinc mineralisation, to Cretaceous or Cainozoic for the haematite mineralisation of the Lake District and adjoining orefields of west and south Cumbria. Recent work has indicated that copper- and magnetite-bearing veins within the Lake District were emplaced prior to the regional cleavage-forming event. Studies of each of the major mineralisation types suggest evidence of different fluid compositions and temperatures for each episode. Metals may have been derived by hydrothermal leaching of wall rocks of the Skiddaw and Borrowdale Volcanic groups with, in some instances, contributions from the batholith. Cooper et al. (1988) demonstrated that Skiddaw Group rocks in the Buttermere area were depleted of several elements, including zinc during the development of the Crummock Water metasomatic aureole, thus indicating the potential of these rocks as sources of some of the metals within vein deposits. In a study of sulphur isotopes, Lowry et al. (1991) suggested that sulphur in sulphides in the major copper and lead–zinc suites of veins was derived from Skiddaw Group rocks: magmatic sulphur is an important constituent of the mineralisation associated with the Shap and Skiddaw granites. Carboniferous sea water is the likely source of sulphur in much of the baryte mineralisation.

The main characteristics of each of the mineralisation types found in the district are outlined below, together with appropriate detailed descriptions of selected individual occurrences. Current views on the likely origin of each group of deposits are also discussed.

CHALCOPYRITE–PYRITE–ARSENOPYRITE

The most abundant base metal-bearing veins within the district are those in which the dominant metal is copper. These have been grouped together as chalcopyrite–pyrite–arsenopyrite-type veins by Stanley and Vaughan (1982a). The district includes the whole of the Coniston and Tilberthwaite mining fields (Figures 64, 65), formerly one of Britain's main copper-producing centres. Though the greatest concentration of such veins is around Coniston, smaller clusters of copper-bearing veins, almost certainly part of the same genetic suite of deposits, occur around Ulpha and in the upper part of the Duddon valley. Virtually all of these occurrences are within volcaniclastic rocks of the Borrowdale Volcanic Group.

Typically, these veins carry abundant quartz as the main gangue, but in some, brecciated country rock is the principal non-metalliferous component. Chlorite is common and dolomite locally important. Chalcopyrite is the characteristic-copper sulphide, though small amounts of tennantite, digenite, djurleite and bornite are found locally. Arsenopyrite is usually abundant, together with pyrite and in places some pyrrhotite. Magnetite is present in two veins and this is discussed further below in the section describing magnetite–haematite mineralisation. Galena and sphalerite are also common in small amounts. The galena in these veins may be distinguished from that in the later lead–zinc-dominated veins by the rarity of bismuth sulphosalt inclusions and apparent absence of antimony-bearing inclusions. The sphalerite invariably has a significant iron content (over 5 mole per cent FeS) compared with that in the later lead–zinc veins (Stanley

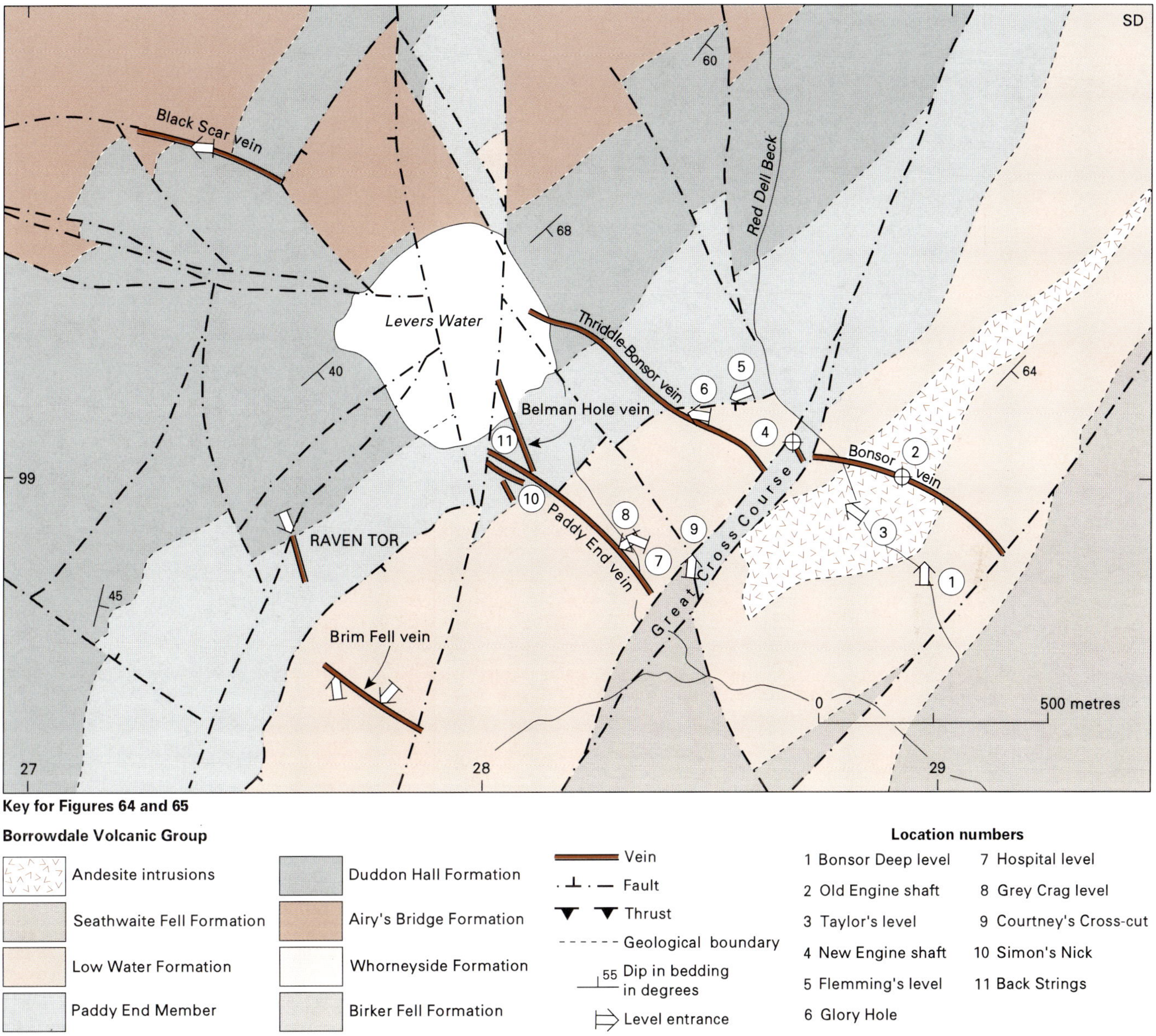

Figure 64 Veins and mines of the Coniston and Levers Water area.

and Vaughan, 1982a). Minor metalliferous constituents found locally include native bismuth, bismuthinite, bismuth sulphoselenides and sulphotellurides, cobalt and nickel minerals, and rare gold.

Work by various authors (e.g. Dagger, 1977; Stanley and Criddle, 1979; Stanley and Vaughan, 1982a, b) on the paragenesis of these veins indicates that many are the result of several mineralising episodes. It has been claimed that the Coniston veins, some of which were worked over a vertical interval of over 500 m, exhibited a vertical zonation of constituent primary minerals with pyrite and chalcopyrite near the surface, arsenopyrite increasing in depth and magnetite instead of arsenopyrite in the deepest workings (Dagger, 1977, p.199). This suggestion must be viewed with some caution, as only the Bonsor and Paddy End veins (Figure 64) have been systematically explored over a significant vertical interval. In neither instance is reliable mineralogical observations known to have been made in situ during mine development. Magnetite is present at depth only in Bonsor Vein, but it occurs at a high level in Long Crag Vein (Figure 65). This magnetite is the earliest mineralisation known in the Coniston veins and may be an expression of a more widespread episode of iron oxide mineralisation pre-dating the main copper mineralisation. It may therefore not be a reliable indicator of increasing depth.

The copper sulphides digenite, djurleite and bornite appear to be restricted to veins now seen at relatively high structural levels. It is possible that the first two named minerals, which include much of what was previously described by Dewey and Eastwood (1925, p.69) as 'black sulphide of copper' may be an expression of a higher

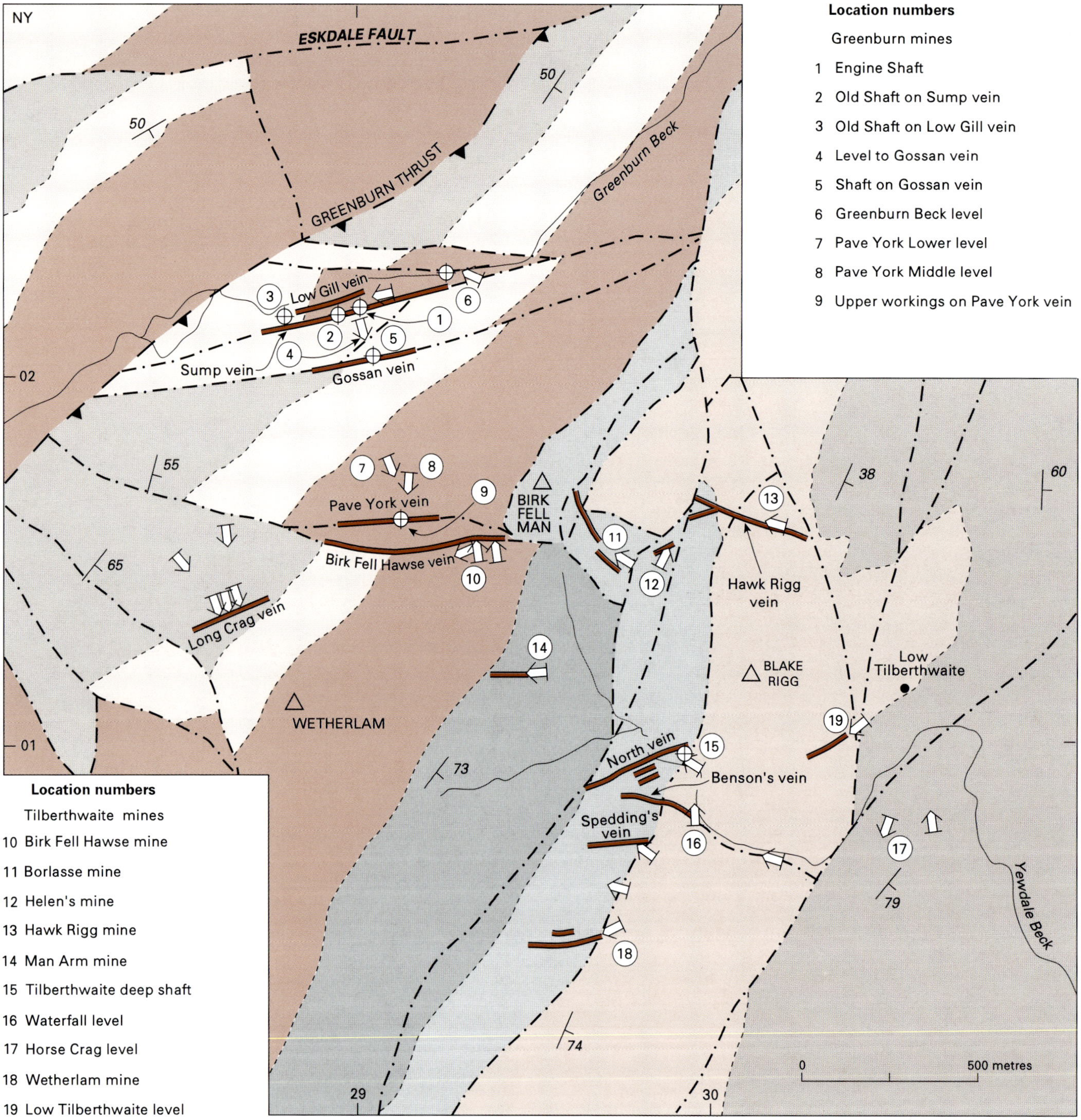

Figure 65 Veins and mines of the Greenburn and Tilberthwaite areas.

zone of sulphide mineralisation. Bornite occurs in a few veins in a situation consistent with it being the product of secondary enrichment.

Supergene assemblages of copper and associated minerals seem to be confined to the topmost few metres of the veins. Surface workings are likely to have long since removed most of such assemblages, but in the few remaining vein outcrops the range of supergene minerals is very restricted, especially compared with the vein deposits of the Caldbeck Fells in the northern Lake District. Removal of the uppermost supergene zones by glaciation may, at least in part, account for this.

It has been observed for many years that most of the copper-bearing veins of the Lake District occupy faults with a north-west–south-east to west–east strike and southerly hade, though there are many exceptions to this (Eastwood, 1921). This orientation is seen in many of the Coniston and Tilberthwaite veins (Figures 64, 65). More

striking is the relationship between vein productivity and wall rock. Veins occur within most lithologies of the upper part of the Borrowdale Volcanic Group. However, in the Coniston and Tilberthwaite areas economic copper mineralisation is in most places restricted to veins within silicic ignimbrite wall rocks (Figures 64, 65) (rhyolites of Mitchell, 1940). Dewey and Eastwood (1925, p.62) noted that the 'slates' worked at Tilberthwaite Gill and south-west of Paddy End effectively limit the area of productive veins to the east of Coniston. In his structural study of the Coniston veins, Dagger (1977, p.197) explained this relationship in terms of the greater tendency of the hard rhyolitic rocks to form open fractures in response to faulting than the associated volcaniclastic sandstone.

In the Bonsor Vein, replacement of original haematite by magnetite, and deposition of early arsenopyrite, which postdates the magnetite, may have occurred at 350 to 400°C (Stanley and Vaughan, 1982b). Quartz, chlorite, stilpnomelane, calcite, dolomite, pyrrhotite, chalcopyrite, sphalerite and late arsenopyrite were probably deposited at about 240°C, and later minerals, including pyrite, native bismuth, bismuthinite, laitakarite, joseite, galena and cosalite are considered to have been formed at temperatures as low as 200°C (Stanley and Vaughan, 1980; 1982a, b). No fluid inclusion data are available for the copper-bearing veins of the district. However, evidence from inclusions in quartz co-existing with chalcopyrite in the Castle Nook Vein in the Vale of Newlands, and likely to be a member of the same genetic suite of veins, indicates that the mineralising fluids were moderately saline brines (~ 5–10 equiv. wt. per cent NaCl, T Shepherd, personal communication in Stanley and Vaughan, 1982a).

Volcanogenic copper mineralisation occurs in close association with felsic volcanic rocks elsewhere in the southern Caledonides at Avoca in the Republic of Ireland, Parys Mountain in Anglesey and Snowdonia, North Wales. The deposits at Avoca and Parys Mountain are volcanogenic massive sulphides, but those in Snowdonia are veins with many similarities to those in the Lake District. The broad geological setting and mineralogy of the Welsh deposits is comparable to the chalcopyrite–pyrite–arsenopyrite veins of the Lake District which are hosted either by rocks of the Borrowdale Volcanic Group or within the Skiddaw Group close to the junction with these rocks. The Borrowdale Volcanic Group has been proposed as a source of the introduced metals (Firman, 1978b, p.239; Stanley and Vaughan, 1982a, p.575). Recently, Lowry et al. (1991) demonstrated from sulphur isotope studies that Skiddaw Group rocks were the most likely source of the sulphide sulphur in the Coniston veins. These authors have attempted, though with some difficulty, to relate the distribution of copper mineralisation to features of the underlying granitic batholith. However, the recent mapping of the Borrowdale Volcanic Group within the district indicates that these copper-bearing veins are almost entirely restricted to the upper part of the Borrowdale Volcanic Group affected by synvolcanic faulting and caldera collapse. Similarities exist between these deposits and those within the broadly contemporaneous Snowdon Volcanic Group of North Wales which Reedman et al. (1985) considered to be the products of hydrothermal convective cells late in the evolution of the caldera. There are thus some grounds for considering a closer genetic link than hitherto considered between the wall-rock geology and mineralisation in the Cumbrian deposits, with vein emplacement late in the volcanic episode, or shortly after its close. Veins formed at this time would predate the main regional cleavage. The discovery of cleavage fabrics in copper veins at several localities throughout the Lake District is entirely consistent with this.

Devonian ages (390–370 Ma; K-Ar; Ineson and Mitchell, 1974) obtained from wall-rock minerals of several Lake District copper veins (including some from the Ambleside district) have been cited as evidence of the age of the main copper mineralisation. The recent recognition of Acadian cleavage fabrics cutting copper veins precludes this. The K-Ar age must therefore derive from an hydrothermal event which postdates the main copper mineralisation.

Polyphase fracture-controlled vein mineralisation which occurs sporadically in Borrowdale Volcanic Group rocks north of Scafell Pike is included here with the chalcopyrite–pyrite–arsenopyrite suite of mineralisation. Individual veins are usually narrow (less than 20 cm), commonly lensoid and impersistent. Locally, mineralised breccias reach 4 m in width. The vein mineralisation is associated with K-feldspar, chlorite, epidote and carbonate alteration. The veins crop out in the bottom of deeply incised fracture-controlled stream courses such as Piers Gill [NY 212 081] north of Scafell Pike, where they are seen to dip steeply to the north-east or north-west and follow the strike of the fracture.

Individual veins and lenses in the Sca Fell area commonly have a simple mineralogy, but several phases of mineralisation may be present within the same fracture zone. The earliest mineralisation recognised comprises one or more quartz–sulphide phases containing very variable proportions of quartz, arsenopyrite, chalcopyrite, galena and sphalerite. Small amounts of carbonate gangue may also occur, and chemical analyses suggest the presence locally of bismuth minerals and gold. At one locality, gold and the silver minerals, freibergite, native silver, electrum, argentite and pyrargyrite have been found associated with pyrite, arsenopyrite, chalcopyrite, sphalerite and galena in a complex quartz–carbonate (ferroan dolomite and calcite) vein structure (Bland, 1988). There is at least one well-developed later phase of carbonate veining which locally contains small amounts of pyrite, arsenopyrite, chalcopyrite, galena and sphalerite. These are postdated by haematite, some of which has a botryoidal habit and is locally accompanied by quartz and carbonate. Barren quartz and carbonate veins are also present in the fractures, some of which postdate haematite.

Brief outlines of the district's major deposits and the most significant minor occurrences follow.

Wasdale and Eskdale

Detailed descriptions of the handful of small workings and trials for copper in the area have been given by Young (1985b). Woodward's (1729) intriguing reference to a fluorite-bearing copper vein in The Screes at Wast Water

is discussed briefly below, though the locality cannot be identified.

In Spothow Gill, near Wha House Farm, short trial levels [NY 2047 0051, NY 2034 0058] explored narrow, east-north-easterly to north-easterly trending fault veins which cut hornfelsed andesite of the Borrowdale Volcanic Group. Mineralisation seen in the veins underground and on the small dumps consists of chalcopyrite with abundant quartz and a little calcite. The uncommon supergene minerals, langite and posnjakite, have been found on the dumps (Young and Johnson, 1985). The date of the trials is unknown and output of ore, if any, would have been very small.

Ulpha area

Several small mines have been worked and trials made on a handful of copper-bearing veins in the valleys of Logan Beck and Holehouse Gill, west of Ulpha. The geology and mineralisation have been described by Cameron et al., (1993), and Adams (1988) has given brief outlines of the history of some of these workings.

Waberthwaite Formation ignimbrite and acid andesite sills exposed in Logan Beck north-east of Plough Fell [SD 1733 9157] are cut by an approximately east–west plexus of quartz–chlorite veins. Similar veinstone containing small amounts of chalcopyrite, specular haematite with traces of magnetite and coatings of green copper supergene minerals occurs on the dump, from a level driven south-west towards this vein. Chalcopyrite-bearing quartz veinstone is also present on the dumps, from a nearby shaft [SD 1729 9160].

On the southern side of Hesk Fell, a quartz–chalcopyrite–arsenopyrite vein occupies a north-east-trending fault which throws rhyodacitic ignimbrite of the Airy's Bridge Formation to the north-west against cleaved andesitic tuff of the Whorneyside Formation to the south-east. Three levels, known collectively as Holehouse Gill or Hesk Fell Mine have been driven on the vein. The dumps from the upper two levels [SD 1758 9427 and SD 1755 9423] contain abundant arsenopyrite, chalcopyrite and, locally, a little tennantite. Pink crusts of erythrite on chloritised wall-rock fragments suggest that one or more primary cobalt minerals may be present, though these have not been identified. Quartz and calcite also occur. The dump from the lowest level [SD 1750 9412] shows only a few thin barren quartz–carbonate veins in country rock. The date of these workings is unknown, but the size of the dumps, especially from the middle level, suggests a moderate sized working. Analyses of vein samples show evidence of As, Bi, Co, Cu, Pb and Zn enrichment (Cameron et al., 1993).

Several levels and an open shaft [SD 1865 9236] in Stonegarth Wood mark the site of Ulpha Copper Mine. In common with other mines in this area no plans have been located. A north-east-trending mineralised fault has been mapped cutting interbedded andesitic tuff and lava of the Duddon Hall Formation. The vein exposed in the sides of the collapsed shaft collar is up to 1.0 m wide and composed mainly of massive white quartz. Veinstone in the adjacent dumps contains much quartz with scattered chalcopyrite, pyrite, arsenopyrite and chlorite. Traces of brown sphalerite and a few specks of galena are also present. Adams (1988, p.125) records that the shaft is about 12 m deep from the foot of which a level has been driven westwards for at least 338 m. The mine is known to have been active at some time between 1860 and 1888. Several levels driven on the south side of Holehouse Gill in Stonegarth Wood appear to be crosscuts to the vein and may connect with the shaft. Similar veinstone to that from the shaft may be seen on some of the dumps.

A small opencast trial [SD 1842 9221] has been made on the vein in porphyritic andesite wall rocks north-east of Long Garth farm. Though the vein is not seen in situ, spoil contains scattered chalcopyrite in quartz in which bladed crystals of bismuthinite are locally conspicuous.

Approximately 600 m east of Hall Dunnerdale, Old Park Vein is a mineralised north-north-east-trending fault which throws down andesite of the Ulpha Formation on the east against tuff of the Duddon Hall Formation. The vein, which is up to 10 m wide, occurs in a zone of rusty weathering in these rocks suggesting the presence of disseminated iron sulphides in the wall rocks. Spoil from four small bell pits or shallow shafts along the course of the vein [SD 2214 9552 to SD 2218 9562] reveals that the vein filling consists of quartz and chlorite with small amounts of chalcopyrite and pyrite. The workings, of unknown age, are small and very little ore can have been obtained.

Copper mineralisation, present within the westernmost 1 km of the Cockley Beck Fault within andesite and breccia of the Birker Fell Formation was worked at the Cockley Beck Copper Mine [NY 2473 0128]. The vein is exposed in breccia wall rocks in Cockley Beck [NY 2502 0129] as a 2 m-wide mineralised zone with numerous veinlets of calcite carrying small amounts of pyrite and chalcopyrite. The Cockley Beck Vein is unusual amongst the district's copper veins in being dominated by a calcite gangue, whereas quartz is scarce. Pyrite and chalcopyrite occur scattered through the calcite. Adams' (1988, p.146) observation of abundant malachite fragments on the floor of one of the ruined mine buildings suggests that this mineral also may have been worked here. Production is likely to have been very small.

Coniston and Levers Water area

The veins and principal mine workings of this area are shown on Figure 64. The area includes two of the Lake District's largest copper mines, Paddy End and Bonsor, as well as numerous smaller workings and trials. Only comparatively brief outlines of the workings are given here. The history of mining at Coniston, which has been described in detail by Holland (1986) is outlined briefly in Chapter 2.

The most westerly group of deposits within the Coniston area are those of the Seathwaite Mines, north-east of Seathwaite Tarn. Mineralisation occurs in at least two veins within a westerly extension of the complex fault belt which hosts the large deposits of the Paddy End and Bonsor mines. The recent survey does not support the views of Dewey and Eastwood (1925, p.69) nor Stanley

and Criddle (1979, p.103) that four east–west veins are present. Dumps from the upper and middle levels [SD 2662 9976 and SD 2653 9941] show quartz veinstone with chlorite, 'grey copper sulphide' and traces of chalcopyrite. Stanley and Criddle (1979) record that the 'grey copper sulphides' at Seathwaite include both digenite and djurleite. They also note the presence in the ore of arsenopyrite, cobaltite, covelline, crystalline haematite, pyrite, and, in veinstone from the upper level dump, small amounts of the rare copper bismuth sulphide, wittichenite (Cu_3BiS_3) and a single grain of gold (see also Criddle and Stanley, 1979). A level driven north [from SD 2611 9933] through andesite of the Birker Fell Formation from the base of the scree slope at the eastern end of Tarn Head Moss appears to have been an unsuccessful trial, though Dewey and Eastwood (1925, pp.69–70) suggest that a vein up to 0.9 m wide was cut: no mineralisation is present on the dump.

The roughly north-west-trending Bonsor Vein is one of the Coniston area's major veins (Figure 64). It may be traced for 0.6 km from the eastern shore of Levers Water [SD 2818 9932] to Kernal Crag [SD 2845 9913]. Contemporary plans and reports commonly refer to this length of the vein as the Thriddle (or Triddle) Vein. East of Kernal Crag a north-north-east-trending fault belt cuts the vein; it is known variously as the Kernal, South, Triddle Foot, or Great Cross-course. This cross-course, which takes the form of a horst between Red Dell Beck and Paddy End Mines, displaces the Bonsor Vein by an aggregate distance of about 80 m sinistrally. The vein may be traced for a farther 0.5 km east of the cross-course.

For much of its course between Levers Water and the Great Cross-course, the vein at outcrop cuts the Paddy End Member. Most exposures show up to 1.5 m of brecciated wall rock with quartz veining. However, in a large cave-like excavation, known as the Glory Hole [SD 2840 9917], the vein is up to 2.7 m wide, and consists of quartz, chlorite, pyrite and chalcopyrite in crude bands parallel to the vein walls. Abundant 'gossany' iron oxides encrust much of this exposure. East of the open stopes [SD 2867 9907] the vein passes into volcaniclastic sandstone within the Low Water Formation and abruptly narrows into a belt of haematite-stained fractures with very little mineralisation. This wall-rock control of mineralisation has long been observed in the Coniston Mines: the veins typically are wide and productive within acid wall rock, but are characteristically poor or barren in andesite, andesitic tuff and volcaniclastic sandstone. East of Red Dell Beck the vein lies within andesite wall rocks and appears to have been generally unproductive.

Bonsor Vein has been worked from the outcrop down to at least 457 m below the surface over a strike length of about 0.5 km. Vein widths vary from about 0.3 m to several metres. From the surface down to Deep Level, the main working level [portal at SD 2898 9882], the vein hades to the south. It is roughly vertical for the succeeding 120 m, below which the hade is to the north. Chalcopyrite is the main copper-bearing ore mineral. It is accompanied by abundant arsenopyrite, some pyrite and marcasite and a little pyrrhotite, galena and sphalerite. Non-metalliferous gangue minerals in Bonsor Vein include quartz, chlorite, stilpnomelane, calcite and dolomite. Magnetite is an important constituent of the vein, especially in the deeper levels where it became progressively more abundant, apparently at the expense of chalcopyrite (Shaw, 1959, p.220). Other metallic minerals, commonly associated with the magnetite, are native bismuth, bismuthinite, joseite, cosalite and laitakarite. The nickel and cobalt minerals niccolite, nickel skutterudite, rammelsbergite and safflorite were found locally in Bonsor Vein or in a branch of it (Russell, 1925; Stanley, 1979; Young, 1987). Production of a small tonnage of these ores is recorded from both the Bonsor and Paddy End workings.

At the extremity of workings from Hospital Level, Bonsor Vein, or a parallel branch from it, was cut beneath Levers Water. The vein was examined at the mine forehead during the present survey and found to be up to 0.65 m wide and filled with quartz, chlorite, wall-rock fragments and only small amounts of chalcopyrite. There the vein lies within the Duddon Hall Formation. The well developed cleavage in the bedded tuff wall rocks passes through the vein. Since the survey a severe collapse of old workings has rendered this section of the mine inaccessible. A similar relationship of cleavage to wall rock is seen in Courtney's Crosscut [SD 2835 9885], where a strong cleavage fabric passes through the Low Water Formation wall rocks and the quartz-chlorite-chalcopyrite vein.

Dewey and Eastwood (1925, p.65) recorded a vein 'several feet in width' carrying galena and sphalerite, in an exploratory drive from the Bonsor Deep Level. The exact position of this vein is unknown though Eastwood (1959, p.170) suggested that these minerals were found in the eastern part of the mine. Firman (1978b, p.235) noted that the copper-bearing veins at Coniston are cut and shifted by barren cross-courses or by lead-bearing veins which strike north–south or north-east–south-west. This lead–zinc mineralisation may be either a part of the main copper mineralisation episode or perhaps part of the lead–zinc mineralisation of Early Carboniferous age which is widespread in the northern part of the Lake District.

Roughly parallel to Bonsor Vein and approximately 300 m to the south-west lies the Paddy End Vein, the second major vein of the Coniston field, and said to have yielded richer ores (Dewey and Eastwood, 1925, p.64). The surface geology of the Paddy End area resembles that of the Bonsor Vein (Figure 64). At its western extremity, south of Levers Water, Paddy End Vein splits into a number of roughly parallel branches which have been worked out to surface within Paddy End Member wall rocks in deep open stopes known as the 'Back Strings' [SD 2805 9900]. About 120 m south-east of Levers Water, a branch vein, known as Belman Hole Vein, diverges northwards, also through the Paddy End Member. The vein is said to have been particularly rich at the junction which was known to the miners as the 'Californian Bunch' (Dewey and Eastwood, 1925, p.64). A few metres east of here the Paddy End Vein has been worked out leaving the prominent gully known as 'Simon's Nick'. Contemporary plans show a number of small veins subparallel to

Paddy End Vein near Levers Water. Few details of the wall-rock geology of the veins within the Paddy End workings have been recorded. All of the outcrop workings near Levers Water are contained within the Paddy End Member. The contact between the Paddy End Member and the overlying Low Water Formation is interpreted as a fault. Though the geometry is not clear, field evidence is consistent with this faulted contact dipping south-east at a shallower angle than the sandstone and dacitic ignimbrite of the Low Water Formation. It is thus likely that much of the payable mineralisation was present within the Paddy End Member though some may have occurred within the Low Water Formation.

Paddy End Vein has a similar change in hade to the Bonsor Vein in depth. From the surface down to the Middle Level of the mine it hades north-north-east; below that it hades to the south. Dewey and Eastwood (1925, p.64) recorded an average vein width of 0.9 to 1.2 m throughout the workings. Between Levers Water and the entrance to Grey Crag Level [SD 2829 9885] the vein has been removed almost entirely by stoping to an average depth of 91 m below the level of Levers Water and good quality ore is said to have been worked to a depth of at least 110 m below this level (Dewey and Eastwood, 1925).

As at Bonsor, the main copper ore mineral was chalcopyrite, though small amounts of tennantite were also present, accompanied by arsenopyrite and pyrite. Traces of native bismuth and bismuthinite have also been found (Stanley and Criddle, 1979). Stanley (1979) recorded traces of gold associated with a bismuth telluride, tentatively identified as wehrlite. Ores of nickel and cobalt were also found locally and traces of the pink cobalt supergene mineral erythrite are relatively common on parts of the Paddy End dumps. Gangue minerals include quartz, chlorite, calcite and dolomite.

South of Little How Crag, west of Levers Water, a prominent quartz vein, up to about 1 m wide, occupies a roughly west-north-west-orientated fault and carries copper mineralisation; this has been worked opencast and by at least one short level, probably at a very early date, at Black Scar Workings [SD 2738 9970]. Sulphides include chalcopyrite, chalcocite, covelline and bornite. Supergene minerals, which are more abundant there than elsewhere in the Coniston field, include malachite, some of which may have been worked as an ore, and chrysocolla. Several trial levels have been driven on Brim Fell Vein on the south-east flank of Brim Fell [for example at SD 2773 9855 and SD 2759 9888]. In the upper reaches of Red Dell Beck, ignimbrite of the Airy's Bridge Formation is cut by small copper-bearing veins which have been explored, possibly in Elizabethan or earlier times at the God's Blessing Mine [NY 2853 0048]. According to Shaw (1970, p.117) the ore from this mine was rich and contained silver and a little gold.

Tilberthwaite area

The largest, and presumably most productive, mine of this area was the Tilberthwaite Mine where at least seven veins were worked. The principal veins are named; from north to south they are North, Shaft, Benson's, Benson's South and Speddings (Figure 65). As with the veins at Bonsor and Paddy End, the available evidence suggests that payable mineralisation was virtually restricted to veins where they cut Paddy End Member wall rocks. In addition to the surface workings and shallow adits already noted, the Tilberthwaite veins were worked at depth from the Deep Shaft on North Vein [NY 2993 0076] and from the Horse Crag Level which was driven westwards from Horse Crag Quarry [NY 3057 0070]. Abundant spoil on the Tilberthwaite dumps suggests that the veins were composed mainly of broken country rock with only relatively small amounts of quartz in thin strings. Chalcopyrite was the main copper mineral accompanied by pyrite and arsenopyrite.

At the foot of Steel Edge, south-west of Tilberthwaite Mine is the adit of Wetherlam Mine [NY 2968 0048] driven on a roughly west–east vein within the Paddy End Member. Nearby Man Arm Mine [NY 2946 0118] is a trial adit driven west along a roughly west–east vein within lithic-rich andesitic to dacitic lapilli-tuff of the Duddon Hall Formation.

The uppermost andesite sheets of the Birker Fell Formation on the north-west flank of Wetherlam are cut by an east-north-east-trending vein known as the Long Crag Vein. Three levels [NY 2862 0133; NY 2865 0135; NY 2867 0137] have tested the vein immediately north of its outcrop, and cross-cut levels [NY 2853 0146 and NY 2865 0154] have also been driven to it from the lower slopes of the hill. Chalcopyrite is present locally in the spoil; on the dumps from the upper workings magnetite, accompanied by traces of native bismuth, and bismuthinite are present. Stanley and Vaughan (1982b) commented on the similarity of the magnetite to that from the deeper levels in the Bonsor Vein.

Birk Fell Hawse, the ridge between the Tilberthwaite and Greenburn valleys, is cut by a roughly west–east set of locally mineralised faults throwing down to the north (Figure 65). The northernmost of these mineralised faults is known as the Pave York Vein which is described with the veins of the Greenburn valley. The southernmost of the faults is mineralised in silicic lapilli-tuff of the Airy's Bridge Formation on the summit of Birk Fell Hawse where it was worked at Birk Fell Hawse Mine from at least three short levels [NY 2929 0153; NY 2934 0154; NY 2938 0155]. Spoil from these show chalcopyrite in a gangue of broken country rock, quartz and chlorite. An opencut along the vein on the summit of the Hawse [NY 2933 0155] exposes portions of the vein up to at least 0.6 m wide in which the most abundant copper mineral is bornite accompanied by a little chalcopyrite, tennantite and arsenopyrite.

Other veins tried, without success, at the head of the Tilberthwaite valley include those at Borlasse Mine [NY 295 016], Hellen's Mine [NY 2988 0152] and at Hawk Rigg near the head of Blake Rigg Gill [NY 301 016] (Figure 65).

Greenburn valley

Greenburn, otherwise known as New or Great Coniston, Mine worked several approximately west–east fault veins which cut various lithologies of the Whorneyside and

Airy's Bridge formations (Figure 65). No good plans of the workings are known and the vein outcrops are more heavily drift covered than most in the Coniston area. Sump Vein was the most productive of the Greenburn veins. The Engine Shaft [NY 2900 0218] was sunk to the 120 fathom (219 m) level entirely in the vein along the faulted contact between andesitic tuff of the Whorneyside Formation on the south and rhyodacitic lapilli-tuff of the Airy's Bridge Formation thrown down to the north. Trials have also been made on the Gossan Vein [NY 2903 0204 and NY 2902 0209], Low Gill Vein [NY 2880 0216] and a subparallel vein on the north bank of Greenburn Beck [NY 2929 0228].

The Pave York Vein occupies the northern boundary of the fault belt which crosses Birk Fell Hawse. Trials have been made on the vein within silicic lapilli-tuff of the Airy's Bridge Formation at three levels on the hill side above Greenburn [NY 2912 0161, NY 2914 0169 and NY 2910 0174]. The vein is said to have been up to 3.7 m wide. Ore from the Pave York workings was transported to the Greenburn site via an inclined surface tramway.

Mineralisation in the Greenburn veins consisted of quartz accompanied by chlorite. Chalcopyrite was the main copper mineral, though Stanley (1979) described small amounts of tennantite with bornite in samples from the Pave York dumps. Arsenopyrite and pyrite are also present and in places in the Sump Vein, dark brown sphalerite was common.

MAGNETITE–HAEMATITE

Veins in which the principal metalliferous mineral is magnetite and/or specular haematite are known from two localities in the district. Also, magnetite is abundant as an early phase in the filling of a few of the major copper-bearing veins in the Coniston area. Specular haematite–quartz mineralisation is present in a few localities within the Eskdale granite.

At the foot of Water Crag, Wasdale [NY 1532 0601] a basic dyke, up to about 2 m wide and striking almost north–south, cuts basaltic andesite of the Birker Fell Formation. For an exposed strike length of several metres the dyke is cut by quartz veins in an anastomosing plexus, up to 1.5 m wide, along the centre of the dyke. Individual quartz veins, up to 0.4 m wide, carry pockets and streaks of dark green scaly chlorite with, in places, crude bands of a few centimetres width of magnetite and platy, specular haematite, some of which appears to be partly replaced by magnetite. A few rusty, in-weathered pockets of medium to coarsely crystalline pyrite occur within some veins. No signs of copper-bearing or other metalliferous minerals have been observed here. Cleavage within the andesite and dyke also cuts the quartz–magnetite veins.

Near Birdhow, Eskdale, basaltic andesite of the Birker Fell Formation within the contact aureole of the Eskdale granite, is cut by a roughly east–west vein emplaced along a fault (Young, 1985b). The vein is exposed above the entrance to a trial level [NY 2069 0131] driven westwards along it from the end of a shallow opencut. It is up to 1 m wide and composed of sheared fragments of hornfels accompanied by quartz and chlorite with masses of hard, compact, crystalline specular haematite intergrown with magnetite. The haematite occurs as bladed anhedral crystals up to about 2 mm across, commonly in radial aggregates which form sinuous bands within the veinstone. The precise proportion of magnetite has not been determined, but it is subordinate to haematite. Massive white quartz overgrows and locally fills vugs in the magnetite and haematite. Small streaks and pockets of chalcopyrite, up to 1 cm across, occur locally within the quartz. Also present on the Birdhow dump are a few blocks of brecciated and reddened Borrowdale Volcanic Group rocks cemented by calcite and dolomite. The relationships of this material to the haematite–magnetite-bearing ore is unknown. No evidence of cleavage cutting the mineralisation has been found at Birdhow.

This vein may be traced westwards to a higher trial level [NY 2051 0131] driven westwards on the vein which there occupies the faulted contact between coarse-grained granite on the north with hornfelsed andesites on the south. The vein is exposed above the level portal as a 2 m-wide rib of massive quartz with included clasts of silicified and greisenised granite. There is much haematite staining and some pockets of pyrite may be seen on the south side of the vein. The small dump shows quartz veinstone with abundant pyrite and haematite. The latter occurs both as micaceous specularite and as red earthy haematite. Textural relationships suggest two generations of haematite. The specularite appears to have been an early component along with the pyrite and much, if not all, of the earthy haematite appears to have formed by alteration of pyrite. No solid, massive haematite or 'kidney ore' has been seen. A farther 200 m to the west, the vein appears as a belt of shattered, greisenised granite in which numerous quartz veinlets locally carry pyrite.

A little magnetite and specular haematite are present in quartz–chalcopyrite veinstone at Logan Beck [SD 1733 9157], north-east of Plough Fell near Ulpha.

Magnetite, commonly accompanied by chlorite, is an important constituent of the Bonsor Vein at Coniston Copper Mines [SD 290 988]. Contemporary records indicate that the proportion of this mineral increased with depth. The inability to separate magnetite from chalcopyrite by the gravity processing then in use was a major contributory factor in the decision to abandon the lowest levels of the mine in 1895 (Shaw, 1970, p.114). At Bonsor, magnetite occurs in two distinct habits. One consists of bladed, and in some instances radiating, grains up to 2 mm across. The other consists of anhedral to subhedral grains. Dagger (1977) commented on the bladed habit, suggesting that it indicated replacement of haematite and Stanley and Vaughan (1982b) also explained the otherwise uncharacteristic lack of typical crystal form of the Bonsor magnetite as pseudomorphous after early haematite.

Magnetite is also abundant in veinstone on the dump from upper workings on Long Crag Vein in the Greenburn valley [NY 2865 0135] (Stanley, 1979). Magnetite-bearing veinstone is not present in the dumps from lower workings on this vein. Stanley and Vaughan (1982b, p.346) observed that the magnetite is accompanied by haematite,

some of which is partly replaced by the latter mineral. The form of some of the bladed magnetite at both Bonsor and Long Crag resembles the bladed form of much of the specular haematite at Water Crag, Wasdale and Birdhow. Limited textural studies suggest that some replacement of specular haematite by magnetite seems likely at these latter sites.

As discussed above, the increasing abundance of magnetite with depth in the Bonsor Vein has led some authors (e.g. Dagger, 1977) to conclude that it is characteristic of the deeper zone of mineralisation. The relative abundance of other metalliferous minerals with depth in the vein suggest a vertical zonation. However, the presence of an extremely similar magnetite-rich assemblage only in the higher levels of the Long Crag Vein, and the presence of very similar vein fillings in Wasdale and Eskdale, lends support to the hypothesis that this magnetite-rich mineralisation may indicate no such simple depth relationship.

Colman and Appleby (1991) described a suite of quartz–magnetite–haematite veins within the Ordovician Snowdon Volcanic Group of North Wales. The iron minerals in part occupy networks of fracture-hosted quartz veins within basaltic tuff and silicic welded tuff. Magnetite commonly occurs as pseudomorphs after early bladed specular haematite. Chlorite and a little pyrite and chalcopyrite locally accompany the iron oxide minerals. The Snowdon quartz–magnetite–haematite veins are a suite of deposits of distinctive mineralogy and structure within the overall setting of the Snowdon caldera mineralisation (Colman and Appleby, 1991). The mineralisation predates the cleavage and is considered to have formed as a result of late-stage hydrothermal activity during caldera evolution (Reedman et al., 1985). Mineralogical, textural and structural similarities between these Welsh and Cumbrian veins is sufficiently close to invite comparison and suggestion of a similar origin.

Dagger (1977, p.199) suggested that introduction of original haematite, now replaced by magnetite, could have followed closely the formation of the fractures now occupied by the Coniston veins and regional metamorphism of the volcanic rocks which accompanied folding. An early date for this mineralisation is thus implied. Stanley and Vaughan (1982b, p.346) confirmed that magnetite is the earliest introduced ore mineral now seen in the Bonsor Vein assemblage. Haematite, accompanied by quartz and chlorite, has been shown by Stanley and Criddle (1979, p.106) to have been the earliest episode of mineralisation in the Seathwaite veins to the west of Coniston, though no subsequent alteration to magnetite has occurred there. Phase relationships of the minerals within the Bonsor Vein imply that magnetite formed at temperatures of 350 to 400°C. The presence of cleavage fabrics within magnetite- and specular haematite-bearing veins at Wasdale and at Honister confirm the early emplacement of this phase of mineralisation. Though the magnetite-bearing portions of the Bonsor and Long Crag veins cannot be seen in situ, examination of veinstone samples reveals textures consistent with a well-developed cleavage fabric. It thus seems likely that these veins do indeed predate the regional cleavage. The K–Ar isotopic age of 388 Ma obtained by Ineson and Mitchell (1974) from Coniston veinstone is likely to reflect a later hydrothermal event.

TUNGSTEN–BISMUTH–MOLYBDENUM

Within the Eskdale granite high-temperature metasomatism, mainly concentrated adjacent to the wall and roof rocks of the intrusion produced a suite of greisens (Young et al., 1988). Many of the greisens carry abundant topaz and in some instances fluorite, indicating that fluorine was at least locally an important component of these late-stage fluids. The large body of topaz–greisen adjacent to the contact of the Eskdale granite with Skiddaw Group rocks at Water Crag, Devoke Water [SD 1525 9723], locally contains irregular pockets, up to 1 cm across, composed of arsenopyrite, bismuthinite, native bismuth and molybdenite. The last mineral is also found there as rare disseminated flakes up to 0.5 mm across. Traces of the supergene minerals scorodite and ferrimolybdite have also been identified, and very small specimens have been obtained which contain minerals showing affinities with zavaritskite (BiOF) and rooseveltite ($BiAsO_4$) (Young, 1985b, p.25). Ansari (1983) noted relatively high, whole-rock concentrations of tin and tungsten (up to 111 ppm and 12 ppm respectively) in the Devoke Water greisens, though no tin or tungsten minerals have yet been found there.

Tungsten minerals have been reported from narrow quartz veins in the Eskdale granodiorite at Buckbarrow Beck, Corney Fell (Young et al., 1986). The principal tungsten-bearing vein at Buckbarrow Beck [SD 1366 9097] which lies a few metres south of the southern margin of the Ambleside district carries conspicuous chalcopyrite together with smaller amounts of scheelite and traces of ferberite. The vein is also noteworthy for the occurrence within it of the rare supergene minerals bismutoferrite ($Bi_2O_3.2Fe_2O_3.4SiO_2.H_2O$), russellite ($Bi_3(V,Fe,Cu,As,U)_1W_1O_9(SO_3, etc.)_x$), cuprotungstite ($Cu_2WO_4(OH_2)$), eulytite ($Bi_4(SiO_4)_3$) and namibite ($CuBi_2VO_6$) (Young et al., 1991; Neall et al., 1993). Geochemical anomalies strongly suggest that further veins of this type may be present elsewhere within the granodiorite (Cameron et al., 1993).

Though tungsten minerals have not been found in the arsenic–bismuth–molybdenite–fluorite mineralisation in the greisen at Water Crag, Devoke Water, the association of tungsten with such an assemblage is well known (for example Pouliot et al., 1978). It is therefore possible, as Young et al. (1986, p.20) suggested, that a genetic link exists between the Water Crag and Buckbarrow Beck mineralisation. The widespread argillic alteration of the granodiorite, present adjacent to the Buckbarrow Beck copper vein and elsewhere may represent metasomatism associated with this mineralising event. Young et al. (1986) commented on similarities between the Buckbarrow Beck tungsten mineralisation and that at Carrock Fell in the northern Lake District, concluding, on balance, that the two occurrences may be coeval and therefore of late Silurian or Early Devonian age. This view was supported by Cameron et al. (1993) who suggested that a concealed

evolved granite underlying the Eskdale granodiorite may be the source of the tungsten mineralisation. However, the possibility of the Eskdale mineralisation being closely linked to the later stages of Eskdale granite emplacement and therefore late Ordovician, cannot be discounted.

Buckley (1984) noted anomalous levels of arsenic and molybdenum in stream sediments derived from the Wasdale Head granite, though no separate minerals of these elements were identified. Whereas hitherto undetected mineralisation similar to that at Water Crag may be present there, it is perhaps more likely that these anomalies are related to mineralisation in the Piers Gill area.

URANIUM

Geochemical investigations of stream sediments collected from the outcrop of the Eskdale granite by McAllister and Firman (1980) showed locally anomalous uranium contents. Compared with an average uranium content of 4.3 ppm for stream sediments throughout the south-western Lake District, 30 samples from Eskdale yielded a mean value of 23.5 ppm with local concentrations of more than 38.3 ppm. McAllister and Firman (1980, p.B179) found that within this area uranium concentrations were generally greater in sediments from north-east-trending streams than from those orientated north-west to south-east. The highest recorded concentration was from Brockshaw Beck [NY 1840 0218] close to its junction with Whillan Beck. If this distribution reflects a mineralised fault pattern within the granite these authors suggested that perhaps two phases of mineralisation were indicated. If so, haematite mineralisation appears to be dominant along north-west to south-east fractures, with possible uranium mineralisation along north-east-trending faults. However, no uranium mineralisation has been observed anywhere within the Eskdale pluton and McAllister and Firman showed that the uranium concentration in quartz veins and the associated granites was less than in unmineralised granite. These authors noted that the Eskdale granite has a mean uranium content appreciably lower than all other major intrusions in northern England, which could either be a primary feature of the granite or reflect leaching at some stage by mineralising fluids. The high uranium contents observed in stream sediments probably result from preferential concentration by scavenging from solution by iron oxides and organic matter. Both McAllister and Firman (1980) and Ansari (1983) failed to locate uranium mineralisation and a satisfactory explanation for the observed uranium values has yet to be found. Subsequent radiometric surveys and rock analyses, the results of which are as yet unpublished, reveal low uranium enrichments of up to 15 ppm in marginal granodiorite. No uranium mineralisation has been identified and scavenging of this element is still considered the most likely explanation for these relatively high values.

Analysis of the fine fraction (less than 150 μm) of boulder clay and stream sediments collected from the outcrop of the Wasdale Head granite showed anomalous levels of uranium, arsenic and molybdenum (Buckley, 1984). Buckley concluded that, whereas the observed levels of these elements may reflect scavenging by manganese oxides and organic matter, this process has probably only enhanced a primary anomaly due to concealed mineralisation currently being leached by circulating groundwater.

FLUORITE

The presence of fluorite as a minor constituent of greisens associated with the Eskdale granite has been discussed in Chapter 8. Fluorite is a very uncommon mineral in Lake District veins and is not known from any copper-dominated vein. Woodward (1729) referred to the presence of fluorite in a copper-bearing vein in The Screes at Wast Water and a specimen from there, consisting of colourless cubes stained with malachite, is in Woodward's collection in the Sedgwick Museum, Cambridge (Sir Arthur Russell manuscript notes in Natural History Museum). No more detailed locality information is preserved and no trace of the occurrence was found during this survey.

LEAD–ZINC

Veins which carry lead and zinc minerals as the major metallic constituents form, like the copper veins, an important group of deposits within the Lake District and were also formerly mined on a considerable scale. Lead–zinc veins are widespread within the northern part of the Lake District, but within the Ambleside district they are represented only by the veins of the Grasmere lead mine and perhaps by a cross-cutting galena–sphalerite vein in the Bonsor Mine, Coniston.

Galena and sphalerite are the main ore minerals, accompanied by minor amounts of chalcopyrite. Silver is an important impurity in much of the Lake District galena with up to 30 ozs of silver per ton of lead recorded from parts of the northern Lake District (Eastwood, 1921). Stanley and Vaughan (1981) have described the widespread occurrence of native antimony and antimony sulphosalts as inclusions within the galena of these deposits. In contrast to the sphalerite of the chalcopyrite–pyrite–arsenopyrite suite, the sphalerite of the lead–zinc veins is typically low in iron (5 mole per cent Fe) (Lowry et al., 1991, p.995). These veins formed at temperatures of 110 to 130°C from highly saline brines (Lowry et al., 1991). Metals may have been derived from the Skiddaw Group or the batholith, perhaps in part involving convective leaching by Carboniferous sea water. Stanley and Vaughan (1982a) noted that these veins occur above the roof region and both the north and south walls of the underlying batholith with no well-defined relationship to known features of the batholith. K–Ar ages suggest that they are of Early Carboniferous age (360 to 330 Ma; Ineson and Mitchell, 1974; Stanley and Vaughan, 1982a). Lead–zinc veins are known locally to cut veins of the chalcopyrite–pyrite–arsenopyrite veins suite.

Similarities between the lead–zinc veins of the Lake District and those of the northern Pennines have been noted by Vaughan and Ixer (1980) and Stanley and

Vaughan (1982a), though in the latter paper the authors highlighted striking similarities in both mineralogy and age with the mineralisation of the Carboniferous limestones of Ireland.

The east-north-east-trending fault that cuts breccia within the Seathwaite Fell Formation near the head of Greenhead Gill, north-east of Grasmere village, is mineralised with quartz, galena, sphalerite, pyrite and minor amounts of chalcopyrite. Small overgrown opencast workings [NY 3495 0870] and an adit [NY 3492 0860] mark the site of Grasmere Mine which is known to have been worked between 1564 and 1573: no more recent working is recorded (Shaw, 1959). The site is thus one of the Lake District's few surviving original Elizabethan mine workings. An adit [NY 3457 0853], about 350 m farther west explored the vein though apparently without success.

HAEMATITE

Veins composed dominantly of haematite are common in the Eskdale granite and at a handful of localities within rocks of the Borrowdale Volcanic Group. These veins are typically filled with haematite or a mixture of haematite and brecciated country rock. Gangue minerals, including dolomite, quartz and calcite, are generally scarce. The haematite is usually the mamillated fibrous crystalline variety known as 'kidney ore', though compact massive and crystalline (specular) varieties are also found. The characteristic form of the haematite and the almost monomineralic nature of the veins serves to link them genetically with the very large replacement bodies of haematite in the Dinantian limestones of west and south Cumbria. The age of iron mineralisation has been disputed for many years. In the most recent studies Shepherd (1973) favoured a Mid- or Late Triassic age, whereas Dunham (1984) argued that the structural and stratigraphical relationship requires a post-Triassic age.

The origin of the Cumbrian haematite deposits has been the subject of much speculation and controversy. Shepherd (1973) proposed, on structural and geochemical grounds, that late-stage alteration of constituents of the Eskdale granite released iron which was locally precipitated as haematite. Remobilisation of this scattered haematite by circulating subsurface brines lead to redeposition of haematite at higher structural levels as the epigenetic deposits seen today. Rose and Dunham (1977) and Dunham (1984) favoured a model for the west and south Cumbrian deposits whereby hypersaline brines, which removed iron from the Sherwood Sandstones or possibly Lower Palaeozoic granites or slates by convective leaching, were driven up-dip by tectonic pressure or by convection driven by a heat source within the East Irish Sea Basin. The mineralising fluids were around 100°C or above (Shepherd, 1973). Where the fluids gained access to the Dinantian limestone west and south of the Ambleside district they produced large-scale metasomatic replacement ore bodies. In the Lower Palaeozoic rocks of the Lake District they produced true fissure veins.

The role of Permo-Triassic rocks is of crucial importance in this mineralising process. In west Cumbria, where these rocks rest directly on Dinantian limestone, ore bodies are present; but where thick shales of Namurian, Westphalian or Late Permian age intervene between the permeable Permo-Triassic rocks and the limestones, no orebodies are found. Over much, if not all, of the Lake District the Permo-Triassic sequence is likely to have begun with a variable thickness of breccias or 'brockrams', which were succeeded by sandstone of the Sherwood Sandstone Group. A permeable route for mineralising fluids is thus likely to have been present over much of the area. The presence of vein deposits of haematite within the Lower Palaeozoic rocks of the Lake District is therefore likely to be related to the ease with which mineralising fluids from the west could gain access via the sub-Permo-Triassic unconformity. The presence of known haematite veins at relatively high topographical levels, their impoverishment downwards which is observed locally, and their increasing frequency towards the western part of the district are consistent with this model.

The distribution of mineralisation within the Eskdale pluton is shown in Figure 66. Detailed descriptions of the deposits are contained in Young (1985b) and only brief comments on the larger or most significant of the deposits are repeated here.

On the north side of Eskdale a north-north-west-trending fault with an easterly hade, known as the Ban Garth Vein, cuts coarse-grained granite on the hillside above Fisherground Farm, east of Eskdale Green. The fault carries haematite mineralisation which was formerly worked in a large open pit on the hill top [NY 1535 0085]. The quarry is today up to 5 m deep and, though considerably overgrown and obscured by collapsed granite debris, parts of the vein may still be seen. In the southern face of the pit up to 4 m of breccia are exposed, consisting of quartz and silicified granite in a matrix of massive purplish grey haematite. From the size of the opencut and the remaining exposures, the vein clearly exceeded 5 m in width at this point.

The north-north-west-trending Blea Tarn Vein may be traced up the hillside between Stanley Ghyll House, near Beckfoot, and Blea Tarn. The vein cuts coarse-grained granite and microgranite outcrops, but no displacement can be demonstrated, though a west-north-west-trending barren quartz vein which branches from the Blea Tarn Vein near the hilltop clearly occupies a fault with a small southerly throw. Numerous levels have been driven into the Blea Tarn Vein [NY 1670 0073, NY 1673 0072, NY 1675 0065, NY 1678 0063, NY 1677 0062, NY 1679 0057 and NY 1679 0052]. The small dumps from these show mainly brecciated and haematite-stained granite with some massive haematite, small 'kidney ore', vuggy quartz and locally some dolomite and manganese oxides.

The prominent hill of Great Barrow [NY 1847 0162] occupies a small graben bounded by strong, northerly trending faults. Between these fractures coarse-grained granite is overlain in turn by porphyritic microgranite and Devoke Water Member lapilli-tuff on the summit of the hill. Both of the major bounding faults and at least one minor fracture within the graben are mineralised with quartz and haematite on the south side of the hill. The western fault, known as the Little Barrow Vein, has

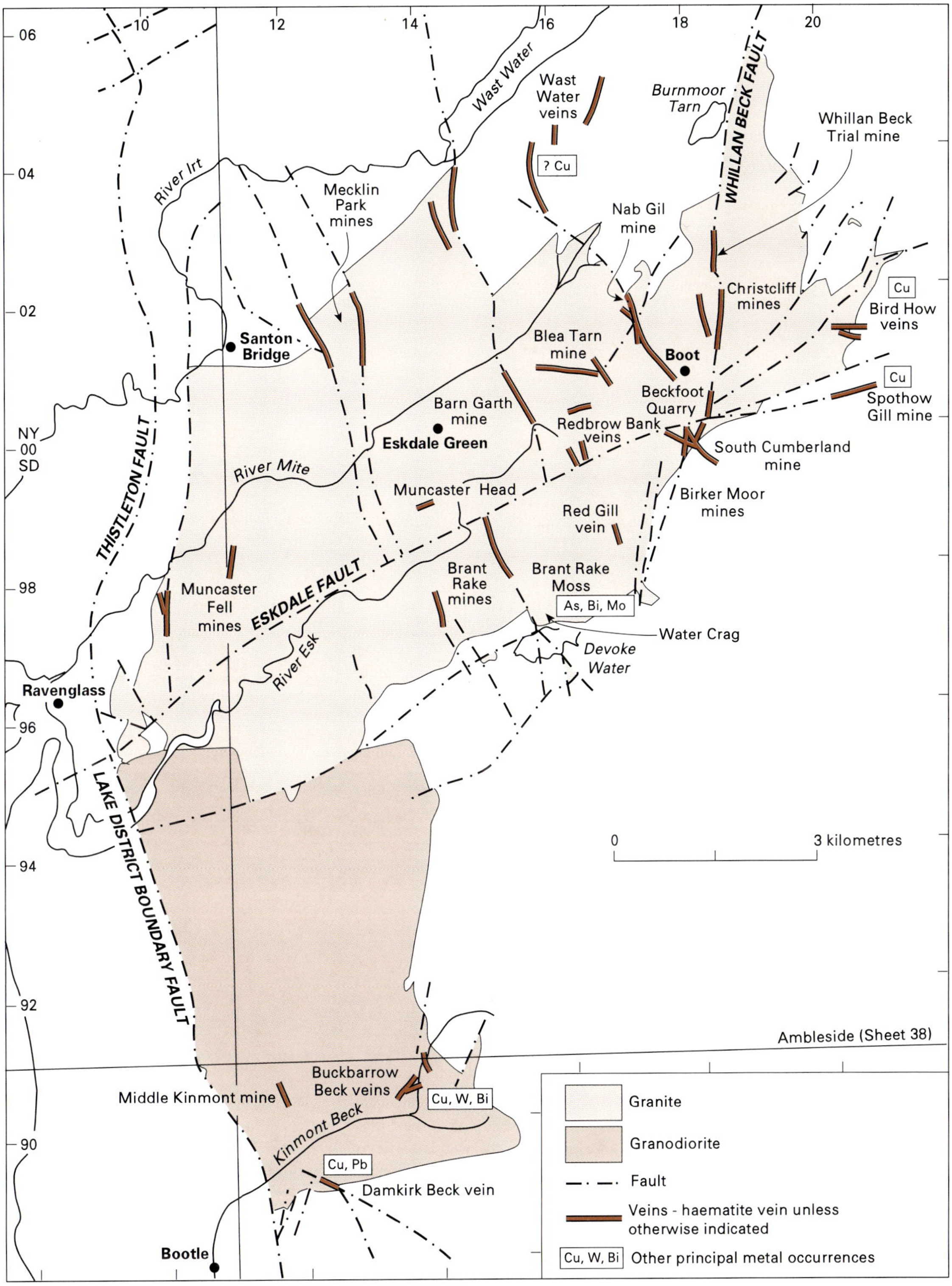

Figure 66 Distribution of mineralisation within the Eskdale intrusion.

been tested by three levels in the wood north-east of Paddock Wray [NY 1834 0134, NY 1835 0130 and NY 1837 0126]. The small dumps are mainly composed of unaltered, though commonly haematite-stained, coarse-grained granite with some quartz, massive haematite, a little small 'kidney ore' and some pale fawn dolomite. Immediately north of the wood [NY 1833 0139] the vein may be seen as a belt of brecciated and silicified granite up to 5 m wide with a network of quartz veinlets penetrating the granite on its eastern side, but without haematite. The major north–south fault which bounds the eastern side of Great Barrow, and known as the Great Barrow Vein, forms

a belt of brecciation and silicification up to 30 m wide. This is particularly well seen in a series of crags north of Christcliff Farm [NY 1853 0131], east of Paddock Wray. Angular clasts of porphyritic microgranite are cemented by white vuggy quartz. No haematite mineralisation is seen at outcrop, though spoil from four levels driven into the vein clearly indicates that haematite is present in places at depth. Haematite-stained granite is the main component of most of the dumps though massive siliceous haematite, some with a brecciated appearance, is present together with blocks of pale fawn, coarsely crystalline dolomite and a little calcite. The mine plans indicate that all of the levels were very short. The remains of an inclined tramway from the middle levels on the Great Barrow Lode to a loading bay at Christcliff Farm may still be seen.

At Nab Gill Mine, Nab Gill Vein and its adjacent branch veins provided the main source of haematite ore in Eskdale. It was the encouraging results of exploration work there in the 1870s that led to the construction of the Ravenglass and Eskdale Railway in order to provide an easy means of transporting ore to the main-line railway at Ravenglass. A brief account of the mine's history has been published (Young, 1984). The main Nab Gill Vein occupies a north-north-west-trending fault which cuts and displaces outcrops of coarse-grained granite and microgranite on the steep fell side north of Boot village. The fault throws down to the east and hades in this direction at up to 25°. The stream known as Nab Gill has been eroded along the outcrop of the vein which is now largely obscured by subsequent mining activities. Near the fell top several branches diverge from the main vein. The main fracture has been mapped across Miterdale where it determines the course of one of the many gullies on The Screes at Wast Water.

The vein filling at Nab Gill Mine consists of brecciated granite wall rock and haematite. Payable ore appears to have been concentrated in two shoots separated by an interval of barren ground (Hibbert et al., 1940, p.82). Brecciated haematite-stained granite, presumably representing this barren section of the vein, can be seen in the old opencut [NY 1731 0142] above the entrance to No. 1 Level. Contemporary reports suggest that the vein attained its maximum width near the fell top where it was up to 6.1 m wide, though this figure included up to 3.4 m of 'horse' or 'rider' of granite. According to Smith (1924, p.211) the average width was between 0.76 and 1.06 m, though in the deepest levels of the mine, below No. 5 Level, the vein thinned to as little as 0.15 m. Apart from brecciated granite, the main vein filling was haematite, much of which consisted of 'kidney ore'. A characteristic feature of the Nab Gill ore is its brecciated texture. The most intensely brecciated ore consists of disorientated fragments of 'kidney ore' or 'pencil ore' in a matrix of crushed, rather earthy haematite. Inclusions of unaltered granite are common. Good examples of this type of ore are common on the dumps, especially those from the lowest, No. 5, level adjacent to the derelict Boot Station [NY 1750 0115]. Similar ore may be seen in situ in the old opencast workings on a subparallel branch vein [NY 1725 0148]. 'Kidney ore', forming bands parallel to the vein walls and with the mamillated surfaces facing the centre of the vein, may be seen in the open stopes [NY 1726 0146] on the same branch vein adjacent to the collapsed portal of No. 2 Level [NY 1738 0133]. Coarsely crystalline, yellowish fawn dolomite, and a little colourless 'nail head' calcite is also present. Manganese oxides, including romanechite, are found locally, especially on a small dump from shaft workings on the fell top [NY 1728 0149].

Opencast workings were made along the outcrop of the main vein in Nab Gill and on a branch vein on the fell top [NY 1725 0148]. Five adit levels were driven on the vein from Nab Gill [No. 1 at NY 1734 0138; No. 2 at NY 1738 0134; No. 3 at NY 1742 0129; No. 4 at NY 1747 0122; and No. 5 at NY 1750 0115]. The bed of the inclined tramway, built to carry ore down the hill side to Boot Station, is still conspicuous.

On the south side of Eskdale at least two veins have been worked for haematite and trials made on other veins at South Cumberland Mine. Haematite mineralisation in the granite dies out as the veins pass upwards into the hornfelsed Birker Fell Formation rocks. The principal vein worked appears to have been the Gill Force Vein. Where exposed in the River Esk [NY 1794 0005], the vein consists of a belt, at least 2 m wide, of shattered and haematite-stained coarse-grained granite with some thin quartz veinlets: the full width of the vein is not exposed. Levels have been driven on Gill Force Vein and the nearby Gate Crag Vein [NY 1791 0004, SD 1792 9995, SD 1792 9993, SD 1793 9993, SD 1817 9984, SD 1819 9982]. The dumps contain much haematite-stained granite with, in places, distinctive stalactitic and specular haematite. Dolomite and quartz are also present. The small dump from the highest, No. 7, level [SD 1822 9980] contains much haematite-stained granite, haematite pseudomorphs after scalenohedral calcite and, in a few specimens, pseudomorphs after a cubic mineral, perhaps either fluorite or pyrite (Young, 1985b, p.19).

A north-north-west-trending fault which locally carries haematite mineralisation may be traced across Birker Moor to the south of Devoke Water where it forms the western boundary of the small granite inlier. An opencast, now much overgrown and collapsed, was opened in the vein near the hill top and small levels have been driven into the vein on the steep slopes above Brantrake Farm [SD 1485 9875].

The major north-west-trending Rossett Gill Fault that cuts rocks of the Esk Pike, Lincomb Tarns and Seathwaite Fell formations between Great End and Allen Crags locally carries haematite mineralisation [NY 2310 0860]. Haematite is present also in the north-north-east-trending fault which crosses Ore Gap [NY 241 072], the col between Esk Pike and Bow Fell, and determines the course of Yeastyrigg Gill to the south. According to Postlethwaite (1913, p.128), two veins of haematite, each about 15 m wide and with a large proportion of 'kidney ore', occur on each side of Yeastyrigg Gill and unite near the gap. Fragments of haematite, including 'kidney ore' and

psilomelane, are present in the weathered exposure of the vein on Ore Gap.

A fault which cuts rocks of the Birker Fell Formation on the east side of Red Tarn [NY 2680 0389] is mineralised with compact haematite and some 'kidney ore', fragments of which occur on the dumps from several old pits along the course of the vein. Ore is known to have been worked in 1700 and smelted in Langdale; and some mining was attempted late in the 19th century (Smith, 1924, p.219). Manuscript notes by W C C Rose in the British Geological Survey archive suggest that some exploration was carried out here in the 1930s.

Haematite occurs in at least three veins in the two Tongue Gills north-east of Grasmere village. The easternmost of these workings, known as Fairfield Mine [NY 3415 0984], worked ore from two levels in the north-west-trending fault which throws breccias and sandstones of the Seathwaite Fell Formation on the south against lapilli-tuff of the Lincomb Tarn Formation on the north. North-west of this, the Providence Mine [NY 3395 1040] explored two veins within the Lincomb Tarns Formation from three short levels on the west side of Little Tongue Gill. Mining is known to have taken place in the early 17th century and between 1873 and 1876, but production was very small.

At Carter Ground on the Dunnerdale Fells, a north-north-west-trending fault, which cuts the Paddy End Member and overlying Caw Formation, carries haematite mineralisation that has been worked on a small scale from a number or surface pits as well as from an adit and shaft [SD 2267 9263]. The spoil heaps consist mainly of country rock with a few fragments of haematite. The mine was active in 1874 (Smith, 1924, p.291), but total production is likely to have been very small. Haematite is present in a parallel fault, also in the Paddy End Member, north-west of Ball Hall where a small trial shaft has been sunk [SD 2260 9213].

Rocks of the Dent Group are cut by a north-north-west-trending fault which carries haematite mineralisation on the west side of Appletree Worth Beck [SD 2460 9268]. The dumps from two levels driven on the vein contain a little haematite, mostly 'kidney ore', together with quartz and some pink baryte. Leviston (1982) noted the presence of chalcopyrite and azurite.

DISSEMINATED MINERALISATION IN THE AUREOLE OF THE ESKDALE PLUTON AND ELSEWHERE

Disseminated sulphide mineralisation has been noted within hornfelsed rocks of the Borrowdale Volcanic Group at several places around the margins of the Eskdale pluton. On the north side of Gate Crag [SD 1840 9980], widely disseminated pyrite, pyrrhotite and chalcopyrite occur in places in the hornfelsed andesites which overlie the granite contact. Near the top of the crags on the east side of Birker Force [SD 1880 9999], several prominent anastomosing veins of dark green hornblende contain a little disseminated pyrrhotite and chalcopyrite together with epidote and chlorite. A manuscript note on the primary geological survey six-inch sheet (Cumberland 79 SE) indicates that fragments of "... magnetic iron ore..." were found there. During this survey several loose blocks of hornblende veinstone were recovered containing masses of fine-grained magnetite up to 2 cm across. However, no magnetite has been found in situ in any of these veins. Similar amphibole veins, though without obvious sulphides or magnetite, have been observed in hornfelsed Borrowdale Volcanic Group rocks close to the granite contact at several other places, for example on the east side of Brat's Hill [NY 1791 0243] north of Boot in Eskdale, Water Crag [SD 1535 9738] and at the eastern end of Devoke Water [SD 1630 9728].

Near the south-eastern margin of the granite, the hornfelsed Borrowdale Volcanic Group rocks locally contain a little disseminated pyrrhotite. This is well seen in a small isolated outcrop [SD 1426 9684] on the east bank of Black Beck, Birkby Fell, where pyrrhotite is prominent as joint coatings and as patches up to 1.5 cm across, perhaps replacing amygdales. Small amounts of disseminated pyrrhotite and pyrite are seen commonly in the Borrowdale Volcanic Group hornfels on the eastern margins of the granodiorite, on Waberthwaite Fell at Red Gill [SD 1408 9336] and Charlesground Gill [SD 1364 9241].

Patchy, widespread silicification of the Borrowdale Volcanic Group rocks of the Wasdale Head and Piers Gill area is accompanied by disseminated sulphides. Pyrite and pyrrhotite are most abundant, but chemical analyses suggest that small amounts of arsenopyrite may be present locally.

TWELVE

Quaternary

The growth of the Antarctic Ice Sheet from about 15 Ma (mid-Miocene) and the Greenland Ice Sheet by three million years ago (late Pliocene) suggests that the climate in northern Britain would have begun to deteriorate long before the beginning of the Quaternary about 1.8 million years ago. During the early Pleistocene there was probably periodic growth of glaciers in the mountains of the Lake District, but there is no evidence that the region was fully glaciated until some 750 000 years ago. The deep ocean sedimentary record indicates that since then growth and decay of large ice sheets occurred periodically in the middle latitudes of Europe (Ruddiman et al., 1980; Bowen, 1991), and ice sheets must have covered the Lake District on several occasions. Cold episodes of ice-sheet growth (glacials) were separated by relatively short, warmer periods (interglacials) when climatic conditions were broadly similar to those of the present day (Table 24). The last five glacial–interglacial cycles have each spanned about 100 000 years. The coldest periods within glacials are known as stadials and the milder interludes as interstadials; both probably became colder as each glacial episode progressed.

A GLACIAL INTERPRETATION FOR THE LAKE DISTRICT

> 'As a huge stone is sometimes seen to lie
> Couched on the bald top of an eminence;
> Wonder to all who do the same espy,
> By what means it could thither come, and whence.'
>
> William Wordsworth
> Resolution and Independence, IX.

The striking glaciated scenery of the Lake District (Plate 27) inspired literary interpretation long before its scientific origin was understood. In the above quotation Wordsworth, working from his home at Grasmere, described perched erratic blocks, stressing the mystery of their origin. As early as 1810 he had commented on the radial (superimposed) drainage pattern in the introduction of his 'Guide to the Lakes'. Sedgwick wrote a section on the geology of the district for later editions of the guide (Wordsworth, 1846) which discussed the then new and revolutionary theory that glaciers were more extensive in the past than they are at present, applied by Agassiz (1841) to northern Britain after its Alpine formulation in the previous year. This theory was also applied to the Lake District by Buckland (1841) who attributed the distribution of Shap granite boulders to the action of ice and described moraines and rounded, polished rock surfaces, including one at Fox Howe near Ambleside.

Tiddeman (1872) introduced the concept that the area had been buried beneath an ice sheet, but the first comprehensive descriptions of the glaciation of the Lake District, including detailed maps showing directions of ice movement and areas covered by glacial deposits, arose from the primary geological survey of the region by Ward (1873; 1874; 1875). Ward debated whether the ice originated from within or outside the district and whether the glaciation was associated with floating ice, a system of local glaciers or an unbroken ice cap. He noted that one of the most striking features of the glaciation was the distribution of Borrowdale Volcanic Group erratics over the outcrops of the Skiddaw Group and Windermere Supergroup strata, concluding that it revealed a general radial movement of the ice away from the central part of the district.

Marr (1916) subsequently used many observations of the primary surveyors in his apposite appraisal of the regional glacial history. He recognised that drainage was superimposed on an elliptical dome with seven major valleys originating in the Scafell area, and an east–west orientated ice divide farther east separating northward- and southward-moving ice. The accumulation and degradation phases of the ice sheet were described as the 'waxing and waning stages' of a glacial cycle. Glacial overflow valleys, oscillations of ice movement and postglacial changes were also described. Marr's conclusion, that the Lake District was covered by a small ice dome which made only a minor contribution to the regional ice sheet, was accepted and modified by later workers (Raistrick, 1925; Trotter, 1929; Hollingworth, 1931; Trotter and Hollingworth, 1932); the model is still considered appropriate (compare with King, 1976; Catt, 1991).

Research into the Quaternary history of the Lake District carried out since the 1930s is listed in bibliographies by Smith (1974, 1990). A developing consensus recognises several episodes within the main Late Devensian glaciation (Huddart, 1972; Evans and Arthurton, 1973; Huddart et al., 1977; Tooley, 1977). The study of lake-bottom sediments in Windermere led to the designation of a site therein at Low Wray Bay [NY 377 013] as the type locality for the Windermere Interstadial (Coope and Pennington, 1977). Geomorphological mapping and reconstruction of Loch Lomond Stadial glaciers has been carried out by Sissons (1980), with important work on glacial deposits in the northern part of the Lake District by Boardman (1991). Other work has been reported in multidisciplinary field guides published by the Quaternary Research Association (Huddart and Tooley, 1972; Boardman, 1981; 1985b; Boardman and Walden, 1994).

Table 24 Quaternary deposits of the Ambleside district and neighbouring areas, and their interpretation in terms of climatic variation, principal events and chronostratigraphy.

PERIOD	EPOCH	STAGE	CHRONOZONE (Stadial-interstadial) Europe		CHRONOZONE British Isles	TIME Radiocarbon years B.P. (1950)	DEEP SEA OXYGEN ISOTOPE STAGES (sidereal years)	CLIMATE	PRINCIPAL EVENTS	DEPOSITS Lacustrine	DEPOSITS Terrestrial
QUATERNARY	HOLOCENE	FLANDRIAN				6 000		Climatic optimum	Maximum sea level		Marine deposits (Estuarine alluvium)
						7 000					Fluvial deposits
						8 000	1	Temperate		Silt and clay	Peat
											Landslip
						9 000					Fluvial deposits
							11 000	Cool temperate		Sand and gravel deltas	
	PLEISTOCENE	LATE DEVENSIAN	Younger Dryas	UPPER WEICHSELIAN	Loch Lomond Stadial	10 000		Arctic	Corrie glaciers	Upper Laminated Clay	Head Scree Moraine (Hummocky Glacial Deposits)
			Alleröd		Windermere Interstadial	11 000		Cool temperate	Declining temperature and increased snowfall	Organic silt	
			Older Dryas							Minerogenic silt with rock debris	
						12 000		Temperate			
			Bölling					Climatic optimum	Maximum diversity of interstadial woodland	Organic silt	Some fluvial deposits (alluvium, alluvial fan and river terrace deposits)
			Late Pleniglacial		Dimlington Stadial	13 000	2	Cool temperate		Upwards increase in organic content of silt	Head
						14 000			Melting of main ice sheet and cutting of meltwater channels	Weakly organic silt	Glaciofluvial deposits (sand and gravel)
						15 000		Arctic	Development of main Late Devensian ice sheet	Minerogenic silt Lower Laminated Clay	(Flow till and melt-out till) Lodgement till
		MID DEVENSIAN	Mid Pleniglacial			26 000	29 000 3	? Cold ? Cool temperate	Corrie glaciers ?	None known	None known
		EARLY DEVENSIAN	(Middle Weichselian)			50 000	61 000 4 73 000	? Cold	Corrie glaciers ?		
			Early Pleniglacial			75 000		? Cool temperate			
			(Lower Weichselian)			80 000	5A-D				
		IPSWICHIAN	Eemian			116 000	115 000 5E 127 000	Warm and humid (probably similar to Holocene climate)			Troutbeck Palaeosol
						132 000	6				

Radiocarbon dates and Deep Sea Oxygen Isotope Stages are given for comparison. Based on Mitchell et al. (1973); Jones and Keen (1993); Shackleton and Opdyke (1976).

Plate 27 Glaciated upland scenery at the head of Great Langdale (D 4686).

GLACIAL HISTORY

Repeated glaciation considerably modified the pre-Quaternary, radial drainage topography of the Lake District by widening, straightening and deepening pre-existing river valleys and breaching watersheds; glacial meltwaters cut drainage channels. Glacial erosion has removed most evidence of Quaternary events prior to the last, Late Devensian, regional ice-sheet glaciation so that only a fragmentary record of earlier Pleistocene events survives locally. The oldest known deposits include a diamicton (Thornsgill Till) of pre-Ipswichian age at Troutbeck (Boardman, 1980; 1985a), and a till below Ipswichian organic-rich strata at Scandal Beck, in the upper Eden valley (Carter et al., 1978).

The last glacial episode began about 120 000 years ago (Table 24), but the Lake District probably was not glaciated fully until the Dimlington Stadial, some 25 000 years ago, when the Late Devensian Ice Sheet covered the entire region. Proglacial deposits and till of local origin accumulated in Lake District valleys but most local ice was deflected southwards by a more powerful flow emanating from the Southern Uplands of Scotland. This ice sheet reached its maximum extent about 18 000 to 20 000 years ago and the landforms and sediments it produced are widespread. Particularly prominent are the extensive drumlin swarms in the Solway Firth lowlands, Eden valley and Furness district. The high ground was first to become free of ice as the climate ameliorated but even the present site of Windermere had been deglaciated by 14 623 ± 360 years before the present (BP) (Pennington, 1978). Sand and gravel were laid down from meltwaters which flowed through, around or were ponded against ice that remained on the lower-lying ground. Ice-marginal moraines and glacial meltwater channels were also formed. The Lake District was almost completely deglaciated by the onset of the Windermere Interstadial (14 000 to 11 000 BP). The dramatic amelioration of the climate has been confirmed by analyses of pollen grains and coleopteran remains found in sediment layers at the bottom of Windermere. By 13 000 BP summer temperatures were little different from those of the present day (Lowe and Walker, 1984) but at about 12 000 BP there was a general cooling of perhaps 3 to 4°C, followed by a further cooling in July temperatures of at least 4°C at about 11 000 BP.

The vegetation in the early part of the Windermere Interstadial was dominated by grasses, sedges and herbaceous plants that thrive on disturbed soils. Open-habitat conditions were succeeded by a shrub vegetation of juniper, willows and crowberry by about 13 000 BP, which later developed into a fairly continuous canopy of birch trees in the Lake District lowlands (Pennington, 1977). The more severe climatic conditions at about 11 000 to 12 000 BP caused woody plants to decline and resulted in a phase of increased clastic inwash into Windermere (Pennington, 1978).

Towards the end of the Windermere Interstadial snow fall increased and corrie glaciers reformed in the Lake District. This restricted phase of glaciation occurred during the Loch Lomond Stadial (11 000 to 10 000 BP) with associated cooling confirmed by pollen evidence (Pennington, 1977). Sissons (1980) has mapped the glacial landforms associated with the stadial, recognising 64 corrie glaciers in the Lake District, but of these only 15 had surface areas greater than 1 km^2. The main features from the Ambleside district are summarised in Figure 67. The former presence of these small corrie glaciers is deduced from the presence of moraines and their age is determined indirectly from their lake sediment pollen and coleopteran content. Cores taken from lakes occupying Loch Lomond Stadial corries contain no record of the biota characteristic of the Windermere Interstadial, whereas cores obtained from lake basins beyond the terminal moraines of these corrie glaciers commonly do include such material. The age of corrie glaciers has been confirmed positively in this manner within the district at two sites in Langdale: in the basin above Red Tarn [NY 267 037] at 515 m above OD (Pennington, 1964) and in Langdale Combe [NY 263 084] at 455 m above OD (Walker, 1965).

The tundra vegetation of the Loch Lomond Stadial was dominated by grasses, sedges, clubmosses, herbaceous plants tolerant of bare or disturbed soils, and willow tree scrub. Mean July temperatures were around 8°C whereas mean January temperatures might have been as low as

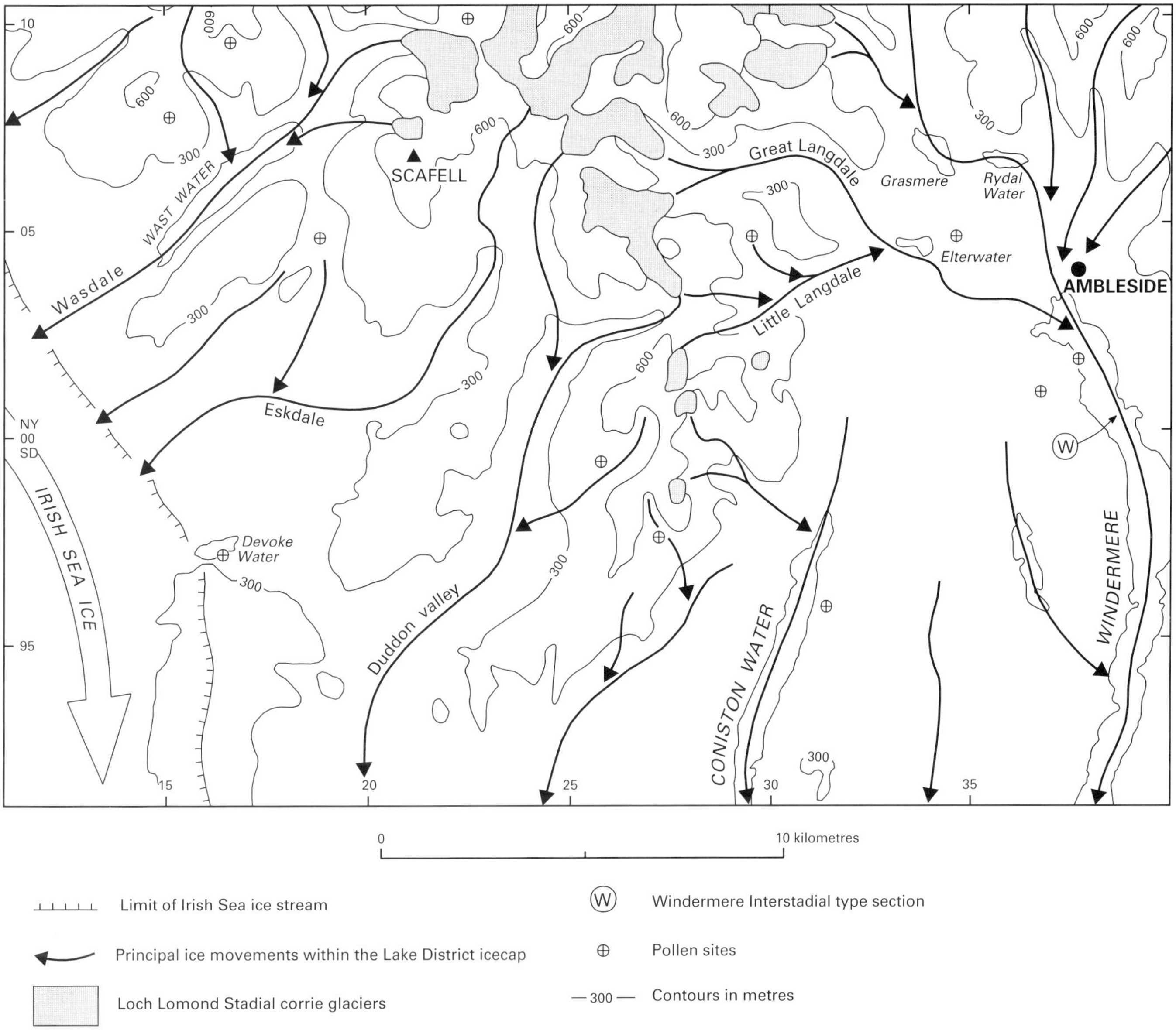

Figure 67 Main features of the Loch Lomond Stadial (after Sissons, 1980).

-20°C (Coope, 1977; Sissons, 1980). Periglacial activity was widespread and included the formation of mountain-top blockfields, solifluction lobes, screes (Plate 28) and protalus ramparts that developed at the feet of perennial mountainside snow patches (Ballantyne and Harris, 1993). The patchy vegetation cover left soils vulnerable to erosion as is indicated by contemporaneous clastic sediments in present and former lake beds (Pennington, 1970). Rivers carried high bed-loads and formed gravelly river terraces.

The corrie glaciers began to retreat at about 10 300 BP, and by 10 000 BP a dramatic amelioration of the climate heralded the Holocene Interglacial. There followed a rapid recolonisation by crowberry heath and juniper scrub, and subsequent expansion of mixed deciduous woodland on lower ground, in which birch and hazel were followed by oak, elm and pine, and then by alder (Pennington, 1964; 1970).

Though the imprint of glaciation remains dominant, postglacial processes have superimposed modifications on the landscape. Steep mountain sides have been affected by rock falls, soil creep and debris flow; on the valley floors, spreads of alluvium and river-terrace deposits have accumulated with alluvial fans developed where tributary streams experienced a rapid change of gradient. Some of these depositional landforms may be relict features that developed either immediately after deglaciation or at the end of the Dimlington Stadial and during the Loch Lomond Stadial (Church and Ryder, 1972; Ballantyne, 1991). Nevertheless, there is good evidence that accelerated erosion and mass transport have occurred within the past few centuries (Macklin et al., 1994) and

Plate 28 The Screes, Wast Water [NY 155 045].

Extensive postglacial scree deposits developed below Illgill Head. The unvegetated parts are active today. A prominent ledge sloping from right to left in the left-hand part of the picture marks a faulted contact between apparently subhorizontal andesite sheets to the right and pyroclastic rocks of the Airy's Bridge Formation to the left (D 4369).

that much of the increase in debris flow may be related to the destabilisation of mountain soils by overgrazing (Harvey et al., 1981), particularly during periods of wetter and cooler climate.

Pollen analyses on cores of peat taken from peat bogs in the Lake District indicate that widespread clearances of the forests began in the late Iron Age (Pennington, 1975a; Barber et al., 1994), when arable agriculture was increasing. The landscape was almost completely cleared of trees during Romano-British times. The later influence of human activity on the environment of the Ambleside district is recorded in the lake-bottom sediments. Atmospheric pollution from the beginning of the industrial revolution increased concentrations of certain heavy metals such as zinc in the sediments (Mackereth, 1966; Haworth, 1985). The biota in Coniston Water is still less diverse than in other lakes following pollution in the 19th century when copper mining in the Coniston area was at its peak (Davison et al., 1985).

GLACIAL EROSION

The central Lake District contains many classic examples of glacial erosional features (e.g. Boardman, 1988) (Plate 27). Most dramatic are the U-shaped valleys with truncated spurs and troughs that are, or were, occupied by ribbon lakes; those within the district include Wasdale, Eskdale, the Duddon valley and Great Langdale. In areas where erosion was most severe, all loose debris has been scoured away to give a 'knock-and-Lochan' landscape (in the sense of Sugden and John, 1976) with abundant glacial striae, for example on the Dunnerdale Fells in the south-west of the district. Loch Lomond Stadial corrie glaciers formed the spectacular corries that now contain tarns, for example, Low Tarn [NY 162 093], Stickle Tarn [NY 287 077], Levers Water [SD 279 994] and Goat's Water [SD 266 977]. Ridges of high ground separate many of the glaciated valleys, for example, Wet Side Edge [NY 237 022] and High Pike [NY 374 087] and sharp arêtes are present locally, as at Prison Band [NY 278 007]. Streamlined landforms such as crag-and-tail features, *roches moutonnées* and drumlins occur mainly within the major valleys and, together with glacial striae, indicate the last direction of ice movement.

Despite the presence of so many features of glacial erosion, it is probable that widespread, ice-sheet glaciations had minimal impact on the landscape (Boardman, 1980). Instead, Boardman (1992) developed a model in which valley glaciers in the early Pleistocene removed weathered Cainozoic deposits, incising and modifying a pre-existing valley system. Corrie development and periglacial slope modification then occurred repeatedly in periods of restricted glaciation, such as the Loch Lomond Stadial. In the west of the district smooth topography is present on the main interfluves, for example those between the intensely glaciated valleys of the Esk and the Duddon, and to the south-west of Birker Fell. This effect is in part a result of more active ice which streamed southwards from the Southern Uplands of Scotland, confining the westward flow of Lake District ice to the main valleys. The directions of Late Devensian ice flow are known to have varied with time (Mitchell and Clark, 1994) but the principal movement directions are summarised in Figure 67.

GLACIAL DEPOSITS

The Lake District glacial deposits were formed during two stadials (Boardman, 1991). The most widespread deposits are associated with the Dimlington Stadial ice-sheet (Boulton et al., 1977) whilst local deposits accompany the spatially limited Loch Lomond Stadial, corrie glaciation (Sissons, 1980). In the active waxing cycle of the Dimlington Stadial, lodgement till was deposited; melt-out till, flow till and glaciofluvial deposits were laid down from the wasting ice-sheet in its waning cycle. Moraine was created by the Loch Lomond Stadial corrie

glaciers. The distribution of glacial deposits in the district is shown on the Solid and Drift map (British Geological Survey, 1991; 1998), though it has proved impracticable in the present study to differentiate their temporal relationships. Hence the till category includes the different types described above, whereas the Moraininc deposits category includes similar moraine deposits from both the Dimlington and Loch Lomond stadials.

Till

Till is the most widely distributed of the glacial deposits, covering much of the district and occurring beneath younger superficial deposits. It consists of ice-transported material laid down either subglacially (lodgement till) or in a supraglacial (meltout till), or paraglacial (flow till) environment. Usually, the meltout and flow tills merged and comprise heterogeneous deposits, a metre or so thick, that are very poorly sorted, crudely stratified, gravelly diamictons intercalated with gravel, silty sand, silt and clay. This till accumulated at the ice front, mainly as debris flows that were modified and redeposited by ephemeral meltwater streams and sheet wash. It is generally permeable (except where an iron pan has developed) and includes large boulders, up to several metres in diameter, originally carried by the ice. Locally these supraglacial and paraglacial deposits form mounds several metres or more in thickness and some may have been classified as Morainic deposits.

In contrast, tills formed in the subglacial environment are generally much thicker (up to 30 m locally) and relatively homogeneous compared with supraglacial and paraglacial tills. They are typically overconsolidated, more clay-rich, and relatively impermeable, with matrix support of clasts and little stratification. Boulders are generally not as large as those occurring in associated supraglacial and paraglacial deposits, and typically they have bevelled and striated surfaces. Most rock fragments in the tills are of local origin, but ice originating in Scotland and thence flowing south through the Irish Sea, penetrated into the westernmost fringe of the district (Figure 67), bringing with it erratics from the Cumbrian coastal plain and southern Scotland. A crude subhorizontal parting in the till may result from pressure release and is particularly common towards the upper surfaces. Other widespread concavo-convex discontinuities originated as subglacial shear planes and erosion surfaces. All types of discontinuity may be lined with silt, clay or indurated silty fine-grained sand, the latter commonly ferruginous, and combine to produce a marked fissility in some till sections.

Extensive, featureless spreads of till occur on major interfluves in the west of the district, but these are generally thin (less than 5 m) and are penetrated by small exposures of solid rock. In the east, the till is generally confined to valley bottoms where thicknesses in excess of 10 m occur locally. In the south-east of the district, mainly over the outcrop of the Windermere Supergroup, southward ice movement has locally moulded the till into drumlins, for example in Grizedale [SD 335 945], west of Esthwaite Water and around the head of Windermere. These drumlins are typically up to 500 m long, 300 m wide and 20 to 30 m high; most are formed entirely of till, but some are moulded around rock outcrops.

Morainic deposits

These distinctive landforms have a highly variable composition and include complex interdigitations of matrix- and clast-supported diamicton, stratified and unstratified silty boulder gravel, as well as beds of sand, silt and clay. Many, but not all, of the deposits are moraines formed primarily by debris-flow and sheet-wash processes. The distribution of Loch Lomond Stadial corrie glaciers is given in Figure 67. A single arcuate ridge commonly defines the former position of a glacier front although it does not necessarily represent the maximum down-valley extent of the glacier. Hummocky moraine occurs on the up-valley side of many arcuate ridges and mainly represents supraglacial debris that was deposited as the glaciers retreated. The moraines commonly comprise an ill-sorted assemblage of angular rock fragments with very little fine-grained matrix. Boulders commonly rest on, or protrude through, the surface of the moraine and in places form the entire feature. The deposits generally are permeable, resulting in water seepage and springs at their margins, particularly on the down-valley side.

Glaciofluvial deposits and glacial meltwater channels

Few glaciofluvial deposits have been identified in the district. Several small deposits of sand and gravel occur near Ulpha [SD 199 939], in Miterdale [SD 117 997] and around Corney Fell [SD 114 915]. Glacial meltwater channels are rare in the central Lake District, but occur in abundance on its periphery. They are commonly several tens of metres wide, generally less than 20 m deep and a few hundred metres in length. A few are present in the west and south-west of the district, good examples occurring on the Dunnerdale Fells [at SD 204 911 and SD 223 922] where most of the channels are cut in solid rock and are now dry. Some of these run obliquely down the side of the Duddon valley and one well-developed channel is followed by the A593 Ambleside to Coniston road north-east of Yew Tree Tarn [NY 322 004].

FLUVIAL DEPOSITS

The oldest fluvial spreads were probably laid down by glacial meltwaters and many river terraces that occur several metres above the present base-level of rivers may have been thus formed. Elsewhere, adjacent to rivers and streams, ***Alluvium*** forms present-day floodplains, ***River terrace deposits*** comprise flat spreads above the floodplain, and ***Alluvial fan deposits*** occur downstream from any sudden decrease in the gradient of a watercourse. These deposits are heterogeneous and consist predominantly of gravel, sand and silt in varying proportions. Locally, organic-rich silt and peat occur both interbedded with and overlying, alluvium and river terrace deposits, particularly adjacent to the more mature streams and rivers. There is little detailed information about the nature

or thickness of the fluvial deposits except in the area west of Ambleside where site investigation boreholes have been drilled on the floodplains of the rivers Rothay and Brathay. The boreholes there encountered up to 20 m of predominantly coarse gravels with cobbles and boulders interbedded with coarse sand and silt, some of which is organic.

Many rivers have built deltas where they enter lakes, for example where the River Brathay and Great Langdale Beck enter Elter Water and the River Rothay enters Windermere. Alluvial fans commonly occur where tributary streams enter the principal valleys, notably in Wasdale, Eskdale, Duddon and Great Langdale. The fans consist predominantly of poorly sorted pebble and cobble gravel.

LACUSTRINE DEPOSITS

Lacustrine deposits occur at the bottom of present-day lakes and tarns as well as within the sites of former lakes and abandoned meanders on river floodplains. The sediments in the present-day lakes and tarns provide much of the evidence of Late Devensian and Flandrian climatic variations since the contained pollen assemblages reflect the contemporary vegetation. On this basis the Windermere sediments are used as the type area for the Late Devensian (Windermere) Interstadial (Coope and Pennington, 1977). Seismic and borehole evidence suggest that the sequence of sediments in Lake Windermere is up to 40 m thick where water depths exceed 30 m (Howell, 1971). The oldest sediments comprise cobble-gravel interbedded with clay. The gravel succession becomes finer upwards, with interbeds of sand and clay towards the top, and then passes upwards into the regularly laminated fine-grained 'Lower Laminated Clay' (Pennington, 1943; 1978) (Table 24). Paired laminae may be varves (i.e. an annual accumulation layer), indicating that the lake was seasonally frozen. The top of the 'Lower Laminated Clay' was taken to be the lithostratigraphical base of the Windermere Interstadial sequence by Coope and Pennington (1977). However, the overlying unlaminated clay is devoid of microfossils and the biostratigraphical base of the Interstadial sequence is taken where microfossils first appear. Thereafter organic detritus becomes increasingly abundant upwards. A bed of silt with angular rock fragments interrupts the organic-rich sequence and possibly represents the cooler Older Dryas chronozone of Scandinavia (Pennington, 1978; Coope, 1977). Above this unit the organic content of the silt declines and the Interstadial sequence is overlain by inorganic varved clays, the 'Upper Laminated Clay', about 0.5 m thick and formed by about 400 silt–clay couplets each of which may represent annual sediment accumulation (Coope and Pennington, 1977; Pennington, 1978).

Glacial meltwaters built out deltas of sand and gravel into the lakes as the Loch Lomond Stadial glaciers decayed, and modern rivers continue the process. Nevertheless, the Holocene sequence of lake bottom sediments is dominated by silt and clay that settled out from suspension. A biostratigraphy based on diatom assemblages (Haworth, 1985) and pollen assemblages (Pennington, 1947; 1964; 1970) is supported by palaeolimnological studies undertaken on sediments in Windermere and other lakes in the district and listed by Haworth and Long (1989). Changes in the composition of the lake sediments have been related to soil dynamics in the catchment areas (Pennington and Lishman, 1971).

MARINE DEPOSITS

The westernmost 2 km of the River Esk within the district is tidal and where the alluvium is considered to have been deposited in an estuarine environment it is classified (British Geological Survey, 1998) as 'undifferentiated' Marine Deposits. These probably contain higher proportions of mud and silt and less gravel than the river alluvium farther upstream. Older marine deposits associated with the Flandrian Transgression in mid-Holocene times may also occur in this area beneath the youngest alluvium. Farther west, in the adjoining Gosforth district, marine transgression reached about 7 m above present Ordnance Datum (King, 1976).

PEAT

Most of the larger spreads of peat formed from about 7500 BP onwards as ombrogenous, upland blankets. The rugged relief and relatively good drainage in central and eastern parts of the district has restricted blanket peat development there, but it covers more of the relatively subdued terrain farther west, notably on the watershed between Eskdale and the Duddon valley. Generally, the thickness increases uphill to a maximum of about 3 m although greater thicknesses occur in some poorly drained depressions on both high and low ground. Pollen analyses indicate that much of this low-ground, basin peat formed in the latter half of the Holocene (Pennington, 1975b). Reed-swamp or fen peat is forming in poorly drained lowland basins today, for example, adjacent to Little Langdale Tarn [NY 305 033] and Elter Water [NY 335 042].

HEAD AND GRAVITY-INDUCED MASS MOVEMENT PHENOMENA

In situ and gravity-transported rock regolith deposits are generally thin and only locally do they exceed 1 m in thickness, for example, North of Hause [NY 108 091], north of Pinnacle Bield [NY 248 098] and on the west side of Swirl How [NY 262 005]. Most head is considered to have formed under periglacial conditions; some may have formed immediately after ice-sheet deglaciation but most probably originated during the Loch Lomond Stadial, when the ground was subjected to permafrost.

The downhill movement of head may be accompanied by cambering in the underlying rockhead surface, particularly on steep slopes where bedding or cleavage planes dip steeply into the hillside. As the beds and cleavage lithons rotate downslope the strata may overturn as slabs of bedrock and gradually become detached, as for example on the south-west side of Wet Side Edge [NY 275 022] and

in the Duddon valley between Stonythwaite [SD 220 969] and Hazel Head [SD 196 941]. Foundered strata occur locally on some of the major crags where rock masses have become detached and moved slightly downslope.

Scree

Aprons of scree are present below most crags of Borrowdale Volcanic Group rock and occur locally beneath crags of the Eskdale granite. A particularly impressive example, The Screes at Wast Water [NY 146 038 to NY 176 064], lies in the north-west of the district (Plate 28). Scree aprons probably began to accumulate during deglaciation with rapid growth during the Loch Lomond Stadial when frost shattering was most intense. Many of the screes have remained active during the Holocene and continue to grow.

Landslip

Gravity-induced slope failures are present locally in till. The largest examples occur adjacent to the River Lickle between Lind End [SD 230 911] and Stephenson Ground [SD 235 931]. On the south-east side of the river, the landslips are superficial and occur on steep slopes; the basal slip-plane coincides with the rockhead surface, the slope failures probably being induced by fluvial erosion undercutting the base of the slope. The landslip on the opposite side of the river appears to have occurred within the till. The fresh appearance of the landslips and fissuring within them indicate that they are still active though initial movement may date from earliest Flandrian times when melting of permafrost weakened the previously frozen till.

MADE GROUND

Most of the larger areas of made ground comprise spoil from slate quarrying in the Seathwaite Fell Formation, for example, on Broughton Moor [SD 255 946], near High Tilberthwaite [NY 317 017 and NY 311 020], Elterwater [NY 325 047] and at Pets Quarry [NY 392 074] north-east of Ambleside. Other areas of made ground consist of mine spoil, for example around Coniston copper mines [SD 286 986] and some small landfills may include domestic waste.

INFORMATION SOURCES

Geological information relevant to the Ambleside district and held by the British Geological Survey at the time of compilation is given in this section. Published and unpublished material in the form of maps, memoirs and reports are included, as are other sources of data held in a number of collections, including borehole records, mine plans, fossils, rock samples, thin sections, hydrogeological data and photographs.

Searches of indexes to some of the collections may be made on the Geoscience Index System in BGS libraries. This is a computer-based system under development which can be used to carry out searches of indexes to collections and digital databases for specified geographical areas. It is based on a geographical information system linked to a relational database management system. Results of the searches are displayed on maps on the screen. At the present time (1996) the datasets are limited and not all are complete. The indexes which are available are listed below:

- Index of boreholes
- Topographical backdrop based on 1:250 000 scale maps
- Outlines of BGS maps at 1:50 000 and 1:10 000 scale and 1:10 560 scale County Series
- Chronostratigraphical boundaries and areas from BGS 1:250 000 maps
- Geochemical sample locations on land
- Aeromagnetic and gravity data recording stations
- Land survey records

MAPS

1:1 000 000 and 1 500 000

Colour shaded relief magnetic map of Britain, Ireland and adjacent areas, 1998
Colour shaded relief gravity map of Britain, Ireland and adjacent areas, 1997
Industrial mineral resources map of Britain, 1996
Metallogenic map of Britain and Ireland, 1996
Tectonic map of Britain, Ireland and adjacent areas, 1996

1:250 000

Lake District 54N 04W, Solid geology, 1980
Lake District 54N 04W, Sea bed sediments and Quaternary geology, 1983
East Irish Sea, Solid, 1994

1:50 000

Sheet 23 Cockermouth, Solid, 1997
Sheet 23 Cockermouth, Solid and Drift, 1997
Sheet 28 Whitehaven, Solid, 1979
Sheet 29 Keswick, Solid, 1999
Sheet 29 Keswick, Solid and Drift, 1999
Sheet 37 Gosforth, Solid, 1998
Sheet 37 Gosforth, Solid and Drift, 1998
Sheet 38 Ambleside, Solid Geology, 1996
Sheet 38 Ambleside, Solid and Drift, 1998
Sheet 47 Bootle, Solid and Drift, 1997
Sheet 48 Ulverston, Solid Geology, 1997
Sheet 48 Ulverston, Solid and Drift, 1997

1:25 000

SD19 Devoke Water and Ulpha, 1991
SD29 and parts of SD39, NY20 and NY30 Coniston special sheet, in press
NY20 and parts of NY10 Sca Fell special sheet, in press
SD18 and part of SD28 Black Coombe, 1998

1:10 000 and 1:10 560

The district covered by the 1:50 000 Series England and Wales Sheet 38 Ambleside was originally surveyed on the six-inch County Series sheets Cumberland 73, 74, 75, 78, 79, 80, 82, 83, 85 and 86, Westmorland 18, 19, 25, 26, 32, and 37, and Lancashire 1, 2, 3, 4, 5, 6, 7 and 8 by W T Aveline, J C Ward, C E de Rance, G H Wollaston, F Rutley, A G Cameron and E J Hebert, and published on one-inch Old Series sheet 98NW in 1882.

The resurvey, upon which this memoir is based, was carried out from 1982 to 1992 by P M Allen, B Beddoe-Stephens, E W Johnson, D Millward, M G Petterson, A J Reedman and B Young, and under NERC contracts to Sheffield and Liverpool universities. The Central Fells area was surveyed by B C Kneller and B J McConnell, under the direction of B P Kokelaar at Liverpool University in 1988–91; the work done by M J Branney, N C Davis and R Smith (1984–89) during the tenure of research studentships at the Universities of Sheffield and Liverpool was incorporated in this contract. The Windermere Supergroup was wholly surveyed by N J Soper, B C Kneller and R W Scott of Sheffield University between 1988 and 1990.

Geological 1:10 000 scale National Grid maps included, wholly or in part, in the 1:50 000 Series Sheet 38 Ambleside are listed below, together with the initials of the surveyors and dates of the survey; in the case of marginal sheets, all surveyors are listed. The maps are not published but are available for inspection in the library, British Geological Survey, Keyworth and Edinburgh. Dyeline copies may be purchased from the Sales Desk of the British Geological Survey.

NY 10 NW	MGP, BBS, AJR	1987–89
NY 10 NE	MGP, BBS, NCD	1987–88
NY 10 SW	BBS, MGP	1987–88
NY 10 SE	BY, BBS, DM	1982–86
NY 11 SW	MGP, BBS	1988–89
NY 11 SE	BCK, BJM, MGP	1988–92
NY 20 NW	NCD, BCK, BJM, MJB, BBS, DM	1983–91
NY 20 NE	MJB, BJM, BCK	1983–91
NY 20 SW	DM, BBS, PMA, EWJ, MJB, BY	1982–86
NY 20 SE	EWJ, MJB, PMA, DM, BBS, MGP	1982–90
NY 21 SW	BJM, NCD, BCK, BPK	1987–92
NY 21 SE	BJM, BCK, BPK	1990–92
NY 30 NW	BBS, EWJ, MGP, AJR	1989–91
NY 30 NE	MGP, BBS, AJR	1989–91
NY 30 SW	DM, EWJ, BCK, RWS	1988–91
NY 30 SE	EWJ, DM, BCK, RWS, NJS	1988–91
NY 31 SW	MGP, BBS	1990–91
NY 31 SE	MGP, BBS	1989–91
NY 40 NW	BBS	1991
SD 19 NW	BY, PMA, DM, BBS	1983–87
SD 19 NE	BY, DM, EWJ, PMA	1983–85
SD 19 SW	BY, DM	1983–85

SD 19 SE	DM, EWJ	1985–86
SD 29 NW	DM, EWJ	1983–87
SD 29 NE	DM, EWJ, BCK, RWS, MGP	1986–90
SD 29 SW	EWJ, RWS, NJS, BCK	1985–89
SD 29 SE	BCK, NJS, RWS, DM, EWJ	1986–90
SD 39 NW	BCK, DM, RWS	1987–90
SD 39 NE	BCK	1988
SD 39 SW	NJS	1988
SD 39 SE	BCK	1988–89

BOOKS AND REPORTS

Regional geochemistry of the Lake District and adjacent areas, 1992.
British regional geology: Northern England, 1971.
Glossary of the minerals of the Lake District and adjoining areas, 1987.
United Kingdom offshore regional report: the geology of the Irish Sea, 1995.
Subsurface memoir: the structure and evolution of the Northumberland–Solway Basin and adjacent areas, 1995.
Memoir: geology of the west Cumbria district, 1997.

Technical reports

Technical reports relevant to the district are referred to in the text and included in the References. These are as follows: Bland (1988); Branney et al. (1993a, b); Branney and Kokelaar (1993); Cameron et al. (1993); Kneller (1990a, b); Kneller and Soper (1990); Kneller and McConnell (1993); Kneller et al. (1993a); Lee (1984a, b; 1986b; 1988; 1989); Lee et al. (1984); McConnell (1993); McConnell et al. (1993); Musson (1987; 1994); Musson et al. (1984); Rundle (1992); Scott and Kneller (1990); Webb and Brown (1984); Wheildon et al. (1984); Wright and Richards (1995) and Young (1985a, b).

Additional to these reports are:

Armstrong, H A, Johnson, E W, and Scott, R W. 1994. Preliminary conodont biostratigraphy of the attenuated Dent Group (Upper Ordovician) at Hartley Ground, Broughton in Furness, Cumbria. *British Geological Survey Technical Report*, WA/94/25.

Beddoe-Stephens, B. 1991. Geochemical variation through the Lower Borrowdale Volcanic Group, Devoke Water area. *British Geological Survey Technical Report*, WG/91/19.

FURTHER INFORMATION

Geophysical surveys

The district is covered by part of the BGS 1:250 000 Bouguer gravity anomaly and Aeromagnetic anomaly maps, Lake District (54°N-04°W). A Geophysical Information Map (GIM) at a scale of 1:50 000 is available for the district. GIMs show information held in the BGS digital databases including Bouguer gravity and aeromagnetic anomalies, distribution of data points, deep borehole sites and locations of detailed geophysical surveys.

Geochemical data

The Geochemical Baseline Survey of the Environment (G-BASE) is based on the collection of stream sediment and stream water samples at an average density of one sample per 1.5 km^2. The fine (minus 150 μm) fractions of stream sediment samples are analysed for a wide range of elements, using automated instrumental methods.

The samples from the Lake District were collected in 1978-80. The results (including Ag, As, Ba, Be, Bi, B, CaO, Cd, Co, Cr, Cu, Fe_2O_3, Ga, K_2O, La, Li, MgO, Mn, Mo, Ni, Pb, Rb, Sb, Sn, Sr, TiO_2, U, V, Y, Zn and Zr in stream sediments, and pH, conductivity, fluoride, bicarbonate and U for stream waters) are published in atlas form:

British Geological Survey. 1992. *Regional geochemistry of the Lake District and adjacent areas.* (Keyworth, Nottingham: British Geological Survey.)

The geochemical data, with location and site information, are available as hard copy for sale or in digital form under licensing agreement. The coloured geochemical atlas is also available in digital form (on CD-ROM or floppy disk) under licensing agreement. British Geological Survey offers a client-based service for interactive GIS interrogation of G-BASE data.

Major and trace element analyses of whole-rock samples from Borrowdale Volcanic Group, Windermere Supergroup and intrusive rocks have been made in laboratories at the BGS, and in the University departments of geological sciences at Royal Holloway (University of London), Nottingham, Leicester, Lancaster and Sheffield. Geochemical data are held by the British Geological Survey.

The following environmental geochemistry reports are available:

Appleton, J D, Hooker, P J, and Smith, N J P. 1995. Methane, carbon dioxide and oil seeps from natural sources and mining areas: characteristics, extent and relevance to planning and development in Great Britain. *British Geological Survey Technical Report*, WP/95/1.

Appleton, J D, and Ball, T K. 1995. Radon and background radioactivity from natural sources: characteristics, extent and relevance to planning and development in Great Britain. *British Geological Survey Technical Report*, WP/95/2.

Appleton, J D. 1995. Potentially harmful elements from natural sources and mining areas: characteristics, extent and relevance to planning and development in Great Britain. *British Geological Survey Technical Report*, WP/95/3.

Appleton, J D. 1995. Radon, methane, carbon dioxide, oil seeps and potentially harmful elements from natural sources and mining areas: characteristics, extent and relevance to planning and development in Great Britain. *British Geological Survey Technical Report*, WP/95/4.

Lexicon

Definitions of the named rock units shown on BGS maps, including those shown on Sheet 38 Ambleside, are held in the Lexicon database. This can be accessed on BGS website (see below). Further information on the database can be obtained from the Lexicon Manager at BGS Keyworth.

Remote sensing data

Interpretation of enhanced Landsat thematic mapper imagery of the Lake District is included in the following report:

Berrange, J P. 1991. Lineament analysis of Landsat Thematic Mapper imagery of the English Lake District and environs. *British Geological Survey Technical Report*, WA/91/30.

Minerals information

There are concentrations of quarries and abandoned metalliferous mines, mine shafts and trial levels in parts of the district. Many of these are shown on the 1:10 000-scale geological maps and some information is held in British Geological Survey files. An indication of the level of these holdings is given in the catalogue below, though all files are now held at the British Geological Survey's Murchison House office.

Young, B, and Millward, D. 1984. Catalogue of mining information (other than coal, fireclay and slate) for the Lake District and South Cumbria held by the Northern England

Office of the BGS. *Lake District Regional Geological Survey,* Report No. 1.

Minerals information, such as the Mines and Quarries database (BRITPITS), can be accessed through the British Geological Survey MINGOL GIS data delivery system. Minerals information is also contained within the following publications:

British Geological Survey. 1996. *United Kingdom Minerals Yearbook 1995.* (Keyworth, Nottingham: British Geological Survey.)

Colman, T B. 1990. *Exploration for metalliferous and related minerals in Britain: a guide.* (Keyworth, Nottingham: British Geological Survey.)

Harris, P M. 1994. *The United Kingdom minerals industry.* (Keyworth, Nottingham: British Geological Survey).

Harris, P M, Highley, D E, Hillier, J A, and Whitwood, A. 1994. *Directory of mines and quarries 1994*; 4th edition. (Keyworth, Nottingham: British Geological Survey.)

Rock physical properties data

The British Geological Survey rock Physical Property Databank contains density (saturated and grain density, porosity), magnetic and sonic velocity data from rocks within the district. Data from the Lake District are summarised in:

Lee, M K. 1988. Density variations within Lake District granites and Lower Palaeozoic rocks. *British Geological Survey Technical Report,* WK/88/9.

MATERIAL COLLECTIONS

Boreholes

British Geological Survey holds a small number of borehole records for the Ambleside district. These are mostly shallow, having been drilled for site investigations.

Geological Survey photographs

About 100 photographs illustrating aspects of the geology of the district are deposited for reference in the BGS libraries. These photographs were taken at various times since 1981 and depict details of the various rocks; there are also some views. Copies of the photographs can be purchased as black and white or colour prints, and 50 × 50 mm transparencies.

Petrological collections

Thin sections of 2840 rocks from the district are registered in the British Geological Survey sliced rock collection. Most of these are of the igneous rocks.

Palaeontological collections

The Ordovician and Silurian rocks of the district have long been visited by fossil collectors, and many major museums have material from the Windermere Supergroup. The most comprehensive collections, particularly of graptolites, are in the Sedgwick Museum, Cambridge which holds the type specimens of Marr and Nicholson, Temple, McNamara, Hutt and Rickards, and information on these is available from the Curator. Important additional material is in the Natural History Museum, London, and the Hunterian Museum, Glasgow.

The biostratigraphy collections of the British Geological Survey contain: (1) collections made for the present survey from a selection of sites including the trace fossils from the Borrowdale Volcanic Group, (2) the bulk of the collections by John Temple and Robin Scott, and (3) collections of such earlier workers as Daniel Sharpe and John Bolton. Enquiries regarding these may be made to the Curator of the Biostratigraphy Collections.

ADDRESSES FOR DATA SOURCES

Enquiries concerning geological data for the district should be addressed to the Manager, National Geosciences Data Centre.

British Geological Survey (Headquarters)
Keyworth
Nottingham NG12 5GG

Telephone 0115 936 3100
Fax 0115 936 3200

London Information Office at the
Natural History Museum Earth Galleries
Exhibition Road
South Kensington
London SW7 2DE

Telephone 0207 589 4090
Fax 0207 584 8270

British Geological Survey
Murchison House
West Mains Road,
Edinburgh EH 93 LA

Telephone 0131 667 1000
Fax 0131 668 2683

British Geological Survey
Website *http://www.bgs.ac.uk*

REFERENCES

Most of the references listed below are held in the Library of the British Geological Survey at Keyworth, Nottingham. Copies of the references can be purchased subject to the current copyright legislation.

Adams, J. 1988. *Mines of the Lake District Fells.* (Clapham (Via Lancaster): Dalesman Books.)

Agassiz, L. 1841. Glaciers, and the evidence of their having once existed in Scotland, Ireland and England. *Proceedings of the Geological Society of London*, Vol. 3, 327–332.

Aissaoui, D M, and Purser, B H. 1983. Nature and origins of internal sediments in Jurassic limestones of Burgundy (France) and Fnoud (Algeria). *Sedimentology*, Vol. 30, 273–283.

Al Jawadi, A F. 1987. Minor igneous intrusions of the Lake District; geochronology, geochemistry and petrology. Unpublished PhD thesis, University of Newcastle.

Allen, J R L. 1982. Sedimentary structures; their character and physical basis. *Developments in Sedimentology*, No. 30.

Allen, P M. 1987. The Solway line is not the Iapetus suture. *Geological Magazine*, Vol. 124, 485–486.

Allen, P M, and Cooper, D C. 1986. The stratigraphy and composition of the Latterbarrow and Redmain sandstones, Lake District, England. *Geological Journal*, Vol. 21, 59–76.

Allen, P M, Cooper, D C, and Fortey, N J. 1987. Composite lava flows of Ordovician age in the English Lake District. *Journal of the Geological Society of London*, Vol. 144, 945–960.

Ansari, S M. 1983. Petrology and petrochemistry of the Eskdale and adjacent intrusions (Cumbria) with special reference to mineralization. Unpublished PhD thesis, University of Nottingham.

Armstrong, H A, Johnson, E W, and Scott, R W. 1994. Conodont biostratigraphy of the attenuated Dent Group (upper Ordovician) at Hartley Ground, Broughton in Furness, Cumbria, UK. *Proceedings of the Yorkshire Geological Society*, Vol. 51, 9–21.

Arthurton, R S, Johnson, E W, and Mundy, D J C. 1988. Geology of the country around Settle. *Memoir of the British Geological Survey*, Sheet 60 (England and Wales).

Aveline, W T. 1872. On the continuity and breaks between the various divisions of the Silurian strata in the Lake District. *Geological Magazine*, Vol. 9, 441–442.

Aveline, W T, and Hughes, T M. 1872. The geology of the country around Kendal, Sedbergh, Bowness and Tebay. *Memoir of the Geological Survey of Great Britain*, Quarter Sheet 98NE.

Bacon, C R. 1990. Calc-alkaline, shoshonitic and primitive tholeiitic lavas from monogenetic volcanoes near Crater Lake, Oregon. *Journal of Petrology*, Vol. 31, 135–166.

Ballantyne, C K. 1991. Holocene geomorphic activity in the Scottish Highlands. *Scottish Geographical Magazine*, Vol. 107, 84–98.

Ballantyne, C K, and Harris, C. 1993. *The periglaciation of Great Britain.* (Cambridge: Cambridge University Press.)

Barber, K E, and five others. 1994. Climatic change and human impact in North Cumbria: peat stratigraphic and pollen evidence from Bolton Fell Moss and Walton Moss. 20–49 in *The Quaternary of Cumbria: field guide.* Boardman, J, and Walden, J J (editors). (Oxford: Quaternary Research Association.)

Barnes, C G. 1992. Petrology of monogenetic volcanoes, Mount Bailey area, Cascade Range, Oregon. *Journal of Volcanology and Geothermal Research*, Vol. 52, 141–156.

Barnes, R P, Lintern, B C, and Stone, P. 1989. Timing and regional implications of deformation in the Southern Uplands of Scotland. *Journal of the Geological Society of London*, Vol. 146, 905–908.

Beddoe-Stephens, B, and Mason, I. 1991. The volcanogenetic significance of garnet-bearing minor intrusions within the Borrowdale Volcanic Group, Eskdale area, Cumbria. *Geological Magazine*, Vol. 128, 505–516.

Beddoe-Stephens, B, Petterson, M G, Millward, D, and Marriner, G F. 1995. Geochemical variation and magmatic cyclicity within an Ordovician continental-arc volcanic field: the lower Borrowdale Volcanic Group, English Lake District. *Journal of Volcanology and Geothermal Research*, Vol. 65, 81–110.

Bell, A M. 1975. A finite strain study of accretionary lapilli tuff in the Borrowdale Volcanic Group, English Lake District. Unpublished PhD thesis, University of Sheffield.

Bergström, S M. 1980. Conodonts as paleotemperature tools in Ordovician rocks of the Caledonides and adjacent areas in Scandinavia and the British Isles. *Geologiska Foreningens i Stockholm Forhandlingar*, Vol. 102, 377–392.

Bland, D J. 1988. An occurrence of silver and gold in Wasdale, Lake District. *British Geological Survey Technical Report*, WG/88/7.

Bloomfield, K. 1975. A late-Quaternary monogenetic volcano field in central Mexico. *Geologische Rundschau*, Vol. 64, 476–497.

Boardman, J. 1980. Evidence for pre-Devensian glaciation in the northeastern Lake District. *Nature, London*, Vol. 286, 599–600.

Boardman, J (editor). 1981. *Field guide to eastern Cumbria.* (Cambridge: Quaternary Research Association.)

Boardman, J. 1985a. The Troutbeck Paleosol, Cumbria, England. 231–260 in *Soils and Quaternary Landscape Evolution.* Boardman, J (editor). (Chichester: Wiley.)

Boardman, J (editor). 1985b. *Field guide to the periglacial landforms of Northern England.* (Cambridge: Quaternary Research Association.)

Boardman, J. 1988. *Classic landforms of the Lake District, classic landform guides.* (Sheffield: Geographical Association.)

Boardman, J. 1991. Glacial deposits in the English Lake District. 175–183 in *Glacial deposits in Great Britain and Ireland.* Ehlers, J, Gibbard, P L, and Rose, J (editors). (Rotterdam: Balkema.)

Boardman, J. 1992. Quaternary landscape evolution in the Lake District — a discussion. *Proceedings of the Cumberland Geological Society*, Vol. 5, 285–315.

Boardman, J, and Walden, J (editor). 1994. *The Quaternary of Cumbria: field guide.* (Oxford: Quaternary Research Association.)

BOTT, M H P. 1974. The geological interpretation of a gravity survey of the English Lake District and the Vale of Eden. *Journal of the Geological Society of London*, Vol. 130, 309–331.

BOTT, M H P. 1978. Deep structure. 25–40 *in* The geology of the Lake District. Moseley, F (editor). *Occasional Publication of the Yorkshire Geological Society*, No. 3.

BOULTER, C A, and SOPER, N J. 1973. Structural relationships of the Shap granite. *Proceedings of the Yorkshire Geological Society*, Vol. 39, 365–369.

BOULTON, G S, JONES, A S, CLAYTON, K M, and KENNING, M J. 1977. A British ice-sheet model and patterns of glacial erosion and deposition in Britain. 231–246 in *British Quaternary studies: recent advances*. Shotton, F W (editor). (Oxford: Claredon Press.)

BOUMA, A H. 1962. Sedimentology of some flysch deposits. (Amsterdam: Elsevier Publishing Company.)

BOWEN, D Q. 1991. Time and space in the glacial sediment systems of the British Isles. 3–11 in *Glacial deposits in Great Britain and Ireland*. EHLERS, J, GIBBARD, P L, and ROSE, J (editors). (Rotterdam: Balkema.)

BRANNEY, M J. 1988a. Subaerial explosive volcanism, intrusion, sedimentation, and collapse in the Borrowdale Volcanic Group, SW Langdale, English Lake District. Unpublished PhD thesis, University of Sheffield.

BRANNEY, M J. 1988b. The subaerial setting of the Ordovician Borrowdale Volcanic Group, English Lake District. *Journal of the Geological Society of London*, Vol. 145, 887–890.

BRANNEY, M J. 1991. Eruption and depositional facies of the Whorneyside Tuff Formation, English Lake District: an exceptionally large-magnitude phreatoplinian eruption. *Geological Society of America Bulletin*, Vol. 103, 886–897.

BRANNEY, M J, DAVIS, N C, KOKELAAR, B P, and MCCONNELL, B J. 1993a. The Airy's Bridge Formation in the Central Fells. *British Geological Survey Technical Report*, WA/93/41.

BRANNEY, M J, KNELLER, B C, and KOKELAAR, B P. 1990. Disordered turbidite facies (DTF): a product of continuous surging density flows (abstract). *13th International Sedimentological Congress.* (Nottingham UK: International Sedimentological Congress.)

BRANNEY, M J, and KOKELAAR, B P. 1992. A reappraisal of ignimbrite emplacement: progressive aggradation and changes from particulate to non-particulate flow during emplacment of high-grade ignimbrite. *Bulletin of Volcanology*, Vol. 54, 504–520.

BRANNEY, M J, and KOKELAAR, B P. 1993. The Whorneyside Tuff Formation in the Central Fells. *British Geological Survey Technical Report*, WA/93/40.

BRANNEY, M J, and KOKELAAR, B P. 1994. Volcanotectonic faulting, soft-state deformation and rheomorphism of tuffs during development of a piecemeal caldera, English Lake District. *Bulletin of the Geological Society of America*, Vol. 106, 507–530.

BRANNEY, M J, KOKELAAR, B P, KNELLER, B C, and DAVIS, N C. 1993b. Structure of the Borrowdale Volcanic Group in the Central Fells. *British Geological Survey Technical Report*, WA/93/46.

BRANNEY, M J, KOKELAAR, B P, and MCCONNELL, B J. 1992. The Bad Step Tuff: a lava-like rheomorphic ignimbrite in a calc-alkaline piecemeal caldera, English Lake District. *Bulletin of Volcanology*, Vol. 54, 187–199.

BRANNEY, M J, and SOPER, N J. 1988. Ordovician volcano-tectonics in the English Lake District. *Journal of the Geological Society of London*, Vol. 145, 367–376.

BRANNEY, M J, and SPARKS, R S J. 1990. Fiamme formed by diagenesis and burial-compaction in soils and subaqueous sediments. *Journal of the Geological Society of London*, Vol. 147, 919–922.

BRANNEY, M J, and SUTHREN, R J. 1988. High-level peperitic sills in the English Lake District: distinction from block lavas and implications for Borrowdale Volcanic Group stratigraphy. *Geological Journal*, Vol. 23, 171–187.

BRENCHLEY, P J. 1988. Environmental changes close to the Ordovician–Silurian boundary. *Bulletin of the British Museum of Natural History (Geology)*, Vol. 43, 377–385.

BRENCHLEY, P J, and CULLEN, B. 1984. The environmental distribution of associations belonging to the *Hirnantia* fauna — evidence from North Wales and Norway. 113–125 *in* Aspects of the Ordovician System. BRUTON, D L (editor). *Palaeontological Contributions from the University of Oslo*, Vol. 295.

BRIDGEWATER, J, FOO, W S, and STEPHENS, D J. 1985. Particle mixing and segregation in failure zones — theory and experiment. *Powder Technology*, Vol. 41, 147–158.

BRITISH GEOLOGICAL SURVEY. 1991. Devoke Water and Ulpha. Sheet SD 19. Solid and Drift geology. 1:25 000. (Southampton: Ordnance Survey for the British Geological Survey.)

BRITISH GEOLOGICAL SURVEY. 1992. *Regional geochemistry of the Lake District and adjacent areas.* (Keyworth, Nottingham: British Geological Survey.)

BRITISH GEOLOGICAL SURVEY. 1996. Ambleside. England and Wales Sheet 38. Solid Geology 1:50 000. (Keyworth, Nottingham: British Geological Survey.)

BRITISH GEOLOGICAL SURVEY. 1998. Ambleside. England and Wales Sheet 38. Solid with Drift geology. 1:50 000. (Keyworth, Nottingham: British Geological Survey.)

BROWN, P E, MILLER, J A, and SOPER, N J. 1964. Age of the principal intrusions of the Lake District. *Proceedings of the Yorkshire Geological Society*, Vol. 34, 331–342.

BUCKLAND, W. 1841. Memoir on the evidences of glaciers in Scotland and the North of England. *Proceedings of the Geological Society of London*, Vol. 3, 332–345.

BUCKLEY, A. 1984. Anomalous uranium in superficial deposits overlying the Wasdale Head Granite, Cumbria. *Proceedings of the Yorkshire Geological Society*, Vol. 44, 431–442.

BUNCH, B, and FELL, C. 1949. A stone axe factory at Pike of Stickle, Great Langdale, Westmorland. *Proceedings of the Prehistoric Society*, Vol. 15, 1–20.

BYERS, C W. 1977. Biofacies patterns in euxinic basins: a general model. 5–17 *in* Deep water carbonate environments. COOK, H E, and ENOS, P (editors). *Society of Economic Palaeontologists and Mineralogists Special Publication*, No. 25.

CAMERON, D G, and six others. 1993. Mineral exploration in the Lower Palaeozoic rocks of south-west Cumbria. Part 1: regional surveys. *British Geological Survey Technical Report (Mineral Reconnaissance Programme Report 128)*, WF/93/4.

CARTER, P A, JOHNSON, G A L, and TURNER, J. 1978. An interglacial deposit at Scandal Beck, N W England. *New Phytologist*, Vol. 81, 785–790.

CAS, R A F, and WRIGHT, J V. 1987. *Volcanic successions modern and ancient: a geological approach to processes, products and successions.* (London: Allen and Unwin.)

CATT, J A. 1991. Late Devensian glacial deposits and glaciations in eastern England and the adjoining offshore region. 61–68 in *Glacial Deposits of Great Britain and Ireland.* EHLERS, J, GIBBARD, P L, and ROSE, J (editors). (Rotterdam: Balkema.)

CHADWICK, R A, HOLLIDAY, D W, HOLLOWAY, S, and HULBERT, A G. 1995. The Northumberland–Solway basin and adjacent areas. *Subsurface Memoir of the British Geological Survey.*

CHADWICK, R A, KIRBY, G A, and BAILY, H E. 1994. The post-Triassic structural evolution of north-west England and adjacent parts of the East Irish Sea. *Proceedings of the Yorkshire Geological Society*, Vol. 50, 91–102.

CHANNELL, J E T, and MCCABE, C. 1992. Palaeomagnetic data from the Borrowdale Volcanic Group: volcano-tectonics and Late Ordovician palaeolatitudes. *Journal of the Geological Society of London*, Vol. 149, 881–888.

CHAPPELL, B W, and WHITE, A J R. 1974. Two contrasting granite types. *Pacific Geology*, Vol. 8, 173–174.

CHAPPELL, S. 1987. Stone axe morphology and distribution in Neolithic Britain. *British Archaeological Reports*, No. British Series 177.

CHURCH, M, and RYDER, J M. 1972. Paraglacial sedimentations: a consideration of fluvial processes conditioned by glaciation. *Geological Society of America Bulletin*, Vol. 83, 3059–3072.

CLARIS, P, QUATERMAINE, J, and WOOLEY, A R. 1989. The Neolithic axe quarries and axe factory sites of Great Langdale and Scafell Pike: a new field survey. *Proceedings of the Prehistoric Society*, Vol. 55, 1–25.

CLARK, L. 1963. The geology and petrology of the Ennerdale Granophyre, its metamorphic aureole and associated mineralization. Unpublished PhD thesis, University of Leeds.

CLARK, L. 1964. The Borrowdale Volcanic Series between Buttermere and Wasdale, Cumberland. *Proceedings of the Yorkshire Geological Society of London*, Vol. 34, 343–356.

COBBING, E J, MALLICK, D I J, PITFIELD, P E J, and TEOH, L H. 1986. The granites of the Southeast Asian Tin Belt. *Journal of the Geological Society of London*, Vol. 143, 537–550.

COCKS, L R M. 1978. A review of British Lower Palaeozoic brachiopods, including a synoptic revision of Davidson's Monograph. *Monographs of the Palaeontographical Society*, No. 549, part of Vol. 131 for 1977.

COCKS, L R M. 1988. Brachiopods across the Ordovician–Silurian boundary. *Bulletin of the British Museum of Natural History (Geology)*, Vol. 43, 311–315.

COCKS, L R M, and FORTEY, R A. 1982. Faunal evidence for oceanic separations in the Palaeozoic of Britain. *Journal of the Geological Society of London*, Vol. 139, 465–478.

COCKS, L R M, and FORTEY, R A. 1990. Biogeography of Ordovician and Silurian faunas. 97–104 *in* Palaeozoic Palaeogeography and Biogeography. MCKERROW, R S, and SCOTESE, C R (editors). *Memoir of the Geological Society of London*, No. 12.

COLMAN, T B, and APPLEBY, A–K. 1991. Volcanogenic quartz-magnetite-hematite veins. Snowdon, North Wales. *Mineralogical Magazine*, Vol. 55, 257–262.

COOPE, G R. 1977. Fossil coleopteran assemblages as sensitive indicators of climatic changes during the Devensian (last) cold stage. *Philosophical Transactions of the Royal Society of London Series B*, Vol. 280, 313–340.

COOPE, G R, and PENNINGTON, W. 1977. Discussion: the Windermere Interstadial of the Late-Devensian. 337–339 *in* Fossil coleopteran assemblages as sensitive indicators of climatic changes during the Devensian (last) cold stage. *Philosophical Transactions of the Royal Society of London Series B*, Vol. 280, 321–340.

COOPER, A H, and HUGHES, R A. 1993. Discussion of MILLWARD, D, and MOLYNEUX, S G. 1992 'Field and biostratigraphic evidence for an unconformity at the base of the Eycott Volcanic Group in the English Lake District'. *Geological Magazine*, Vol. 130, 271–272.

COOPER, A H, MILLWARD, D, JOHNSON, E W, and SOPER, N J. 1993. The early Palaeozoic evolution of northwest England. *Geological Magazine*, Vol. 130, 711–724.

COOPER, A H, and MOLYNEUX, S G. 1990. The age and correlation of the Skiddaw Group (early Ordovician) sediments in the Cross Fell inlier (northern England). *Geological Magazine*, Vol. 127, 147–157.

COOPER, A H, and five others. 1995. The stratigraphy, correlation, provenance and palaeogeography of the Skiddaw Group (Ordovician) in the English Lake District. *Geological Magazine*, Vol. 132, 185–211.

COOPER, D C, and BRADLEY, A D. 1990. The ammonium content of granites in the English Lake District. *Geological Magazine*, Vol. 127, 579–586.

COOPER, D C, and six others. 1988. The Crummock Water aureole: a zone of metasomatism and source of ore metals in the English Lake District. *Journal of the Geological Society of London*, Vol. 145, 523–540.

COPE, J C W, INGHAM, J K, and RAWSON, P F. 1992. Atlas of palaeogeography and lithofacies. *Memoir of the Geological Society of London*, No. 18.

CREWDSON, C. 1915. New fossiliferous horizon in the Coniston Grits of Windermere. *Geological Magazine*, Vol. 2, 169–171.

CRIDDLE, A J, and STANLEY, C J. 1979. New data on wittichenite. *Mineralogical Magazine*, Vol. 43, 109–113.

CUMMINS, W A. 1979. Neolithic stone axes: distribution and trade in England and Wales. 5–12 in *Stone axe studies.* CLOUGH, T H M, and CUMMINS, W A (editors). (London: Council for British Archaeology, Research Report 23.)

DAGGER, G W. 1977. Controls of copper mineralization at Coniston, English Lake District. *Geological Magazine*, Vol. 114, 195–202.

DAVIDSON, J P, FERGUSON, K M, COLUCCI, M T, and DUNGAN, M A. 1988. The origin and evolution of magmas from the San Pedro–Pellado Volcanic Complex, S Chile: multicomponent sources and open system evolution. *Contributions to Mineralogy and Petrology*, Vol. 100, 429–445.

DAVIES, W J K. 1981. *The Ravenglass and Eskdale Railway.* (2nd edition). (Newton Abbot: David and Charles.)

DAVIS, J, and HAWKESWORTH, C J. 1993. The petrogenesis of 30–20 Ma basic and intermediate volcanics from the Mogollon–Datil Volcanic Field, New Mexico, USA. *Contributions to Mineralogy and Petrology*, Vol. 115, 165–183.

DAVIS, N. 1989. The relationship between ignimbrite eruption and caldera collapse in the Borrowdale Volcanic Group of the Central Fells, English Lake District. Unpublished PhD thesis, University of Sheffield.

DAVISON, W, HILTON, J, LISHMAN, J P, and PENNINGTON, W. 1985. Contemporary lake transport processes determined from sedimentary records of copper mining activity. *Environmental Science and Technology*, Vol. 19, 356–360.

DEWEY, H, and EASTWOOD, T. 1925. Copper ores of the Midlands, the Lake District and the Isle of Man. *Special Reports on the Mineral Resources of Great Britain. Memoir of the Geological Survey of Great Britain*, Vol. 30.

DICKINSON, W R, and eight others. 1983. Provenance of North American Phanerozoic sandstones in relation to tectonic setting. *Geological Society of America Bulletin*, Vol. 94, 222–235.

DRUITT, T H, MELLORS, R A, PYLE, D, and SPARKS, R S J. 1989. Explosive volcanism on Santorini, Greece. *Geological Magazine*, Vol. 126, 95–126.

DUNHAM, K C. 1952. Age-relations of the epigenetic mineral deposits of Britain. *Transactions of the Geological Society of Glasgow*, Vol. 21, 395–429.

DUNHAM, K C. 1984. Genesis of the Cumbria hematite deposits. *Proceedings of the Yorkshire Geological Society*, Vol. 45, 130.

DWERRYHOUSE, A R. 1909. On some intrusive rocks in the neighbourhood of Eskdale (Cumberland). *Quarterly Journal of the Geological Society of London*, Vol. 65, 55–80.

EASTWOOD, T. 1921. The lead and zinc ores of the Lake District. *Special Reports on the Mineral Resources of Great Britain. Memoir of the Geological Survey of Great Britain*, Vol. 22.

EASTWOOD, T. 1959. The Lake District mining field. 149–174 in *The future of non-ferrous mining in Great Britain and Ireland.* (London: Institution of Mining and Metallurgy.)

EICHELBERGER, J C. 1975. Origin of andesite and dacite: evidence of mixing at Glass Mountain in California and at other circum-Pacific volcanoes. *Geological Society of America Bulletin*, Vol. 86, 1381–1391

EVANS, D J, ROWLEY, W J, CHADWICK, R A, KIMBELL, G S, and MILLWARD, D. 1994. Seismic reflection data and the internal structure of the Lake District batholith, Cumbria, northern England. *Proceedings of the Yorkshire Geological Society*, Vol. 50, 11–24.

EVANS, D J, ROWLEY, W J, CHADWICK, R A, and MILLWARD, D. 1993. Seismic reflections from within the Lake District batholith, Cumbria, northern England. *Journal of the Geological Society of London*, Vol. 150, 1043–1046.

EVANS, J A, MILLAR, I L, and NOBLE, S R. 1995. Hydration during uplift is recorded by reset Rb-Sr whole-rock ages. *Journal of the Geological Society of London*, Vol. 152, 209–212.

EVANS, W B, and ARTHURTON, R S. 1973. North-west England. 28–36 *in* A correlation of Quaternary deposits in the British Isles. MITCHELL, G F, PENNY, L F, SHOTTON, F W, and WEST, R G (editors). *Special Report of the Geological Society of London*, No. 4.

EWART, A. 1982. The mineralogy and petrology of Tertiary–Recent orogenic volcanic rocks: with special reference to the andesitic–basaltic compositional range. 25–95 in *Andesites.* THORPE, R S (editor). (Chichester: John Wiley.)

FIRMAN, R J. 1957. The Borrowdale Volcanic Series between Wastwater and Duddon valley, Cumberland. *Proceedings of the Yorkshire Geological Society*, Vol. 31, 39–64.

FIRMAN, R J. 1978a. Intrusions. 146–163 *in* The geology of the Lake District. MOSELEY, F (editor). *Occasional Publication of the Yorkshire Geological Society*, No. 3.

FIRMAN, R J. 1978b. Epigenetic mineralisation. 226–241 *in* The geology of the Lake District. MOSELEY, F (editor). *Occasional Publication of the Yorkshire Geological Society*, No. 3.

FIRMAN, R J, and LEE, M K. 1986. The age and structure of the concealed English Lake District batholith and its probable influence on subsequent sedimentation, tectonics and mineralisation. 117–127 in *Geology in the real world — the Kingsley Dunham volume.* NESBITT, R W, and NICHOL, J (editors). (Institution of Mining and Metallurgy.)

FITTON, J G. 1971. The petrogenesis of the calc-alkaline Borrowdale Volcanic Group, Northern England. Unpublished PhD thesis, University of Durham.

FITTON, J G. 1972. The genetic significance of almandine–pyrope phenocrysts in the calc-alkaline Borrowdale Volcanic Group, Northern England. *Contributions to Mineralogy and Petrology*, Vol. 36, 231–248.

FITTON, J G, and HUGHES, D J. 1970. Volcanism and plate tectonics in the British Ordovician. *Earth and Planetary Science Letters*, Vol. 8, 223–228.

FITTON, J G, THIRLWALL, M F, and HUGHES, D J. 1982. Volcanism in the Caledonian orogenic belt of Britain. 611–636 in *Andesites.* THORPE, R S (editor). (New York: Wiley.)

FORDHAM, B G. 1992. Chronometric calibration of mid-Ordovician to Tournaisian conodont zones: a compilation from recent graphic-correlation and isotope studies. *Geological Magazine*, Vol. 129, 709–721.

FORTEY, N J. 1989. Low grade metamorphism in the Lower Ordovician Skiddaw Group of the Lake District, England. *Proceedings of the Yorkshire Geological Society*, Vol. 47, 325–337.

FURNESS, R R. 1965. The petrography and provenance of the Coniston Grits east of the Lune valley, Westmorland. *Geological Magazine*, Vol. 102, 252–260.

FYFE, W S, and MCBIRNEY, A R. 1975. Subduction and the structure of andesitic volcanic belts. *American Journal of Science*, Vol. 275A, 285–297.

GEIKIE, A. 1891. The Anniversary Address of the President. *Quarterly Journal of the Geological Society of London*, Vol. 47, 48–162.

GIBBS, A D. 1989. Structural styles in basin formation. 81–93 *in* Extensional tectonics and stratigraphy of the North Atlantic margins. TANKARD, A J, and BALKWILL, H R (editors). *American Association of Petroleum Geologists Memoir*, No. 46.

GILL, J B. 1981. *Orogenic andesites and plate tectonics.* (Berlin: Springer-Verlag.)

GOUGH, D I. 1973. Dynamic uplift of Andean Mountains and Island Arcs. *Nature, London, Physical Sciences*, Vol. 242, 39–41.

GOVINDA-RAJULU, B V S. 1959. The structural and petrological relations of the Ennerdale Granophyre (Cumberland) to the adjoining rocks. Unpublished PhD thesis, University College, London.

GREEN, J F N. 1915. The garnets and streaky rocks of the English Lake District. *Mineralogical Magazine*, Vol. 17, 207–217.

GREEN, J F N. 1919. The vulcanicity of the Lake District. *Proceedings of the Geologists' Association*, Vol. 30, 153–182.

GREEN, J F N. 1920. The geological structure of the Lake District. *Proceedings of the Geologists' Association*, Vol. 31, 109–126.

GROVE, T L, GERLACH, D C, and SANDO, T W. 1982. Origin of calc-alkaline series lavas at Medicine Lake Volcano by fractionation, assimilation and mixing. *Contributions to Mineralogy and Petrology*, Vol. 80, 160–182.

HALLIDAY, A N. 1984. Coupled Sm–Nd and U–Pb systematics in late Caledonian granites and the basement under northern Britain. *Nature, London*, Vol. 307, 229–233.

HARLAND, W B, COX, A V, CRAIG, L E, SMITH, A G, and SMITH, D G. 1990. *A geologic time scale 1989.* (Cambridge: Cambridge University Press.)

HARMON, R S, and HALLIDAY, A N. 1980. Oxygen and strontium isotope relationships in the British late Caledonian granites. *Nature, London*, Vol. 283, 21–25.

HARTLEY, J J. 1925. The succession and structure of the Borrowdale Volcanic Series as developed in the area lying between the lakes of Grasmere, Windermere and Coniston. *Proceedings of the Geologists' Association*, Vol. 36, 203–226.

HARTLEY, J J. 1932. The volcanic and other igneous rocks of Great and Little Langdale, Westmorland; with notes on the

tectonics of the district. *Proceedings of the Geologists' Association*, Vol. 43, 32–69.

HARVEY, A M, OLDFIELD, F, BARON, A F, and PEARSON, G W. 1981. Dating of post-glacial landforms in the Central Howgills. *Earth Surface Processes and Landforms*, Vol. 6, 401–412.

HAWORTH, E Y. 1985. The highly nervous system of the English Lakes; aquatic ecosystem sensitivity to external changes, as demonstrated by diatoms. *Freshwater Biological Association Annual Report*, No. 53, 60–79.

HAWORTH, E Y, and LONG, J. 1989. *Lake District palaeolimnology: a bibliography.* (Ambleside: Institute of Freshwater Ecology.)

HELM, D G. 1970. Stratigraphy and structure in the Black Combe inlier, English Lake District. *Proceedings of the Yorkshire Geological Society*, Vol. 38, 105–148.

HIBBERT, A, JOHNSON, W H, and ADAMS, J J. 1940. *Survey of iron ore resources of West Cumberland.* (Cumberland: Cumberland Development Council Ltd.)

HIGGINS, M W. 1973. Petrology of Newberry volcano, central Oregon. *Geological Society of America Bulletin*, Vol. 84, 455–488.

HILDRETH, J, and MOORBATH, S. 1988. Crustal contributions to arc magmatism in the Andes of central Chile. *Contributions to Mineralogy and Petrology*, Vol. 98, 455–489.

HILDRETH, W, GRUNDER, A L, and DRAKE, R E. 1984. The Loma Seca Tuff and the Calabozos caldera: a major ash-flow and caldera complex in the southern Andes of central Chile. *Geological Society of America Bulletin*, Vol. 95, 45–54.

HOLLAND, E G. 1986. *Coniston copper.* (Milnthorpe, Cumbria: Cicerone Press.)

HOLLINGWORTH, S E. 1931. Glaciation of western Edenside and adjoining areas and the drumlins of the Edenside and Solway Basin. *Quarterly Journal of the Geological Society of London*, Vol. 87, 281–359.

HOWELL, F T. 1971. A continuous seismic profile survey of Windermere. *Geological Journal*, Vol. 7, 329–334.

HUDDART, D. 1972. Late Devensian glacial history. 3–47 in *The Cumberland Lowland. Handbook.* HUDDART, D, and TOOLEY, M J (editors). (Durham: Quaternary Research Association.)

HUDDART, D, and TOOLEY, M J (editor).1972. *Field Guide to the Cumberland Lowland.* (Cambridge: Quaternary Research Association.)

HUDDART, D, TOOLEY, M J, and CARTER, P. 1977. The coasts of North West England. 119–154 *in* The Quaternary History of the Irish Sea. KIDSON, C, and TOOLEY, M J (editors). *Geological Journal Special Issue.* (7th edition.)

HUGHES, R A, COOPER, A H, and STONE, P. 1993. Structural evolution of the Skiddaw Group (English Lake District) on the northern margin of eastern Avalonia. *Geological Magazine*, Vol. 130, 621–629.

HUGHES, R A, EVANS, J A, NOBLE, S R, and RUNDLE, C C. 1996. U-Pb chronology of the Ennerdale and Eskdale intrusions supports sub-volcanic relationships with the Borrowdale Volcanic Group (Ordovician, English Lake District). *Journal of the Geological Society of London*, Vol. 153, 33–38.

HUTT, J E. 1974. The Llandovery graptolites of the English Lake District, Pt 1. *Palaeontographical Society Monograph*, 1–56.

HUTT, J E. 1975. The Llandovery graptolites of the English Lake District, Pt 2. *Palaeontographical Society Monograph*, 57–137.

INESON, P R, and MITCHELL, J G. 1974. K-Ar isotopic age determinations from some Lake District mineral localities. *Geological Magazine*, Vol. 111, 521–537.

INGHAM, J K. 1966. The Ordovician rocks in the Cautley and Dent districts of Westmorland and Yorkshire. *Proceedings of the Yorkshire Geological Society*, Vol. 35, 455–505.

INGHAM, J K. 1970. The Upper Ordovician trilobites from the Cautley and Dent districts of Westmorland and Yorkshire. *Palaeontographical Society Monograph*, Pt 1, 1–58.

INGHAM, J K, and McNAMARA, K J. 1978. The Coniston Limestone Group. 121–129 *in* The geology of the Lake District. MOSELEY, F (editor). *Occasional Publication of the Yorkshire Geological Society*, No. 3.

INGHAM, J K, McNAMARA, K J, and RICKARDS, R B. 1978. The Upper Ordovician and Silurian rocks. 121–145 *in* The geology of the Lake District. MOSELEY, F (editor). *Occasional Publication of the Yorkshire Geological Society*, No. 3.

JOHNSON, E W, BRIGGS, D E G, SUTHREN, R J, WRIGHT, J L, and TUNNICLIFF, S P. 1994. Non-marine arthropod traces from the subaerial Ordovician Borrowdale Volcanic Group, English Lake District. *Geological Magazine*, Vol. 131, 395–406.

JOHNSON, G A L. 1961. Skiddaw Slates proved in the Teesdale Inlier. *Nature, London*, Vol. 90, 996–997.

JOHNSON, M R W, SANDERSON, D J, and SOPER, N J. 1979. Deformation in the Caledonides of England, Ireland and Scotland. 165–186 *in* The Caledonides of the British Isles — reviewed. HARRIS, A L, HOLLAND, C H, and LEAKE, B E (editors). *Special Publication of the Geological Society of London*, No. 8.

JONES, R L, and KEEN, D H. 1993. *Pleistocene environments in the British Isles.* (London: Chapman and Hall.)

KANARIS-SOTIRIOU, R, MILLWARD, D, and RUNDLE, C C. 1991. The Great Whinscale Dacite — an enigmatic lava flow from the Borrowdale Volcanic Group, English Lake District. *Proceedings of the Yorkshire Geological Society*, Vol. 48, 393–408.

KENDALL, J D. 1884. Mineral veins of the Lake District. *Transactions of the Manchester Geological Society*, Vol. 17, 292–341.

KIMBELL, G S. 1994. The interpretation of gravity and aeromagnetic data for the geological investigation of the Sellafield area, west Cumbria. *Proceedings of the Yorkshire Geological Society*, Vol. 50, 103–112.

KIMBELL, G S, and STONE, P. 1995. Crustal magnetization variations across the Iapetus Suture Zone. *Geological Magazine*, Vol. 132, 599–609.

KING, C A M. 1976. *The geomorphology of the British Isles: Northern England.* (London: Methuen and Company Limited.)

KING, L M. 1992. A basin study of the early Palaeozoic Windermere Group, NW England. Unpublished PhD thesis, University of Cambridge.

KNELLER, B C. 1990a. The Wenlock rocks of Sheet 38 (Ambleside). *British Geological Survey Technical Report*, WA/90/64.

KNELLER, B C. 1990b. The Ludlow rocks of Sheet 38 (Ambleside). *British Geological Survey Technical Report*, WA/90/62.

KNELLER, B C. 1991. A foreland basin on the southern margin of Iapetus. *Journal of the Geological Society of London*, Vol. 148, 207–210.

KNELLER, B C, and BELL, A M. 1993. An Acadian mountain front in the English Lake District: the Westmorland Monocline. *Geological Magazine*, Vol. 130, 203–213.

KNELLER, B C, EDWARDS, E, McCAFFREY, W D, and MOORE, R. 1991. Oblique reflection of turbidity currents. *Geology*, Vol. 19, 250–252.

KNELLER, B C, KING, L M, and BELL, A M. 1993b. Foreland basin development and tectonics on the northwest margin of eastern Avalonia. *Geological Magazine*, Vol. 130, 691–697.

KNELLER, B C, KOKELAAR, B P, and DAVIS, N C. 1993a. The Lingmell Formation. *British Geological Survey Technical Report*, WA/93/42.

KNELLER, B C, and MCCONNELL, B J. 1993. The Seathwaite Fell Formation in the Central Fells. *British Geological Survey Technical Report*, WA/93/43.

KNELLER, B C, SCOTT, R W, SOPER, N J, JOHNSON, E W, and ALLEN, P M. 1994. Lithostratigraphy of the Windermere Supergroup, Northern England. *Geological Journal*, Vol. 29, 219–240.

KNELLER, B C, and SOPER, N J. 1990. Geology of SD39SE and SD39SW. *British Geological Survey Technical Report*, WA/90/65.

KOKELAAR, B P. 1982. Fluidisation of wet sediments during the emplacement and cooling of various igneous bodies. *Journal of the Geological Society of London*, Vol. 139, 21–33.

LAKE DISTRICT NATIONAL PARK. 1994. *Lake District National Park Local Plan incorporating policies in respect of minerals and waste.* (Kendal, Cumbria: Lake District National Park.)

LAUFELD, S, BERGSTRÖM, J, and WARREN, P T. 1975. The boundary between the Silurian Cyrtograptus and Colonus Shales in Skane, southern Sweden. *Geologiska Föreningens i Stockholm Förhandlingar*, Vol. 97, 207–222.

LAWRENCE, D J D, WEBB, B C, YOUNG, B, and WHITE, D E. 1986. The geology of the late Ordovician and Silurian rocks (Windermere Group) in the area around Kentmere and Crook. *British Geological Survey Report*, Vol. 18, No. 5.

LE MAITRE, R W. 1989. *A classification of igneous rocks and glossary of terms. Recommendations of the International Union of Geological Sciences Subcommission on the Systematics of Igneous Rocks.* (Oxford: Blackwell Scientific Publications.)

LEE, M K. 1984a. *The three-dimensional form of the Lake District granite batholith. Investigation of the geothermal potential of the UK.* (Keyworth, Nottingham: British Geological Survey.)

LEE, M K. 1984b. *Analysis of geophysical logs from the Shap, Skiddaw, Cairngorm, Ballater, Mount Battock and Bennachie heat flow boreholes. Investigation of the geothermal potential of the UK.* (Keyworth, Nottingham: British Geological Survey.)

LEE, M K. 1986a. A new gravity survey of the Lake District and three-dimensional model of the granite batholith. *Journal of the Geological Society of London*, Vol. 143, 425–435.

LEE, M K. 1986b. Hot dry rock. 21–42 in *Geothermal energy: the potential in the United Kingdom.* DOWNING, R A, and GRAY, D A (editors). (London: HMSO for the British Geological Survey.)

LEE, M K. 1988. Density variations within Lake District granites and Lower Palaeozoic rocks. *British Geological Survey Technical Report*, WK/88/9.

LEE, M K. 1989. Upper crustal structure of the Lake District from modelling and image processing of potential field data. *British Geological Survey Technical Report*, WK/89/1.

LEE, M K, BROWN, G C, WEBB, P C, WHEILDON, J, and ROLLIN, K E. 1987. Heat flow, heat production and thermo-tectonic setting in mainland UK. *Journal of the Geological Society of London*, Vol. 144, 35–42.

LEE, M K, PHARAOH, T C, and SOPER, N J. 1990. Structural trends in central Britain from images of gravity and aeromagnetic fields. *Journal of the Geological Society of London*, Vol. 147, 241–258.

LEE, M K, and eight others. 1984. *Hot dry rock prospects in Caledonian granites. Investigation of the geothermal potential of the UK.* (Keyworth, Nottingham: British Geological Survey.)

LEVISTON, D. 1982. The Appletreeworth area. 106–114 in *The Lake District.* SHIPP, T (editor). (London: Unwin Paperbacks.)

LEWIS, C L E, GREEN, P F, CARTER, A, and HURFORD, A J. 1992. Elevated K/T palaeotemperatures throughout Northwest England: three kilometres of Tertiary erosion? *Earth and Planetary Science Letters*, Vol. 112, 131–145.

LIOU, J G, MARUYAMA, S, and CHO, M. 1985. Phase equilibria and mineral parageneses of metabasites in low-grade metamorphism. *Mineralogical Magazine*, Vol. 49, 321–333.

LIPMAN, P W. 1984. The roots of ash-flow calderas in western North America: windows into the tops of granitic batholiths. *Journal of Geophysical Research*, Vol. 89 B10, 8801–8841.

LOPEZ-ESCOBAR, L, FREY, F A, and VERGARA, M. 1977. Andesites and high-alumina basalts from the central-south Chile High Andes: geochemical evidence bearing on their petrogenesis. *Contributions to Mineralogy and Petrology*, Vol. 63, 199–228.

LOWE, J J, and WALKER, M J C. 1984. *Reconstructing Quaternary environments.* (Harlow: Longman.)

LOWRY, D, BOYCE, A J, PATTRICK, R A D, FALLICK, A E, and STANLEY, C J. 1991. A sulphur isotopic investigation of the potential sulphur sources for Lower Palaeozoic-hosted vein mineralization in the English Lake District. *Journal of the Geological Society of London*, Vol. 148, 993–1004.

MACDONALD, R, MILLWARD, D, BEDDOE-STEPHENS, B, and LAYBOURN-PARRY, J. 1988. The role of tholeiitic magmatism in the English Lake District: evidence from dykes in Eskdale. *Mineralogical Magazine*, Vol. 52, 459–472.

MACDONALD, R, THORPE, R S, GASGARTH, J W, and GRINDROD, A R. 1985. Multi-component origin of Caledonian lamprophyres of northern England. *Mineralogical Magazine*, Vol. 49, 485–494.

MACKERETH, F J H. 1966. Some chemical observations on post-glacial lake sediments. *Philosophical Transactions of the Royal Society of London Series B*, Vol. 250, 165–213.

MACKLIN, M G, RUMSBY, B T, HEAP, T, and PASSMORE, D G. 1994. Thinhope Burn, Northumberland. 50–57 in *The Quaternary of Cumbria: field guide.* BOARDMAN, J, and WALDEN, J (editors). (Oxford: Quaternary Research Association.)

MALIN, M C, DZURISIN, D, and SHARPE, R P. 1983. Stripping of Keanakakoi tephra on Kilauea Volcano, Hawaii. *Geological Society of America Bulletin*, Vol. 94, 1148–1158.

MANIAR, P D, and PICCOLI, P M. 1989. Tectonic discrimination of granitoids. *Geological Society of America Bulletin*, Vol. 101, 635–643.

MARR, J E. 1878. On some well-defined life-zones in the Lower Part of the Silurian (Sedgwick) of the Lake District. *Quarterly Journal of the Geological Society of London*, Vol. 34, 871–885.

MARR, J E. 1892. On the Wenlock and Ludlow Strata of the Lake District. *Geological Magazine*, Vol. 9, 534–541.

MARR, J E. 1900. Notes on the geology of the English Lake District. *Proceedings of the Geologists' Association*, Vol. 16, 449–483.

MARR, J E. 1916. *The geology of the Lake District and the scenery as influenced by geological structure.* (Cambridge: Cambridge University Press.)

MARR, J E, and NICHOLSON, H A. 1888. The Stockdale Shales. *Quarterly Journal of the Geological Society of London*, Vol. 44, 654–732.

MARTIN-DEL POZZO, A L. 1982. Monogenetic vulcanism in Sierra Chichinautzin, Mexico. *Bulletin Volcanologique*, Vol. 45, 9–24.

MATHESON, I. 1986. Coniston mines in 1849. *The Mine Explorer*, Vol. 2, 41–44.

McAllister, W D N, and Firman, R J. 1980. Stream-sediment uranium anomalies in Eskdale, Cumbria, England. *Transactions of the Institution of Mining and Metallurgy*, Vol. 89, Pt B, 177–181.

McBirney, A R, and White, C M. 1982. The Cascade Province. 115–136 in *Andesites*. Thorpe, R S (editor). (Chichester: John Wiley.)

McCabe, C, and Channell, J E T. 1990. Paleomagnetic results from volcanic rocks of the Shelve Inlier, Wales: evidence for a wide Late Ordovician Iapetus Ocean in Britain. *Earth and Planetary Science Letters*, Vol. 96, 458–468.

McCaffrey, W D. 1991. Silurian turbidite provenance in NW England. Unpublished PhD thesis, University of Leeds.

McCaffrey, W D. 1994. Sm–Nd isotopic characteristics of sedimentary provenance: the Windermere Supergroup of NW England. *Journal of the Geological Society of London*, Vol. 151, 1017–1021.

McCaffrey, W D, and Kneller, B C. 1996. Silurian turbidite provenance on the northern Avalonian margin. *Journal of the Geological Society of London*, Vol. 153, 437–450.

McConnell, B J. 1993. The Lincomb Tarns and Esk Pike Formations. *British Geological Survey Technical Report*, WA/93/44.

McConnell, B J, Kneller, B C, and Davis, N C. 1993. Minor intrusions in the Central Fells. *British Geological Survey Technical Report*, WA/93/45.

McKerrow, W S. 1979. Ordovician and Silurian changes in sea level. *Journal of the Geological Society of London*, Vol. 136, 137–145.

McNamara, K J. 1979a. The age, stratigraphy and genesis of the Coniston Limestone Group in the southern Lake District. *Geological Journal*, Vol. 14, 41–68.

McNamara, K J. 1979b. Trilobites from the Coniston Limestone Group (Ashgill Series) of the Lake District, England. *Palaeontology, Palaeoclimatology, Palaeoecology*, Vol. 22, 53–92.

Merriman, R J, Rex, D C, Soper, N J, and Peacor, D R. 1995. The age of Acadian cleavage in northern England, UK: K–Ar and TEM analysis of a Silurian metabentonite. *Proceedings of the Yorkshire Geological Society*, Vol. 50, 255–265.

Miklius, A, Flower, M F J, Huisjmans, J P P, Mukasa, S B, and Castillo, P. 1991. Geochemistry of lavas from Taal Volcano, southwestern Luzon, Philippines; evidence for multiple magma supply systems and mantle source heterogeneity. *Journal of Petrology*, Vol. 32, 593–627.

Millward, D. 1980. Three ignimbrites from the Borrowdale Volcanic Group. *Proceedings of the Yorkshire Geological Society*, Vol. 42, 595–616.

Millward, D, and Molyneux, S G. 1992. Field and biostratigraphic evidence for an unconformity at the base of the Eycott Volcanic Group in the English Lake District. *Geological Magazine*, Vol. 129, 77–92.

Millward, D, Moseley, F, and Soper, N J. 1978. The Eycott and Borrowdale volcanic rocks. 99–120 *in* The geology of the Lake District. Moseley, F (editor). *Occasional Publication of the Yorkshire Geological Society*, No. 3.

Milodowski, A E, and Zalasiewicz, J A. 1991. The origin and sedimentary, diagenetic and metamorphic evolution of chlorite-mica stacks in Llandovery sediments of central Wales, UK. *Geological Magazine*, Vol. 128, 263–278.

Mitchell, G F, Penny, L F, Shotton, F W, and West, R G. 1973. A correlation of Quaternary deposits in the British Isles. *Special Report of the Geological Society of London*, No. 4.

Mitchell, G H. 1940. The Borrowdale Volcanic Series of Coniston, Lancashire. *Quarterly Journal of the Geological Society of London*, Vol. 96, 301–319.

Mitchell, G H. 1956a. The geological history of the Lake District. *Proceedings of the Yorkshire Geological Society*, Vol. 30, 407–463.

Mitchell, G H. 1956b. The Borrowdale Volcanic Series of the Dunnerdale Fells, Lancashire. *Liverpool and Manchester Geological Journal*, Vol. 1, 428–449.

Mitchell, G H. 1963. The Borrowdale Volcanic rocks of the Seathwaite Fells, Lancashire. *Liverpool and Manchester Geological Journal*, Vol. 3, 289–300.

Mitchell, W A, and Clark, C D. 1994. The last ice sheet in Cumbria. 4–14 in *The Quaternary of Cumbria: field guide*. Boardman, J, and Walden, J (editors). (Oxford: Quaternary Research Association.)

Molyneux, S G. 1988. Micropalaeontological evidence for the age of the Borrowdale Volcanic Group. *Geological Magazine*, Vol. 125, 541–542.

Moseley, F. 1974. Structural relations between the Skiddaw Slates and the Borrowdale Volcanics. *Proceedings of the Cumberland Geological Society*, Vol. 3, 127–145.

Moseley, F. 1983. *The volcanic rocks of the Lake District*. (London: MacMillan Press.)

Moseley, F. 1984. Lower Palaeozoic lithostratigraphical classification in the English Lake District. *Geological Journal*, Vol. 19, 239–247.

Moseley, F. 1990. The Tilberthwaite Tuffs from Langdale to Tilberthwaite. 143–150 in *Geology of the Lake District*. (Geologists' Association Guide). Moseley, F (editor). (London: Geologists' Association.)

Moseley, F. 1993. Field relations and movement history of the Coniston Fault, Cumbria. *Proceedings of the Geologists' Association*, Vol. 104, 1–13.

Moseley, F, and Millward, D. 1982. Ordovician volcanicity in the English Lake District. 93–111 in *Igneous rocks of the British Isles*. Sutherland, D S (editor). (London: Wiley.)

Musson, R M W. 1987. Seismicity of southwest Scotland and northwest England; with a catalogue of earthquakes within 75 km of Chapelcross. *British Geological Survey Global Seismology Report*, No. 316.

Musson, R M W. 1994. A catalogue of British earthquakes. *British Geological Survey Technical Report*, WL/94/04.

Musson, R M W, Neilson, G, and Burton, P W. 1984. Macroseismic reports on historical British earthquakes 1: northwest England and southwest Scotland. *British Geological Survey Global Seismology Report*, Vol. 1 (2 vols), No. 207.

Neall, T, Young, B, Hyslop, E K, and Fakes, R D. 1993. Namibite from Buckbarrow Beck, Cumbria: the first British occurrence. *Mineralogical Magazine*, Vol. 57, 754–755.

Norman, T N. 1961. The geology of the Silurian strata in the Blawith area, Furness. Unpublished PhD thesis, University of Birmingham.

Numan, N M S. 1974. Structure and stratigraphy of the southern part of the Borrowdale Volcanic Group, English Lake District. Unpublished PhD thesis, University of Sheffield.

O'Brien, C, Plant, J A, Simpson, P R, and Tarney, J. 1985. The geochemistry, metasomatism and petrogenesis of the granites of the English Lake District. *Journal of the Geological Society of London*, Vol. 142, 1139–1157.

OLIVER, G J H. 1988. Arenig to Wenlock regional metamorphism in the paratectonic Caledonides of the British Isles: a review. 347–363 *in* The Caledonian–Appalachian Orogen. HARRIS, A L, and FETTES, D J (editors). *Special Publication of the Geological Society of London*, 38.

OLIVER, R L. 1954. Welded tuffs in the Borrowdale Volcanic Series, English Lake District, with a note on similar rocks in Wales. *Geological Magazine*, Vol. 91, 473–483.

OLIVER, R L. 1956a. The origin of garnets in the Borrowdale Volcanic Series and associated rocks, English Lake District. *Geological Magazine*, Vol. 93, 121–139.

OLIVER, R L. 1956b. The origin of garnets in the Borrowdale Volcanic Series. *Geological Magazine*, Vol. 93, 516–517.

OLIVER, R L. 1961. The Borrowdale Volcanic and associated rocks of the Scafell area, English Lake District. *Quarterly Journal of the Geological Society of London*, Vol. 117, 377–417.

ORCHARD, M J. 1980. Upper Ordovician conodonts from England and Wales. *Geologica et Palaeontologica*, Vol. 14, 9–44.

PARKER, W R. 1966. Quicksand structures in the bedded tuffs of Great Langdale, Westmorland. *Nature, London*, Vol. 210, 1247–1248.

PATRICK, C K. 1978. Hydrogeology. 250–261 *in* The geology of the Lake District. MOSELEY, F (editor). *Occasional Publication of the Yorkshire Geological Society*, No. 3.

PEARCE, J A. 1982. Trace element characteristics of lavas from destructive plate boundaries. 525–548 in *Andesites*. THORPE, R S (editor). (New York: Wiley.)

PEARCE, J A. 1983. Role of the sub-continental lithosphere in magma genesis at active continental margins. 230–249 in *Continental basalts and mantle xenoliths*. HAWKESWORTH, C J, and NORRY, M J (editors). (Nantwich, Cheshire: Shiva.)

PEARCE, J A, HARRIS, N B W, and TINDLE, A G. 1984. Trace element discrimination diagrams for the tectonic interpretation of granitic rocks. *Journal of Petrology*, Vol. 25, 956–983.

PEARCE, J A, and PARKINSON, I J. 1993. Trace element models for mantle melting: application to volcanic arc petrogenesis. 373–403 *in* Magmatic processes and plate tectonics. PRITCHARD, H M, ALABASTER, T, HARRIS, N B, and NEARY, C R (editors). *Special Publication of the Geological Society of London*, No. 76.

PENNINGTON, W. 1943. Lake sediments: the bottom deposits of the North Basin of Windermere, with special reference to the Diatom Succession. *New Phytologist*, Vol. 42, 1–27.

PENNINGTON, W. 1947. Studies of the post-glacial history of British vegetation. VII. Lake sediments: pollen diagrams from the the bottom deposits of the North Basin of Windermere. *Philosophical Transactions of the Royal Society of London, Series B*, Vol. 233, 137–175.

PENNINGTON, W. 1964. Pollen analyses from the deposits of six upland tarns in the Lake District. *Philosophical Transactions of the Royal Society of London, Series B*, Vol. 248, 205–244.

PENNINGTON, W. 1970. Vegetation history in the north-west of England: a regional synthesis. 41–79 in *Studies in the vegetational history of the British Isles*. WALKER, D, and WEST, R G (editors). (Cambridge: Cambridge University Press.)

PENNINGTON, W. 1975a. The effect of Neolithic man on the environment in north-west England: the use of absolute pollen diagrams. 74–86 in *The effect of man on the landscape: the Highland Zone. CBA Research Report Number 11*. EVANS, J G, LIMBREY, S, and CLEERE, H (editors). (London: Council for British Archaeology.)

PENNINGTON, W. 1975b. A chronostratigraphic comparison of Late-Weichselian and Late-Devensian subdivisions, illustrated by two radiocarbon-dated profiles from western Britain. *Boreas*, Vol. 4, 157–171.

PENNINGTON, W. 1977. The Late Devensian flora and vegetation of Britain. *Philosophical Transactions of the Royal Society of London, Series B*, Vol. 280, 247–271.

PENNINGTON, W. 1978. Quaternary geology. 207–225 *in* The geology of the Lake District. MOSELEY, F (editor). *Occasional Publication of the Yorkshire Geological Society*, No. 3.

PENNINGTON, W, and LISHMAN, J P. 1971. Iodine in lake sediments in northern England and Scotland. *Biological Reviews*, Vol. 46, 279–313.

PETTERSON, M G, BEDDOE-STEPHENS, B, MILLWARD, D, and JOHNSON, E W. 1992. A pre-caldera plateau-andesite field in the Borrowdale Volcanic Group of the English Lake District. *Journal of the Geological Society of London*, Vol. 149, 889–906.

PICHLER, H, and KUSSMAUL, S. 1980. Comments on the geological map of the Santorini islands. 413–427 in *Thera and the Aegean World II*. DOUMAS, C (editor). (London: Thera and the Aegean World.)

PICKERING, K T, HISCOTT, R N, and HEIN, F J. 1989. *Deep marine environments*. (London: Unwin Hyman.)

PIDGEON, R T, and AFTALION, M. 1978. Co-genetic and inherited zircon U–Pb systems in granites: Palaeozoic granites of Scotland and England. 183–220 *in* Crustal evolution in northwestern Britain and adjacent regions. BOWES, D R, and LEAKE, B E (editors). *Geological Journal Special Issue*, No. 10.

PINKERTON, H, and SPARKS, R S J. 1976. The 1975 sub-terminal lavas, Mount Etna: a case history of the formation of a compound lava field. *Journal of Volcanology and Geothermal Research*, Vol. 1, 167–182.

POSTLETHWAITE, T. 1913. *Mines and mining in the (English) Lake District*. (3rd edition). (Whitehaven: W H Moss (reprinted 1975 by Michael Moon, Whitehaven).)

POULIOT, G, BARONDEAU, B, SAUVE, P, and DAVIS, M. 1978. Distribution of alteration minerals and metals in the Fire Tower Zone at Brunswick Tin Mines Ltd., Mount Pleasant area, New Brunswick. *Canadian Mineralogist*, Vol. 16, 223–237.

PRUEHER, L M, and MCBIRNEY, A R. 1988. Relations of cinder cones to the magmatic evolution of Mount Mazama, Crater Lake National Park, Oregon. *Journal of Volcanology and Geothemal Research*, Vol. 35, 253–268.

RAISTRICK, A. 1925. The glaciation of Borrowdale, Cumberland. *Proceedings of the Yorkshire Geological Society*, Vol. 20, 155–181.

RASTALL, R H. 1906. The Buttermere and Ennerdale Granophyre. *Quarterly Journal of the Geological Society of London*, Vol. 62, 253–274.

RASTALL, R H. 1942. The ore deposits of the Skiddaw district. *Proceedings of the Yorkshire Geological Society*, Vol. 24, 328–343.

REDFEARN, P. 1979. The geology of the Silurian south of the Shap Granite with special reference to the Birkbeck Tunnel. Unpublished PhD thesis, University of Nottingham.

REEDMAN, A J, COLMAN, T B, CAMPBELL, S D G, and HOWELLS, M F. 1985. Volcanogenic mineralization related to the Snowdon Volcanic Group (Ordovician), Gwynedd, North Wales. *Journal of the Geological Society of London*, Vol. 142, 875–888.

RICKARDS, R B. 1964a. The graptolitic mudstone and associated facies in the Silurian strata of the Howgill Fells. *Geological Magazine*, Vol. 101, 435–451.

RICKARDS, R B. 1964b. Two new genera of Silurian phacopid trilobites. *Palaeontology*, Vol. 7, 541–551.

RICKARDS, R B. 1967. The Wenlock and Ludlow succession in the Howgill Fells (north-west Yorkshire and Westmorland). *Quarterly Journal of the Geological Society of London*, Vol. 123, 215–249.

RICKARDS, R B. 1969. Wenlock graptolite zones in the English Lake District. *Proceedings of the Geological Society of London*, Vol. 1654, 61–65.

RICKARDS, R B. 1970. The age of the Middle Coldwell Beds. *Proceedings of the Geological Society of London*, Vol. 1663, 111–114.

RICKARDS, R B. 1978. Silurian. 130–145 *in* The geology of the Lake District. MOSELEY, F (editor). *Occasional Publication of the Yorkshire Geological Society*, No. 3.

ROBERTS, B, MERRIMAN, R J, and PRATT, W. 1991. The influence of strain, lithology and stratigraphical depth on white mica (illite) crystallinity in mudrocks from the vicinity of the Corris Slate Belt, Wales: implications for the timing of metamorphism in the Welsh Basin. *Geological Magazine*, Vol. 128, 633–645.

ROCK, N M S. 1987. The nature and origin of lamprophyres: an overview. 191–226 *in* Alkaline igneous rocks. FITTON, J G, and UPTON, B G J (editors). *Special Publication of the Geological Society of London*, No. 30.

ROSE, W C C, and DUNHAM, K C. 1977. Geology and haematite deposits of south Cumbria. *Memoir of the British Geological Survey.*

RUDDIMAN, W F, MCINTYRE, A, NIEBLER-HUNT, V, and DURAZZI, J T. 1980. Oceanic evidence for the mechanism of rapid northern hemisphere glaciation. *Quaternary Research*, Vol. 13, 33–34.

RUNDLE, C C. 1979. Ordovician intrusions in the English Lake District. *Journal of the Geological Society of London*, Vol. 136, 29–38.

RUNDLE, C C. 1992. Review and assessment of isotopic ages from the English Lake District. *British Geological Survey Technical Report*, No. WA/92/38.

RUSSELL, A. 1925. A notice of the occurrence of native arsenic in Cornwall; of bismuthinite at Shap, Westmorland; and of smaltite and niccolite at Coniston, Lancashire. *Mineralogical Magazine*, Vol. 20, 299–304.

SALTER, J W. 1873. *A catalogue of the collection of Cambrian and Silurian fossils contained in the Geological Museum of the University of Cambridge, 72.* (Cambridge: Cambridge University Press.)

SCOFFIN, T P. 1971. The conditions of growth of the Wenlock reefs of Shropshire (England). *Sedimentology*, Vol. 17, 173–219.

SCOTT, R W. 1992. A palaeoenvironmental analysis of the Coniston Limestone Formation (Ashgill) in northern England. Unpublished PhD thesis, University of Sheffield.

SCOTT, R W, and KNELLER, B C. 1990. A report on the lithostratigraphy of the Ashgill and Llandovery age rocks on Sheet 38 (Ambleside). *British Geological Survey Technical Report*, WA/90/63.

SEDGWICK, A. 1832. On the geological relations of the stratified and unstratified groups of rocks composing the Cumbrian Mountains. *Proceedings of the Geological Society of London*, Vol. 1, 399.

SEDGWICK, A. 1845. On the comparative classification of the fossiliferous strata of North Wales, with the corresponding deposits of Cumberland, Westmoreland and Lancashire. *Quarterly Journal of the Geological Society of London*, Vol. 1, 442–450.

SEDGWICK, A. 1846. On the classification of the fossiliferous slates of Cumberland, Westmoreland and Lancashire. *Quarterly Journal of the Geological Society of London*, Vol. 2, 106–131.

SEGERSTROM, K. 1950. Erosion studies at Paricutin, State of Michoacan, Mexico. *United States Geological Survey Bulletin*, No. 965A, 1–164.

SERVAIS, T. 1993. The Ordovician acritarch *Frankea*. *Special Papers in Palaeontology*, Vol. 48, 79–95.

SHACKLETON, N J, and OPDYKE, N D. 1976. Oxygen isotope and palaeomagnetic stratigraphy of Equatorial Pacific core. V28–239, Late Pliocene to Latest Pleistocene. 449–464 *in* Investigation of late Quaternary palaeoceanography and palaeoclimatology. CLINE, R M, and HAGS, J D (compilers). *Geological Society of America Memoir*, No. 145.

SHAW, W T. 1959. Joint discussion on the Lake District Mining field. 219–223 in *The future of non-ferrous mining in Great Britain and Ireland.* (London: Institution of Mining and Metallurgy.)

SHAW, W T. 1970. *Mining in the Lake Counties.* (Clapham (via Lancaster): Dalesman Books.)

SHEPHERD, T J. 1973. Geochemical evidence for basement control of the West Cumberland haematite mineralization. Unpublished PhD thesis, University of Durham.

SHEPHERD, T J, BECKINSALE, R D, RUNDLE, C C, and DURHAM, J. 1976. Genesis of Carrock Fell tungsten deposits Cumbria: fluid inclusion and isotopic study. *Transactions of the Institution of Mining and Metallurgy*, Vol. 85, B63–73.

SIMPSON, A. 1967. The stratigraphy and tectonics of the Skiddaw Slates and the relationship of the overlying Borrowdale Volcanic Series in part of the Lake District. *Geological Journal*, Vol. 5, 391–418.

SIMPSON, B. 1934. The petrology of the Eskdale (Cumberland) Granite. *Proceedings of the Geologists' Association*, Vol. 45, 17–34.

SISSONS, J B. 1980. The Loch Lomond Advance in the Lake District, northern England. *Transactions of the Royal Society of Edinburgh: Earth Sciences*, Vol. 71, 13–27.

SMITH, B. 1924. Iron ores: haematites of west Cumberland, Lancashire and the Lake District. *Special Reports on the Mineral Resources of Great Britain. Memoir of the Geological Survey of Great Britain*, Vol. 8.

SMITH, R A. 1974. *A bibliography of the geology and geomorphology of Cumbria.* (Penrith: Cumberland Geological Society.)

SMITH, R A. 1990. *A bibliography of the geology and geomorphology of Cumbria Part 2 (1974–1990).* (Penrith: Cumberland Geological Society.)

SMITH, R L. 1979. Ash-flow magmatism. 5–27 *in* Ash-flow tuffs. CHAPIN, C E, and ELSTON, W E (editors). *Geological Society of America Special Paper*, No. 180.

SOPER, N J. 1987. The Ordovician (?) batholith of the English Lake District. Discussion of FIRMAN, R J and LEE, M K. 1986. Age and structure of the concealed English Lake District batholith and its probable influence on subsequent sedimentation, tectonics and mineralisation. 117–127 *in Geology and the Real World — the Kingsley Dunham Volume.* Institution of Mining and Metallurgy. *Geological Magazine*, Vol. 124, 481–482.

SOPER, N J, ENGLAND, R W, SNYDER, D B, and RYAN, P D. 1992. The Iapetus suture zone in England, Scotland and eastern Ireland: a reconciliation of geological and deep seismic data. *Journal of the Geological Society of London*, Vol. 149, 697–700.

SOPER, N J, and KNELLER, B C. 1990. Cleaved microgranite dykes of the Shap swarm in the Silurian of NW England. *Geological Journal*, Vol. 25, 161–170.

SOPER, N J, and MOSELEY, F. 1978. Structure. 45–67 *in* The geology of the Lake District. MOSELEY, F (editor). *Occasional Publication of the Yorkshire Geological Society*, No. 3.

SOPER, N J, WEBB, B C, and WOODCOCK, N H. 1987. Late Caledonian (Acadian) transpression in north-west England: timing, geometry and geotectonic significance. *Proceedings of the Yorkshire Geological Society*, Vol. 46, 175–192.

SOPER, N J, and WOODCOCK, N H. 1990. Silurian collision and sediment dispersal patterns in southern Britain. *Geological Magazine*, Vol. 127, 527–542.

SPARKS, R S J. 1976. Grain size variations in ignimbrites and implications for the transport of pyroclastic flows. *Sedimentology*, Vol. 23, 147–188.

STANLEY, C J. 1979. Mineralogical studies of copper, lead, zinc and cobalt mineralization in the English Lake District. Unpublished PhD thesis, University of Aston in Birmingham.

STANLEY, C J, and CRIDDLE, A G. 1979. Mineralization at Seathwaite Tarn, near Coniston, English Lake District: the first occurrence of wittichenite in Great Britain. *Mineralogical Magazine*, Vol. 43, 103–107.

STANLEY, C J, and VAUGHAN, D J. 1980. Interpretative studies of copper mineralization to the south of Keswick, England. *Transactions of the Institution of Mining and Metallurgy*, Vol. 89, B25–30.

STANLEY, C J, and VAUGHAN, D J. 1981. Native antimony and bournonite intergrowths in galena from the English Lake District. *Mineralogical Magazine*, Vol. 44, 257–260.

STANLEY, C J, and VAUGHAN, D J. 1982a. Copper, lead, zinc and cobalt mineralization in the English Lake District: classification, conditions of formation and genesis. *Journal of the Geological Society of London*, Vol. 139, 569–579.

STANLEY, C J, and VAUGHAN, D J. 1982b. Mineralization in the Bonser vein, Coniston, English Lake District: mineral assemblages, paragenesis, and formation conditions. *Mineralogical Magazine*, Vol. 46, 343–350.

STEPHENS ASSOCIATES. 1988. *Slate quarrying in the Lake District. A cause for concern? A report for Friends of the Lake District.* (Kendal: Stephens Associates.)

SUGDEN, D E, and JOHN, B S. 1976. *Glaciers and landscape: a geomorphological approach.* (London: Arnold.)

SUTHREN, R J. 1977. Volcanic and sedimentary facies of part of the Borrowdale Volcanic Group, Cumbria. Unpublished PhD thesis, University of Keele.

SUTHREN, R J, and FURNES, H. 1980. Origin of some bedded welded tuffs. *Bulletin Volcanologique*, Vol. 43, 61–71.

THIRLWALL, M F, and FITTON, J G. 1983. Sm–Nd garnet age for the Ordovician Borrowdale Volcanic Group, English Lake District. *Journal of the Geological Society of London*, Vol. 140, 511–518.

THOMAS, L J. 1986. Low grade metamorphism of the Lake District, England. Unpublished PhD thesis, University of St Andrews.

THOMAS, L J, HARMON, R D, and OLIVER, G J H. 1985. Stable isotope composition of alteration fluids in low-grade Lower Palaeozoic Rocks, English Lake District. *Mineralogical Magazine*, Vol. 49, 425–434.

THOMPSON, J B, MULLINS, H T, NEWTON, C R, and VERCOUTERE, T L. 1985. Alternative biofacies model for dysaerobic communities. *Lethaia*, Vol. 18, 167–179.

TIDDEMAN, R H. 1872. On the evidence for the ice-sheet in north Lancashire and adjacent parts of Yorkshire and Westmoreland. *Quarterly Journal of the Geological Society of London*, Vol. 28, 471–491.

TOMITA, K, KANAI, T, KOBAYASHI, T, and OBA, N. 1985. Accretionary lapilli formed by the eruption of Sakurajima volcano. *Journal of the Japanese Association of Mineral and Petroleum Economic Geologists*, Vol. 80, 49–54.

TOOLEY, M J (editor). 1977. *The Isle of Man, Lancashire coast and Lake District. Guidebook for Excursion A4.* (INQUA.)

TORSVIK, T H, and TRENCH, A. 1991. The Ordovician history of the Iapetus Ocean in Britain: new palaeomagnetic constraints. *Journal of the Geological Society of London*, Vol. 148, 423–425.

TRENCH, A, TORSVIK, T H, SMETHURST, M A, WOODCOCK, N H, and METCALF, R. 1991. A paleomagnetic study of the Builth Wells–Llandrindod Wells Ordovician Inlier, Wales: palaeogeographic and structural implications. *Geophysical Journal International*, Vol. 105, 477–489.

TROTTER, F M. 1929. The glaciation of the eastern Edenside, the Alston Block, and the Carlisle Plain. *Quarterly Journal of the Geological Society of London*, Vol. 85, 549–612.

TROTTER, F M, and HOLLINGWORTH, S E. 1932. The glacial sequence in the North of England. *Geological Magazine*, Vol. 69, 374–380.

TROTTER, F M, HOLLINGWORTH, S E, EASTWOOD, T, and ROSE, W C C. 1937. Gosforth district. *Memoir of the Geological Survey of the Great Britain*, Sheet 37, New Series (England).

TURNER, R E, and WADGE, A J. 1979. Acritarch dating of Arenig volcanism in the Lake District. *Proceedings of the Yorkshire Geological Society*, Vol. 42, 405–414.

VAUGHAN, D J, and IXER, R A. 1980. Studies of sulphide mineralogy of north Pennine ores and its contribution to genetic models. *Transactions of the Institution of Mining and Metallurgy*, Vol. 89, B99–109.

VERGARA, M, and MUNIZAGA, F. 1974. Age and evolution of the upper Cenozoic andesitic volcanism in central-south Chile. *Geological Society of America Bulletin*, Vol. 85, 603–606.

VESSEL, R K, and DAVIES, D K. 1981. Non-marine sedimentation in an active fore-arc basin. 31–45 *in* Recent and ancient non-marine depositional environments: models for exploration. ETHRIDGE, F G, and FLORES, R M (editors). *Society of Economic Palaeontologists and Mineralogists Special Publication*, No. 31.

WADGE, A J. 1966. Hydrogeology. 23–29 in *Cumbrian Rivers Hydrological Survey, hydrometric areas 75, 76 and part 74. Ministry of housing and local Government.* (London: HMSO.)

WADGE, A J. 1972. Sections through the Skiddaw–Borrowdale unconformity in eastern Lakeland. *Proceedings of the Yorkshire Geological Society*, Vol. 39, 179–198.

WADGE, A J. 1978. Classification and stratigraphical relationships of the Lower Ordovician rocks. 68–78 *in* The geology of the Lake District. MOSELEY, F (editor). *Occasional Publication of the Yorkshire Geological Society*, No. 3.

WADGE, A J, GALE, N H, BECKINSALE, R D, and RUNDLE, C C. 1978. A Rb–Sr isochron age for the Shap Granite. *Proceedings of the Yorkshire Geological Society*, Vol. 42, 297–305.

WALKER, D. 1965. The post-glacial period in the Langdale fells, English Lake District. *New Phytologist*, Vol. 64, 488–510.

WALKER, E E. 1904. Notes on the garnet-bearing and associated rocks of the Borrowdale Volcanic Series. *Quarterly Journal of the Geological Society of London*, Vol. 60, 70–104.

WALKER, G P L. 1970. Compound and simple lava flows and flood basalts. *Bulletin Volcanologique*, Vol. 35, 579–590.

WALKER, J A. 1981. Petrogenesis of lavas from cinder cane fields behind the volcanic front of Central America. *Journal of Geology*, Vol. 89, 721–739.

WALKER, J A, WILLIAMS, S N, KALAMARIDES, R I, and FEIGENSON, M D. 1993. Shallow open-system evolution of basaltic magma

beneath a subduction zone volcano: the Masaya Caldera Complex, Nicaragua. *Journal of Volcanology and Geothermal Research*, Vol. 56, 379–400.

WARD, J C. 1873. The glaciation of the northern part of the Lake District. *Quarterly Journal of the Geological Society of London*, Vol. 29, 422–441.

WARD, J C. 1874. The origin of some of the Lake basins of Cumberland. *Quarterly Journal of the Geological Society of London*, Vol. 30, 96–104.

WARD, J C. 1875. The glaciation of the southern part of the Lake District and the glacial origin of the lake-basins of Cumberland and Westmoreland. *Quarterly Journal of the Geological Society of London*, Vol. 31, 152–166.

WARD, J C. 1876. The geology of the northern part of the English Lake District. *Memoir of the Geological Survey of Great Britain.*

WARREN, P T, PRICE, D, NUTT, M J C, and SMITH, E G. 1984. Geology of the country around Rhyl and Denbigh. *Memoir of the British Geological Survey*, Sheets 95 and 107.

WATSON, C R. 1984. A volcanic vent at Colt Crag near Coniston, Cumbria. *Proceedings of the Cumberland Geological Society*, Vol. 4, 237–244.

WATSON, E B, and HARRISON, T M. 1983. Zircon saturation revisited: temperature and composition effects in a variety of crustal magma types. *Earth and Planetary Science Letters*, Vol. 64, 295–304.

WEBB, B C, MILLWARD, D, JOHNSON, E W, and COOPER, A H. 1987a. The Ordovician (?) batholith of the English Lake District. Discussion of FIRMAN, R J and LEE, M K. 1986. Age and structure of the concealed English Lake District batholith and its probable influence on subsequent sedimentation, tectonics and mineralisation. 117–127 *in Geology and the Real World–the Kingsley Dunham Volume.* Institution of Mining and Metallurgy. *Geological Magazine*, Vol. 124, 482–483.

WEBB, P C, and BROWN, G C. 1984. *The Lake District granites: heat production and related geochemistry. Investigation of the geothermal potential of the UK.* (Keyworth, Nottingham: British Geological Survey.)

WEBB, P C, LEE, M K, and BROWN, G C. 1987b. Heat flow–heat production relationships in the UK and the vertical distribution of heat production in granite batholiths. *Geophysical Research Letters*, Vol. 14, 279–282.

WERNICKE, B, and BIRCHFIELD, B C. 1982. Modes of extensional tectonics. *Journal of Structural Geology*, Vol. 4, 105–115.

WHEILDON, J, KING, G, CROOK, C N, and THOMAS-BETTS, A. 1984. *The Lake District Granites: heat flow, heat production and model studies. Investigation of the geothermal potential of the UK.* (Keyworth, Nottingham: British Geological Survey.)

WILSON, A A, and CORNWELL, J D. 1982. The Institute of Geological Sciences borehole at Beckermonds Scar, North Yorkshire. *Proceedings of the Yorkshire Geological Society*, Vol. 44, 59–88.

WILSON, D W R. 1954. The stratigraphy and palaeontology of Valentian rocks of Cautley (Yorkshire WR). Unpublished PhD thesis, University of Birmingham.

WILSON, M. 1989. *Igneous petrogenesis.* (London: Unwin Hyman Ltd.)

WOODCOCK, N H. 1990. Sequence stratigraphy of the Palaeozoic Welsh Basin. *Journal of the Geological Society of London*, Vol. 147, 537–547.

WOODWARD, J. 1729. *An attempt towards a natural history of the fossils of England: in a catalogue of the English fossils in the collection of J Woodward MD.* (London: F Fayram.)

WORDSWORTH, W. 1846. *The scenery of the Lakes including four letters on the geology of the Lake District by the Reverend Professor Sedgewick (edited by the Publisher).* (3rd edition). (Kendal: J Hudson.)

WRIGHT, A D. 1985. The Ordovician–Silurian boundary at Keisley, northern England. *Geological Magazine*, Vol. 122, 261–273.

WRIGHT, F, and RICHARDS, J A. 1995. The Coniston Earthquake of 18 July 1994 (2.2ML). *British Geological Survey Technical Report*, WL/95/34.

YOUNG, B. 1984. Geology and history of Nab Gill Mine, Eskdale, Cumbria. *Proceedings of the Cumberland Geological Society*, Vol. 4, 269–275.

YOUNG, B. 1985a. Greisens and related rocks associated with the Eskdale Granite, Cumbria. *British Geological Survey Report*, PDA2/85/2.

YOUNG, B. 1985b. Mineralization associated with the Eskdale intrusion, Cumbria. *Report of Programme Directorate A. British Geological Survey*, No. PDA2 85/3.

YOUNG, B. 1987. *Glossary of the minerals of the Lake District and adjoining area.* (Newcastle Upon Tyne: British Geological Survey.)

YOUNG, B, ANSARI, S M, and FIRMAN, R J. 1988. Field relationships, mineralogy and chemistry of the greisens and related rocks associated with the Eskdale Granite, Cumbria. *Proceedings of the Yorkshire Geological Society*, Vol. 47, 109–123.

YOUNG, B, FORTEY, N J, and NANCARROW, P H A. 1986. An occurrence of tungsten mineralisation in the Eskdale intrusion, west Cumbria. *Proceedings of the Yorkshire Geological Society*, Vol. 46, 15–21.

YOUNG, B, FORTEY, N J, and NANCARROW, P H A. 1991. Russellite from Buckbarrow Beck, Cumbria, England. *Journal of the Russell Society*, Vol. 4, 3–7.

YOUNG, B, and JOHNSON, E W. 1985. Langite and posnjakite from the Lake District. *Journal of the Russell Society*, Vol. 1, 80.

FOSSIL INVENTORY

To satisfy the rules and recommendations of the international codes of botanical and zoological nomenclature, authors of species cited in the text are listed below.

Bracteoleptaena polonica (Temple, 1965)
Calymene s.l. *subdiademata* McCoy, 1851
Cardiola interrupta J de C Sowerby, 1839
Cyphaspis brachypygus (Marr & Nicholson, 1888)
Dalmanites cf. *tenuimucronatus* Whittard, 1938
Decoroproetus piriceps (Ingham, 1970)
Dedzetina cf. *microstoma* Havlíček, 1977
Delops nobilis marri? Rickards, 1965
Delops obtusicaudatus (Salter, 1849)
Diacanthaspis decacantha (Angelin, 1854)
Eophacops glaber (Marr & Nicholson, 1888)
Eostropheodonta hirnantensis (McCoy, 1851)
Erratencrinurus cornutus (Ingham, 1974)
Fardenia (?) cf. *transversaria* Bancroft, 1949
Frankea sartbernardensis (Martin) Colbath, 1986
Hirnantia sagittifera (McCoy, 1851)
Leonaspis erinaceus (Marr & Nicholson, 1888)
Mucronaspis mucronata (Brongniart, 1822)
Mucronaspis olini (Temple, 1952)
Phillipsinella parabola aquilonia Ingham, 1970
Plaesiomys inflata (Salter, 1848)
Plaesiomys porcata (McCoy, 1846)
Plectothyrella crassicostis (Dalman, 1828)
Plectatrypa flexuosa (Marr & Nicholson, 1888)
Proromma acanthodes (Marr & Nicholson, 1888)
Raphiophorus aloniensis (Marr & Nicholson, 1888)
Scotoharpes judex (Marr & Nicholson, 888)
Staurocephalus cf. *clavifrons* Angelin, 1854
Toxochasmops marri (Reed, 1894)
Tretaspis hadelandica cf. *brachystichus* Ingham, 1970
Tretaspis hadelandica deliqua McNamara, 1979
Uhakiella strangulata (McCoy, 1851)
Veryhachium irroratum Loeblich & Tappan, 1969
Youngia moroides (Marr & Nicholson, 1888)

INDEX

See Contents (pp.v–vii) for principal lithostratigraphical and lithodemic units

BRITISH GEOLOGICAL SURVEY

Keyworth, Nottingham NG12 5GG
0115 936 3100

Murchison House, West Mains Road, Edinburgh EH9 3LA
0131 667 1000

London Information Office, Natural History Museum
Earth Galleries, Exhibition Road, London SW7 2DE
020 7589 4090

The full range of Survey publications is available through the Sales Desks at Keyworth and at Murchison House, Edinburgh, and in the BGS London Information Office in the Natural History Museum (Earth Galleries). The adjacent bookshop stocks the more popular books for sale over the counter. Most BGS books and reports can be bought from The Stationery Office and through Stationery Office agents and retailers. Maps are listed in the BGS Map Catalogue, and can be bought together with books and reports through BGS-approved stockists and agents as well as direct from BGS.

The British Geological Survey carries out the geological survey of Great Britain and Northern Ireland (the latter as an agency service for the government of Northern Ireland), and of the surrounding continental shelf, as well as its basic research projects. It also undertakes programmes of British technical aid in geology in developing countries as arranged by the Department for International Development and other agencies.

The British Geological Survey is a component body of the Natural Environment Research Council.

Published by The Stationery Office and available from:

The Publications Centre
(mail, telephone and fax orders only)
PO Box 276, London SW8 5DT
Telephone orders/General enquiries 0870 600 5522
Fax orders 0870 600 5533

www.tso-online.co.uk

The Stationery Office Bookshops
123 Kingsway, London WC2B 6PQ
020 7242 6393 Fax 020 7242 6412
68–69 Bull Street, Birmingham B4 6AD
0121 236 9696 Fax 0121 236 9699
33 Wine Street, Bristol BS1 2BQ
0117 926 4306 Fax 0117 929 4515
9–21 Princess Street, Manchester M60 8AS
0161 834 7201 Fax 0161 833 0634
16 Arthur Street, Belfast BT1 4GD
028 9023 8451 Fax 028 9023 5401
The Stationery Office Oriel Bookshop
18–19 High Street, Cardiff CF1 2BZ
029 2039 5548 Fax 029 2038 4347
71 Lothian Road, Edinburgh EH3 9AZ
0870 606 5566 Fax 0870 606 5588

The Stationery Office's Accredited Agents
(see Yellow Pages)

and through good booksellers

Acknowledgements

The authors thank their many colleagues in BGS and academia for their help and encouragement in the field, for access to unpublished data and for extensive discussions. Geochemical work in this project has benefitted from collaboration with Dr G F Marriner (Royal Holloway, University of London), Professor R Macdonald (University of Lancaster), Dr R Kanaris-Sotiriou (University of Sheffield), Dr A D Saunders (University of Leicester), and Drs P K Harvey and B P Atkin (University of Nottingham).

The resurvey of the Ambleside district forms part of the Lake District Regional Geological Survey, at various times led by Drs P M Allen, B C Webb, A J Reedman and P Stone, under the programme management of Dr D B Smith, Mr D H Land and Dr D J Fettes. Parts of the district were resurveyed under NERC contract to Sheffield and Leeds universities (1988–90), led by Dr N J Soper, and to Liverpool University (1988–91), led by Dr B P Kokelaar.

Chapters in this memoir have been written by the following authors.

One	Introduction: D Millward
Two	Applied geology: B Young, R J Marks, E W Johnson
Three	Concealed geology: M K Lee
Four	Ordovician: Borrowdale Volcanic Group: D Millward and B Beddoe-Stephens
Five	Borrowdale Volcanic Group: Birker Fell Formation: D Millward, M G Petterson, B Beddoe-Stephens, E W Johnson, P M Allen and M J Branney
Six	Borrowdale Volcanic Group: volcaniclastic successions: E W Johnson, D Millward, B C Kneller, M J Branney, B J McConnell, B Beddoe-Stephens, M G Petterson and S G Molyneux
Seven	Ordovician–Silurian: Windermere Supergroup: B C Kneller, A W A Rushton, R W Scott, N J Soper, N J Fortey and P Stone
Eight	Intrusive igneous rocks: B Beddoe-Stephens, P M Allen, D C Cooper, D Millward, M G Petterson, C C Rundle and B Young
Nine	Structure: D Millward, P Stone, M J Branney, B C Kneller, B P Kokelaar, E W Johnson and N J Soper
Ten	Metamorphism: N J Fortey, B Roberts and S Hirons
Eleven	Metalliferous mineralisation: B Young, D Millward, D C Cooper and E W Johnson
Twelve	Quaternary: E W Johnson, J W Merritt, D Millward and B Beddoe-Stephens

The memoir was compiled by Dr D Millward, and edited by Drs P Stone and A A Jackson. It was produced under the programme management of Dr D J Fettes.

The farmers, landowners, National Trust and the Lake District National Park are thanked for their co-operation and assistance during the resurvey. Burlington Slate Limited and Kirkstone Quarries Limited kindly provided information and production figures on the slate industry.

Notes

Throughout the memoir the word 'district' refers to the area covered by British Geological Survey 1:50 000 Series England and Wales Sheet 38 Ambleside.

National Grid references are within the 100 km squares NY and SD and are given in square brackets.

Numbers preceded by the letter D, L, or MNS refer to the BGS collection of photographs; those preceded by the letter E refer to registered numbers in the BGS sliced rock collection.

The authorship of fossil names is given in the fossil inventory.

PREFACE

Geology underpins a wide range of activities vital to the creation of wealth, particularly in relation to the exploration for, and exploitation of, resources. It is also vital that we have the best possible understanding of the geology of the United Kingdom if we are to maintain the quality of life, whether through the identification of potential hazards prior to development, or through helping to ameliorate the problems caused by future developments. The British Geological Survey is funded by central Government to improve our understanding of the three-dimensional geology of the UK through a national programme of geoscience surveying, data collection, interpretation, publication and archiving. One aim of this programme is to provide coverage of the UK land area by modern 1:50 000 scale geological maps, together with explanatory memoirs, by the year 2005. This memoir on the Ambleside district of the English Lake District is part of the output from that programme. It is best read in conjunction with the colour-printed 1:50 000 scale geological map. The comprehensive studies summarised in the memoir incorporate the results of detailed geological mapping by the British Geological Survey and teams from the Universities of Sheffield, Leeds and Liverpool, funded by University Mapping Contracts awarded by NERC.

The Ambleside district lies within the scenically beautiful Lake District National Park. The main settlements are the villages of Grasmere, Ambleside and Coniston. The high fells contrast with lower, rolling countryside around Lake Windermere and Coniston Water. The economy is dependent upon tourism, but hill farming remains the main land use. The Lake District has attracted geological interest for many years both for research and training, but this memoir is the first comprehensive account of the geology of the Ambleside district.

The Cumbrian mountains have been sculpted by glaciers out of the 460 Ma-old subaerial Borrowdale Volcanic Group. Preservation of subaerial volcanoes in the geological record is unusual because of their low preservation potential, and in many relatively young volcanic fields the internal nature of caldera systems is rendered inaccessible by the lack of erosion and the extensive blankets of silicic ignimbrite. In the Lake District tectonic deformation and dissection of the terrain have produced a unique insight into the deep structure of a subaerial caldera volcano.

Resurvey of the Windermere Supergroup has produced a major revision of the lithostratigraphical nomenclature. These rocks are interpreted to have been deposited largely during the final stages of closure of the Iapetus Ocean in a foreland basin ahead of a south-eastwards migrating thrust belt.

David A Falvey, PhD
Director

British Geological Survey
Kingsley Dunham Centre
Keyworth
Nottingham
NG12 5GG

PLATES

TABLES

FIGURES

CONTENTS